APPLICATIONS OF TRANSITION METAL CATALYSIS IN DRUG DISCOVERY AND DEVELOPMENT

APPLICATIONS OF TRANSITION METAL CATALYSIS IN DRUG DISCOVERY AND DEVELOPMENT

An Industrial Perspective

Edited by

MATTHEW L. CRAWLEY
Main Line Health
Berwyn, Pennsylvania, USA

BARRY M. TROST
Stanford University
Stanford, California, USA

A JOHN WILEY & SONS, INC., PUBLICATION

Published by John Wiley & Sons, Inc., Hoboken, New Jersey
Published simultaneously in Canada

For general information on our other products and services or for technical support, please contact
our Customer Care Department within the United States at (800) 762-2974, outside the United States
at (317) 572-3993 or fax (317) 572-4002.

Wiley also publishes its books in a variety of electronic formats. Some content that appears in
print may not be available in electronic formats. For more information about Wiley products,
visit our web site at www.wiley.com.

Library of Congress Cataloging-in-Publication Data:

Applications of transition metal catalysis in drug discovery and development : an industrial perspective /
edited by Matthew L. Crawley, Barry M. Trost.
    p. ; cm.
    Includes bibliographical references and index.
    ISBN 978-0-470-63132-4 (cloth)
    I. Crawley, Matthew L. II. Trost, Barry M.
    [DNLM: 1. Drug Discovery. 2. Catalysis. 3. Pharmaceutical Preparations–chemistry.
    4. Transition Elements. QV 744]
    615.1′9–dc23
    2011047552

Printed in the United States of America

ISBN: 9780470631324

10 9 8 7 6 5 4 3 2 1

CONTENTS

PREFACE

Designing structure for function is a key activity of a chemist approaching problems in fields ranging from material science to medicine. A critical aspect of such an endeavor is the ability to access the designed structure in a time-efficient manner. This requirement puts a major constraint on the level of complexity of the designed structure, which, in turn, may limit the performance of the compound especially in terms of selectivity. Another component of this issue is the ability to synthesize any compound that does prove to have desirable properties in a practical way. At the heart of both of these requirements is the ability to obtain the required compound in as few steps as possible from readily available commercial materials.

The strategic route to produce the desired structure depends upon the toolbox of synthetic reactions. To be as efficient as possible, the synthetic reactions must be chemoselective (i.e., differentiate among various bond types including between multiple functional groups of the same type), regioselective (i.e., control orientation of approach of two reactants), and, where applicable, diastereoselective (i.e., control relative stereochemistry/geometry) and enantioselective (i.e., control absolute stereochemistry). In addition, practical synthetic reactions must maximize the generation of raw materials and minimize the generation of waste—an aspect referred to as being atom economic. In the ideal condition, the reaction should be a simple addition in an intermolecular process or an isomerization in an intramolecular process. In any event, any stoichiometric by-product should be as small and innocuous as possible. To the extent another reactant is necessary, it should be needed only catalytically.

While some synthetic reactions do meet these requirements, such as the Diels–Alder and more recently the Aldol reaction, most synthetic reactions do not. Thus, a huge need for new synthetic methodology to meet these challenges exists. No application is more impacted by these issues as pharmaceutical research.

To obtain the rigorous performance requirements of pharmaceuticals, ever more complex structures will be needed. However, our toolbox remains limited. Catalysis has rapidly emerged as a critical approach to meet these challenges. In the first instance, improving the ability to perform existing reactions is a more efficient and atom-economic fashion is one approach. Good examples are the aforementioned Diels–Alder and Aldol approaches. In the Diels–Alder reaction, catalysis becomes the vehicle to deal with issues of regio-, diastereo-, and enantioselectivity. In the Aldol addition, catalysis becomes critical to perform this important reaction in its most atom-economic fashion as well as to address the issues of regio-, diastereo-, and enantioselectivity.

In the second instance, generating new patterns of reactivity is even more powerful in meeting these goals. New reactivity allows the creation of strategies that previously did not exist. A good illustration is the transition metal catalyzed cross-coupling reaction. No single reaction has had such an immense impact in pharmaceutical research and development than this process in the last 40 years, the reaction being first disclosed in the 1970s. While aryl coupling reactions such as the Ullmann coupling did exist, the thought that two different aryl groups could couple in a chemoselective manner was only a pipe dream. This could evolve into being so powerful that almost any two subunits can be coupled in this C–C bond-forming process that it has become was unimaginable. In a very real sense, a major change in the practice of the science has occurred and has allowed the design of new structures for pharmaceutical applications that has already borne such fruit. How many more such reactions may exist is impossible to estimate. However, it is reasonable to deduce that many remain undiscovered as strong incentive for the future generations of scientists.

This monograph provides an overview of where we are in meeting these challenges. Leading scientists who are themselves confronted with these problems provide a description of how far we have come. At the same time, their chapters reveal that, in spite of the progress, we have a very long way to go. Yes, we have gotten better, but that does not mean that we are anywhere near where we need to be. It is fair to predict that we are still in our infancy in making the synthesis of pharmaceuticals more empowering. While the imagination, determination, and skill of future efforts will be required for us to move forward in meeting these challenges, the immensity of the opportunities undoubtedly will mean that these activities will be critical for a very long time.

BARRY M. TROST

Stanford University
Stanford, California, USA

CONTRIBUTORS

Hans-Ulrich Blaser, Solvias AG, Basel, Switzerland (ret.)

Carl A. Busacca, Chemical Development, Boehringer Ingelheim Pharmaceuticals, Inc., Ridgefield, CT, USA

Cheol K. Chung, Merck Research Laboratories, Merck & Co., Inc., Rahway, NJ, USA

Matthew L. Crawley, Main Line Health, Berwyn, PA, USA

Daniel R. Fandrick, Chemical Development, Boehringer Ingelheim Pharmaceuticals, Inc., Ridgefield, CT, USA

Chris H. Senanayake, Chemical Development, Boehringer Ingelheim Pharmaceuticals, Inc., Ridgefield, CT, USA

Hong C. Shen, Department of Medicinal Chemistry, Roche R&D Center Ltd., Pudong, Shanghai, China

Jinhua J. Song, Chemical Development, Boehringer Ingelheim Pharmaceuticals, Inc., Ridgefield, CT, USA

William A. Szabo, Consultant in Drug Development, San Diego, CA, USA

Lamont Terrell, Heart Failure DPU, GlaxoSmithKline, King of Prussia, PA, USA

Oliver R. Thiel, Chemical Process Research & Development, Amgen, Inc., Thousand Oaks, CA, USA

Vince Yeh, Genomics Institute of the Novartis Research Foundation, San Diego, CA, USA

Jingjun Yin, Department of Process Chemistry, Merck Research Laboratories, Merck & Co., Inc., Rahway, NJ, USA

ABOUT THE AUTHORS

EDITORS

Professor Barry M. Trost

Born in Philadelphia, Pennsylvania in 1941 where he began his university training at the University of Pennsylvania (BA, 1962), he obtained a Ph.D. degree in Chemistry just three years later at the Massachusetts Institute of Technology (1965). He directly moved to the University of Wisconsin where he was promoted to Professor of Chemistry in 1969 and subsequently became the Vilas Research Professor in 1982. He joined the faculty at Stanford as Professor of Chemistry in 1987 and became Tamaki Professor of Humanities and Sciences in 1990. In addition, he has been Visiting Professor of Chemistry in Denmark (University of Copenhagen), France (Universities of Paris VI and Paris-Sud), Germany (Universities of Marburg, Hamburg and Munich), Italy (University of Pisa), Spain (Universities of Barcelona and Santiago de Compostela) and the UK (Cambridge University). In 1994 he was presented with a *Docteur Honoris Causa* of the Université Claude-Bernard (Lyon I), France, and in 1997 a *Doctor Scientiarum Honoris Causa* of the Technion, Haifa, Israel. In 2006, he was appointed Honorary Professor of the Shanghai Institute of Organic Chemistry.

Professor Trost's work has been characterized by a very high order of imagination, innovation and scholarship. He has ranged over the entire field of organic synthesis, particularly emphasizing extraordinarily novel methodology. Further, he has repeatedly demonstrated how his innovative methodology allows for the simplification of many complex target oriented syntheses leading to natural products of high biological activity. Early in his career, Professor Trost participated in the isolation and structure determination of the Cecropia Juvenile hormone. These studies helped to promulgate the concept that insect growth regulators can serve as ecologically preferable alternatives to pesticides. The long term goal and defining mission of Professor Trost's career has been toward enhancing synthetic effectiveness. He has virtually created whole new methodologies and strategies in organic synthesis. Among the areas which he has pioneered are the use of sulfur based reagents and transition metal, most notably palladium, and more recently ruthenium, catalysts in complex setting. More recently, his work has focused on enantioselective catalysts via the rational design of "chiral space." This work requires a detailed grasp of the mechanistic nuances of organometallic reactions. As part of this program, he designed a new class of ligands that spontaneously forms dinuclear complexes that are proving to be especially effective at asymmetric catalysis. Another continuing theme, directed to the realities of organic synthesis in fostering commercial goals, is the strategy of "atom economy." Thus, synthetic building blocks are combined to produce complex targets in high yield with a bare minimum of debris under the guidance of carefully crafted catalysts. In this way, organic synthesis can be brought to bear in process settings to reach substances of high complexity in a commercially feasible way which minimizes environmental impact. Barry Trost's triumphs in total synthesis include complex terpenoids, steroids, alkaloids, vitamins, antibiotics, nucleosides and macrolides. His syntheses are conducted with characteristic flair and serve as learning resources in addressing the limits of the methodology which he tends to develop independently.

In recognition of his many contributions, Professor Trost has received a number of awards, including the ACS Award in Pure Chemistry (1977), the ACS Award for Creative Work in Synthetic Organic Chemistry (1981), the Baekeland Award (1981), the first Allan R. Day Award of the Philadelphia Organic Chemists' Club (1983), the Chemical Pioneer Award of the American Institute of Chemists (1983), the Alexander von Humboldt Stiftung Award (1984), MERIT Award of NIH (1988), Hamilton Award (1988), Arthur C. Cope Scholar Award (1989), Guenther Award in the Chemistry of Essential Oils and Related Products (1990), the Dr. Paul Janssen Prize (1990), the ASSU Graduate Teaching Award (1991), Pfizer Senior Faculty Award (1992), Bing Teaching Award (1993), the ACS Roger Adams Award (1995), the Presidential Green Chemistry Challenge Award (1998), the Herbert C. Brown Award for Creative Research in Synthetic Methods (1999), the Belgian Organic Synthesis Symposium Elsevier Award (2000), the Nichols Medal (2000), the Yamada Prize (2001), the ACS Nobel Laureate Signature Award for Graduate Education in Chemistry (2002), the ACS Cope Award (2004), the City of Philadelphia John Scott Award (2004), Thomson Scientific Laureate (2007), the Kitasato Microbial Chemistry Medal, and the Nagoya Medal (2008). He has held a Sloan Fellowship, a Camille and Henry Dreyfus Teacher-Scholar grant and an American-Swiss Foundation Fellowship as well as having been the Julius Stieglitz Memorial Lecturer of the ACS-Chicago section (1980–81) and Centenary Lecturer of the Royal Society of Chemistry (1981–82). Professor Trost has been elected a Fellow of the American Academy of Sciences (1982) and a member of the National Academy of Sciences (1980). He has served as editor and on the editorial board of many books and journals, including being Associate Editor of the Journal of the American Chemical Society

(1974–80). He has served as a member of many panels and scientific delegations, and served as Chairman of the NIH Medicinal Chemistry Study Section. He has held over 120 special university lectureships and presented over 260 Plenary Lectures at national and international meetings. He has published two books and over 840 scientific articles. He edited a major compendium entitled Comprehensive Organic Synthesis consisting of nine volumes and serves as editor for ChemTracts/Organic Chemistry.

Dr. Matthew L. Crawley

Matthew Lantz Crawley obtained his B.A. degree from Williams College with a double major in political economy and chemistry under the guidance of Professor J. Hodge Markgraf. In 1998, he started his graduate studies with Professor Barry M. Trost at Stanford University, where his work focused on the development of asymmetric palladium-catalyzed reactions with applications in total synthesis. After completion of his Ph.D., Matthew joined the medicinal chemistry department of Incyte Corporation in Delaware, where he worked for several years. From late 2005 through 2010, Matthew worked at Wyeth Research in Pennsylvania (now Pfizer), first as a Senior Research Scientist and then as a Team Leader and Principal Research Scientist doing drug discovery in an array of therapeutic areas, including neuroscience, inflammation, and cardiovascular and metabolic disease. Most recently, while remaining active in chemistry through consulting, his work has focused on the healthcare sector where he is the Director of Electronic Medical Records for the Philadelphia based Main Line Health system. Matthew has authored or coauthored dozens of publications and is an inventor on several non-provisional and issued patents.

AUTHORS

Dr. Chris H. Senanayake

Dr. Chris H. Senanayake was born in Sri Lanka and received a BS degree (First Class) in Sri Lanka. After coming to the United States, he completed his MS at Bowling Green State University with Professor Thomas Kinstle in synthetic chemistry. He obtained his Ph.D. under the guidance of Professor James H. Rigby at Wayne State University in 1987 where he worked on the total synthesis of complex natural products such as, ophiobolanes, and completed the first total synthesis of grosshemin in the guaianolide family. He then undertook a postdoctoral fellow with Professor Carl R. Johnson and worked on the total synthesis of polyol systems such as amphotericin B and compactin analogous, and the synthesis of C-nucleoside precursors.

In 1989, he joined the Department of Process Development at Dow Chemical Co. In 1990, he joined the Merck Process Research Group. After 6 years at Merck, he accepted a position at Sepracor, Inc. in 1996 where he was promoted to Executive

Director of Chemical Process Research. In 2002, he joined Boehringer Ingelheim Pharmaceuticals. Currently, he is the Vice President of Chemical Development and leading a group of highly talented scientists, engineers, and administrative staff located in Ridgefield, CT.

Senanayake's research interests focus on the development of new asymmetric methods for the synthesis of bioactive molecules and heterocycles and on catalytic, enzymatic, and mechanistic studies. He has published and lectured in the area of practical asymmetric synthesis and many disciplines of organic chemistry how to develop drugs on an economical, greener and practical manner in large-scale operation for rapid development of drugs.

Senanayake demonstrates the ability to define and optimize chemical research and development strategies and tactics. He is able to "connect the dots" between the purely scientific and commercial perspectives and set up creative and effective strategies for new and proprietary products in ways that build value for the organization and create a competitive advantage. He is an Editorial Advisory Board member of the Organic Process Research & Development Journal. In 2008, he was the chairperson of Stereochemistry Gordon Conference. In 2010, he received the prestigious Siegfried gold medal award for development of practical processes for APIs and Process Chemistry. In 2011, He was appointed as an editorial board member of the Advance Synthesis and Catalysis Journal.

Dr. Daniel R. Fandrick

Dr. Daniel R. Fandrick received his B.S. degree with a major in chemistry from the University of California, San Diego under the guidance of Professor Joseph M. O'Conner. In 2006, Daniel earned his PhD degree in organic chemistry at Stanford University under the mentorship of Professor Barry M. Trost. His graduate studies focused on the development of the dynamic kinetic asymmetric transformations of vinyl aziridines and allenes and their applications to total synthesis. After graduation, he joined the chemical development group at Boehringer-Ingelheim Pharmaceuticals Inc. in Ridgefield, Connecticut where he is a principle scientist. Dr. Fandrick has published over 30 papers and numerous patents. His research interests are in the development of transition metal catalyzed and sustainable methodologies to provide efficient assess to pharmaceutically useful scaffolds and chiral centers. At Boehringer-Ingelheim Pharmaceuticals, Inc. Dr. Fandrick developed the novel asymmetric propargylation methodologies which provide general access to chiral homopropargylic alcohols and amines.

Dr. Jinhua Jeff Song

Dr. Jinhua Jeff Song received his undergraduate education at Nankai University in Tianjin, China (1989 to 1992). After a brief stay at Rice University in Houston, TX, he moved to the Massachusetts Institute of Technology in 1993 and obtained his Ph.D. degree in 1998 under the supervision of Prof. Satoru Masamune. Subsequently, he joined the Department of Chemical Development at Boehringer

Ingelheim Pharmaceuticals in Ridgefield, CT, where he is currently a Senior Associate Director in Process Research.

Dr. Song's research areas encompass natural product synthesis, asymmetric synthesis of chiral biologically active compounds, efficient methodologies for heterocycle synthesis, and novel N-heterocyclic carbene catalyzed reactions. He has published >40 research papers, review articles and book chapters including some "Most-Cited" and "Most-Accessed" papers. Over the years, Dr. Song has delivered invited lectures at various international conferences as well as academic institutions. Some of his work also received media attention and has been highlighted in the *Chemical and Engineering News*. Additionally, Dr. Song holds >15 patents on efficient synthesis of pharmaceutical agents.

Dr. Carl Busacca

Dr. Carl Busacca is a native of Milwaukee, Wisconsin, who graduated from high school in North Carolina, and then attended North Carolina State University, receiving his BS in Chemistry in 1982. He performed some undergraduate research in both laser resonance Raman spectroscopy, and the ^{60}Co radiolysis of fluorocarbons. He then worked three years in Research Triangle Park, North Carolina doing organic synthesis for Union Carbide, before returning to graduate school, at Colorado State University. At CSU, he studied under Professor A.I. Meyers, earning his Ph.D. in 1989 following research into asymmetric Diels-Alder cycloadditions. In 1990, he entered the Pharmaceutical Industry working for five years as a medicinal chemist at Sterling Winthrop in Rensselaer, New York. At Sterling, he worked on novel anti-arrhythmics and Thrombin inhibitors, and carried out research supporting the existence of palladium carbenes as intermediates in cross-coupling reactions. He joined the Department of Chemical Development at Boehringer-Ingelheim Pharmaceuticals in Ridgefield, Connecticut as a process chemist in 1994. At Boehringer, Dr. Busacca worked first for Dr. Vittorio Farina, and then for Dr. Chris Senanayake, and he is currently Distinguished Research Fellow. He has worked extensively in the anti-viral area, developing inhibitors of various parts of the HIV and HCV machinery. Dr. Busacca's principal research interests include mechanistic organopalladium chemistry, ligand design, the development of new organophosphorus chemistry, asymmetric catalysis, applications of NMR spectroscopy to Process Research, and the design of efficient chemical processes. He is deeply interested in the nucleo-synthesis of transition metals in supernovae.

Dr. Hong C. Shen

Hong C. Shen received his B.S. degree in chemistry from Peking University under the direction of Professor Yunhua Ye in 1997. He subsequently moved to University of Minnesota, where he developed a formal [3 + 3] cycloaddition with Professor Richard Hsung, and obtained his M.S. degree in 1998. Hong Shen then joined the research group of Professor Barry Trost at Stanford University. His work spanned from Ru- and Pd-catalyzed reactions to their applications in total syntheses of natural products. After obtaining his Ph.D. in 2003, Hong Shen assumed a senior research chemist position at Merck Research Laboratories, Rahway, New Jersey. He then took on an exploratory chemistry team lead working in the areas of cardiovascular, thrombosis, and metabolic diseases. Most recently he returned to China to pursue an exceptional opportunity as the section head for medicinal chemistry at Roche. Hong has authorship on more than 55 scientific publications and is an inventor on 17 patent applications. Hong is currently a visiting professor at Tianjin University.

Dr. Jingjun Yin

Jingjun Yin obtained a B.S. degree from the University of Science and Technology of China in 1994 and then went on to earn his Ph.D. degree in Organic Chemistry under the direction of Prof. Lanny Liebeskind at Emory University in 1999. After that, he moved to Massachusetts Institute of Technology as a postdoctoral fellow with Prof. Stephen Buchwald. In 2001, he joined the Process Research Department of Merck at Rahway, New Jersey focusing on designing efficient and practical syntheses of complex drug candidates with a special interest in transition-metal catalyzed reactions.

Dr. William A. Szabo

Bill Szabo received his undergraduate degree in Chemistry from Lehigh University, worked for 2 years as an R&D chemist at Johnson & Johnson's McNeil Laboratories, and earned a Ph.D. degree in Heterocyclic and Medicinal Chemistry from the University of Florida. He held a 2-year postdoctoral fellowship in natural product synthesis at Wesleyan University with Professor Max Tishler, former President of Merck Research Laboratories and an early pioneer in pharmaceutical process development. Bill was next recruited by Alfred Bader, cofounder of the Aldrich Chemical Company, and worked for 18 years at Aldrich in Milwaukee in various management positions in R&D, production, and advertising. He then relocated to St. Louis and spent 4 years in the sales and marketing of reagents and bulk pharma-ceutical intermediates for Sigma-Aldrich, positions which included Sales Director for North America and Vice President of International Sales. In 1998, Bill took an early retirement and moved to San Diego. He has since been consulting in drug and business development. Bill has been active in the 2800-member San Diego Section of the American Chemical Society. He served as Chairman of its executive board in 2011.

Dr. Vince Yeh

Vince Yeh completed his B. S. degree from the University of British Columbia in 1994. Subsequently, he earned his Ph.D. in organic chemistry from University of Alberta in 2001 under the guidance of Professor Derrick Clive. He then moved to Stanford University for post-doctoral studies under Professor Barry Trost. During that time he contributed to the development of asymmetric direct aldol catalysts invented in the Trost labs. In 2003 he moved to Chicago to work for Abbott Laboratories in their Metabolic Diseases unit where he invented an advanced development candidate. In 2007, he worked for Astellas Pharmaceuticals where he established their first American preclinical chemistry research lab. Later that year, he moved back to California (San Diego) to pursue medicinal chemistry research at the Genomic Institute of Novartis Research Foundation. Vince holds numerous patents and has written a number of original research papers, reviews and book chapters in chemistry and medicinal chemistry.

Dr. Oliver R. Thiel

Oliver R. Thiel received his undergrad education in chemistry at the Technical University in Munich, Germany, completing a thesis in the labs of Prof. Matthias Beller, focusing on rhodium catalyzed hydroaminations. He then pursued his Ph.D. at the Max-Planck-Institut für Kohlenforschung in Mülheim, Germany under the guidance of Prof. Alois Fürstner, graduating with a thesis on the application of ring-closing metathesis for the synthesis of complex natural products. From 2001 to 2003 he was a Feodor Lynen Postdoctoral Research fellow with Prof. Barry M. Trost at Stanford University, applying palladium-catalyzed reactions for the synthesis of Furaquinocins and a Bryostatin analogue. Subsequently he joined the Chemical Process Research and Development group at Amgen in Thousand Oaks, CA where he currently holds the position of Principal Scientist. His main areas of interest are the development of robust processes for active pharmaceutical ingredients, the application of catalytic reactions in organic synthesis and the development of hybrid modalities. Oliver has authored more than 30 publications.

Dr. Cheol K. Chung

Cheol Chung obtained his B.S. and M.S. degree from Seoul National University. After a brief stint at LG Chemical working as a researcher in the pharmaceutical division, he joined the laboratory of Professor Barry Trost at Stanford University where he studied the transition metal catalyzed transformations of alkynes and their application to natural products synthesis. After the completion of his Ph.D. in 2006, he moved to California Institute of Technology for postdoctoral training to further his knowledge in organometallic catalysis under the supervision of Professor Robert Grubbs. Since 2008, Cheol Chung has been senior research chemist in Process Research at the Merck Research Laboratories in Rahway, NJ.

Dr. Lamont Terrell

Lamont Terrell earned his B.S. degree in Chemistry at Texas Southern University in 1995. Under the direction and guidance of Professor Robert Maleczka, Jr. at Michigan State University, he studied organic synthesis, completed the total synthesis of the antileukemic natural product amphidinolide A and earned his Ph.D. in 2001. Upon completion of his graduate studies at MSU, he continued synthetic training with a two-year postdoctoral stint with Professor Barry Trost at Stanford University. The focus of his postdoctoral studies was the development of a catalytic dinuclear zinc asymmetric Mannich reaction. After completion of his postdoctoral studies, Lamont obtained a position with GlaxoSmithKline in the cardiovascular medicinal chemistry group, where he has been for the last 9 years.

Dr. Hans-Ulrich Blaser

Hans-Ulrich Blaser carried out his doctoral research with A. Eschenmoser at the Federal Institute of Technology (ETH) Zürich, where he received the Ph.D. degree in 1971. Between 1971 and 1975 he held postdoctoral positions at the University of Chicago (J. Halpern), Harvard University (J.A. Osborn), and Monsanto (Zürich). During 20 years at Ciba-Geigy (1976–1996) he gained practical experience at R&D in the fine chemicals and pharmaceutical industry, which continued at Novartis (1996–1999) and at Solvias where he was chief technology officer until 2009. Presently he acts a scientific advisor. His main interest is selective catalysis with emphasis on enantioselective catalysts. During his industrial carrier he has developed and implemented numerous catalytic routes for agrochemicals, pharmaceuticals and fine chemicals both as project leader and section head. He and his team received several awards for their contributions to industrial catalysis in general and enantio-selective catalysis in particular, notably the Sandmeyer Prize of the Swiss Chemical Society in 1998, the Horst Pracejus Prize of the German Chemical Society (2009) and the Paul Rylander Award of the Organic Reaction Catalysis Society (2010).

1

TRANSITION METAL CATALYSIS IN THE PHARMACEUTICAL INDUSTRY

CARL A. BUSACCA, DANIEL R. FANDRICK, JINHUA J. SONG, AND CHRIS H. SENANAYAKE

1.1 OVERVIEW OF CATALYSIS

Catalysis typically provides the technology to enable the efficient and cost-effective synthesis of pharmaceutical products. By definition, catalysis increases the reaction rate by lowering the activation energy of the reaction, therefore allowing the chemical transformation to take place under much milder conditions over the uncatalyzed process. Furthermore, the catalyst typically imparts chemo-, regio-, or stereoselectivities over the course of the reaction to enable highly efficient syntheses of target molecules.

Applications of Transition Metal Catalysis in Drug Discovery and Development: An Industrial Perspective, First Edition. Edited by Matthew L. Crawley and Barry M. Trost.
© 2012 John Wiley & Sons, Inc. Published 2012 by John Wiley & Sons, Inc.

Catalysis is one of the principle drivers for the modern economy. Catalysis-based industries contribute more than 35% of the global GDP [1]. It has been estimated that about 90% of the chemicals are derived in some fashion from catalytic processes [2]. The annual worldwide demand for catalysts is approaching one million metric tons, and further growth in this sector was projected to continue [3]. Furthermore, catalysis is one of the 12 green chemistry principles [4]. The use of catalysis can significantly reduce waste streams, simplify synthetic processes, and reduce both cycle times and volume requirements, especially in chemical manufacturing. Catalysis often enables a business to enhance the value of the product while minimizing the overall carbon-footprint of their activities.

The significance of catalysis and its proven impact on the advancement of science was recognized by several Nobel Prizes in Chemistry. In 1909, Wilhelm Ostwald won the Nobel Prize for "his work on catalysis and for his investigation into the fundamental principles governing chemical equilibria and rates of reaction." During the first decade of this century, four transition-metal catalyzed reactions were honored with Nobel Prizes in Chemistry: asymmetric hydrogenation and oxidation (2001; Knowles, Noyori, and Sharpless), metathesis (2005, Chauvin, Grubbs, and Schrock), and cross-coupling reactions (2010; Heck, Negishi, and Suzuki). These reactions not only have academic significance but also proved to be critical for the production of industrially important products.

Noyori's BINAP-Rh-catalyzed asymmetric allylic amine isomerization reaction was used to develop an industrial process for menthol (Scheme 1.1) [5]. Menthol is one of the most widely utilized natural products. In 2007, the total world production of menthol was $> 19,000$ tons, over a quarter of which was used for pharmaceutical purposes, while the remainder was used for consumer products such as toothpaste,

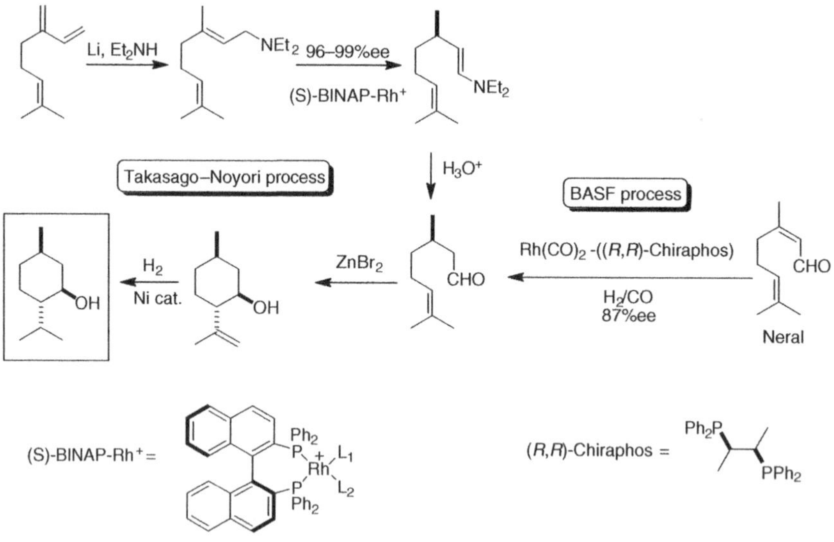

SCHEME 1.1   Industrial menthol processes.

cosmetics, confectionary, and tobacco products [6]. Natural menthol is supplied via isolation from mint cultivated primarily in Asian countries. However, the market demand greatly exceeded the natural supply. In addition, the reliability of natural supply is affected by weather and climate of the mint-growing region. A need existed for an efficient and economical method for synthetic menthol to close the supply gap and also to alleviate the volatility of price on the market.

The new Takasago–Noyori menthol process commenced with the conversion of myrcene to geranyldiethylamine by treatment with lithium and diethylamine. Then asymmetric isomerization of the allylic amine with a cationic BINAP-Rh catalyst afforded a chiral enamine, which was hydrolyzed to (R)-citronellal (96–99%ee). Elaboration of (R)-citronellal to (−)-menthol was accomplished in two additional straightforward steps. This new process allowed Takasago to produce 1000–3000 tons of synthetic menthol every year for the past 30 years.

BASF recently disclosed a new menthol process using Chiraphos-Rh-catalyzed hydrogenation reaction as the key step [7]. They were able to achieve the direct asymmetric hydrogenation of neral to give (R)-citronellal with 87%ee. The projected production capacity of the BASF menthol process was 3000–5000 tons/year [8]. This menthol process described here clearly underscored the importance of catalysis to our everyday life.

1.2 TRANSITION METAL CATALYSIS IN THE PHARMACEUTICAL INDUSTRY

Transition metal catalyzed processes have been extensively utilized in the pharmaceutical industry for over the past 30 years. They have been employed for library preparations, discovery syntheses, and large-scale preparation of active pharmaceutical ingredients. This use relates to the efficiency to conduct a large number of chemical transformations with tolerance of numerous functional groups, and high enantio-, diastereo-, and chemoselectivities. The most commonly applied transition metal catalyzed applications relate to the transformations that result in a cross-coupling for the formation of carbon–carbon and carbon–heteroatom bonds, asymmetric hydrogenation, oxidation, asymmetric addition, and metathesis. The emergence of each technology, evolution into its current status, impact, and recent advances that are projected to provide additional value to the pharmaceutical industry deserve further discussion.

1.2.1 Cross-Couplings for the Formation of Carbon–Carbon Bonds

The importance of cross-couplings for the formation of carbon–carbon bonds to the chemical industry is best appreciated by awarding the 2010 Nobel Prize to Heck, Negishi, and Suzuki for "palladium-catalyzed cross-couplings in organic synthesis." The basis of cross-coupling is the reductive elimination of two organic components from a high valent late transition metal for the formation of a C–C bond (Scheme 1.2) [9]. The utility of this reaction was realized by the development of

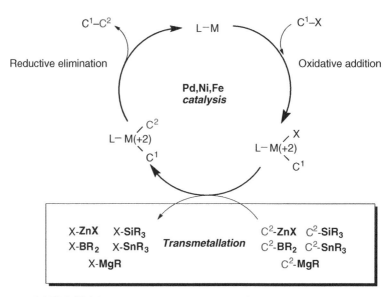

SCHEME 1.2   General cross-coupling mechanism and extensions.

suitable components for the selective formation of the mixed bis-organometallic intermediate. In 1971, Kochi demonstrated that a Fe(III) complex can catalyze the coupling of organo-magnesium reagents with haloalkenes [10]. The following year in 1972, Kumada, Tamao, and Corriu independently reported the cross-coupling of organo-magnesium reagents with alkenyl or aryl halides catalyzed by a Ni(II) complex [11]. Since these seminal reports, palladium and nickel complexes have emerged as the mainstream catalysts employing organo-boronates, silicon, tin, magnesium, and zinc reagents as the nucleophilic components wherein the corresponding cross-couplings are referred to as Suzuki-Miyaura [12], Hiyama [13], Stille [14], Kumada [15], and Negishi [16] couplings. The transmetallation operation can also be replaced by a migratory insertion with an olefin or carbon monoxide to achieve a Heck coupling [17] or carbonylation [18]. Since the advent of these technologies and by proper choice of reaction components, catalyst, and conditions, most carbon–carbon single bonds can be constructed through this process.

The utility of cross-couplings for the accessibility of bi-aryl, aryl-alkenyl, and aryl-alkynyl moieties has made these structures common synthetic intermediates for APIs and as pharmacophores rationally designed into numerous drugs and clinical candidates as exemplified by Losartan, Naratriptan, and Singulair (Fig. 1.1) [19]. A survey of reactions scaled in Pfizer's GMP facility at the Groton site showed a steady increase in the use of cross-couplings over the past two decades [20]. Of the 14% of reactions that generate a C–C bond, 4.3% were cross-couplings from 1985 to 1996, which increased to 14.5% for the period between 1997 and 2007. Further utility of cross-couplings will be due to advances in broadening the substrate scope for incorporation into a cross coupling. Extension to less reactive electrophiles such as aryl chlorides [21], phenolates [22], carbon-nitriles [23], and aryl ammonium

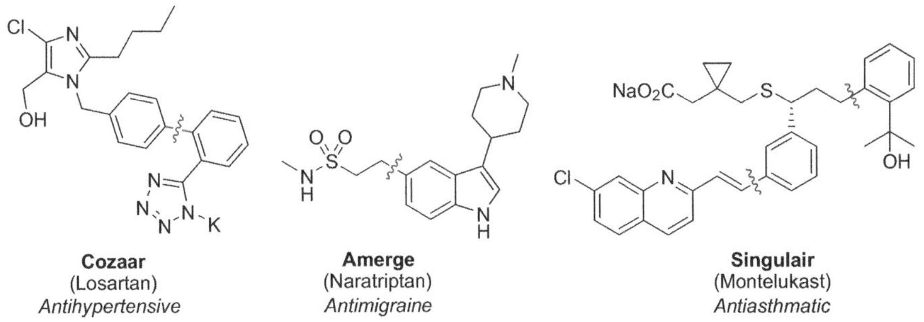

FIGURE 1.1 Selected examples for application of C–C cross-couplings.

salts [24] has greatly increased the flexibility for incorporating a larger pool of commercially available materials into a synthesis. Recently, some progress for the cross-coupling of aryl fluorides has been achieved [25]. Complementary to the advances with the electrophile, the nucleophile scope has also expanded to include aldol equivalents [26], carboxylic acids (decarboxylative couplings) [27], perfluorinated alkanes [28], and C–H insertions [29]. Progress has also been made for enantioselective cross-couplings [30] to provide access to atropisomers, which are emerging as pharmacophores [31]. More promising are the developments of enantioselective cross-couplings for the generation of classical carbon stereocenters, which provide general access to otherwise difficult structures [32]. These advances in cross-couplings are growing exponentially, a trend that will add additional value to the pharmaceutical industry.

1.2.2 Cross-Couplings for the Formation of Carbon–Heteroatom Bonds

The impact of cross-couplings for the formation of carbon–heteroatom bonds is more significant to the pharmaceutical industry than cross-couplings for the formation of C–C bonds. The process for heteroatom coupling is based on principles similar to those used for carbon–carbon bonds but varies with mechanism due to the influences of metal, ligand, and nucleophilic component (Scheme 1.3) [33]. Although the copper mediated C–N coupling has been known for over a century, that is, Ullmann [34] and

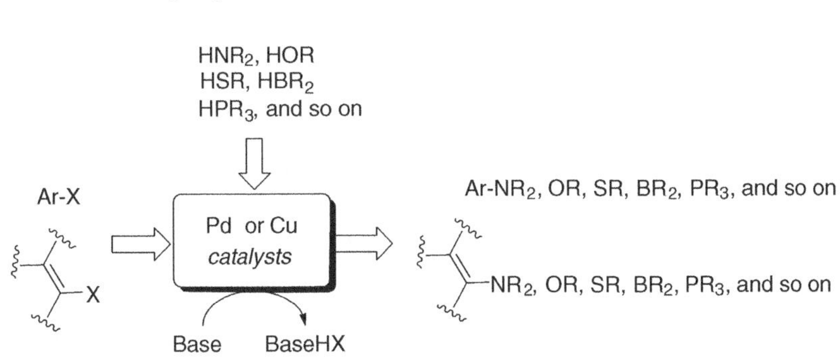

SCHEME 1.3 General scope of C–X bond formation through a cross-coupling.

Goldberg [35] reactions, the use of harsh conditions prevented general application to complex molecule synthesis. The discovery and development of ligand-mediated copper catalyzed couplings by Buchwald and coworkers [36] and Goodbrand et al. [37] allowed for much milder conditions and tolerance of numerous functional groups. These reactions have been extended to the coupling of nitrogen, phosphorous, oxygen, and sulfur providing general access to the respective carbon sp^2–X bond. Complementary to copper-catalyzed reactions, the palladium-catalyzed process has been known since the pioneering research of Migita and coworkers with amide-tin reagents [38]. The expansion in the palladium process originated from the independent developments of tin-free processes by Buchwald [39], Louie and Hartwig [40]. The palladium-catalyzed process was then extended to the formation of C–O, C–P, and C–B bonds. The recent application for the formation of a C–B bond by Ishiyama and coworkers [41] presented a valuable process for the preparation of organo-boronates for Suzuki couplings with significant functional group tolerance that simply cannot be achieved by traditional Grignard or organo-lithium-mediated approaches. The breadth of bond formations that are possible with cross-couplings for the formation of carbon–heteroatom bonds is nearly unmatched when compared to other methodologies.

The impact of cross-couplings for carbon–heteroatom bond formation for the pharmaceutical industry is considerable. These methodologies provide the robustness required for the formation of a large diversity of structures that are necessary for discovery and development (Fig. 1.2). Some notable examples are the multibillion dollar drugs Gleevec from Novartis [42], Abilify from Otsuka Pharmaceuticals [43], and Pfizer's Phase III candidate Torcetrapid [44]. The advances in broadening the electrophile scope for cross-couplings for carbon–carbon bonds are often extended to the formation of carbon–heteroatom bonds. The use of aryl chlorides [45] and aryl ethers [46] are routinely depicted in the recent literature. Other significant developments in cross-couplings are related to C–H activations and numerous reports are published on the efficient amination of C–H bonds [47]. One promising methodology reported by Hartwig is the C–H borolation promoted by an iridium catalyst, which provides interesting regioselectivities that are not achievable by existing methodologies [48]. These seminal advances will further increase the substrate scope and applicability for pharmaceutical synthesis.

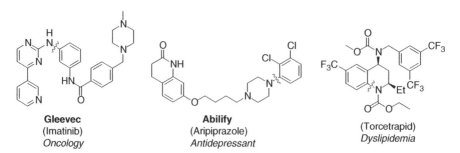

Gleevec
(Imatinib)
Oncology

Abilify
(Aripiprazole)
Antidepressant

(Torcetrapid)
Dyslipidemia

FIGURE 1.2   Selected examples for application of heteroatom cross-couplings.

1.2.3 Asymmetric Hydrogenation

Asymmetric hydrogenation is the most significant asymmetric technology utilized to establish chirality in pharmaceutical products. Half of the 2001 Nobel Prize was awarded to William Knowles and Ryoji Noyori "for their work on chirally catalyzed hydrogenation reactions." A hydrogenation is the transfer of a molecule of hydrogen to a pi-bond resulting in a formal reduction, typically catalyzed by a transition metal (Scheme 1.4) [49]. The utility of the process is the ability to control reduction of one enantiotopic face of the pi-bond by a chiral ligand resulting in asymmetric induction. Since the advent of transition metal catalyzed asymmetric hydrogenations by Knowles and Noyori, an exceedingly large pool of chiral ligands have been invented typically based on the class of biaryl diphosphines, monophosphines, pholanes, ferrocenyl-based diphosphines, and miscellaneous phosphines, which have comparable performance [50]. As coined by Jacobsen, the selected few that are routinely applied to large-scale production of commercially relevant molecules are referred to as "privileged ligands [51]." Intellectual protection for the initial structures has expired, and the industry has free access to utilize the technology. Some structures are still protected by patents. However, since the variation based on the ligand structure can be considerable while retaining similar performance to known systems, modification of the initial ligand design to circumvent IP protection is often achievable [52]. Asymmetric hydrogenation has been applied to the saturation of C=C, C=O, and C=N bonds most typically with Rh, Ru, and Ir-based catalysts. Due to the high levels of discrimination exhibited for these processes, multiple stereocenters can often be established in a single transformation through a hydrogenation that proceeds through a dynamic kinetic asymmetric transformation.

Asymmetric hydrogenation can be considered as the powerhouse for installation of chirality into a drug substance [53]. Numerous examples attest to this importance (Fig. 1.3). The classical example of multiton applications are the synthesis of L-Dopa by Monsanto as well as the menthol processes described earlier in this chapter [54]. Sitagliptin from Merck [44b] and a Cathepsin S inhibitor from Boehringer Ingelheim [55] highlight the usefulness of this technology for the synthesis of human pharmaceuticals. The value of asymmetric hydrogenation for the establishment of chirality is due to the low cost for hydrogen, high levels of asymmetric induction, and most importantly the typical high turnover allowing for exceedingly low catalyst loadings. Achievement of 10,000–1,000,000 turnovers is common and expected for production processes. Due to the large pool of available ligands, pharmaceutical companies and specialized CROs built catalysis or automation groups that can conduct high throughput optimization to establish the optimal ligand and conditions for a particular transformation.

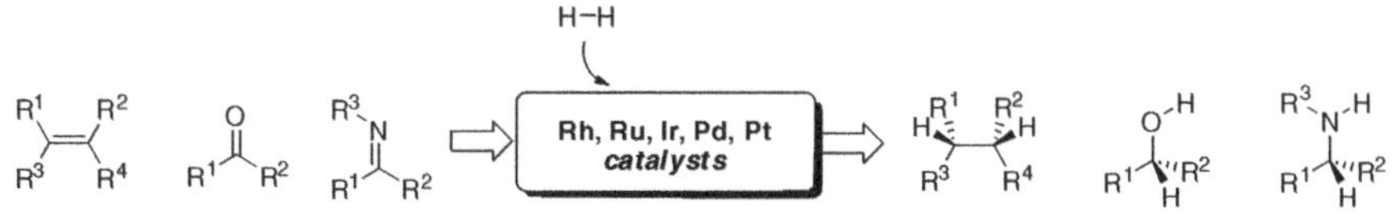

SCHEME 1.4 General hydrogenation scope.

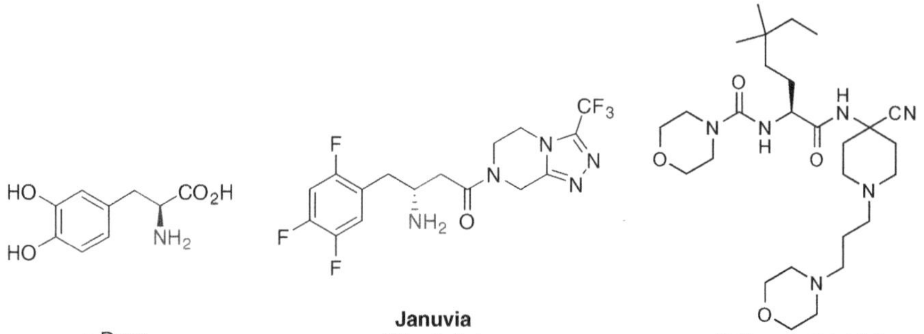

L-Dopa
Antiparkinsonian

Januvia
(Sitagliptin)
Antidiabetic

Cathepsin S Inhibitor
Anti-inflammatory

FIGURE 1.3 Selected examples for application of asymmetric hydrogenation.

Clearly asymmetric hydrogenation is one of the most mature of all asymmetric catalytic technologies employed by the pharmaceutical and agrochemical industries. There are several reasons why this methodology has been so fully embraced. From an operations perspective, asymmetric hydrogenation requires no significant changes to equipment that has been used for many years for standard hydrogenations at scale. Thus, minimal capital investments and limited operator retraining are needed to use the technology. The steady march of chiral ligand development in the past 35 years has led to reasonably good scope with respect to the substrate classes which can be successfully reduced. In parallel, there have been dramatic improvements in the efficiency of these chiral catalysts, leading to ever higher TON's (turnover numbers) and TOF's (turnover frequencies) and lower catalyst loadings. Hydrogen, particularly on a mole basis, is also by far the cheapest reducing agent available. When these advantages are coupled with a transformation that is inherently "atom economical," which also provides value-added (chiral) products the broad acceptance of asymmetric hydrogenation can be understood.

However, asymmetric hydrogenation of certain substrates is still challenging. For example, only recently has a general process for the asymmetric hydrogenation of some tetra-substituted and unfunctionalized olefins been reported by Pfaltz and coworkers [56]. These catalysts normally require the relatively expensive BAr^F counterion, which also has a high molecular weight. There is also nearly exclusive use of iridium for these hydrogenations, rather than cheaper precious metals such as ruthenium. Achieving very low catalyst loadings thus becomes a critical part of the cost analysis in these processes as well. These outstanding issues of expanding the substrate scope and improving the economics of asymmetric hydrogenation will likely form the basis of future academic and industrial research in this critical area.

1.2.4 Oxidative Catalysis

Oxidative catalysis, the complementary process to hydrogenation, has also impacted pharmaceutical research and development albeit to a lesser degree than for academic total synthesis. Oxidations catalyzed by transition metals involve a broad class of transformations to generate pro-chiral, racemic, and chiral products. The significance of asymmetric oxidations was recognized by awarding half of the 2001 Nobel Prize to Sharpless "for his work on chirally catalyzed oxidation reactions." The general utility for transition metals to catalyze an oxidation is through a metal-mediated oxidation wherein the oxidized metal is regenerated by a stoichiometric oxidant (Scheme 1.5). Use of the transition metal allows for use of less reactive and desirable oxidants as well as providing the chemo- and stereoselectivity necessary for an efficient process. The classical example for this mechanism was developed by Ley for the oxidation of alcohols and aldehydes with a ruthenium oxide catalyst [57]. Some extensions have been applied to allylic C–H oxidations [58], but the vast utility of transition metal oxidations was realized by pi-bond oxidations. The Wacker process enables access to aldehydes and ketones by the oxidation of olefins; however, utilization of the methodology on highly functionalized compounds is restricted due to limited chemoselectivity [59]. Alternatively, stereoselective olefin oxidation can efficiently provide chiral epoxides, through a Sharpless [60] or Jacobsen–Katsuki [61] asymmetric epoxidation, as well as chiral diols or amino-alcohols through a Sharpless dihydroxylation [62] or amino-hydroxylation [63], respectively. Application of a slight modification of the Sharpless asymmetric epoxidation conditions to the oxidation of sulfides has provided a general access to the chiral sulfoxide pharmacophore [64].

Transition metal catalyzed oxidation has had limited applications in the pharmaceutical industry to date. The decreased use of oxidations in API syntheses in relation to academic total syntheses [9] is due to the inefficiency of introducing elements in an oxidation state that requires adjustment, associated waste, and safety concerns. The typical use of nongreen solvents, such as chlorinated hydrocarbons, further limits the utility for oxidations to the pharmaceutical industry. This issue is reflected by one goal of the American Chemical Society Green Initiative for oxidations without a chlorinated solvent [65]. The notable application to the synthesis of esomepraxol [66], rosavastatin [67], and indinavir [68] underscores the value of transition metal catalyzed oxidations (Fig. 1.4). The utility of asymmetric oxidations for the formation of C–O bonds has also been extended to the synthesis of other more complicated stereocenters. The use of a Sharpless asymmetric epoxidation by Eisai provided the stereochemistry to construct a quaternary carbon stereocenter through a

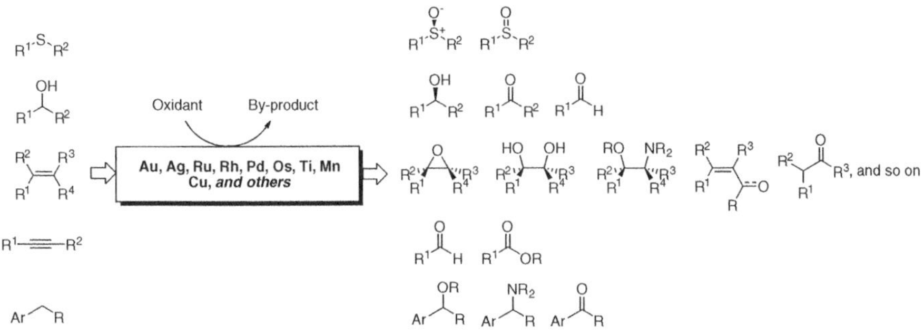

SCHEME 1.5   Transition metal catalyzed oxidation and extensions.

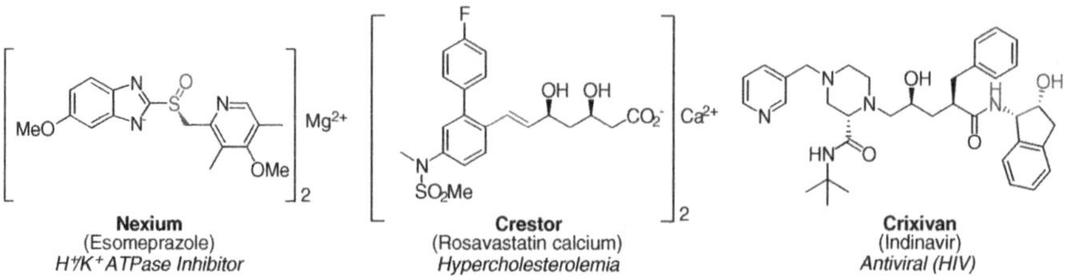

Nexium
(Esomeprazole)
H⁺/K⁺ATPase Inhibitor

Crestor
(Rosavastatin calcium)
Hypercholesterolemia

Crixivan
(Indinavir)
Antiviral (HIV)

FIGURE 1.4   Selected examples for application of oxidations.

subsequent Pinacol rearrangement (Scheme 1.6) [69]. This type of stereocenter cannot be readily accessed by other more traditional asymmetric transformations thereby demonstrating the significant utility of asymmetric oxidations toward API synthesis. Recent advances in transition metal catalyzed oxidations toward asymmetric allylic oxidation [70], oxidative coupling [71], and C–H oxidations [72] have provided increased utility for complex molecule synthesis. Adaptation of these new methodologies by medicinal chemists for the design and discovery of new APIs and the utilization of these new methods by process chemists will increase the use of these valuable transition metal catalyzed oxidations.

1.2.5   Asymmetric Addition Reactions

Asymmetric addition reactions comprise a wide variety of transformations, which due to their versatility provide access to chiral structures that are not otherwise efficiently accessible by other technologies. The utility of transition metals for these additions stems from the ability to use a substoichiometric amount of a chiral ligand that provides a more efficient asymmetric transformation relative to more traditional chiral auxiliary–based technologies. Although some reports on enantioselective additions to aldehydes date back to the 1940s [73], the first reproducible reaction was the addition of Grignard reagents to carbonyl species with a chiral ethereal solvent reported by Cohen and Wright in 1953 [74]. Oguni et al. in 1984 reported the (*S*)-leucinol catalyzed addition of diethylzinc to benzaldehyde with 49%ee [75]. This seminal achievement sparked the dramatic developments in transition metal catalyzed asymmetric additions with hundreds of methodological developments and thousands of applications. These types of transition metal–based addition reactions can be crudely divided into four types of transformations (Scheme 1.7). The first involves addition to C=O and C=N bonds. These additions provide chiral alcohol or amine products by the addition of a carbon-based nucleophile such as through an asymmetric organo-zinc-mediated, aldol, and cyanide addition reactions. The second

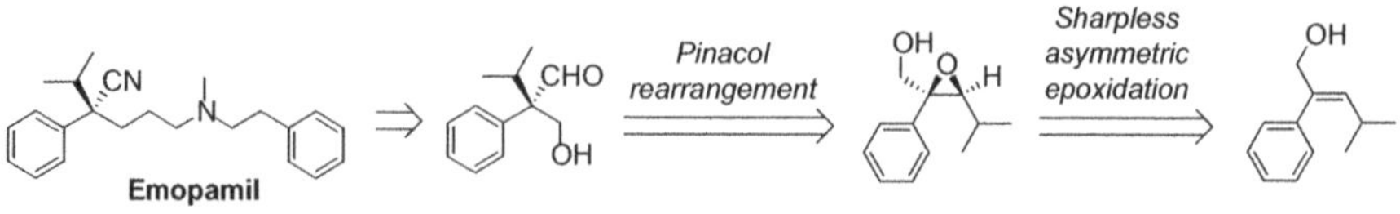

Emopamil

Pinacol rearrangement

Sharpless asymmetric epoxidation

SCHEME 1.6   Eisai SAE and application to a quaternary carbon stereocenter.

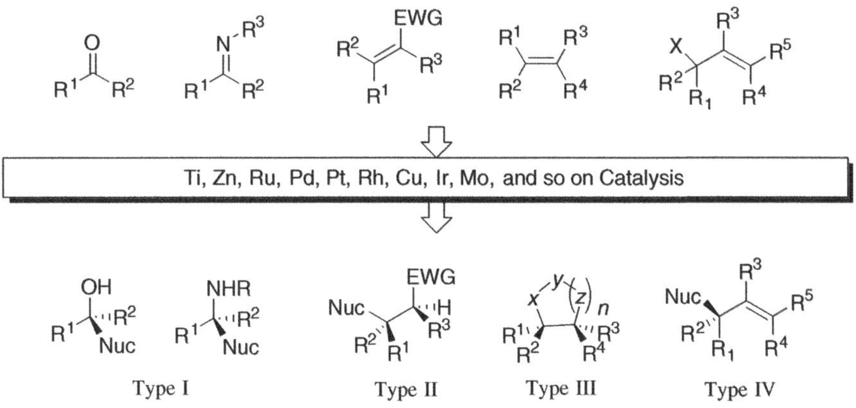

SCHEME 1.7 General types of asymmetric addition reactions.

reaction type is the asymmetric addition to C=C bonds through a conjugate (1,4) addition. The third class of additions is cycloadditions achieved by cyclopropanations, dipolar [3 + 2] cycloadditions, and more traditional Diels–Alder [4 + 2] reactions. The final addition type involves asymmetric allylic alkylations that provide chiral allylic stereocenters.

The utility of asymmetric addition reactions for pharmaceutical companies comes from the ability to access stereocenters that are not amenable to efficient construction through asymmetric hydrogenation. This general trend is due to the higher catalyst loadings and costs typically associated with implementing an asymmetric addition. The suggestion of Pfizer's Hawkins that the chances of using an asymmetric transformation on commercial scale is greater if the process was implemented early in development and applied to a late synthetic step [76] is more relevant for asymmetric addition reactions than for hydrogenations. The costs associated with resolution of an early intermediate can be lower than implementing the asymmetric addition reaction. Several applications of asymmetric additions have been reported by pharmaceutical companies (Fig. 1.5). The applications to GSK 3082 [77], oseltamivir [78], and tipranavir [79] highlight the utility of the technology. Significant advances in asymmetric additions have recently emerged that can change these perceptions. C–H activation for the stereoselective insertion of a heteroatom provides

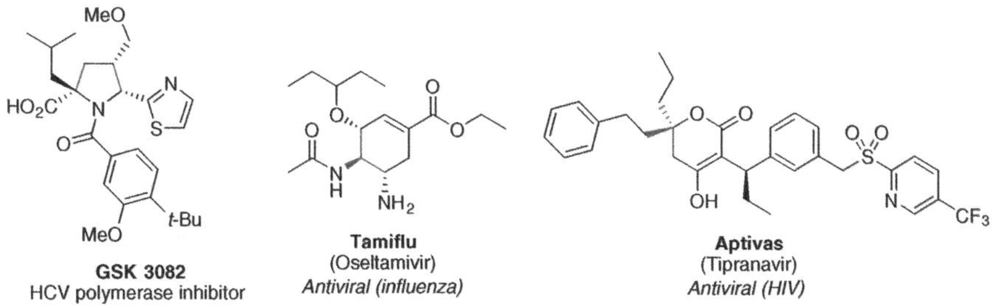

FIGURE 1.5 Selected examples of the application of asymmetric additions.

the opportunity to establish stereocenters from readily available materials [80]. The asymmetric addition to ketones has efficiently afforded chiral tertiary alcohols [81]. Recent advances in asymmetric allylations [82] and propargylations [83] provide access to chiral homoallylic and propargylic alcohols that are not readily accessed by hydrogenation of the corresponding 2,3-unsaturated ketones, due to their propensity to isomerize the pi-bond into conjugation. The ability to access a large number of chiral moieties and scaffolds through these advances will increase the value of asymmetric addition reactions in pharmaceutical production.

1.2.6 Metathesis

Metathesis has also impacted academic research more significantly than for the pharmaceutical industry. However, the scientific impact of metathesis technology cannot be overstated, as exemplified by the awarding of the 2005 Nobel Prize to Chauvin, Grubbs, and Schrock "for the development of the metathesis method in organic synthesis." Industrial application of the metathesis reaction dates back to the 1950s as exemplified by the Shell Higher Olefin Process (SHOP) [84]. Over the past several decades, the emergence of highly active and well-defined molybdenum and ruthenium metathesis catalysts allowed the chemoselectivity required for fine chemical and complex organic molecule synthesis. The notable catalysts are those developed by Grubbs and coworkers [85], Hoveyda and coworkers [86], Nolan and coworkers [87], Grela and coworkers [88], Schrock et al. [89], and Zhan [90]. Applications to the pharmaceutical industry are generally limited to the ring closing metathesis (RCM) for the construction of medium to large rings and cross metathesis (Scheme 1.8). The well-known tolerance of numerous functional groups to the metathesis conditions and the ability to utilize unactivated olefins presents significant utility for the incorporation into complex molecule synthesis.

The use of metathesis is more significant in early discovery with limited examples in development. Some notable examples are depicted in Figure 1.6, which include BILN 2061 [91], telcagepant [92], and SB-462795 [93]. The drawback to the incorporation of metathesis into a multikilogram batch is often the limited catalyst turnover and typical high dilution necessary to prevent dimerization. When conditions and the substrate are thoroughly optimized, < 1 mol% catalysts loadings with reasonable concentrations can often be achieved. The importance for substrate optimization is exemplified by the process reported by Boehringer Ingelheim for the synthesis of the macrocycle BILN 2061. By derivatizing an amide nitrogen, the molecule was shown to adopt a more favorable conformation for cyclization, which enabled the reaction to be conducted at 20 times the initial reaction concentration. Achieving a concentration of 0.2 M for the RCM equates to a 20-fold improvement

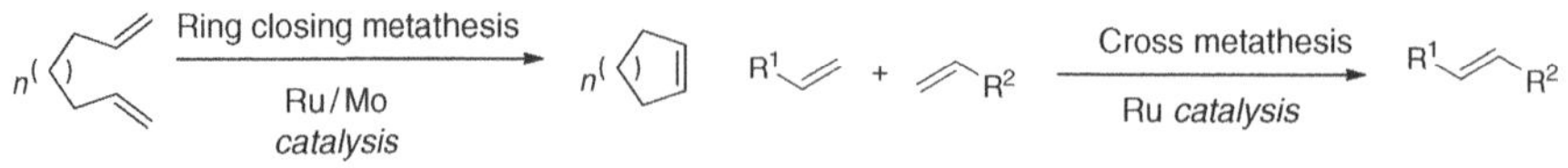

SCHEME 1.8   General types of metathesis utilized in the pharmaceutical industry.

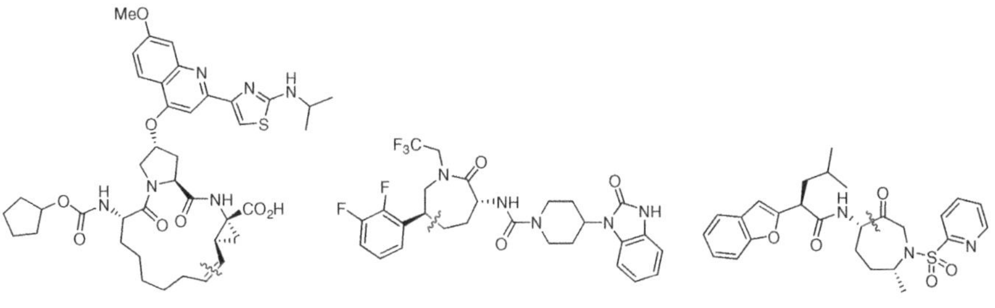

BILN 2061
Antiviral (HIV)

Telcagepant
CGRP receptor antagonist

SB-462795
Cathepsin K inhibitor

FIGURE 1.6 Selected examples for application of metathesis.

in the volume–time factor for production (*vida infra*). Similar to other transition metal catalysts processes, recent advances in metathesis chemistry have enabled formation of tetra-substituted olefins [94], asymmetric metatheses [95], and the metatheses of alkynes [96]. This research has directly increased the ability to construct the increasingly complex biologically active molecules that are being discovered and developed as the next generation of medicines.

1.3 CHALLENGES IN TAKING CATALYSIS TO INDUSTRIAL SCALES

In recent years, transition-metal catalysis has become a truly indispensible technology for industrial scale production of APIs. This is due to several inter-related factors (a) the constantly increasing regulatory requirements, for example, the strong regulatory pressure to develop a single-enantiomer of a drug as well as more stringent environmental protection legislations; (b) the pressure to reduce cost and time to market (a single day delay to the market will result in >$1 million loss for a $400 million annual revenue drug); and (c) the discovery of more efficient catalytic reactions from both academia and industry. The interplay of all these elements has resulted in the uptake of many of these catalytic methods for industrial scale API production. Successful examples of catalytic reactions on commercial scales are now routinely reported in the literature, yet there still exists a time delay between the initial (usually academic) reports on new catalysis and their implementation in industry.

The development and implementation of a catalytic reaction for industrial scale production is not trivial and many considerations have to be taken into account. The major factors that might hamper the technology transfer from lab-scale to large-scale production involve (a) the limited availability of the catalysts; (b) high cost of catalysts; (c) nonscalable operations (such as column chromatography, high dilution, extreme temperatures/pressures, unsafe reagents, etc.); and (d) the intellectual property (IP) issues (Fig. 1.7). Thus, a significant amount of process research and evaluation is almost always required to translate the chemistry demonstrated in initial reports into a "process-friendly" state.

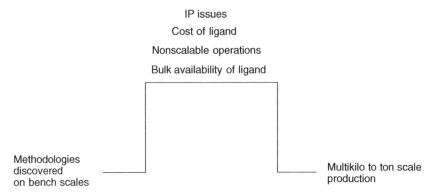

IP issues

Cost of ligand

Nonscalable operations

Bulk availability of ligand

Methodologies discovered on bench scales

Multikilo to ton scale production

FIGURE 1.7   Taking Benchtop reactions to large scales.

Perhaps most importantly though, it is generally in industry where the new catalysis is truly "stressed" with complex substrates containing multiple potentially interfering functional groups. At this time one often finds that the chemo-, stereo-, or regioselectivity of a catalyst is not as good as the performance shown for a limited number of substrates lacking true steric and electronic breadth. When catalysis performs poorly at this early stage, the reasons for this are rarely apparent. The substrate, the product, or the catalyst itself might be responsible, and significant effort is often required to elucidate these effects. This is likely the critical juncture where the speed with which new methodologies are embraced by industry is decided. When time and resource investments are not made for new chemistries that are capable of being truly transformative to API synthesis, their acceptance and use will be clearly delayed and an early opportunity missed.

To overcome these hurdles, it therefore takes the efforts and collaboration from both industry and academia. From the industry perspective, more efforts should be directed at enhancing collaboration with academia on catalysis research and pro-actively increasing investment in strategic technologies. The companies should encourage forward thinking and a long-term vision. They should ideally provide incentives and rewards for industrial scientists to take smart risks to innovate and explore new chemistry.

Along this line, several major pharmaceutical companies have already invested heavily in setting up new technological platforms such as high throughput catalysis screening. Boehringer Ingelheim, Merck, and Pfizer all have their own in-house catalysis groups dedicated to catalyst development and these organizations have proven to be very successful [97].

Academic researchers by contrast should become more aware of and appreciate the constraints of industry. The research programs should be designed in such a way that barriers for technology transfer of the newly developed chemistry into the industrial setting are minimized. Environmental friendly solvents and reagents should be included in the routine optimization for new methodologies. Another encouraging trend is that more research groups partner with chemical suppliers to make their new catalysts commercially available for initial testing. Simple steps like these can greatly facilitate the uptake of new catalytic methods for pharmaceutical applications.

To better understand the complexity of technology transfer to large-scale production and how the factors described above play out in the pharmaceutical industry, it is instructive to further consider Boehringer Ingelheim's BILN 2061 process. BILN 2061 (Ciluprevir, Fig. 1.8) was discovered by Boehringer Ingelheim (BI) as an HCV NS3 protease inhibitor [98] and entered the development phase in the late 2000 as the first small molecule therapy for hepatitis C infection. The most unique structural feature of the molecule was the 15-membered macrocycle ring containing a *cis*-double bond. The macrocyclization was achieved via a ruthenium-catalyzed RCM reaction. Back in 2000, RCM was largely an academic exercise [99] and there were no industrial applications of this reaction to large-scale API synthesis. In order to support clinical studies and the projected market supplies, BI required the development of a manufacturing process suitable for multiton production. A strategic decision was, therefore, made to invest in the development of the metathesis reaction for production scales [100].

At the outset of the project, Grubbs' first generation catalyst (**1**) [101] was used to cyclize the tripeptide diene **2** (Scheme 1.9). When the acyclic precursor was treated with >5 mol% catalyst in refluxing dichloromethane for 24 h, the desired macrocycle **3** was formed along with up to 50% of the *epi*-RCM product (epi-**3**). The extent of epimerization varied from batch to batch and was scale-dependent. In addition to the stereochemical concerns, other critical scale up issues for this key RCM reaction were identified, which included: (1) the high dilution (0.01 M) required to minimize the unproductive intermolecular metathesis side reactions, (2) high catalyst loadings, (3) a projected multimillion-dollar capital investment due to the high solvent volume, (4) the uncertain supply of the catalyst due to cost and availability concerns, and (5) a complicated IP situation. All of these issues need to be resolved before the RCM reaction could be reliably scaled up to support the clinical development of BILN 2061.

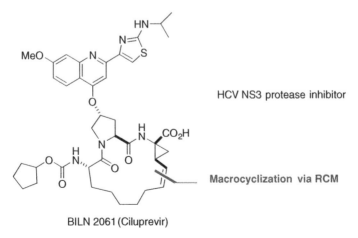

HCV NS3 protease inhibitor

Macrocyclization via RCM

BILN 2061 (Ciluprevir)

FIGURE 1.8   BILN 2061.

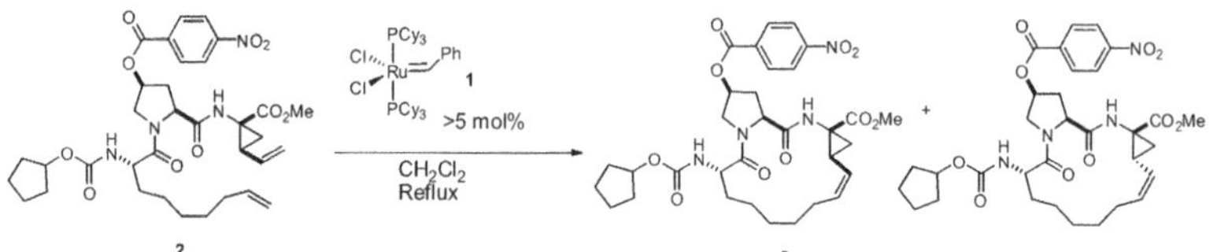

Reaction 2 3 Epi-3

SCHEME 1.9 Initial results for RCM reaction.

In order to better understand the RCM reaction, additional catalysts were screened (Fig. 1.9). With the second generation, NHC-containing catalysts from Grubbs and coworkers [102], Hoveyda and coworkers [103], and Grela and coworkers [104] (**4**, **5**, and **6**), very little isomerization occurred in the RCM reaction. However, these more reactive catalysts led to the formation of 8–10 mol% of the cyclic dimer, which not only resulted in lower yields but also complicated product purification. In contrast, the first generation Hoveyda catalyst **7** [105] that has somewhat attenuated activity, gave a much cleaner reaction with almost no dimer formation and no epimerization. With 3–5 mol% catalyst at 0.01 M diene concentration, 90–95% yields of compound **3** were obtained after 24 h in refluxing dichloromethane. More than 100 kg API was manufactured by using these conditions. Several critical scale up concerns still remained. The most serious issue was the high dilution required to effect an efficient macrocyclization. When the initial diene concentration was increased from 0.01 to 0.1 M, the yield of **3** dropped from >90% to 40%. This challenge is certainly not unique to the Boehringer's substrate. In fact, most of the RCM-based macrocyclizations required a substrate concentration ranging between 0.2 and 8.5 mM, and a catalyst loading at 2–10 mol% [106].

From a business perspective to secure the projected market supplies, the large reaction volume requirement of the RCM would necessitate a multimillion dollar capital investment to build a new production facility at Boehringer's manufacturing site. Although the RCM reaction itself is intrinsically green by virtue of being a catalytic process with excellent atom economy [107], the need for high dilution resulted in an E-factor [108] of 370 kg/kg for this step alone. This means that to

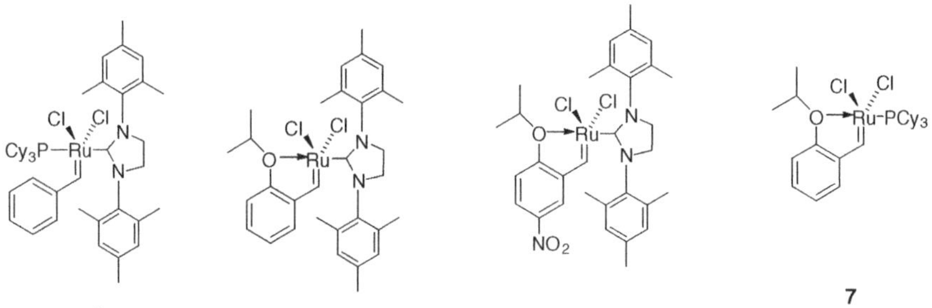

4 2nd generation Grubbs catalyst

5 2nd generation Hoveyda catalyst

6 Grela catalyst

7 1st generation Hoveyda catalyst

FIGURE 1.9 RCM catalysts.

cyclize 1 kg diene, ~370 kg of chemical waste would be generated. An average synthetic step should have an E-factor in the range of 5–30 kg/kg. In addition, it was highly desirable to replace dichloromethane with a greener solvent.

During the early work on the RCM reaction, a small but detectable effect on the RCM rate by varying the substituent on the remote C-4 position was noticed. Based on this observation, it was hypothesized that by changing the substituents on the acyclic precursor backbone, it might be possible to coil the molecule into a conformation that is more favorable for intramolecular cyclization. Indeed, it was found that the protecting group on the C-4 amide nitrogen has a profound influence on the conformation of the acyclic precursor and consequently the concentration requirement for the intramolecular macrocyclization.

A series of diene precursors with different substituents on the C-4 amide nitrogen were synthesized. Using Grela's catalyst, when R = H (**2**), an 82% yield of **3** was obtained at 0.01 M concentration (Scheme 1.10). The reaction was not only faster with Boc as a protecting group on the nitrogen as in compound **8** but also cleaner even at 0.2 M with lower catalyst loading (93% yield of **9**). Presumably, the introduction of the Boc group on the C-4 amide nitrogen relieved the ring strain of the acyclic precursor, thereby facilitating the desired intramolecular metathesis pathway. Detailed mechanistic studies using NMR techniques also showed that by changing from the unprotected substrate **2** to the Boc-protected diene **8**, the initiation site of metathesis shifted from the vinylcyclopropane to the nonenoic acid moiety (Scheme 1.11). It was hypothesized that initiation at the vinylcyclopropane leads to a slower reaction due to stabilization of the Ru carbene by chelation, whereas initiation at the nonenoic acid site accelerates the ring-closing step.

With this breakthrough discovery, the RCM reaction could be accomplished at more than 20-fold higher substrate concentration with the use of 0.05 mol% catalyst, to give 93% yield of the RCM product within a short reaction time. To prepare 1 kg of macrocycle intermediate **3**, only 0.5 g catalyst and 7.5 L of solvent were needed. As a result, the E-factor of this step was reduced from 370 to 52 kg/kg. The new RCM process thus improved the greenness of the reaction by about one order of magnitude. In addition, the reaction solvent was changed from dichloromethane to toluene without negatively impacting the performance of the RCM. More importantly, the

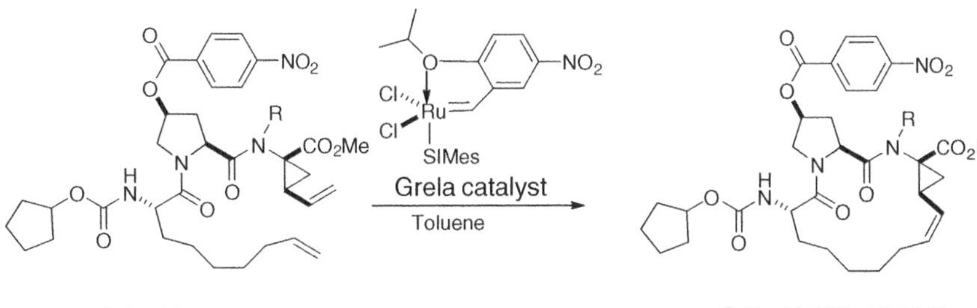

2: R = H
8: R = Boc

3: R = H: 82% at 0.01 M
    R = H: 40% at 0.1 M
9: R = Boc: 93% at 0.2 M

SCHEME 1.10 RCM reaction with modified substrates.

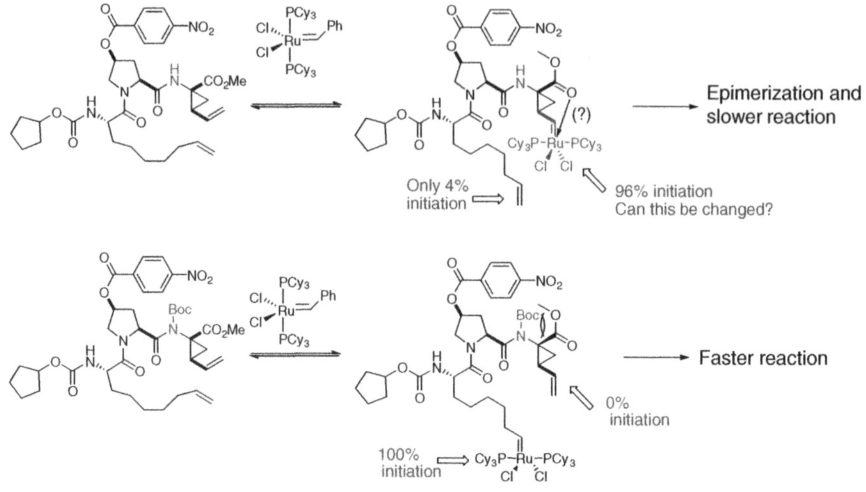

SCHEME 1.11   Initiation of RCM reaction with modified substrates.

new RCM process could be readily accommodated in the existing standard multi-purpose reactors in chemical production, and the need for a multimillion dollar capital investment was averted.

Securing the IP rights for various RCM catalysts was required to use this metathesis chemistry to supply the market with the HCV protease inhibitor. This "freedom-to-operate" is necessary to ensure the reliability of the drug supply once the compound is marketed. Any disruption of this supply could potentially affect patient safety, and was therefore unacceptable. The IP issue was taken very seriously and was handled with care throughout process development. In parallel to the scientific program, BI engaged in extensive contract negotiations with several technology owners and eventually was able to reach an agreement of collaboration with Professor Grela.

Through innovative chemical research and development, a breakthrough was achieved for the RCM reaction for BILN 2061 production. The new macrocyclization process runs at improved concentrations with a low catalyst loading (ca. 0.05%) to give the RCM product in >90% yield. Overall, these improvements significantly reduced the API cost. This achievement was highlighted in professor Grubbs' Nobel Prize Speech [109] as well as in Chemical & Engineering News [110]. The success story of BILN 2061 showed that in order to develop a commercially viable catalytic process, a multitude of inter-related factors need to be considered including reaction optimization, green chemistry, IP, the manufacturing facility, catalyst cost, and raw material sourcing. These challenges can only be met through the collaborative work of a multidisciplinary team.

1.4   SUMMARY AND FUTURE OUTLOOK

The rapid growth of transition-metal catalysis has enabled the pharmaceutical industry to accelerate both drug discovery and drug development. Future catalysis research will undoubtedly bring more exciting chemistry and allow novel transformations to take place with ultra low catalyst loadings and good functional group compatibility. Collaboration of synthetic organic, physical organic, as well as organometallic, analytical and computational chemists provides the foundation to elucidate mechanistic pathways, rationalize catalyst behavior, invent new methodologies, and optimize and implement processes into chemical production. The ultimate goal of this effort is to develop catalysts that impart the selectivities, reactivity, and atom economy needed for cost-effective approaches to a variety of critical chemical transformations. Finally, a close collaboration between academia and industry is vitally important for the expedient implementation of catalysis on scale. It is our hope that this book will not only serve as an overview of the fast-growing field of transition-metal catalysis but also provide impetus and inspirations for the future discovery of more efficient and practical catalysts to help us tackle the economical and ecological challenges in the twenty-first century.

REFERENCES

[1] North American Catalysis, Society. Available at http://www.nacatsoc.org/what.asp

[2] Recognizing the Best in Innovation: Breakthrough Catalyst. *R&D Magazine*, September 2005, p. 20.

[3] *World Catalyst Market*, June, 2008.

[4] Anastas, P. T.; Warner, J. C., Eds., *Green Chemistry Theory and Practice*; Oxford University Press: Oxford, 1998.

[5] Noyori, R. *Adv. Syn. Cat.* **2003**, *345*, 15.

[6] Clark, G. S. *Perfumer Flavorist* **2007**, *32*(12), 38.

[7] Jäkel, C.; Paciello, R. U.S. Patent 7534921 (05/19/2009).

[8] McCoy, M. *Chem. Eng. News* **2010**, *88*(35), 15.

[9] For a general review, see: Slagt, V. F.; Vries, A. H. M.; Vries, J. G.; Kellogg, R. M. *Org. Process Res. Dev.* **2010**, *14*, 30–47.

[10] Tamura, M.; Kochi, J. K. *J. Am. Chem. Soc.* **1971**, *93*, 1487.

[11] (a) Tamao, K.; Sumitani, K.; Kumada, M. *J. Am. Chem. Soc.* **1972**, *94*, 4374; (b) Corriu, R. J. P.; Masse, J. P. *J. Chem. Soc. Chem. Commun.* **1972**, 144.

[12] Miyaura, N.; Suzuki, A. *Chem. Rev.* **1995**, *95*, 2457.

[13] (a) Denmark, S. E.; Regens, C. S. *Acc. Chem. Res.* **2008**, *41*, 1486; (b) Denmark, S. E. *J. Org. Chem.* **2009**, *74*, 2915.

[14] (a) Stille, J. *Angew. Chem. Int. Ed.* **1986**, *25*, 508; (b) Farina, V.; Krishnamurthy, D.; Scott, W. J. *Org. React.* **1997**, *50*, 1.

[15] Tamao, K. In *Comprehensive Organic Synthesis*; Trost, B., Ed., Pergamon: Oxford, 1991; Vol. 3, p. 435.

[16] (a) Negishi, E.; King, A.; Okukado, N. *J. Org. Chem.* **1977**, *42*, 1821; (b) Knochel, P.; Singer, R. *Chem. Rev.* **1993**, *93*, 2117.

[17] Heck, R. F. *Org. React.* **1982**, *27*, 345.

[18] Bryndza, H. E.; Tam, W. *Chem. Rev.* **1988**, *88*, 1163.

[19] (a) Duncia, J. V.; Carini, D. J.; Chiu, A. T.; Johnson, A. L.; Price, W. A.; Wong, P. C.; Wexler, R. R.; Timmermans, P. B. M. W. M. *Med. Res. Rev.* **1992**, *12*, 149; (b) Blatcher, P.; Carter, M. W.; Butina, D.; Owen, M. R. U.S. Patent 4997841 (1991); (c) Labelle, M.; Belley, M.; Gareau, Y.; Gauthier, J. Y.; Guay, D.; Gordon, R.; Grossman, S. G.; Jones, T. R.; Leblanc, Y.; McAuliffe, M.; MaFarlane, C. S.; Masson, P.; Metters, K. M.; Quimet, N.; Patrick, D. H.; Piechuta, H.; Rochette, C.; Sawyer, N.; Xiang, Y. B.; Pickett, C. B.; Ford-Hutchinson, A. W.; Zamboni, R. J.; Young, R. N. *Bioorg. Med. Chem. Lett.* **1995**, *5*, 283.

[20] Dugger, R. W.; Ragan, J. A.; Ripin, D. H. B. *Org. Process Res. Dev.* **2005**, *9*, 253.

[21] For a selected example, see: Littke, A. F.; Fu, G. C. *J. Am. Chem. Soc.* **2001**, *123*, 6989.

[22] Yu, D.-G.; Li, B.-J.; Zheng, S.-F.; Guan, B.-T.; Wang, B.-Q.; Shi, Z.-J. *Angew. Chem. Int. Ed.* **2010**, *49*, 4566.

[23] (a) Watson, M. P.; Jacobsen, E. N. *J. Am. Chem. Soc.* **2008**, *130*, 12594; (b) Nakao, Y.; Ebata, S.; Yada, A.; Hiyama, T.; Ikawa, M.; Ogoshi, S. *J. Am. Chem. Soc.* **2008**, *130*, 12874; (c) Najera, C.; Sansano, J. M. *Angew. Chem. Int. Ed.* **2009**, *48*, 2452.

[24] Reeves, J. T.; Fandrick, D. R.; Tan, Z.; Song, J. J.; Lee, H.; Yee, N. K.; Senenayake, C. H. *Org. Lett.* **2010**, *12*, 4388.

[25] (a) Littke, A. F. *Modern Arylation Methods*; Wiley-VCH: Weinheim, 2009; pp. 25–67; (b) Furuya, T.; Benitez, D.; Tkatchouk, E.; Strom, A. E.; Tang, P.; Goddard, W. A.; Ritter, T. *J. Am. Chem. Soc.* **2010**, *132*, 5922; (c) Ohashi, M.; Kambara, T.; Hatanaka, T.; Saijo, H.; Doi, R.; Ogoshi, S. *J. Am. Chem. Soc.* **2011**, *133*, 3256.

[26] (a) Hama, T.; Liu, X.; Culkin, D. A.; Hartwig, J. F. *J. Am. Chem. Soc.* **2003**, *125*, 11176; (b) Kundig, E. P.; Seidel, T. M.; Jia, Y.; Bernardinelli, G. *Angew. Chem. Int. Ed.* **2007**, *46*, 8484; (c) Garcia-Fortanet, J.; Buchwald, S. L. *Angew. Chem. Int. Ed.* **2008**, *47*, 8108.

[27] (a) Gooben, L. J.; Zimmermann, B.; Knauber, T. *Angew. Chem. Int. Ed.* **2008**, *47*, 7103; (b) Goossen, L. J.; Rodriguez, N.; Linder, C. *J. Am. Chem. Soc.* **2008**, *130*, 15248; (c) Gooben, L. J.; Rodriguez, N.; Lange, P. P.; Linder, C. *Angew. Chem. Int. Ed.* **2010**, *49*, 1111.

[28] (a) Senecal, T. D.; Parsons, A. T.; Buchwald, S. L. *J. Org. Chem.* **2011**, *76*, 1174; (b) Zhang, C.-P.; Wang, Z.-L.; Chen, Q.-Y.; Zhang, C.-T.; Gu, Y.-C.; Xiao, J.-C. *Angew. Chem. Int. Ed.* **2011**, *50*, 1896.

[29] (a) Chen, X.; Engle, K. M.; Wang, D.-H.; Yu, J.-Q. *Angew. Chem. Int. Ed.* **2009**, *48*, 5094; (b) Peng, H. M.; Dai, L.-X.; You, S. L. *Angew. Chem. Int. Ed.* **2010**, *49*, 5826; (c) Rousseaux, S.; Gorelsky, S. L.; Chung, B. K. W.; Fagnou, K. *J. Am. Chem. Soc.* **2010**, *132*, 10692.

[30] (a) Bermejo, A.; Ros, A.; Fernandez, R.; Lassaletta, J. M. *J. Am. Chem. Soc.* **2008**, *130*, 15798; (b) Shen X.; Jones, G. O.; Watson, D. A.; Bhayana, B.; Buchwald, S. L. *J. Am. Chem. Soc.* **2010**, *132*, 11278.

[31] LaPlante, S. R.; Edwards, P. J.; Fader, L. D.; Jakalian, A.; Hucke, O. *Chem. Med. Chem.* **2011**, *6*, 505.

[32] (a) For a selected review, see: Glorius, F. *Angew. Chem. Int. Ed.* **2008**, *47*, 8347; (b) Owston, N. A.; Fu, G. C. *J. Am. Chem. Soc.* **2010**, *132*, 11908–11909; (c) Lou, S.; Fu, G. C. *J. Am. Chem. Soc.* **2010**, *132*, 1264.

[33] (a) For a comprehensive review, see: Ley, S. V.; Thomas, A. W. *Angew. Chem. Int. Ed.* **2003**, *42*, 5400; (b) Beletskaya, I. P.; Cheprakov, A. V. *Chem. Rev.* **2004**, 2337.

[34] Ullmann, F. *Ber. Dtsch. Chem. Ges.* **1903**, *36*, 2382.

[35] Goldberg, I. *Ber. Dtsch. Chem. Ges.* **1906**, *39*, 1691.

[36] Kiyomori, A.; Marcoux, J.-F.; Buchwald, S. L. *Tetrahedron Lett.* **1999**, *40*, 2657.

[37] Goodbrand, H. B.; Hu, N.-X. *J. Org. Chem.* **1999**, *64*, 670.

[38] Kosugi, M.; Kameyama, M.; Migita, T. *Chem. Lett.* **1983**, 927.

[39] Guram, A. S.; Rennels, R. A.; Buchwald, S. L. *Angew. Chem. Int. Ed.* **1995**, *34*, 1348.

[40] Louie, J.; Hartwig, J. F. *Tetrahedron Lett.* **1995**, *36*, 3609.

[41] (a) Ishiyama, T.; Matsuda, N.; Miyaura, N.; Suzuki, A. *J. Am.Chem. Soc.* **1993**, *115*, 11018; (b) Chemler, S. R.; Trauner, D.; Danishefsky, S. J. *Angew. Chem. Int. Ed.* **2001**, *40*, 4544.

[42] Loiseleur, O.; Kaufmann, D.; Abel, S.; Buerger, H. M.; Meisenbach, M.; Schmitz, B.; Sedelmeier, G.W.O. Patent 03/066613 (2003).

[43] Morita, S.; Kitano, K.; Matsubara, J.; Ohtani, T.; Kawano, Y.; Otsubo, K.; Uchida, M. *Tetrahedron* **1998**, *54*, 4811.

[44] Damon, D. B.; Dugger, R. W.; Hubbs, S. E.; Scott, J. M.; Scott, R. W. *Org. Process Res. Dev.* **2006**, *10*, 472.

[45] (a) Shen, Q.; Hartwig, J. F. *Org. Lett.* **2008**, *10*, 4109; (b) Vo, G. D.; Hartwig, J. F. *J. Am. Chem. Soc.* **2009**, *131*, 11049.

[46] Hartwig, J. F. *Acc. Chem. Res.* **2008**, *41*, 1534.

[47] Sun, K.; Li, Y.; Xiong, T.; Zhange, J.; Zhang, Q. *J. Am. Chem. Soc.* **2011**, *133*, 1694.

[48] (a) Boebel, T. A.; Hartwig, J. F. *J. Am. Chem. Soc.* **2008**, *130*, 7534; (b) Boebel, T. A.; Hartwig, J. F. *Organometallics* **2008**, *27*, 6013; (c) Kawamorita, S.; Ohmiya, H.; Hara, K.; Fukuoka, A.; Sawamura, M. *J. Am. Chem. Soc.* **2009**, *131*, 5058.

[49] (a) W. Tang, X. Zhang. *Chem. Rev.* **2003**, *103*, 3029; (b) Schenkel, L. B.; Ellman, J. A. *J. Org. Chem.* **2004**, *69*, 1800; (c) Barbaro, P.; Bianchini, C.; Giambastiani, G.; Parisel, S. L. *Coord. Chem. Rev.* **2004**. *248*, 2131; (d) Blaser, H.-U.; Studer, M. *Acc. Chem. Res.* **2007**, *40*, 1348.

[50] Blaser, H. U.; Malan, C.; Pugin, B.; Spindler, F.; Steiner, H.; Studer, M. *Adv. Synth. Cat.* **2003**, *343*, 103.

[51] Yoon, T. P.; Jacobsen, E. N. *Science* **2003**, *299*, 1691.

[52] Farina, V.; Reeves, J. T.; Senanayake, C. H.; Song, J. J. *Chem. Rev.*, **2006**, *106*, 2734.

[53] (a) Shimizu, H.; Nagasaki, I.; Matsumura, K.; Sayo, N.; Saito, T. *Acc. Chem. Res.* **2007**, *40*, 1385; (b) Schultz, C. S.; Krska, S. W. *Acc. Chem. Res.* **2007**, *40*, 1320.

[54] (a) Knowles, W. S.In *Large Scale Asymmetric Catalysis*, Blaser, H.U.; Schmidt, E., Eds., Wiley-VCH: Weinheim, 2003; p. 23; (b) Selke, R. In *Large Scale Asymmetric Catalysis*, Blaser, H. U.; Schmidt, E., Eds., Wiley-VCH: Weinheim, **2003**; p. 39.

[55] Lorenz, J. C.; Busacca, C. A.; Feng, X. W.; Grinberg, N.; Haddad, N.; Johnson, J.; Kapadia, S.; Lee, H.; Saha, A.; Sarvestani, M.; Spinelli, E. M.; Varsolona, R.; Wei, X.; Zeng, X.; Senanayake, C. H. *J. Org. Chem.* **2010**, *75*, 1155.

[56] Bell, S.; Wustenberg, B.; Kaiser, S.; Menges, F.; Netscher, T.; Pfaltz, A. *Science* **2006**, *311*, 642.

[57] Ley, S. V.; Norman, J.; Griffith, W. P.; Marsden, S. P. *Synthesis* **1994**, 639.

[58] For selected recent examples, see: (a) Stang, E. M.; White, M. C. *Nat. Chem.* **2009**, *1*, 547; (b) McLaughlin, E. C.; Choi, H.; Wang, K.; Chiou, G.; Doyle, M. P. *J. Org. Chem.* **2009**, *74*, 730; (c) Henderson, W. H.; Check, C. T.; Proust, N.; Stambuli, J. P. *Org. Lett.*

2010, *12*, 824; (d) Pilarski, L. T.; Janson, P. G.; Szabo, K. J. *J. Org. Chem.* **2011**, *76*, 1503.

[59] (a) Tsuji, J. *Synthesis* **1984**, 369; (b) Feringa, B. L. Wacker oxidation. In *Transition Metals for Organic Synthesis*; Wiley-VCH: Weinheim, 1998; Vol. 2, p. 307.

[60] (a) Pfenninger, A. *Synthesis* **1986**, 89; (b) Katsuki, T.; Martin, V. S. *Org. React.* **1996**, *48*, 1.

[61] Jacobsen, E. N. Asymmetric catalytic epoxidation of unfunctionalized olefins. In *Catalytic Asymmetric Synthesis*; Ojima, I., Ed., Wiley-VCH: New York, 1993.

[62] Kolb, H. C.; VanNieuwenhze, M. S.; Sharpless, K. B. *Chem. Rev.* **1994**, *94*, 2483.

[63] (a) Li, G.; Chang, H.-T.; Sharpless, K. B. *Angew. Chem. Int. Ed.* **1996**, *35*, 451. (b) Reiser, O. *Angew. Chem. Int. Ed.* **1996**, *35*, 1308.

[64] (a) Zhao, S. H.; Samuel, O.; Kagan, H. B. *Tetrahedron* **1987**, *43*, 5135; (b) Donnoli, M. I.; Superchi, S.; Rosini, C. *J. Org. Chem.* **1998**, *63*, 9392.

[65] Constable, D. J. C.; Dunn, P. J.; Hayler, J. D.; Humphrey, G. R.; Leazer, J. L., Jr.; Linderman, R. J.; Lorenz, K.; Manley, J.; Pearlman, B. A.; Wells, A.; Zaks, A.; Zhang, T. Y. *Green Chem.* **2007**, *9*, 411.

[66] Wang, F.; Montemayor, L. K.; Che, D.; Horne, S. E. U.S. Patent 2007-797921 20070509 (2010).

[67] Watanabe, M.; Koike, H.; Ishiba, T.; Okada, T.; Seo, S.; Hirai, K. *Bio Med. Chem. Lett.* **1997**, *5*, 437.

[68] Senanayake, C. H.; Smith, G. B.; Ryan, K. M.; Fredenburgh, L. E.; Liu, J. *Tetrahedron* **1996**, *37*, 3271.

[69] Kimura, T.; Yamamoto, N.; Suzuki, Y.; Kawano, K.; Norimine, Y.; Ito, K.; Nagato, S.; Iimura, Y.; Yonaga, M. *J. Org. Chem.* **2002**, *67*, 6228.

[70] Covell, D. C.; White, M. C. *Angew. Chem. Int. Ed.* **2008**, *47*, 6448.

[71] Xie, J.; Huang, Z.-Z. *Angew. Chem. Int. Ed.* **2010**, *49*, 10181.

[72] (a) Zhang, Y.-H.; Yu, J.-Q. *J. Am. Chem. Soc.* **2009**, *131*, 14654; (b) Chen, M. S.; White, M. C. *Science* **2010**, *327*, 566.

[73] (a) Betti, M.; Lucchi, E. *Boll. Sci. Fac. Chim. Ind. Bologna* **1940**, 2–5; *Chem. Abstr.* **1940**, *34*, 2354; (b) Tarbell, D. S.; Paulson, M. C. *J. Am. Chem. Soc.* **1942**, *64*, 2842.

[74] Cohen, H. L.; Wright, G. F. *J. Org. Chem.* **1953**, *18*, 432.

[75] (a) Oguni, N.; Omi, T.; Yamomoto, Y.; Nakamura, A. *Chem. Lett.* **1983**, 841; (b) Oguni, N.; Omi, T. *Tetrahedron Lett.* **1984**, *25*, 2823.

[76] Hawkins, J. M.; Watson, T. J. N. *Angew. Chem. Int. Ed.* **2004**, *43*, 3224.

[77] Agbodjan, A. A.; Cooley, B. E.; Copley, R. C. B.; Corfield, J. A.; Flanagan, R. C.; Glover, B. N.; Guidetti, R.; Haigh, D.; Howes, P. D.; Jackson, M. M.; Matsuoka, R. T.; Medhurst, K. J.; Millar, A.; Sharp, M. J.; Slater, M. J.; Toczko, J. F.; Xie, S. *J. Org. Chem.* **2008**, *73*, 3094.

[78] Trost, B. M.; Zhang, T. *Eur. J. Org. Chem.* **2011**, *17*, 3630.

[79] Trost, B. M.; Andersen, N. G. *J. Am. Chem. Soc.* **2002**, *124*, 14320.

[80] (a) Maier, T. C.; Fu, G. C. *J. Am. Chem. Soc.* **2006**, *128*, 4594; (b) Lee, E. C.; Fu, G. C. *J. Am. Chem. Soc.* **2007**, *129*, 12066.

[81] Hatano, M.; Ishihara, K.; *Synthesis* **2008**, 1647.

[82] (a) Wada, R.; Oisaki, K.; Kanai, M.; Shibasaki, M. *J. Am. Chem. Soc.* **2004**, *126*, 8910; (b) Shi, S.-L.; Xu, L.-W.; Oisaki, K.; Kanai, M.; Shibasaki, M. *J. Am. Chem. Soc.* **2010**, *132*, 6638.

[83] Fandrick, D. R.; Fandrick, K. R.; Reeves, J. T.; Tan, Z.; Tang, W.; Capacci, A. G.; Rodriguez, S.; Song, J. J.; Lee, H.; Yee, N. K.; Senenayake, C. H. *J. Am. Chem. Soc.* **2010**, *132*, 7600.

[84] Ivin, K. J.; Mol, J. C. *Olefin Metathesis and Metathesis Polymerisation*; Academic Press: San Diego, 1997.

[85] Scholl, M.; Ding, S.; Lee, S. W.; Grubbs, R. H. *Org. Lett.* **1999**, *1*, 953.

[86] Garber, S. B.; Kingsbury, J. S.; Gray, B. L.; Hoveyda, A. H. *J. Am. Chem. Soc.* **2000**, *122*, 8168.

[87] Huang, J.; Stevens, E. D.; Nolan, S. P.; Peterson, J. L. *J. Am. Chem. Soc.* **1999**, *121*, 2674.

[88] Michrowska, A.; Bujok, R.; Harutyunyan, S.; Sashuk, V.; Dolgonos, G.; Grela, K. *J. Am. Chem. Soc.* **2004**, *126*, 9318.

[89] Schrock, R. R., Murdzek, J. S.; Bazan, G. C.; Robbins, J.; DiMare, M.; O'Regan, M. *J. Am. Chem. Soc.* **1990**, *112*, 3875.

[90] Zhan, Z.-Y. U.S. Patent 0043180 (2007).

[91] (a) Shu, C.; Zeng, X.; Hao, M.-H.; Wei, X.; Yee, N. K.; Busacca, C. A.; Han, Z.; Farina, V.; Senanayake, C. H. *Org. Lett.* **2008**, *10*, 1303; (b) Farina, V.; Shu, C.; Zeng, X.; Wei, X.; Han, Z.; Yee, N. K.; Senanayake, C. H. *Org. Process Res. Dev.* **2009**, *13*, 250.

[92] Burgey, C. S.; Paone, D. V.; Shaw, A. W.; Deng, J. Z.; Nguyen, D. N.; Potteiger, C. M.; Graham, S., Vacca, J. P.; Williams, T. M. *Org. Lett.* **2008**, *10*, 3235.

[93] Marquis, R. W.; Ru, Y.; LoCastro, S. M.; Zeng, J.; Yamashita, D. S.; Oh, H.-J.; Erhard, K. F.; Davis, L. D.; Tomaszek, T. A.; Tew, D.; Salyers, K.; Proksch, J.; Ward, K.; Smith, B.; Levy, M.; Cummings, M. D.; Haltiwanger, R. C.; Trescher, G.; Wang, B.; Hemling, M. E.; Quinn, C. J.; Cheng, H.-Y.; Lin, F.; Smith, W. W.; Janson, C. A.; Zhao, B.; McQueney, M. S.; D'Alessio, K.; Lee, C.-P.; Marzulli, A.; Dodds, R. A.; Blake, S.; Hwang, S.-M.; James, I. E.; Gress, C. J.; Bradley, B. R.; Lark, M. W.; Gowen, M.; Veber, D. F. *J. Med. Chem.* **2001**, *44*, 1380.

[94] (a) Stewart, I. C.; Ung, T.; Pletnev, A. A.; Berlin, J. M; Grubbs, R. H.; Schrodi, Y. *Org. Lett.* **2007**. *9*, 1589; (b) Vorfalt, T.; Leuthausser, S.; Plenio, H. *Angew. Chem. Int. Ed.* **2009**, *48*, 5191.

[95] Sattely, E. S.; Meek, S. J.; Malcolmson, S. J.; Schrock, R. R.; Hoveyda, A. H. *J. Am. Chem. Soc.* **2009**, *131*, 943.

[96] Heppekausen, J.; Stade, R.; Goddard, R.; Furstner, A. *J. Am. Chem. Soc.* **2010**, *132*, 11045.

[97] (a) Tang, W.; Capacci, A. G.; Wei, X.; Li, W.; White, A.; Patel, N. D.; Savoie, J.; Gao, J. J.; Rodriguez, S.; Qu, B.; Haddad, N.; Lu, B. Z.; Krishnamurthy, D.; Yee, N. K.; Senanayake, C. H. *Angew. Chem. Int. Ed. Engl.* **2010**, *49*, 5879; (b) Fandrick, D. R.; Fandrick, K. R.; Reeves, J. T.; Tan, Z.; Tang, W.; Capacci, A. G.; Rodriguez, S.; Song, J. J.; Lee, H.; Yee, N. K.; Senanayake, C. H. *J. Am. Chem. Soc.* **2010**, *132*, 7600; (c) Tang, W.; Capacci, A. G.; White, A.; Ma, S.; Rodriguez, S.; Qu, B.; Savoie, J.; Patel, N. D.; Wei, X.; Haddad, N.; Grinberg, N.; Yee, N. K.; Krishnamurthy, D.; Senanayake, C. H. *Org. Lett.* **2010**, *12*, 1104; (d) Lorenz, J. C.; Busacca, C. A.; Feng, X.; Grinberg, N.; Haddad, N.; Johnson, J.; Kapadia, S.; Lee, H.; Saha, A.; Sarvestani, M.; Spinelli, E. M.; Varsolona, R.; Wei, X.; Zeng, X.; Senanayake, C. H. *J. Org. Chem.* **2010**, *75*, 1155; (e) Busacca, C. A.; Lorenz, J. C.; Grinberg, N.; Haddad, N.; Lee, H.; Li, Z.; Liang, M.; Reeves, D.; Saha, A.; Varsolona, R.; Senanayake, C. H. *Org. Lett.* **2008**, *10*, 341.

[98] (a) Lamarre, D.; Anderson, P. C.; Bailey, M.; Beaulieu, P.; Bolger, G.; Bonneau, P.; Boes, M.; Cameron, D. R.; Cartier, M.; Cordingley, M. G.; Faucher, A.-M.; Goudreau, N.; Kawai, S. H.; Kukolj, G.; Lagacé L.; LaPlante, S. R.; Narjes, H.; Poupart, M.-A.; Rancourt, J.; Sentjens, R. E.; St. George, R.; Simoneau, B.; Steinmann, G.; Thibeault, D.; Tsantrizos, Y. S.; Weldon, S. M.; Yong, C.-L.; Llinás-Brunet, M. *Nature* **2003**, *426*, 186; (b) Llinás-Brunet, M.; Bailey, M. D.; Bolger, G.; Brochu, C.; Faucher, A.-M.; Ferland, J. M.; Garneau, M.; Ghiro, E.; Gorys, V.; Grand-Maître, C.; Halmos, T.; Lapeyre-Paquette, N.; Liard, F.; Poirier, M.; Rhéaume, M.; Tsantrizos, Y. S.; Lamarre, D. *J. Med. Chem.* **2004**, *47*, 1605; (c) Tsantrizos, Y.; Bolger, G.; Bonneau, P.; Cameron, D. R.; Goudreau, N.; Kukolj, G.; LaPlante, S. R.; Llinás-Brunet, M.; Nar, H.; Lamarre, D. *Angew. Chem. Int. Ed.* **2003**, *42*, 1356; (d) Goudreau, N.; Cameron, D. R.; Bonneau, P.; Gorys, V.; Plouffe, C.; Poirier, M.; Lamarre, D.; Llinás-Brunet, M. *J. Med. Chem.* **2004**, *47*, 123; (e) Rancourt, J.; Cameron, D. R.; Gorys, V.; Lamarre, D.; Poirier, M.; Thibeault, D.; Llinás-Brunet, M. *J. Med. Chem.* **2004**, *47*, 2511.

[99] (a) Grubbs, R. H., Ed. *Handbook of Metathesis*; Wiley-VCH: Weinheim, Germany, 2003; (b) Grubbs, R. H. *Tetrahedron* **2004**, *60*, 7117.

[100] (a) Medicinal Chemistry Synthesis: Faucher, A.-M.; Bailey, M. D.; Beaulieu, P. L.; Brochu, C.; Duceppe, J.-S.; Ferland, J.-M.; Ghiro, E.; Gorys, V.; Halmos, T.; Kawai, S. H.; Poirier, M.; Simoneau, B.; Tsantrizos, Y. S.; Llinás-Brunet, M. *Org. Lett.* **2004**, *4*, 2901; (b) First Large-Scale Synthesis: Yee, N. K.; Farina, V.; Houpis, I. N.; Haddad, N.; Frutos, R. P.; Gallou, F.; Wang, X.-J.; Wei, X.; Simpson, R. D.; Feng, X.; Fuchs, V.; Xu, Y.; Tan, J.; Zhang, L.; Xu, J.; Smith-Keenan, L. L.; Vitous, J.; Ridges, M. D.; Spinelli, E. M.; Johnson, M.; Donsbach, K.; Nicola, T.; Brenner, M.; Winter, E.; Kreye, P.; Samstag, W. *J. Org. Chem.* **2006**, *71*, 7133; (c) Nicola, T.; Brenner, M.; Donsbach, K.; Kreye, P. *Org. Proc. Res. Dev.* **2005**, *9*, 513; (d) Highly Practical Large-Scale Synthesis: Shu, C. et al. *Org. Lett.* **2008**, *10*, 1303; (e) Farina, V.; Shu, C.; Zeng, X.; Wei, X.; Han, Z.; Yee, N. K.; Senanyake, C. H. *Org. Proc. Res. Dev.* **2009**, *13*, 250.

[101] Schwab, P.; Grubbs, R. H.; Ziller, J. W. *J. Am. Chem. Soc.* **1996**, *118*, 100.

[102] Scholl, M.; Trnka, T. M.; Morgan, J. P.; Grubbs, R. H. *Tetrahedron Lett.* **1999**, *40*, 2247.

[103] Garber, S. B.; Kingsbury, J. S.; Gray, B. L.; Hoveyda, A. H. *J. Am. Chem. Soc.* **2000**, *122*, 8168.

[104] Michrowska, A.; Bujok, R.; Harutyunyan, S.; Sashuk, V.; Dolgonos, G.; Grela, K. *J. Am. Chem. Soc.* **2004**, *126*, 9318.

[105] Kingsbury, J. S.; Harrity, J. P. A.; Bonitatebus, P. J., Jr., Hoveyda, A. H. *J. Am. Chem. Soc.* **1999**, *121*, 791.

[106] Gradillas, A.; Perez-Castells, J. *Angew. Chem. Int. Ed. Engl.* **2006**, *45*, 6086.

[107] Anastas, P. T.; Warner, J. C., Eds., *Green Chemistry Theory and Practice*; Oxford University Press: Oxford, 1998.

[108] E-Factor Definition: (a) Sheldon, R. A. *Green Chem.* **2007**, *9*, 1273; (b) Sheldon, R. A. *Chem. Ind.* **1992**, 903.

[109] Grubbs, R. H. *Angew. Chem. Int. Ed. Engl.* **2006**, *45*, 3760.

[110] (a) "Sustainable Syntheses," Thayer, A. M. in *Chemical & Engineering News* June 8, 2009; (b) "Drug Production: Boehringer-Ingelheim blazes trail in Scaling Up Metathesis Reaction" Thayer, A. M. in *Chemical & Engineering News* Feb. 12, 2007.

2

SELECTED APPLICATIONS OF TRANSITION METAL-CATALYZED CARBON–CARBON CROSS-COUPLING REACTIONS IN THE PHARMACEUTICAL INDUSTRY

Hong C. Shen

Applications of Transition Metal Catalysis in Drug Discovery and Development: An Industrial Perspective, First Edition. Edited by Matthew L. Crawley and Barry M. Trost.
© 2012 John Wiley & Sons, Inc. Published 2012 by John Wiley & Sons, Inc.

2.1   INTRODUCTION

In the past few decades, myriads of transition metal-catalyzed carbon–carbon cross-coupling reactions have been discovered and developed [1]. Many of these powerful methods have been successfully harnessed in the pharmaceutical industry for large-scale synthesis of small molecule drugs. In part, as a result of their synthetic accessibility, biaryl moieties have become privileged pharmacophores to be incorporated into numerous drugs and clinical candidates. The most well-known biaryl drugs (Fig. 2.1) include Cozaar™ (Losartan, Dupont-Merck, angiotensin II (AngII) receptor antagonist) [2], Atarcand™ (Candesartan, Takeda, AngII receptor antagonist) [3], Avapro™ (Irbesartan, Bristol-Myers Squibb and Sanofi, AngII receptor antagonist) [4], Dolobid™ (Diflunisal, Merck, cyclooxygenase-1 (COX-1) and cyclooxygenase-2 (COX-2) dual inhibitors) [5], Vioxx™ (Rofecoxib, Merck, COX-2 inhibitor) [6], Arcoxia™ (Etoricoxib, Merck, COX-2 inhibitor) [7], Celebrex™ (Celecoxib, Pfizer, COX-1 and COX-2 dual inhibitors) [8], Reyataz™ (Atazanavir, Bristol-Myers Squibb, HIV protease inhibitor) [9], Evista™ (Raloxifene, Eli Lilly, selective estrogen receptor modulator) [10], and Gleevec™ (Imatinib, Novartis, tyrosine kinase inhibitor) [11]. In addition, drugs possessing alkynyl, vinyl, or alkyl aryl moieties could also be accessed via C–C cross-coupling reactions. The examples include Tazorac™ (Tazarotene, Allergan, receptor-selective retinoid) [12], Maxalt™ (rizatriptan benzoate, Merck, 5-hydroxytryptamine (5-HT) subtype 1B and 1D agonist) [13], Singulair™ (Montelukast sodium, Merck, cysteinyl leukotriene antagonist) [14], Advil™/Motrin™ (Ibuprofen, COX-1 and COX-2 dual inhibitors) [15], Lamisil™ (Terbinafine, Sandoz, squalene epoxide inhibitor) [16], Relpax™ (Eletriptan, Pfizer, 5-HT$_{1B}$ and 5-HT$_{1D}$ agonist) [17], Nolvadex™ (Tamoxifen, AstraZeneca, a prodrug of a potent estrogen receptor antagonist) [18], and many others.

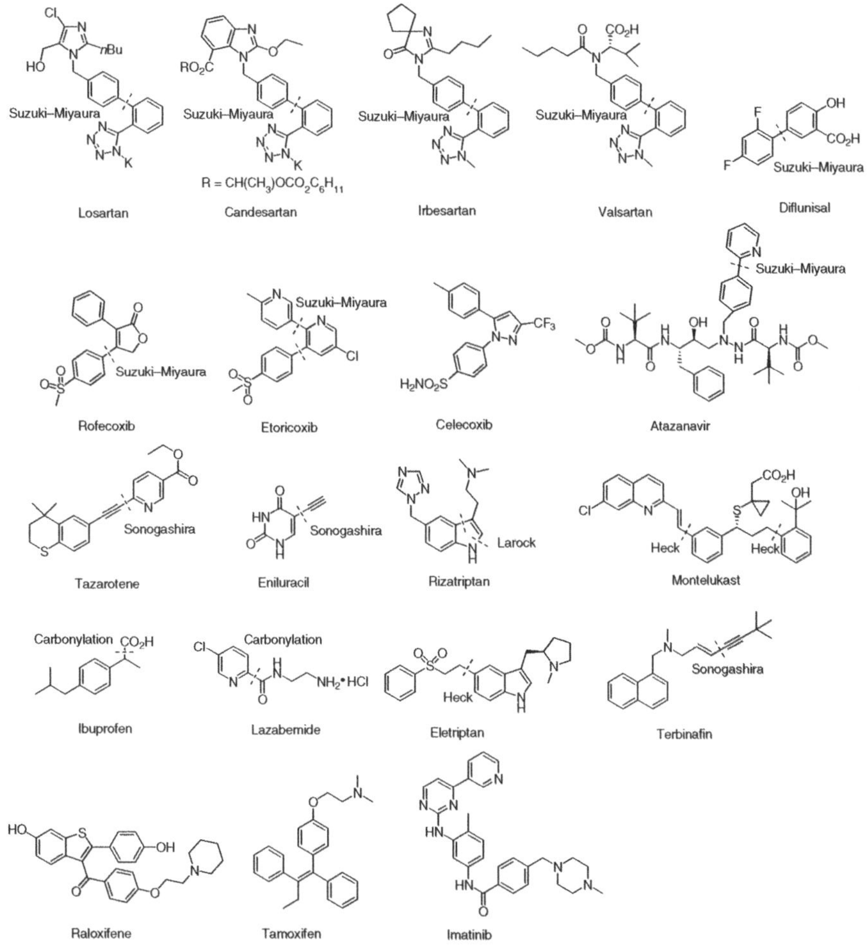

FIGURE 2.1 Representative biaryl drugs.

2.2 CARBON–CARBON CROSS-COUPLINGS

Depending on the types of organometallic species or nucleophiles, C–C cross-coupling reactions can be divided into Suzuki–Miyaura coupling [19], Negishi coupling [20], Kumada–Corriu–Tamao coupling [21], Hiyama coupling [22], Kosugi–Migita–Stille coupling [23], Heck–Mizoroki coupling [24], Sonogashira coupling [25], Trost–Tsuji coupling [26], conjugated C–C coupling [27], α-arylation of carbonyl compounds [28], C–H activation/direct arylation [29], cyanation [30], and carbonylation [31] reactions (Scheme 2.1). For the purpose of clarity, only one key inventor of each type of cross-coupling reaction is referred in this chapter. Among these transformations, Suzuki coupling, Negishi coupling, Sonogashira coupling, Kumada coupling, and Heck coupling reactions have been extensively

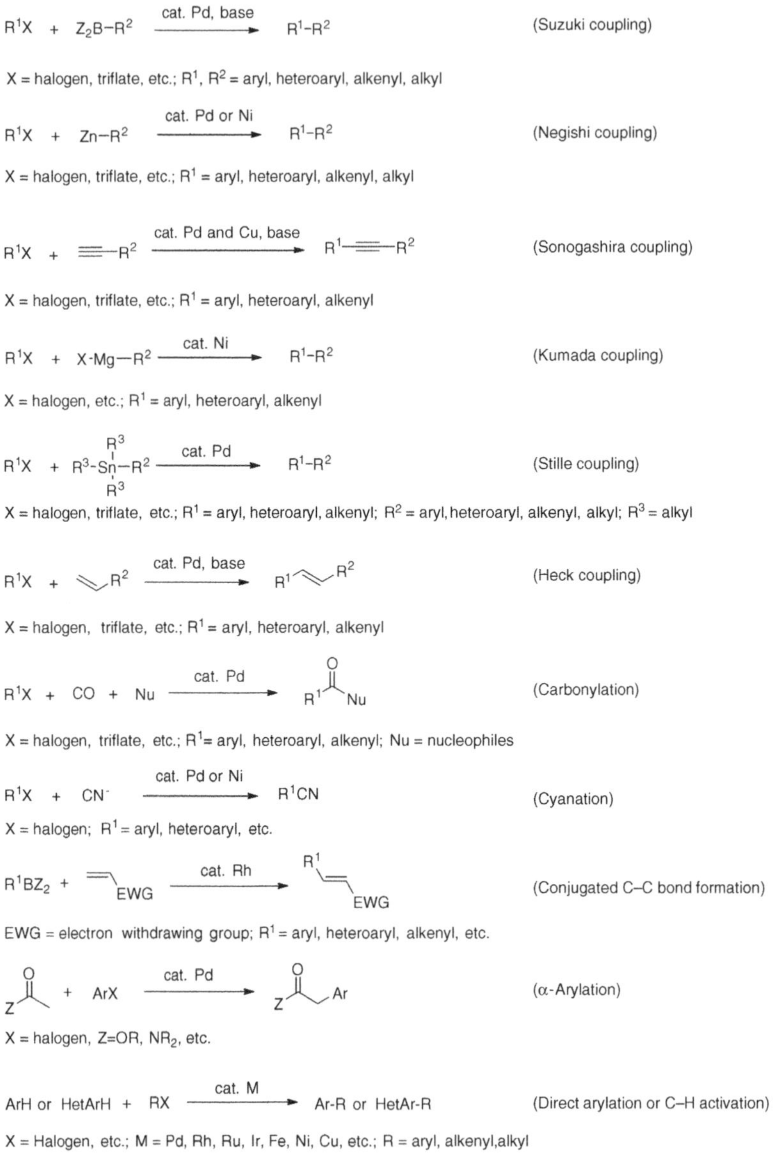

R^1X + Z_2B-R^2 $\xrightarrow{\text{cat. Pd, base}}$ R^1-R^2 (Suzuki coupling)

X = halogen, triflate, etc.; R^1, R^2 = aryl, heteroaryl, alkenyl, alkyl

R^1X + $Zn-R^2$ $\xrightarrow{\text{cat. Pd or Ni}}$ R^1-R^2 (Negishi coupling)

X = halogen, triflate, etc.; R^1 = aryl, heteroaryl, alkenyl, alkyl

R^1X + $\equiv\!\!-R^2$ $\xrightarrow{\text{cat. Pd and Cu, base}}$ $R^1\!\!-\!\!\equiv\!\!-R^2$ (Sonogashira coupling)

X = halogen, triflate, etc.; R^1 = aryl, heteroaryl, alkenyl

R^1X + $X-Mg-R^2$ $\xrightarrow{\text{cat. Ni}}$ R^1-R^2 (Kumada coupling)

X = halogen, etc.; R^1 = aryl, heteroaryl, alkenyl

R^1X + $R^3\text{-}\!\!\overset{\displaystyle R^3}{\underset{\displaystyle R^3}{Sn}}\!\!-R^2$ $\xrightarrow{\text{cat. Pd}}$ R^1-R^2 (Stille coupling)

X = halogen, triflate, etc.; R^1 = aryl, heteroaryl, alkenyl; R^2 = aryl, heteroaryl, alkenyl, alkyl; R^3 = alkyl

R^1X + $\diagdown\!\!=\!\!R^2$ $\xrightarrow{\text{cat. Pd, base}}$ $R^1\diagup\!\!\diagdown\!\!R^2$ (Heck coupling)

X = halogen, triflate, etc.; R^1 = aryl, heteroaryl, alkenyl

R^1X + CO + Nu $\xrightarrow{\text{cat. Pd}}$ $R^1\!\!\overset{\displaystyle O}{\diagup\!\!\diagdown}\!\!Nu$ (Carbonylation)

X = halogen, triflate, etc.; R^1 = aryl, heteroaryl, alkenyl; Nu = nucleophiles

R^1X + CN^- $\xrightarrow{\text{cat. Pd or Ni}}$ R^1CN (Cyanation)

X = halogen; R^1 = aryl, heteroaryl, etc.

R^1BZ_2 + $\diagup\!\!=\!\!\diagdown_{EWG}$ $\xrightarrow{\text{cat. Rh}}$ $\overset{\displaystyle R^1}{\diagup\!\!\diagdown\!\!\diagup}_{EWG}$ (Conjugated C–C bond formation)

EWG = electron withdrawing group; R^1 = aryl, heteroaryl, alkenyl, etc.

$Z\!\!\overset{\displaystyle O}{\diagup\!\!\diagdown}\!\!\diagdown$ + ArX $\xrightarrow{\text{cat. Pd}}$ $Z\!\!\overset{\displaystyle O}{\diagup\!\!\diagdown}\!\!\diagup\!\!Ar$ (α-Arylation)

X = halogen, Z=OR, NR_2, etc.

ArH or HetArH + RX $\xrightarrow{\text{cat. M}}$ Ar-R or HetAr-R   (Direct arylation or C–H activation)

X = Halogen, etc.; M = Pd, Rh, Ru, Ir, Fe, Ni, Cu, etc.; R = aryl, alkenyl, alkyl

SCHEME 2.1   C–C bond cross-coupling reactions.

used in industry. The Stille coupling is often unsuitable for pilot and plant production mainly due to the toxicity and price issues related to organostannane reagents. The Hiyama coupling is underutilized presumably as a result of the poor availability of silicon-containing coupling partners.

Besides various materials in electronics, numerous drugs and drug candidates were prepared on large scales from the aforementioned transition metal-catalyzed coupling reactions in the pharmaceutical industry, which manifest the impact of these reactions. As such, the 2010 Nobel Prize in Chemistry was awarded to three giants in this field: Richard Heck, Ei-ichi Negishi, and Akira Suzuki. These pioneers, together with many others, solved selectivity issues in conventional C–C bond forming reactions and provided efficient and precise tools to enable industrial chemists to build molecules that have preserved and improved people's lives.

From 1997 to 2002, there was a significant increase in the use of transition metal-catalyzed C–C cross-coupling reactions according to a recent survey [32]. This survey disclosed that during this period 14% of all good manufacturing practice (GMP) bulk reactions fall into the category of C–C bond formation, which are largely C–C bond cross-coupling reactions.

To develop a large-scale transition metal-catalyzed process, one needs to judiciously consider factors such as the reactivity of the catalytic system, efficiency characterized by turnover number (TON), selectivity (chemo- and regioselectivity and in some cases enantioselectivity), scalability, cost of both catalysts and ligands, waste management and metal contamination in product, and so on.

With respect to catalyst selection, several relatively inexpensive catalysts derived from Cu [33], Ni [34], or Fe [35], are found to be effective for some of the aforementioned processes. Nevertheless, Pd catalysts are most commonly used in the pharmaceutical industry despite their higher cost in general. This is due in part to the fact that once optimized, Pd catalysts are often more reactive and catalytically more efficient with higher TONs than their Cu, Ni, or Fe counterparts [36]. In some cases, Pd-catalyzed reactions can proceed at room temperature [37]. Efficient reactions with large TONs of highly active Pd catalysts, represented primarily by palladacycles and coordinatively unsaturated Pd catalysts using electron-rich bulky phosphines, have been developed and reviewed [38]. It appears that the new "PdL$_1$" catalysts (where L$_1$ is a monodentate bulky P ligand of high donating ability) bodes well for large-scale, wide scope, and highly efficient cross-coupling reactions.

Despite the significant development of ligands for Pd-catalyzed reactions [39], newly developed ligands such as heterocyclic carbenes and sterically encumbered electron-rich phosphines are not commonly used in process chemistry mainly due to their cost.

The extent of metal contamination in active pharmaceutical ingredients (APIs) needs to be in compliance with regulations (http://www.emea.europa.eu/pdfs/human/swp/444600en.pdf.) and general guidelines of each company. Thus, extensive research has been conducted to address this issue and numerous solutions have emerged [40]. These include the treatment by resin, metal sequestrant, crystallization, and another ligand, to name a few. Some of these approaches will be described by examples later. Novel approaches that have not been extensively adopted by industry include the use of switchable polarity solvents [41], nanofiltration [42], and temperature-dependent multicomponent solvent systems [43]. The toxic metal level in final products should generally be less than 10 ppm as a rule of thumb.

Although an excellent patent review for cross-coupling reactions by Corbet and Mignani was published in 2006 1d, this chapter aims to provide selected industrial applications of each cross-coupling reaction type based on recent publications, in particular *Organic Process Research & Development*. Novel technologies developed during the industrial applications are also discussed. It is not, however, the intention to conduct a comprehensive survey of all transition metal-catalyzed C–C bond cross-coupling reaction examples used in the pharmaceutical and chemical industry.

2.2.1 Suzuki Coupling

Owing to the functional group compatibility, good accessibility of boron reagents, remarkable versatility in forming various types of C–C bonds, and high success rate when carried out in a parallel synthesis format, Suzuki coupling is one of the most widely applied coupling reactions in the pharmaceutical industry. It is particularly useful for sp^2 (aryl)–sp^2 (aryl) carbon–carbon bond formation, thus allowing the discovery and development of numerous drugs and clinical candidates containing a biaryl motif. Suzuki coupling can also be applied to construct sp^2 (aryl)–sp^2 (vinyl), sp^2 (vinyl)–sp^2 (vinyl), sp^2 (aryl)–sp^3, and even sp^3–sp^3 C–C bonds.

2.2.1.1 sp^2 (Aryl)–sp^2 (Aryl) Coupling   The most common Suzuki coupling reactions used by medicinal chemists are perhaps sp^2–sp^2 coupling, in particular sp^2 (aryl)–sp^2 (aryl) coupling reactions.

An early classical example of the use of Suzuki coupling in the pharmaceutical industry was Merck's synthesis of losartan **2**, an angiotensin II receptor antagonist (Scheme 2.2) [44]. The boronic acid **1** was prepared via ortho-directed lithiation, followed by the treatment of triisopropryl borate and water. The subsequent Suzuki coupling proceeded using Pd(0) species derived *in situ* from the relatively inexpensive Pd(OAc)$_2$ and triphenylphosphine instead of tetrakis(triphenylphosphine)palladium(0). The amount of water was crucial for the success of the reaction as an excess caused agglomeration of K$_2$CO$_3$, whereas a deficit led to a sluggish reaction. A novel method of removing residual Pd from the product involved an addition

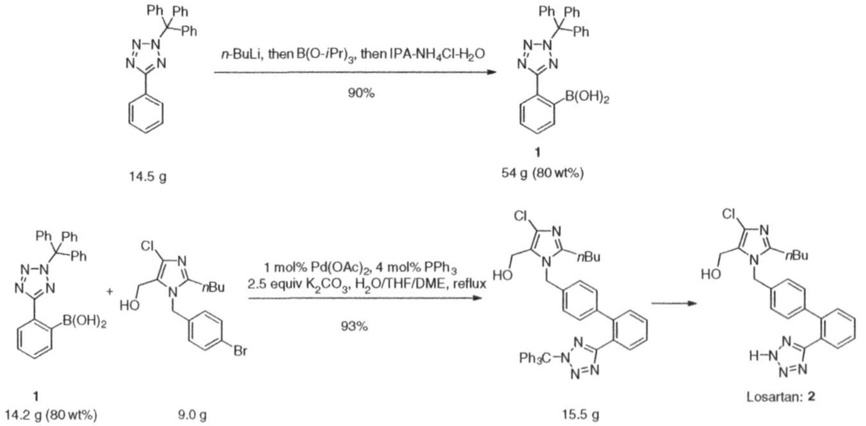

SCHEME 2.2   Merck's synthesis of losartan via a Suzuki coupling.

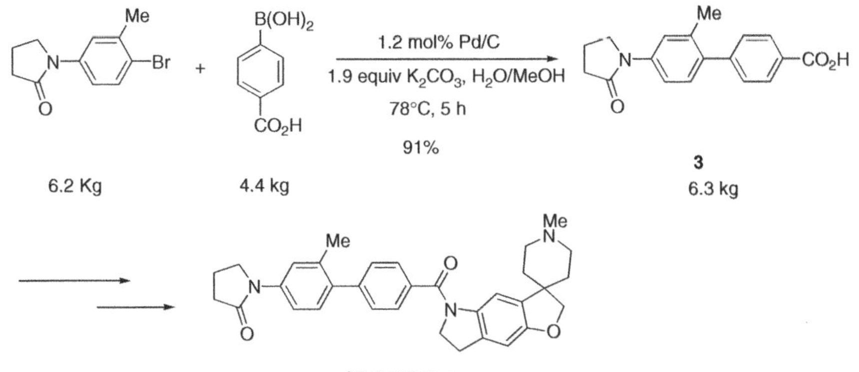

SCHEME 2.3   GSK's synthesis of SB-245570 via a Suzuki coupling.

of tributylphosphine (10 mol%) to the organic layer after aqueous extraction. As a result, Pd(0) species was stabilized and < 50 ppm of Pd level was detected in the trityl losartan intermediate.

Another example of biaryl Suzuki coupling was reported by GSK in the synthesis of SB-245570 (**4**), a compound in development for the treatment of depression (Scheme 2.3) [45]. One nice feature of the reaction is that inexpensive Pd/C in the absence of phosphine ligand smoothly catalyzed the reaction to provide an excellent yield of the biaryl product **3**. Furthermore, the residual Pd level was considerably low (<6 ppm). In contrast, Pd(PPh$_3$)$_4$, Na$_2$CO$_3$, DME/H$_2$O gave only 64% yield of the desired product with 40–80 ppm of Pd level. Another issue associated with the use of Pd(PPh$_3$)$_4$ was that the quality of catalyst varied from batch to batch causing inconsistent results.

The Lilly process group also used a ligandless Pd black-catalyzed Suzuki coupling to provide biaryl product **5** in good yield (Scheme 2.4). The use of 0.99 equiv of boronic acid was intended to avoid the difficult removal of extra boronic acid if it was

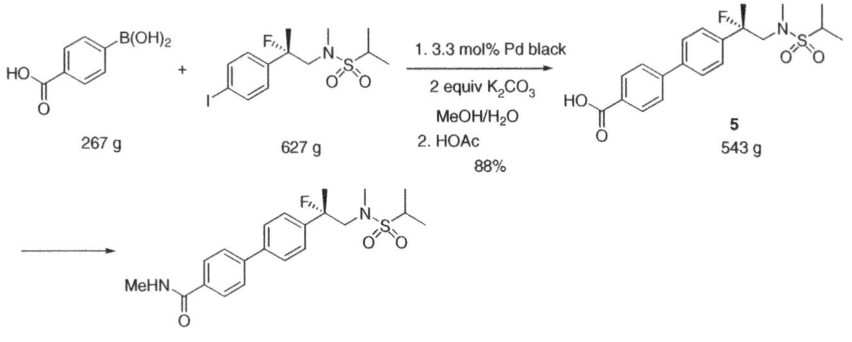

SCHEME 2.4   Lilly's synthesis of LY503430 via a Suzuki coupling.

not completely consumed. After the completion of the reaction, treatment with acetic acid induced crystallization of the product with excellent purity, and the crystalline product could be easily filtered off with only 3–8 ppm of Pd. It should be noted that compound **5** is a key intermediate leading to LY503430 (**6**), a potential therapeutic agent for Parkinson's disease [46].

To prepare LY451395 (**10**), an AMPA potentiator developed for Alzheimer's disease, Eli Lilly process chemists reported another ligandless protocol to form not only pinacol boronic ester **8** from aryl iodide **7**, but also compound **10** from the subsequent Suzuki coupling of boronic acid **9** (Scheme 2.5) [47]. It was found that deoxygenation, which was conducted by a subsurface sparge of nitrogen, was required to minimize the homocoupling of the boronic ester, presumably through a Pd(II) species in the Suzuki coupling step. In addition, potassium formate was applied to suppress the concentration of Pd(II) in the mixture, thereby further decreasing the dimer formation. Finally, the use of heterogeneous Pd black catalyst allowed the efficient separation of catalyst from the reaction mixture by a simple glass fiber filter.

In the synthesis of the potent $D_3/D_2/5-HT_2$ receptor ligand **13**, a Suzuki coupling of 2-cyanoboronic acid and aryl bromide **11** generated the biaryl product **12** on a kilogram scale (Scheme 2.6) [48]. In this reaction, high chemoselectivity of bromide versus chloride was observed. One particular challenge of the reaction was the high Pd contamination (>2000 ppm) in the product. A DIAION CR20 polyamine resin was then selected to treat the reaction mixture, thus removing most of the Pd impurity. A subsequent crystallization using ($+$)-di-p-toluoyl-D-tartaric acid (DTTA) generated a salt of **12** with only 35 ppm of Pd. Intermediate **12** was finally converted to E2040 (**13**) containing less than 10 ppm of the Pd content, which met the requirement for API.

To support the development of $5HC_{2c}$ receptor agonist **15**, a potential treatment for obesity [49], a group of BMS chemists developed a multikilogram synthesis (Scheme 2.7) [50]. The use of isopropyl acetate/water to replace DME/water proved to be effective for the reaction and thus eliminated a solvent exchange during workup. Although the reaction proceeded well under the given conditions, the coupling

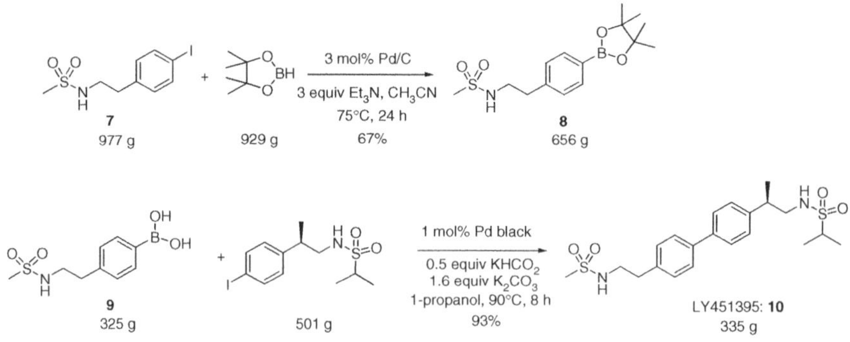

SCHEME 2.5 Lilly's synthesis of LY451395 via a Suzuki coupling.

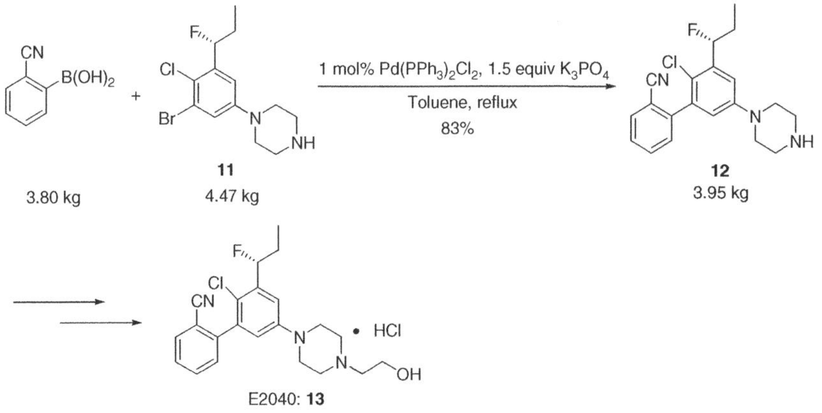

SCHEME 2.6   Eisai's process for the synthesis of E2040 via a Suzuki coupling.

product **14** was not a solid and it contained a high level of Pd (2500–3500 ppm). While the treatment of the reaction mixture with Picachem carbon or a solution of tris (hydroxymethyl)aminomethane was insufficient to remove the Pd residue, the use of a 20% sodium carbonate solution of trithiocyanuric acid and the ensuing treatment of Picachem carbon 80PN eventually reduced the Pd level to less than 100 ppm.

In the synthesis of biaryl compound **18**, a microsomal prostaglandin E synthase I (mPGES-1) inhibitor developed by Merck, a directed lithiation of **16** followed by treatment with triisopropoxyborate provided the lithium triisopropyl borate **17** with full conversion (Scheme 2.8) [51]. The high water solubility of the corresponding boronic acid precluded its isolation. It was found that the crude mixture could be used directly for the cross-coupling reaction after the addition of water and catalyst. The addition of base was not needed and the reaction produced the biaryl product **18** in excellent yield.

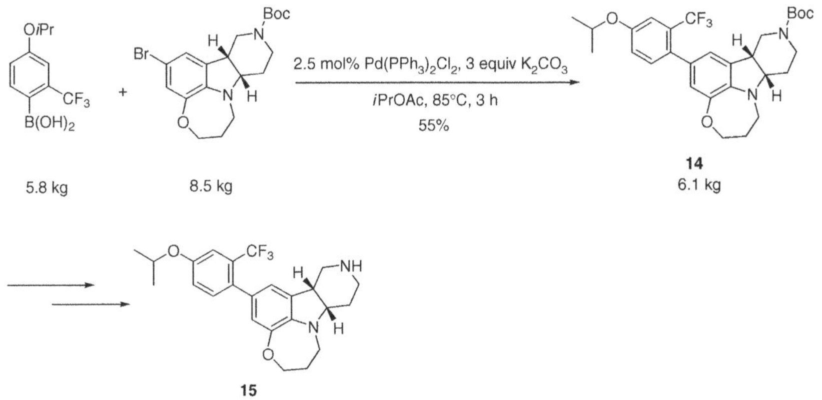

SCHEME 2.7   BMS's synthesis of a 5-HT$_{2c}$ agonist via a Suzuki coupling.

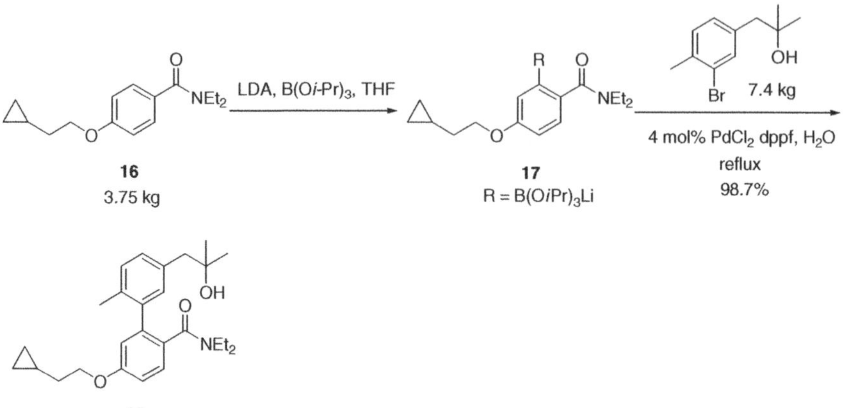

SCHEME 2.8   Merck's synthesis of an mPGES-1 inhibitor via a Suzuki coupling.

To address the challenge of the removal of residual palladium in cross-coupling reactions, the pharmaceutical industry has also been engaged in developing solid-support catalysts based on various supporting materials, such as carbon [52], silica [53], zeolites [54], sepiolites [55], polyionic gels [56], hydrotalcite [57], triphenylpho-sphinated polystyrene beads [58], core–shell block copolymers [59], and palladium-containing perovskites [60]. Recently, a new catalyst called Pd-Smopex-111, in which Smopex-111 is a fiber-based benzenethiol, was applied by Novartis scientists for Suzuki coupling reactions (Scheme 2.9) [61]. By grafting metal binding functionality onto fibers, the unique metal scavenging system Smopex allows effective recovery of transition metals from process streams. For example, the Suzuki coupling of phenyl boronic acid **19** and bromide **20** in the presence of 2.5% of Pd-Smopex-111 afforded the desired coupling product **21** with nearly quantitative yield and < 1 ppm of leached Pd. The catalyst can be recycled and no noticeable loss of catalyst activity was observed after four cycles.

2.2.1.2 *sp*² (Aryl)–*sp*² (Heteroaryl) Coupling   An efficient multikilogram synthesis of ABT-963 (**25**), a selective COX-2 inhibitor developed by Abbott, involved a Suzuki coupling as one of the key steps (Scheme 2.10) [62]. This reaction of boronic acid **22** and aryl bromide **23** rendered the desired triaryl intermediate **24** in

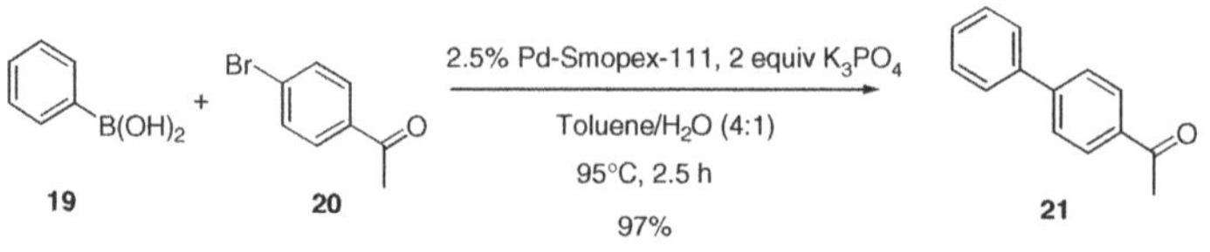

SCHEME 2.9   Novartis's synthesis of a biaryl intermediate via a Suzuki coupling using a solid-support catalyst.

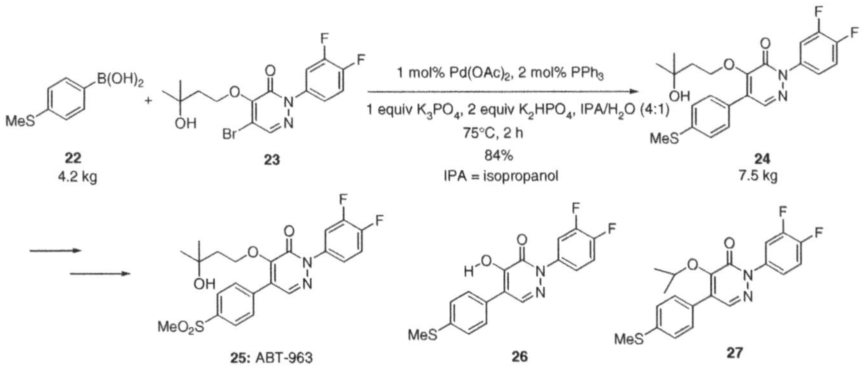

SCHEME 2.10   Abbott's synthesis of ABT-963 via a Suzuki coupling.

95–98% HPLC yield and 84–88% isolated yield. During the reaction optimization, it was found that the degassing of the reaction mixture was essential for high yields. In addition, 1.1 equiv of boronic acid **22** was required to drive the reaction to completion. Regarding reaction solvents, isopropanol (IPA) or THF was superior to EtOH or t-BuOH. The choice of base (a mixture of K_3PO_4 and K_2HPO_4) was critical in that other bases led to more solvolysis by-products (**26** and **27**). Interestingly, the solvolysis by-products were found to increase as the stirring rate increased. Other catalysts such as $Pd(PPh_3)_2Cl_2$, Pd/C, and $Pd(OAc)_2/P(o\text{-tol})_3$ provided slightly lower yields of **24** than $Pd(OAc)_2/PPh_3$. After the reaction, the mixture was passed through a silica cartridge filter and then crystallized in 4:1 heptane/EtOAc.

A Merck process group used Pd/C as a cheap and quasi-heterogeneous catalyst system to enable a kilogram-scale Suzuki coupling (Scheme 2.11) [63]. The initial screening of catalysts led to the identification of $Pd(dppf)Cl_2$ (dppf = 1,1'-bis(diphenylphosphino)ferrocene) for this reaction. However, large amounts of Pd residue (3500 ppm) and Fe residue (1160 ppm) were found in the product. It turned out that Pd/C (5% or 10% Pd) was a preferable catalyst affording the coupling product **28** in excellent yields and with less Pd contamination (18–80 ppm) after recrystallization. Intermediate **28** was then converted to **29**, a potent PDE4 inhibitor with the potential to treat pulmonary diseases such as asthma and chronic obstructive pulmonary disease (COPD) [64].

To improve a Suzuki coupling reaction with $Pd(OAc)_2$, dppf, and organic base Et_3N in dimethylacetamide (DMAc)/H_2O, which has already been established on a large scale (Scheme 2.12), a ligand–solvent–base screening led to further optimized conditions employing $Pd(OAc)_2$, $P(o\text{-tol})_3$, $KHCO_3$, and IPA/H_2O. As a result, a superior yield of the desired biaryl product (82% versus 66% control) and a reduced formation of by-product **30** (<0.1% versus 0.68% control) were observed [65]. Furthermore, since $P(o\text{-tol})_3$ was used instead of dppf, the residual iron in the control conditions (471 ppm) was not an issue for the optimized route. To remove the residual palladium from the multikilogram Suzuki coupling reaction, solid-phase Pd absorption was explored. The best treatment was the triamine-derived silica that reduced the

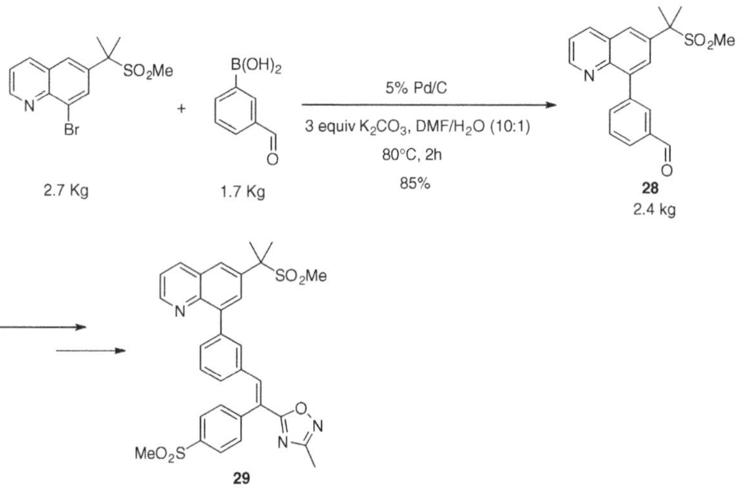

SCHEME 2.11 Merck's synthesis of a PDE4 inhibitor via a Suzuki coupling.

palladium level to 285 ppm. The treatment of the reaction mixture with toluene and 20% aqueous $NaHSO_3$ further reduced the palladium level from ~8000 to 100 ppm or less. Finally, this cost-effective, efficient, and facile process was demonstrated on a 20 L scale.

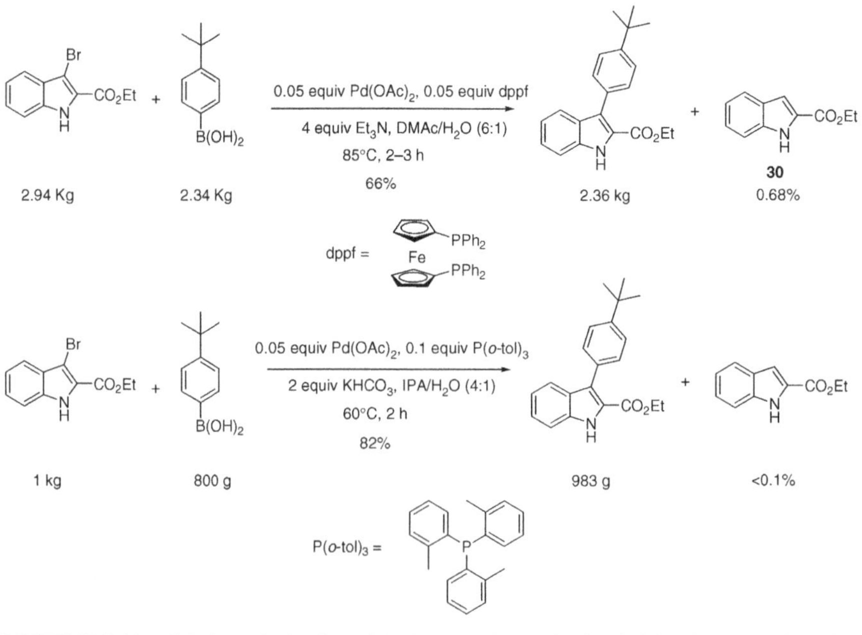

SCHEME 2.12 GSK's optimization of the large-scale synthesis of a biaryl intermediate via a Suzuki coupling.

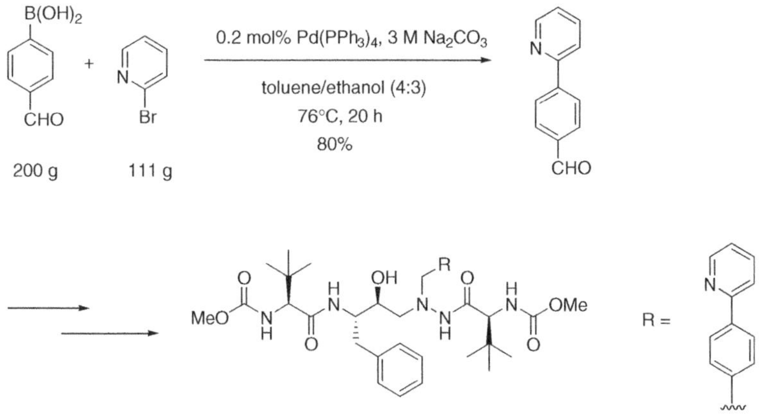

Atazanavir: **31**

SCHEME 2.13 BMS's synthesis of atazanavir via a Suzuki coupling.

The BMS process research group disclosed an efficient synthesis of the HIV protease inhibitor atazanavir™ (**31**) (Scheme 2.13) [66]. The biaryl moiety of this drug was prepared via a Kumada coupling in the original process, which required the protection and deprotection of an aldehyde group and the use of DIBALH to generate the activated Ni catalyst. To circumvent these issues, a Suzuki coupling was explored in the second process [67]. Owing to superior functional group compatibility of the Suzuki coupling, the reaction proceeded smoothly without the need for aldehyde protection. The 4:3 ratio of toluene:ethanol solvent mixture was critical for substrate solubility, thus affording an efficient reaction. This process successfully delivered 3.5 kg of the final product.

A Pfizer process research group reported the synthesis of a potent inhibitor of p38α via a Suzuki coupling (Scheme 2.14) [68]. In this reaction, the 4,5-disubstituted oxazole **34** was prepared using a bromooxazole **32** and boronic acid **33**. The weak basicity of the product led to a unique solubility of the product so that it was partitioned primarily in the aqueous layer when distributed between aqueous

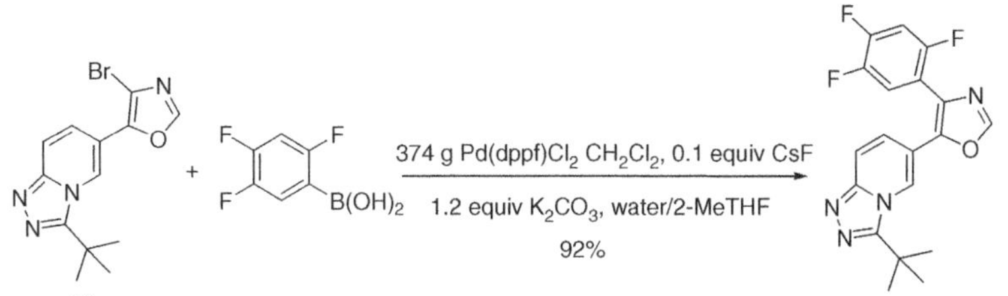

32 3 kg; **33** 1.8 kg; **34** 2.8 kg

SCHEME 2.14 Pfizer's synthesis of a p38α inhibitor via a Suzuki coupling.

HCl solution and toluene. To remove the Pd residue from the final API, a basic amine (e.g., triethylamine) was used to displace Pd from the product by competitive binding. Remarkably, after a triethylamine treatment, the Pd and Fe levels were dropped from 2100 and 3200 ppm to 9 and 8.9 ppm, respectively.

In the synthesis of a Nav1.8 sodium channel modulator (**38**) for treatment of pain, a Pfizer team developed a room-temperature Suzuki coupling to form the key biaryl intermediate **37** (Scheme 2.15) (dba = dibenzylidene acetone) [69]. The use of a slightly high catalyst loading (3 mol%) and the addition of 1 equiv of water allowed a reproducible reaction and minimized the protodeboronation. The use of KF as the base presumably promoted the formation of the potassium boronate salt, which reacted with aryl bromide **36** at a significantly faster rate than the corresponding boronic acid **35**, thereby minimizing the protodeboronation side reaction. An adequate addition of water may assist the dissolution of KF in the reaction mixture and thus facilitate the formation of the fluoroboronate salt.

In the cases where boronic acids/esters are not readily available, excellent scalable procedures were reported to prepare boronic acids/esters *in situ*, which could then react with aryl halides to form the corresponding biaryl product (Scheme 2.16). For example, in the process development for compound **41**, a potent bradykinin 1 antagonist used for potential treatment of inflammatory pain [70], an efficient two-step one-pot process was established. This reaction involved the Pd-catalyzed boronic ester formation *in situ* using bromopyridine **39** and pinacol diborane, as well as a subsequent Suzuki coupling [71]. In both steps, Pd(dppf)Cl$_2$ was used as the catalyst. The first step was quite robust and it provided a quantitative assay yield of the boronic ester intermediate **40** [72]. An alternative route using trifluoroboronate was also explored and then abandoned due to the liberation of HF that corroded the stainless steel and glassware.

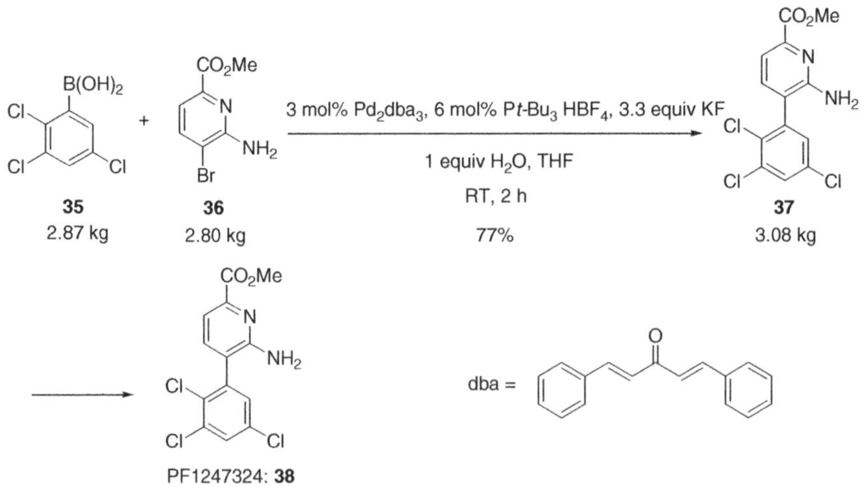

SCHEME 2.15 Pfizer's synthesis of a Nav1.8 modulator via a Suzuki coupling.

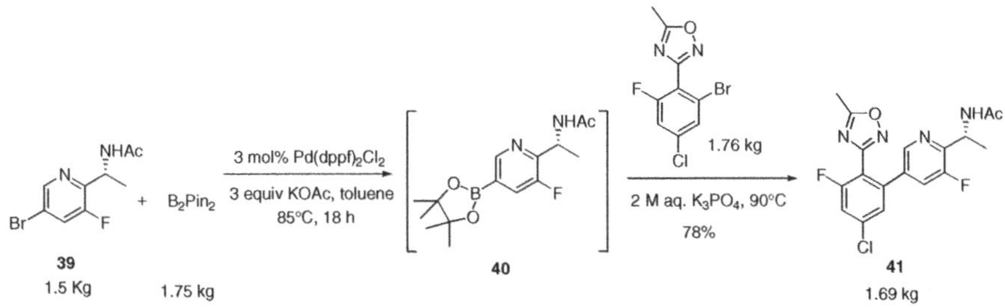

SCHEME 2.16 Merck's synthesis of a bradykinin 1 antagonist via a Suzuki coupling.

Besides boronic acids and esters, boranes can also be applied in Suzuki coupling reactions. For example, a group from Dow and the former Pharmacia developed a large-scale process route to access OSU 6162 (**45**), a potential central nervous system (CNS) agent (Scheme 2.17) [73]. The starting borane was readily prepared on a 100 kg scale via a lithium halogen exchange, followed by the treatment of B(OMe)Et₂. The resulting borane **42** was then coupled with ethylsulfonyl bromobenzene **43** in the presence of a base, Pd(PPh₃)₄, and a phase transfer catalyst tetrabutylammonium bromide (TBAB) to give the desired biaryl product **44** in good yield. Attempts to use heterogeneous Pd catalysts were unsuccessful. To reduce the cost of the high

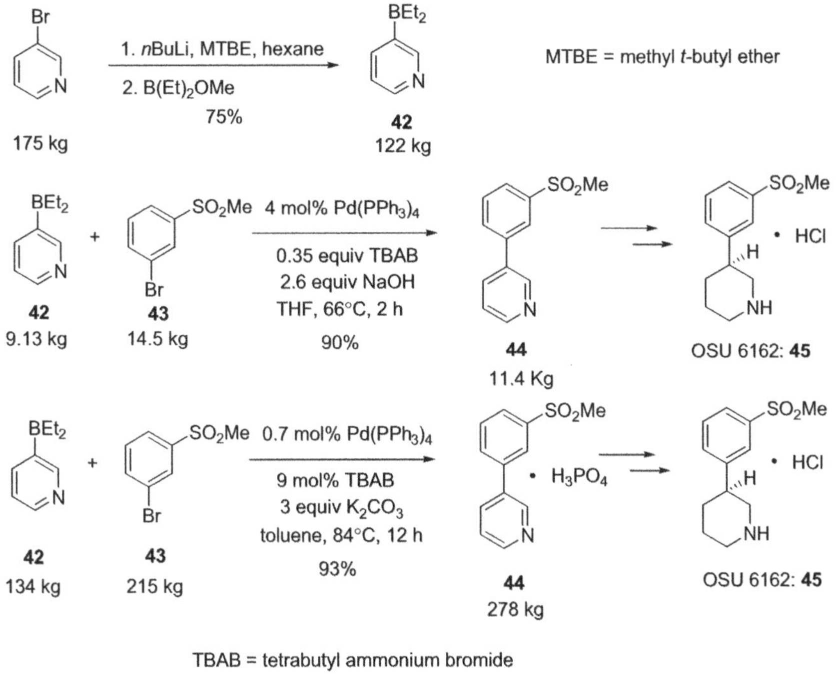

TBAB = tetrabutyl ammonium bromide

SCHEME 2.17 Dow and Pharmacia's synthesis of OSU 6162 via a Suzuki coupling.

catalyst loading, a nonpolar variant of this reaction using toluene was developed. In this case, only 0.7 mol% of Pd(PPh₃)₄ and a much lower loading of TBAB were used. This revised protocol also allowed a more convenient workup and a slightly higher yield.

In the preparation of ABT-869 (**49**), a potent tyrosine kinase receptor inhibitor for the treatment of cancer [74], an Abbott process team developed a second-generation Suzuki coupling reaction using pinacol boronic ester **46** and less reactive chloride **47**, as opposed to the corresponding iodide used in their first-generation delivery (Scheme 2.18). In addition, the new route entailed the use of dbpf (1,1'-bis(di-*tert*-butylphosphino)ferrocene) (**48**) instead of dppf [75]. The advantage of using dbpf lies in its superior reactivity, which allowed the reaction to proceed under 60°C. Since higher reaction temperature caused significant degradation of the product, dppf was employed only under 60°C, resulting in sluggish conversion. One issue of the Pd(OAc)₂/dbpf catalyst system was its sensitivity to oxygen. The reaction mixture was thus sparged with nitrogen (<100 ppm oxygen) prior to the addition of the catalyst solution (<10 ppm oxygen). An oxygen detector was used to monitor the oxygen level to ensure reproducibility on a large scale.

2.2.1.3 sp² (Heteroaryl)–sp² (Heteroaryl) Coupling   Obatoclax **53**, a novel indolylprodigiosin derivative, is used as a B-cell lymphoma (Bcl) antiapoptotic protein inhibitor [76]. This compound is currently in phase II clinical trials for treatment of multiple hematological and solid tumor malignancies. A scalable Suzuki coupling for the synthesis of intermediate **52** was developed using a methyl carbamate protected 2-indole boronic acid **50** and bromoenamine **51** (Scheme 2.19) [77]. The use of methyl carbamate as a protecting group was claimed to be the first in a Suzuki reaction setting and it led to higher yields and a reduced amount of the protodeboronation by-product compared to the use of commercially available Boc-protected 2-indole boronic acid.

2.2.1.4 sp² (Aryl)–sp² (Vinyl) Coupling   The first generation of COX-2 selective inhibitor, rofecoxib (Vioxx, **56**), could be prepared in a straightforward fashion using

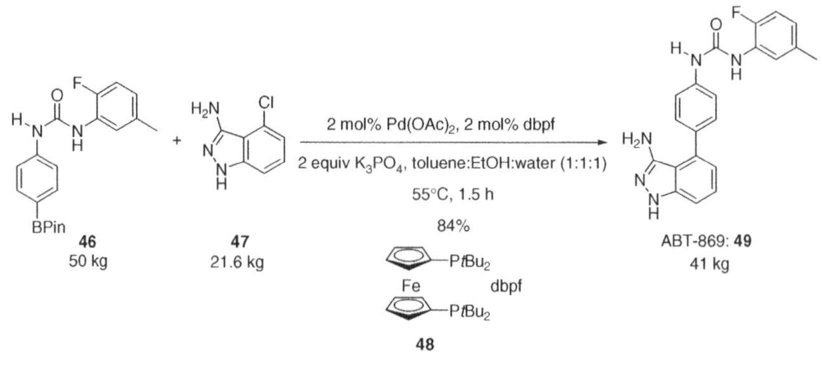

SCHEME 2.18   Abbott's synthesis of ABT-869 via a Suzuki coupling.

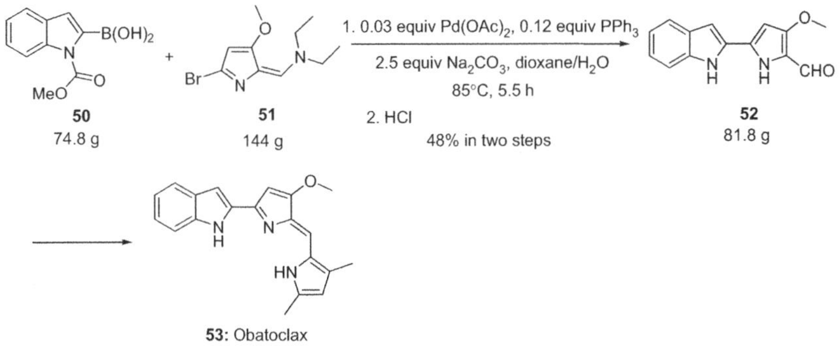

SCHEME 2.19 Synthesis of obatoclax via a Suzuki coupling by Gemin X Biotechnologies.

a Suzuki coupling between vinyl bromide **54** and boronic acid **55** (Scheme 2.20), followed by the oxidation of sulfide to sulfone **56** [78]. This protocol represents one of the most widely used Suzuki coupling procedures in medicinal chemistry.

Highly substituted β-lactams, as potent antibiotics, elicited considerable research activities for their synthesis. One such example is anti-MRS carbapenem candidate **59** developed by Merck. In the synthesis, triflate **57** reacted with boronic acid **58** to create a pivotal C–C bond present in the final candidate (Scheme 2.21) [79]. This versatile approach was also adopted by Shionogi in their synthesis of a variety of other carbapenems [80].

A Pfizer group reported a multikilogram-scale synthesis of CI-1034 (**64**) [81], a novel endothelin ET_A receptor antagonist used for the treatment of primary pulmonary hypertension and congestive heart failure [82]. One of the key steps was the Suzuki coupling of boronic acid **61** and triflate **62** (Scheme 2.22). Boronic acid **61** was derived from the corresponding bromide **60** via the lithium halogen exchange, triisopropyl borate quench, and the ensuing treatment with acetic acid and water. Originally, triflate **62**, Pd(PPh$_3$)$_2$Cl$_2$, and triphenylphosphine were used in the

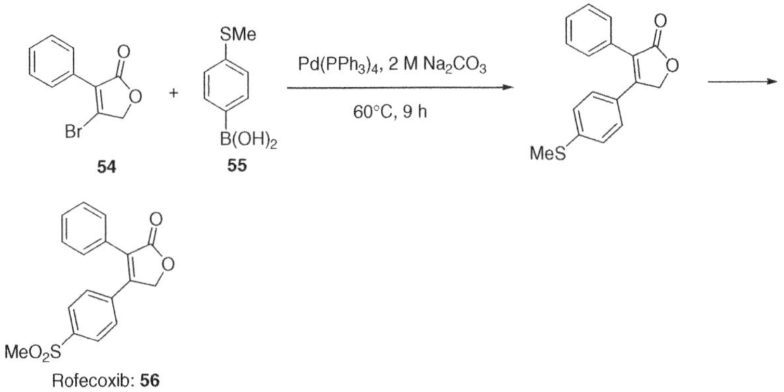

SCHEME 2.20 Merck's synthesis of rofecoxib via a Suzuki coupling.

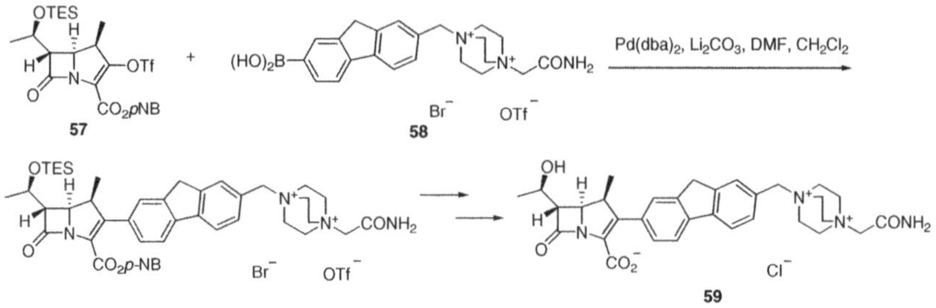

SCHEME 2.21 Merck's synthesis of a carbapenem candidate via a Suzuki coupling.

kilogram coupling reaction that provided an excellent yield of intermediate **63**. Neither recrystallization nor the treatment with carbon was effective in removing the Pd impurity from the product. Several equivalents of trithiocyanuric acid successfully reduced the Pd level to less than 10 ppm. Despite the success of this reaction, the formation of the triflate intermediate required the use of expensive triflic anhydride. Therefore, the screening of alternative electrophiles including phosphates and sulfonates led to the identification of tosylate and 4-fluorotosylate as the preferred electrophiles. In addition, the reaction catalyst and ligand were optimized to provide the desired coupling product in good yields on a multigram scale.

In a closely related target called PD 0182783 (**66**), which was also used as an endothelin A antagonist, Pfizer chemists reported the use of diaryl boronic acid (**65**) for the Suzuki coupling (Scheme 2.23) [83]. Both aryl groups of the diaryl boronic acid (**65**) could be transferred, thus increasing the atom economy of this reaction.

2.2.1.5 sp² (Vinyl)–sp³ Coupling Besides sp^2–sp^2 Suzuki coupling, sp^2–sp^3 coupling has also been attempted in large-scale transformations, although examples are fewer. As illustrated in Scheme 2.24, DSM Pharma Chemicals reported the hydroboration of alkyne **67** to form vinyl borane **68**, which was then coupled with benzyl chloride **69** to furnish **70**, an intermediate used for the synthesis of HIV protease or renin inhibitors [84].

The second example is in the total synthesis of (+)-discodermolide (**71**), an advanced cancer drug candidate developed by Novartis (Scheme 2.25) [85]. After exploring a variation of the Negishi coupling, which provided reasonable results on a 10 g scale, the Novartis group eventually adopted Marshall's strategy [86] by exploiting Suzuki coupling as a key step to construct the C_9–C_{21} unit. This approach was successfully adapted on a 1 kg scale.

2.2.1.6 sp² (Aryl)–sp³ Coupling Boehringer Ingelheim's pilot plant process for an HIV nonnucleoside reverse transcriptase inhibitor (NNRTI) (**76**) relies on the Pd-catalyzed sp^2 (aryl)–sp^3 Suzuki coupling (Scheme 2.26) [87]. In this case, vinylsilane **72** underwent a highly regioselective hydroboration using 9-borabicyclo[3.3.1] nonane (9-BBN) followed by the treatment of NaOH, leading to *in situ* formation of borate. The resulting borate then coupled with bromide **73** in the presence of palladium acetate and triphenylphosphine. It was found that the use of a nonpolar

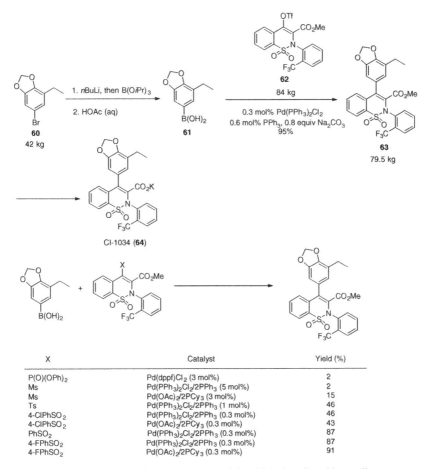

SCHEME 2.22 Pfizer's synthesis of CI-1034 via a Suzuki coupling.

X	Catalyst	Yield (%)
P(O)(OPh)$_2$	Pd(dppf)Cl$_2$ (3 mol%)	2
Ms	Pd(PPh$_3$)$_2$Cl$_2$/2PPh$_3$ (5 mol%)	2
Ms	Pd(OAc)$_2$/2PCy$_3$ (3 mol%)	15
Ts	Pd(PPh$_3$)$_2$Cl$_2$/2PPh$_3$ (1 mol%)	46
4-ClPhSO$_2$	Pd(PPh$_3$)$_2$Cl$_2$/2PPh$_3$ (0.3 mol%)	46
4-ClPhSO$_2$	Pd(OAc)$_2$/2PCy$_3$ (0.3 mol%)	43
PhSO$_2$	Pd(PPh$_3$)$_2$Cl$_2$/2PPh$_3$ (0.3 mol%)	87
4-FPhSO$_2$	Pd(PPh$_3$)$_2$Cl$_2$/2PPh$_3$ (0.3 mol%)	87
4-FPhSO$_2$	Pd(OAc)$_2$/2PCy$_3$ (0.3 mol%)	91

solvent such as toluene could minimize the protodeboronation side reaction. The resulting silane **74** as a crude product was subjected to Tamao–Fleming oxidation to yield 9.15 kg of alcohol **75** in 57% overall yield.

2.2.2 Negishi Coupling

Compared to the Kumada coupling reaction, the Negishi coupling has a broader substrate scope due to the better functional group compatibility of organozinc reagents [88]. For example, Negishi coupling reactions can proceed smoothly in the presence of ester, nitrile, or various heteroaryls. In particular, 2-heteroaryl zinc reagents are useful to install a C–C bond containing a 2-heteroaryl group.

Three methods were applied to prepare organozinc reagents. The first involves the use of zinc metal activated by TMSCl and 1,2-dibromoethane. Alternatively, highly reactive Rieke zinc is employed to insert to a C–X bond [89]. The third method takes

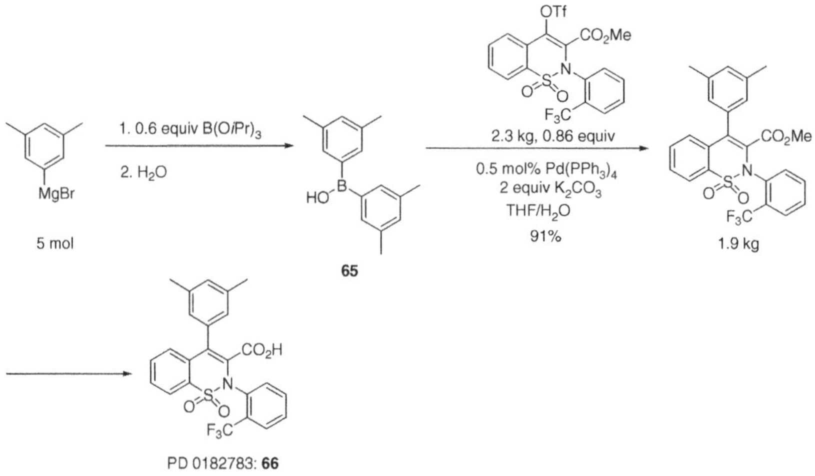

SCHEME 2.23   Pfizer's synthesis of PD 0182783 via a Suzuki coupling.

advantage of the transmetalation from organolithium, Grignard, organoboranes, and organozirconium reagents using $ZnCl_2$.

2.2.2.1 sp^2 (Aryl)–sp^2 (Heteroaryl) Coupling

A Merck group discovered triazole **79** as a potent and selective mGluR antagonist that could be potentially used to treat schizophrenia (Scheme 2.27) [90]. The concise and efficient synthesis of **79** involved a [3 + 2] cycloaddition between azide **77** and 1-propynyl magnesium bromide, adopting a protocol developed by Sharpless and coworkers [91]. The resulting triazole magnesium species **78** underwent a transmetalation followed by a Negishi coupling employing dimethylbisdiphenylphosphinoxanthene (Xantphos) as the ligand to furnish the target **79** in one pot [92]. After a crystallization and a treatment of PBu_3 to remove Pd, a recrystallization gave the final product with >99% purity and <18 ppm of Pd.

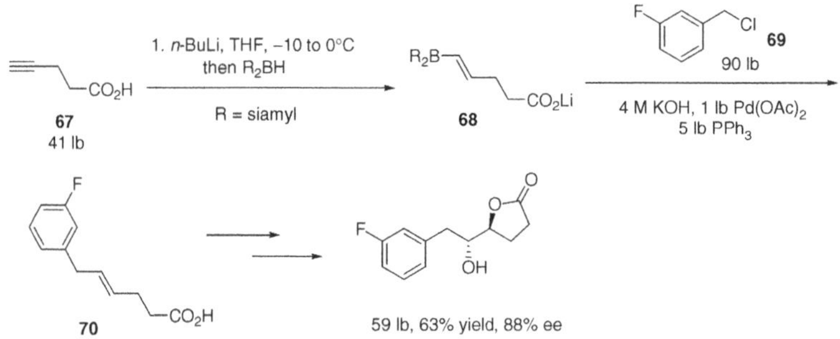

SCHEME 2.24   A large-scale sp^2 (vinyl)–sp^3 Suzuki coupling by DSM Pharma Chemicals.

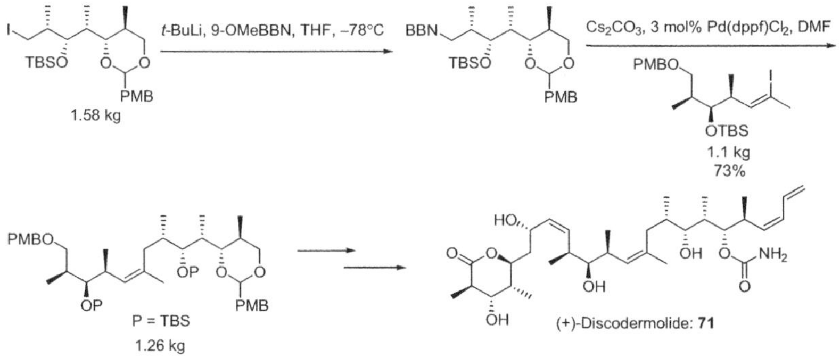

SCHEME 2.25 Novartis's synthesis of (+)-discodermolide via a sp^2 (vinyl)–sp^3 Suzuki coupling.

To develop a Pd-catalyzed Negishi coupling of oxazol-2-yl zinc **80** with aryl bromides, a Pfizer team used solid $ZnCl_2$ without additional purification for transmetalation, resulting in the corresponding zinc reagent (Scheme 2.28) [93]. It appeared that this protocol offered higher yields of the subsequent Negishi coupling reactions than the cases in which $ZnCl_2$ solution was used. Interestingly, as many as 3 equiv of $ZnCl_2$ was required for good coupling reaction yields, and this process was proven to be robust on a 1 kg scale.

2.2.2.2 sp^2 (Heteroaryl)–sp^2 (Heteroaryl) Coupling Negishi coupling proved to be particularly useful for sp^2 (heteroaryl)–sp^2 (heteroaryl) coupling due to its good functional compatibility and excellent reactivity.

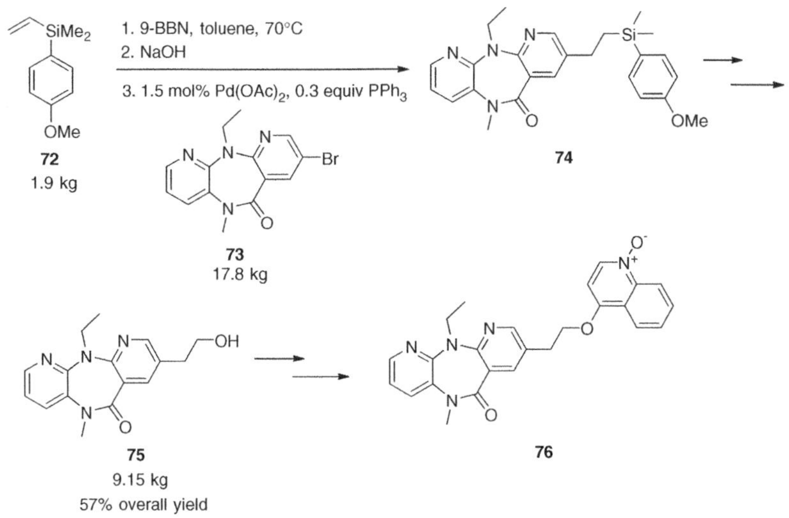

SCHEME 2.26 Boehringer Ingelheim's synthesis of an HIV NNRTI (76) via a sp^2 (aryl)–sp^3 Suzuki coupling.

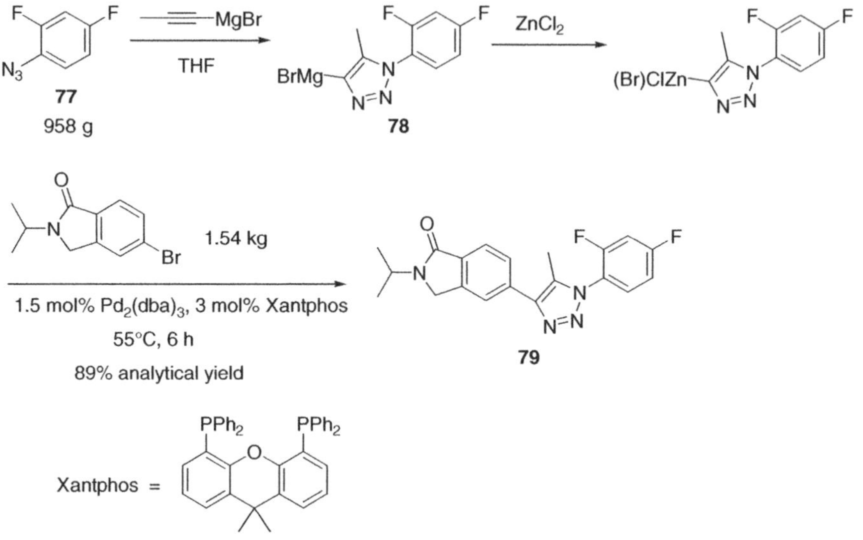

SCHEME 2.27 Merck's synthesis of an mGluR antagonist via a Negishi coupling.

Scheme 2.29 outlines a difficult Negishi coupling reported by Pfizer to construct the bis-heteroaryl moiety of AG-28262 (**84**), a vascular endothelial growth factor receptor (VEGFR) inhibitor used as an antiangiogenesis agent for clinical evaluation (Scheme 2.29) [94]. The synthesis of AG-28262 commenced with the lithiation of methylimidazole **81** to generate the corresponding lithium species, which then underwent a transmetalation to form zincate **82**. The pivotal Negishi coupling proceeded on over 100 g scale to yield the desired product **83** in 63% yield. A total of eight batches of Negishi coupling reaction provided 1.3 kg of intermediate with >95% purity. Although the zincate was generated in THF, the subsequent addition of acetonitrile was critical to solubilize the zincate, thus allowing the Negishi reaction to proceed with good yields.

A Novartis group used the Negishi coupling reaction to rapidly assemble the biaryl pharmacophore of a drug candidate (**86**) for the treatment of melanoma [95]. The mechanism of the action of this molecule is based on the discovery that B-Raf kinase inhibitors could play a role in regulating proliferation, survival, and angiogenesis in various animal models [96]. As shown in Scheme 2.30, the lithium–bromide exchange proceeded without the concomitant lithium–chloride exchange at −75°C. The temperature control was critical as higher temperatures

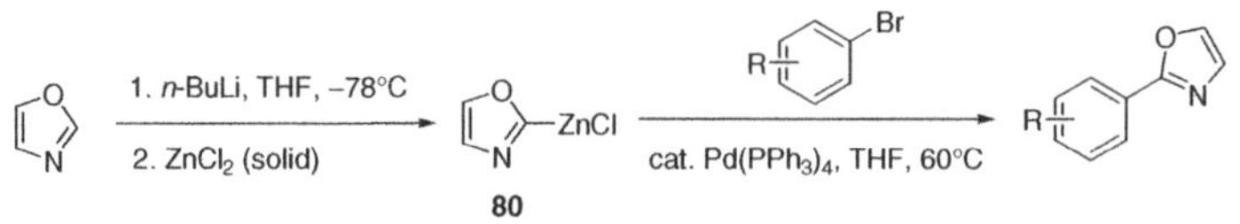

SCHEME 2.28 Pfizer's synthesis of a biaryl intermediate via a Negishi coupling.

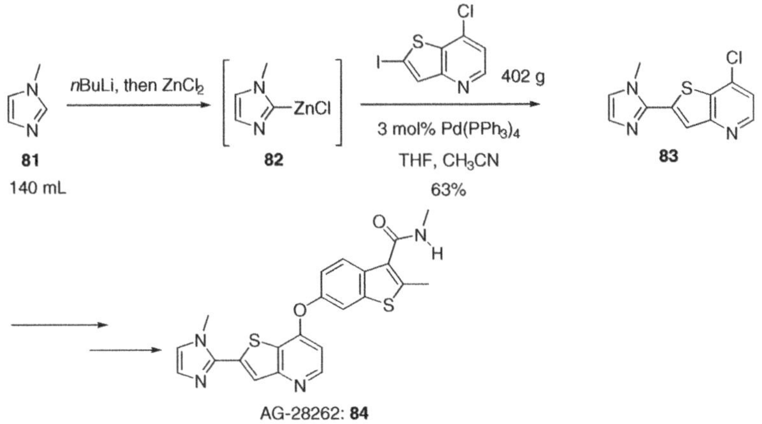

SCHEME 2.29 Pfizer's synthesis of AG-28262 via a Negishi coupling.

(more than −50°C) led to the decomposition of the lithium pyridyl species. After the transmetalation, the resulting zincate was unstable above 30°C. Hence, the Negishi coupling of the zincate with triflate **85** was preformed at 0–5°C. As the quality of Pd(PPh$_3$)$_4$ determined the activity of the catalyst, only new batches of catalyst stored in argon were used for the coupling reaction.

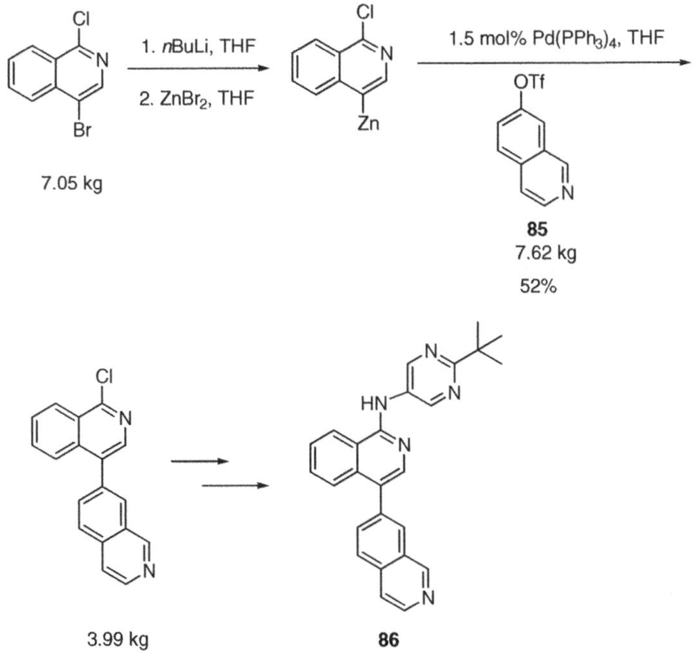

SCHEME 2.30 Novartis's synthesis of a melanoma drug candidate via a Negishi coupling.

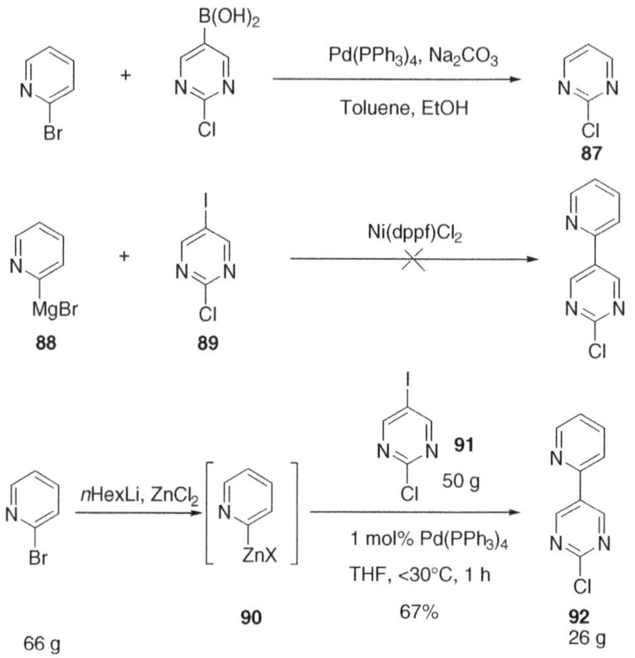

SCHEME 2.31   Johnson & Johnson's synthesis of a PDE5 inhibitor via a Negishi coupling.

A Johnson & Johnson process team demonstrated the unique utility of Negishi coupling in a challenging bis-heteroaryl coupling reaction (Scheme 2.31) [97]. The initial attempt to form the C–C bond via a Suzuki coupling failed because only protodeboronation product **87** was observed. The Kumada coupling of magnesium-pyridyl bromide **86** and pyrimidyl iodide **89** gave a mixture of degradation products. Finally, a Negishi coupling of zincate **90**, generated *in situ* via lithium bromide exchange and zinc chloride treatment, and iodopyrimidine **91** uneventfully provided a good yield of the desired coupling product **92** on a medium scale. This intermediate was then used for the scale-up of a PDE5 inhibitor for further development.

2.2.3 Kumada Coupling

The Kumada coupling was first reported nearly four decades ago [98]. It can be catalyzed by Pd or Ni catalysts. The limitation of the utility of this reaction lies in the functional group compatibility of Grignard reagents, and thus chemoselectivity of reaction could become an issue. At the time of its discovery, the application of transition metal-catalyzed C–C formation in the pharmaceutical industry was rare. Nevertheless, this method has been used in industry to prepare biaryl products in large scale when functional group compatibility was not an issue.

For example, a group in Zambon developed a kilogram-scale Kumada coupling in the synthesis of diflunisal (**93**) (Scheme 2.32) [99]. It should be noted that this

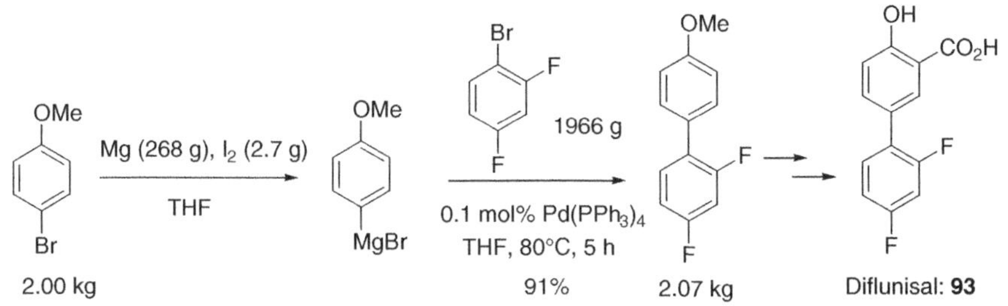

SCHEME 2.32 Zambon's synthesis of diflunisal via a Kumada coupling.

protocol required the use of only 0.1 mol% of Pd(PPh$_3$)$_4$, and the reaction could also be promoted by nickel catalysts.

Another example of the Kumada coupling was reported by a group from Agouron Pharmaceuticals in the synthesis of AG341 (**96**), an inhibitor of thymidylate synthase (Scheme 2.33) [100]. The purpose of the reaction was to introduce a methyl group to the naphthyl unit of the molecule as shown in intermediate **95**. The original trial of the methylation reaction of aryl iodide **94** employed a Stille coupling using tetramethyltin. However, the moderate yield (51%), and the need for an extractive workup and flash chromatography to remove the tin waste, rendered the procedure unsuitable for scale-up. Using the Kumada coupling conditions with only 1 mol% of Pd catalyst, the desired methylation product was obtained with moderate yield via an easy workup. It is worth noting that the lactam group present in the coupling partner did not interfere with the reaction even if excess methyl magnesium bromide was used [101].

2.2.4 Stille Coupling

The Migita–Kosugi group and the Stille group simultaneously reported the coupling of organostannes with carbon electrophiles in the 1970s [102]. This type of transformation is now referred to as the Stille coupling. Trimethylstannyl or tributylstannyl compounds are typically used for Stille coupling and they are synthesized from the corresponding organolithium, magnesium, aluminum, zinc, and boron reagents via a transmetalation reaction. Although trimethylstannyl compounds are more reactive than the corresponding tributylstannyl compounds, the toxicity of the former is about 1000 times higher than that of the latter. Owing to the significant covalent nature of the

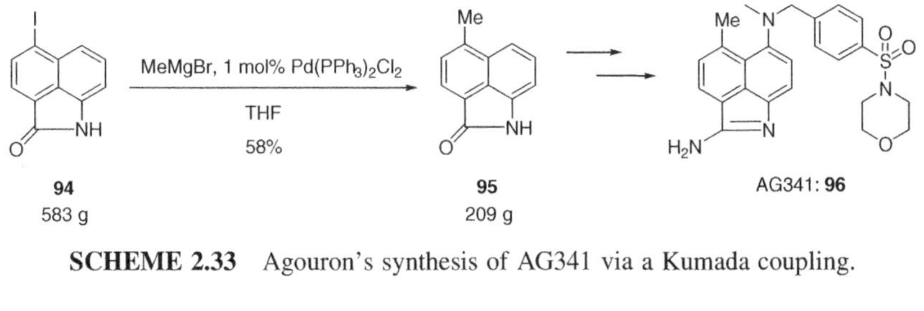

SCHEME 2.33 Agouron's synthesis of AG341 via a Kumada coupling.

C–Sn bond, alkyl stannanes are not very reactive, but the low reactivity can be remedied by using highly polar solvents. Another issue of the Stille coupling is that sometimes butyl and methyl groups can also be transferred in the coupling reaction causing the formation of side products, despite their low transfer potential relative to sp^2 hybridized centers.

An early application of the Stille coupling was reported by Merck chemists in their synthesis of carbapenem candidate **100** (Scheme 2.34) [103]. The use of stannatrane nucleophile **97** facilitated the Stille coupling to give an excellent yield of interme-diate **98**, accomplishing the desired sp^2–sp^3 C–C bond formation. Meanwhile, a quantitative recovery of the stannatrane as chloride (**101**) could be achieved by simply treating the resulting stannatrane triflate **99** with 20% aqueous NaCl. It is noteworthy that numerous Stille coupling reactions have been applied in carbapenem and cephalosporin synthesis due to the remarkable functional group tolerability and neutral reaction conditions [104].

The use of Stille coupling for large-scale syntheses in the pharmaceutical industry is rare due to toxicity issue related to organostannanes. However, when there are no other options, the Stille coupling can indeed be applied even in GMP bulk campaigns.

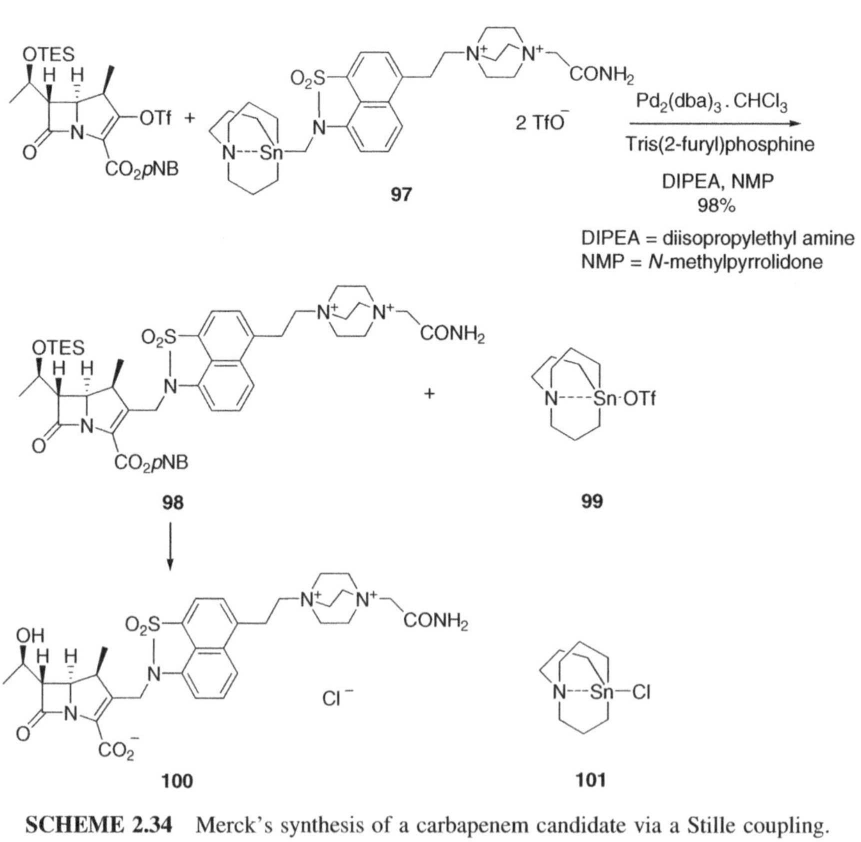

SCHEME 2.34   Merck's synthesis of a carbapenem candidate via a Stille coupling.

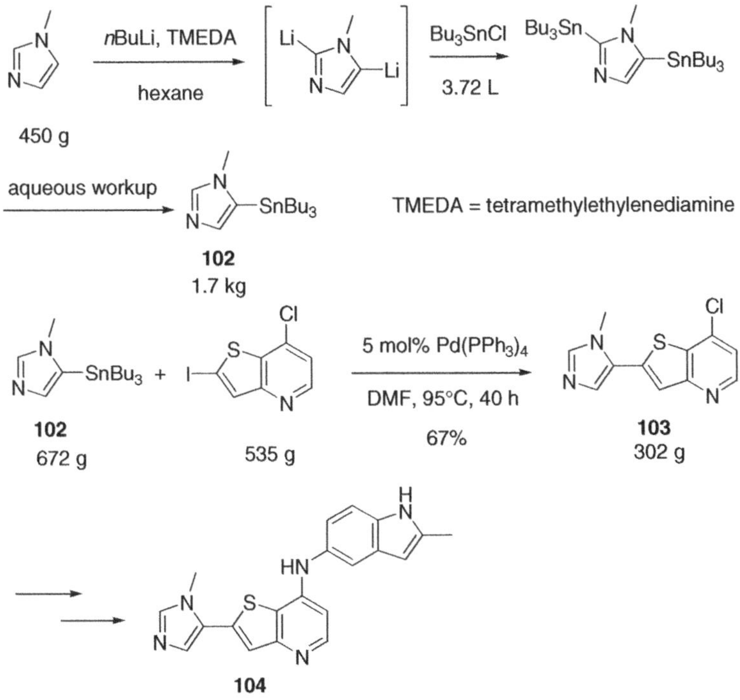

SCHEME 2.35   Pfizer's synthesis of a VEGFR-2 antagonist via a Stille coupling.

This was demonstrated by Pfizer's work on the preparation of **104**, a VEGFR-2 antagonist (Scheme 2.35) [105]. The attempts to assemble the bis-heteroaryl core of **104** via coupling strategies including Suzuki, Negishi, Heck, Hiyama, Kumada, Kobayashi, and DeShong coupling reactions failed on small scale or above a 50 g scale. The only scalable and high-yielding route toward intermediate **103** entailed the Stille coupling. The stannane starting material **102** could be easily prepared on large scale by double lithiation of methylimidazole and quenching with tributyltin chloride, followed by an aqueous workup. The tributyltin chloride-derived Sn impurities could be removed by a hexane–acetonitrile partition. The resulting stannane was then used for the subsequent Stille coupling to provide a good yield of the desired product **103**. The major issues were tank and drug substance contaminations by the tin residue. The tank contamination issue was easily solved by adopting a large glass vessel disposed of after the campaign. To assess the tin level in product, an inductively coupled plasma emission spectroscopy (ICP) analysis was used. This analytical tool allowed the detection limit of Sn and Pd to less than 2 ppm, below the required upper limit of Sn level (~20 ppm) by toxicologists. A simple reslurry of the Stille coupling product removed the bulk of the Sn residue (from 170,000 to 154 ppm). The material was then recrystallized to provide product containing only 19 ppm of Sn residue. Alternatively, without recrystallization, the product could proceed to the next step to give the final product **104**, which was crystallized upon treatment of camphorsulfonic acid to yield material with less than 3 ppm of Pd and Sn accomplishing the cGMP campaign.

2.2.5 Heck Coupling

The Heck coupling reaction, discovered by R. F. Heck in the late 1960s [106], has demonstrated wide utility since the 1980s in both total syntheses of natural products in academia and synthesis in pharmaceutical and agrochemical industry [107]. In addition to water and air tolerability in most Heck coupling reactions, the important virtues of the Heck reaction also include its functional group compatibility and wide olefin substrate scope [108]. This powerful transformation can be applied both inter- and intramolecularly. The intramolecular Heck coupling is particularly useful for forming carbocycles and heterocycles. Thus, it has been applied to numerous natural product total synthesis and sometimes in an enantioselective fashion [109].

2.2.5.1 Intermolecular Heck Coupling   In an efficient synthesis of a leukotriene LTD$_4$ antagonist L-699392 (**107**) [110], a Merck group applied a highly regiose-lective Heck reaction of aryl iodide and allylic alcohol to form intermediate **106** (Scheme 2.36) [111]. The reaction was conducted without phase transfer catalysts, and the Pd catalyst loading was low (0.5 mol%). A similar Heck reaction was applied in the synthesis of the closely related Merck asthma drug Montelukast Sodium [112]. It was found that a 10% excess of methyl *o*-iodobenzoate **105** could help to stabilize the Pd in the form of soluble complex in the crystallization solvent, thus allowing complete rejection of the Pd during the crystallization event.

Eletriptan (**108**), Pfizer's 5-HT$_{1D}$-like partial agonist for the treatment of migraine, was also prepared by a Heck coupling, followed by hydrogenation (Scheme 2.37) [113].

In a chemoselective Heck coupling reaction, the Pfizer process chemists differen-tiated the reactivity of bromide and iodide in substrate **109** by controlling the reaction temperature and selecting the appropriate solvent and base (Scheme 2.38) [114]. As such, excellent chemoselectivity was achieved: 79% of the desired intermediate

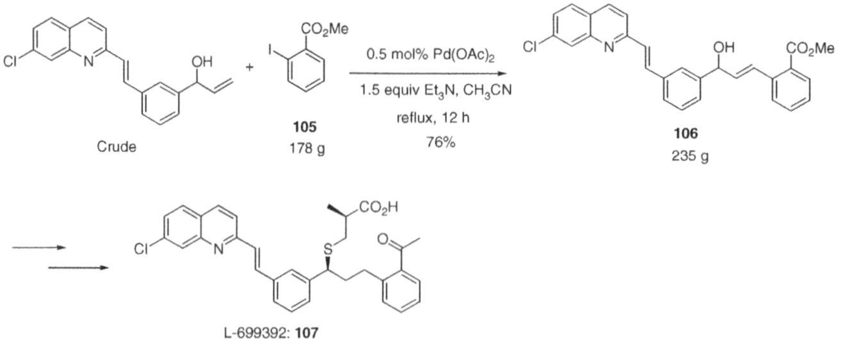

SCHEME 2.36   Merck's synthesis of a LTD$_4$ antagonist via a Heck coupling.

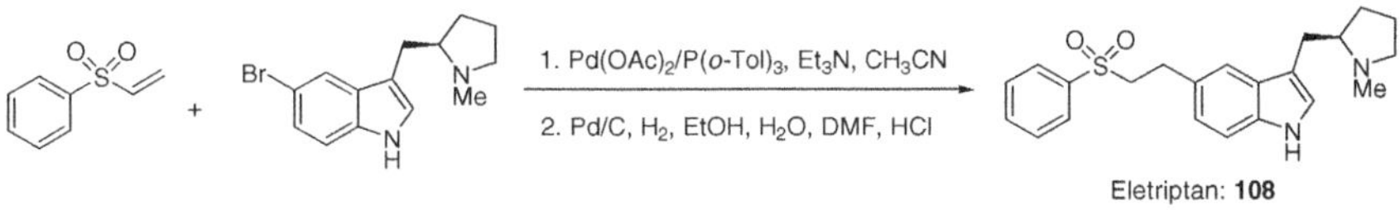

SCHEME 2.37 Pfizer's synthesis of eletriptan via a Heck coupling.

110 was obtained with only 0.4% of the corresponding bis-Heck coupling product. The ensuing second Heck coupling then introduced another carbon unit of the drug in high boiling point xylene at an elevated temperature. This sequence resulted in key intermediate **111** that was eventually converted to **112**, a thromboxane receptor antagonist targeted to prevent restenosis after balloon angioplasty or artery bypass grafts [115].

In a manufacturing route toward ketone alcohol **113**, a Heck coupling was adopted by an AstraZeneca group (Scheme 2.39) [116]. The screening of a range of bases, catalysts, ligands, and solvents led to the identification of a combination of Pd(OAc)$_2$, P(o-tol)$_3$, and Et$_3$N as the optimal reagents. To avoid the workup complication associated with the precipitation of triethylamine HBr salt, Bu$_3$N was selected as its HBr salt is soluble in toluene. The catalyst loading can be reduced to 0.01 mol%. Despite the low yield (33%) that may be due to the catalyst decomposition, this reaction was scaled up to a 3 kg scale.

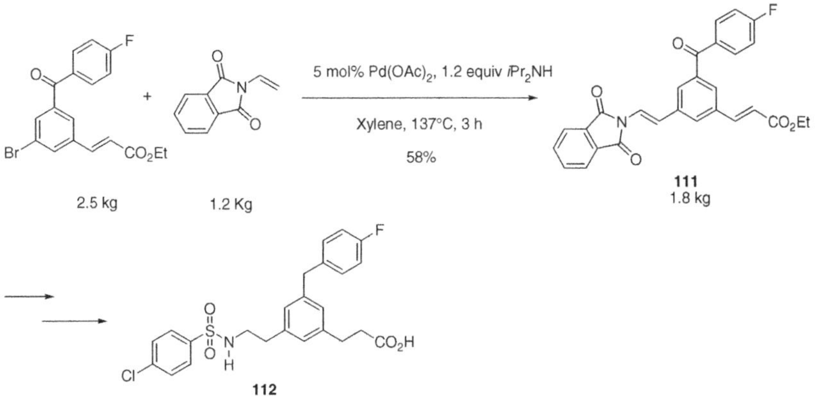

SCHEME 2.38 Pfizer's synthesis of a thromboxane receptor antagonist via a chemoselective Heck coupling.

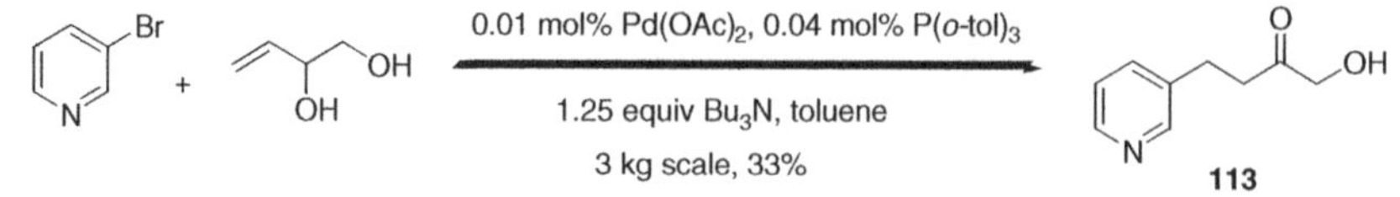

SCHEME 2.39 AstraZeneca's synthesis of an α-hydroxy ketone via a Heck coupling.

Oncology candidate CP-724714 (**116**) is a selective ErbB2 angiogenesis inhibitor that has been evaluated for breast, ovarian, and other types of cancers. The demand of bulk material for clinical studies prompted the Pfizer process group to explore an alternative route from their discovery synthesis that involved a Sonogashira coupling [117]. After exploring the Suzuki and Heck coupling reactions, the Heck coupling of aryl iodide **114** and bis-Boc protected allyl amine **115** proved to be a robust step on a multikilogram scale (Scheme 2.40). Furthermore, the key metrics comparison of the Sonogashira, Suzuki, and Heck coupling revealed that the Heck reaction generated substantially less organic and metal wastes than other strategies. Therefore, the Heck coupling route was selected as the ultimate route to prepare this clinical candidate.

Ligandless Heck coupling could also be useful as demonstrated by the synthesis of rilpivirine™ (TMC278, **118**), a novel NNRTI (Scheme 2.41) [118]. A key step in the commercial route of TMC278 is a Heck coupling between a halogenated aniline and acrylonitrile [119]. This process has been successfully transferred into production on a 2400 mol scale (6000 L reactor). The use of more reactive iodoaniline **117** was preferred over the corresponding bromoaniline for two major reasons. First, iodoaniline substrate allowed the reaction to proceed with low catalyst loading (0.5 mol% Pd/C) in the absence of any ligand, whereas the phosphine ligand with Pd catalyst used for the Heck reaction of bromoaniline was significantly more expensive. Thus, the choice of iodoaniline resulted in tremendous cost-effectiveness. Second, the reaction with iodoaniline gave more reliable control of the Pd residue (58 ppm) versus that with bromoaniline (1196 ppm). After the reaction was complete, a simple treatment

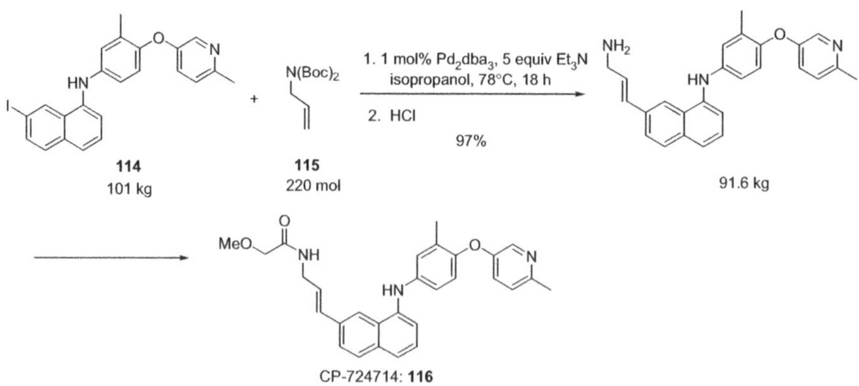

SCHEME 2.40 Pfizer's synthesis of CP-724714 via a Heck coupling.

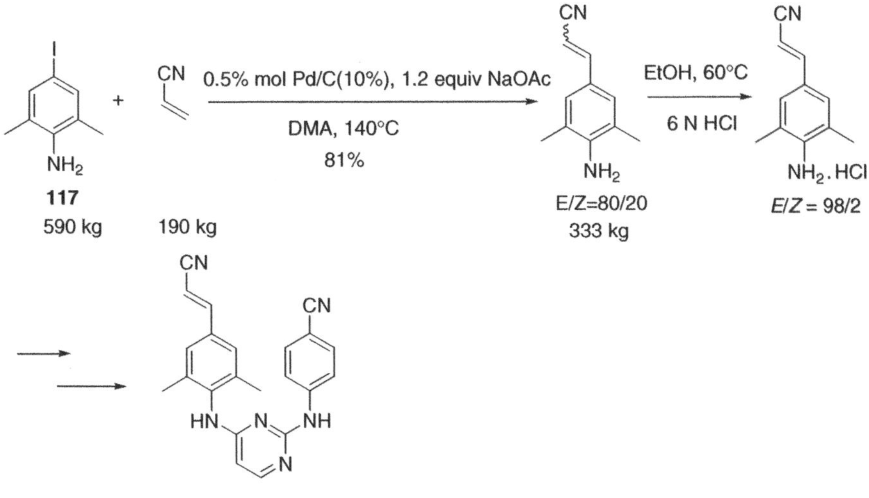

Rilpivirine (TMC278): **118**

SCHEME 2.41 Johnson & Johnson's synthesis of rilpivirine via a ligandless Heck coupling.

of activated charcoal resulted in less than 5 ppm of Pd level in the product, which met the criteria for a commercial route.

A multikilogram-scale vinylation of bromopyridine was successfully executed using 1 mol% of Pd(OAc)$_2$ and a combination of 2,2'-bis(diphenylphosphino)-1,1'-binaphthyl (BINAP) and P(o-tol)$_3$ (Scheme 2.42) [120]. In an earlier process, in the absence of the BINAP additive, Pd(OAc)$_2$ was reduced during the reaction to generate a significant amount of Pd black that was difficult to remove from the reaction vessel. In addition, the reaction was lengthy. By using BINAP to promote the reaction and slow down the reduction of Pd(II) to Pd(0), the reaction was complete in 16 h as opposed to 3 days without BINAP.

GSK reported a process route to prepare a potent vitronectin receptor αvβ3 antagonist SB-273005 (**121**), which may be used for treatment of inflammation, cardiovascular disorders, cancer, and osteoporosis (Scheme 2.43) [121]. The Heck

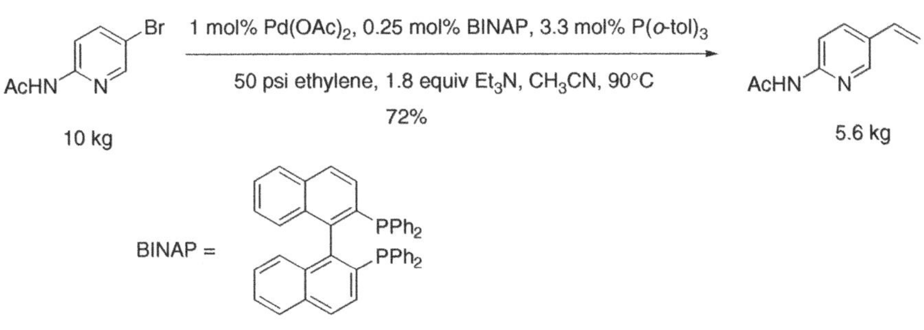

SCHEME 2.42 Pfizer's synthesis of a vinyl pyridine via a Heck coupling.

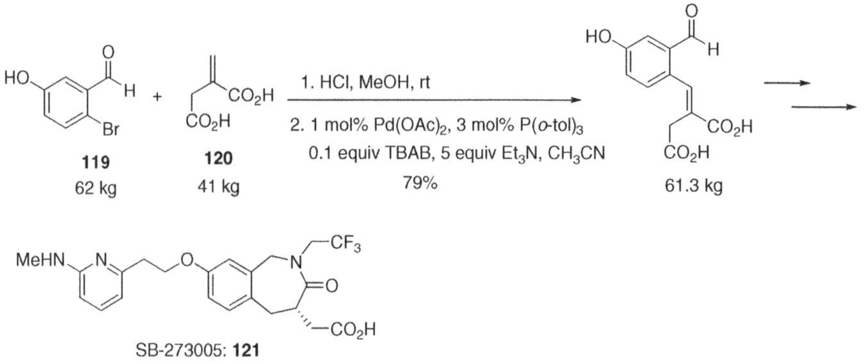

SCHEME 2.43   GSK's synthesis of SB-273005 via a Heck coupling.

reaction between bromobenzaldehyde **119** and itaconic acid **120** proceeded with intramolecular aldol reaction of the product as a side reaction [122]. Therefore, the aldehyde was converted to acetal *in situ* prior to the Heck coupling reaction.

A Pfizer group also applied Heck reaction to generate ketone **122** as a key intermediate for the preparation of a hepatitis C viral polymerase (HCVP) inhibitor (**123**) (Scheme 2.44) [123]. In this reaction, the addition of chloride ligands and ligating bases, such as triethylamine, led to enhanced reactivity and lowered catalyst loading. As the reaction was exothermic and rapid (completed in 15–60 min), triethylamine was dosed in portions. Owing to the generation of HBr that then quenched triethylamine, a non-amine co-base (LiOAc) was also added to stabilize the catalyst system, allowing the reaction to complete with a single initial charge of Pd catalyst.

Solid-support Pd-Smopex-111 catalyst was also applied in a Heck coupling reaction of *t*-butyl acrylate **124** and 4-acetylphenyl bromide **125** (Scheme 2.45). This provided the desired product **126** in good yield and only 8.7 ppm of leached Pd [62].

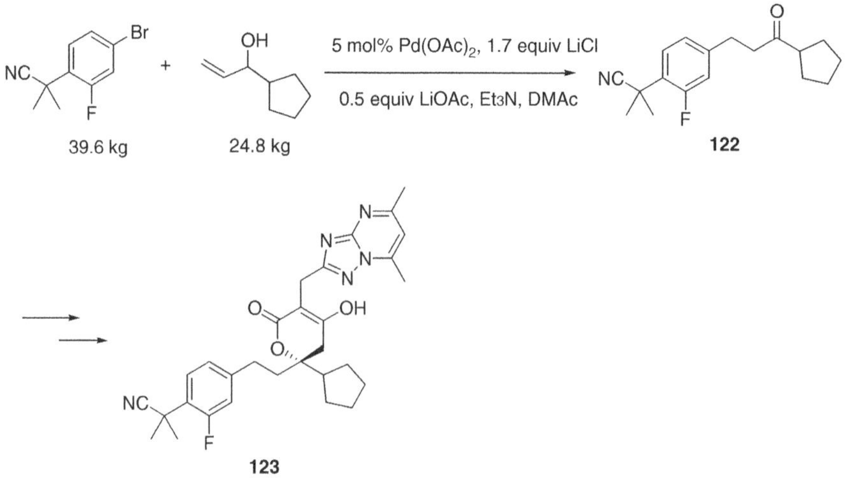

SCHEME 2.44   Pfizer's synthesis of a HCVP inhibitor via a Heck coupling.

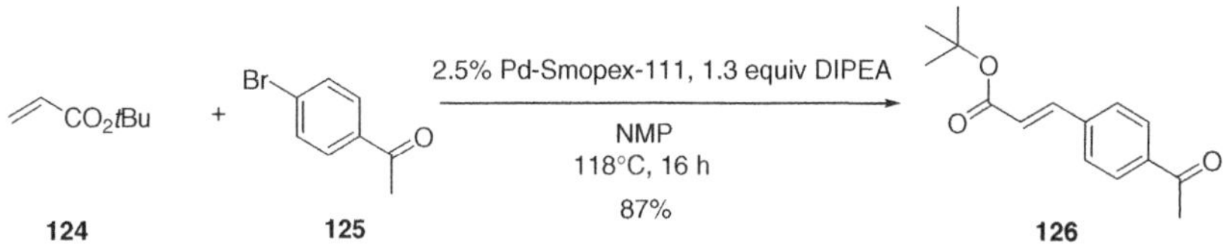

SCHEME 2.45 Novartis's synthesis of a biaryl intermediate via a Heck coupling using a solid-support catalyst.

2.2.5.2 *Intramolecular Heck Coupling*

In the development of a glycine antagonist (**129**) potentially useful for the treatment of nicotine craving [124], a GSK group used an intramolecular Heck reaction to form tetrahydroisoquinoline **127** (Scheme 2.46) [125]. The major side products were the endo olefin isomer **128** resulting from the lactam β-hydride elimination and the protodehalogenation by-product. The portion of the desired product was significantly increased with the addition of triphenylphosphine. It should be noted that the use of excess amount of triethylamine did not cause racemization even while refluxing in toluene. To remove residual Pd from the product, reflux of the reaction mixture with 6 mol% of trimercaptotriazine at the end of the reaction reduced the Pd level to < 15 ppm.

2.2.6 Sonogashira Coupling

As a significant advance over the related Stephens–Castro coupling [126], Sonogashira coupling was developed to form a sp and sp^2 C–C bond under much milder conditions [127]. Although alkyne is not prevalent in drug molecules, the resulting

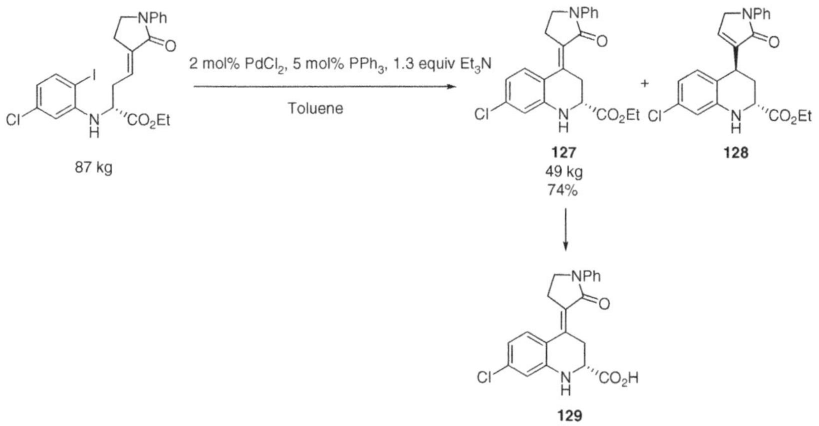

SCHEME 2.46 GSK's synthesis of a glycine antagonist via an intramolecular Heck coupling.

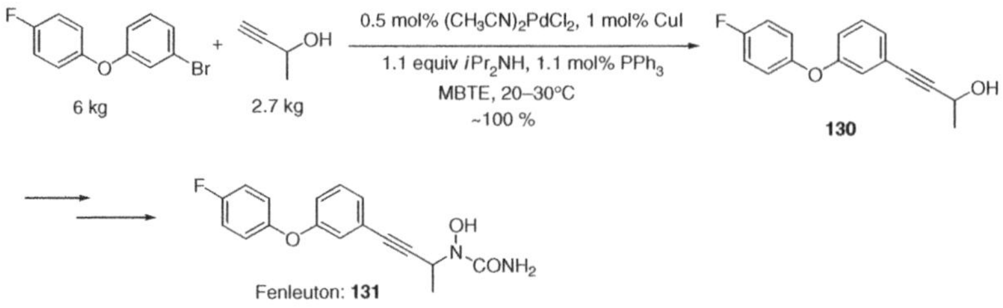

SCHEME 2.47 Abbott's synthesis of fenleuton via a Sonogashira coupling.

alkyne from Sonogashira coupling reactions can be functionalized to alkenes, alkanes, carbocycles, heteroaryls, and heterocycles. As a result, this versatile process has enabled the preparation of multiple drug candidates.

2.2.6.1 Aryl Alkynyl Coupling

In a manufacturing route to fenleuton (**131**), a 5-lipoxygenase (5-LO) inhibitor for the treatment of osteoarthritis and related inflammatory diseases, the Abbott process group employed a Sonogashira coupling to assemble the alkynyl aryl intermediate **130**, which was then converted to fenleuton (**131**) via a nucleophilic displacement (Scheme 2.47) [128]. The Sonogashira coupling was exothermic, but it was well controlled by slowly metering the isopropylamine and keeping the temperature below 35°C. The reaction was essentially quantitative and easily scalable.

p38 MAP kinase inhibitors have been used as antiinflammatory agents. Eli Lilly developed one such compound referred to as LY479754 (**134**) containing a triaryl-substituted imidazole core (Scheme 2.48) [129]. One of the key steps is the Sonogashira coupling of aryl iodide **132** and phenylacetylene [130]. It was found that the Pd catalyst loading needed to be 4 mol% to allow the reaction to be completed within 15 h. Lower catalyst loading resulted in a sluggish reaction. DMSO was found to be a good solvent. Upon the addition of an initial portion of water during workup, a dark precipitate consisting of primarily catalyst was formed and filtered off.

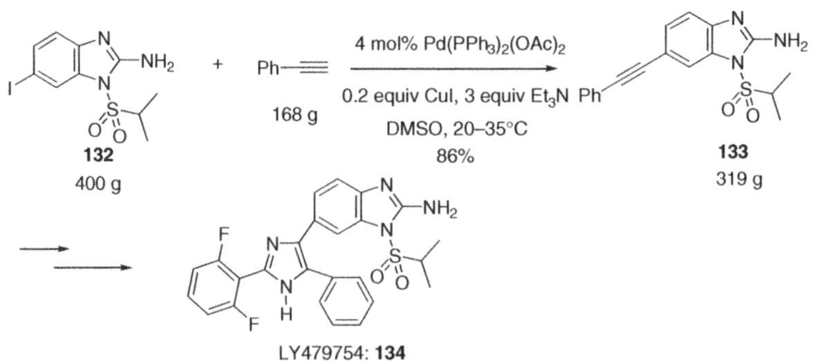

SCHEME 2.48 Lilly's synthesis of LY479754 via a Sonogashira coupling.

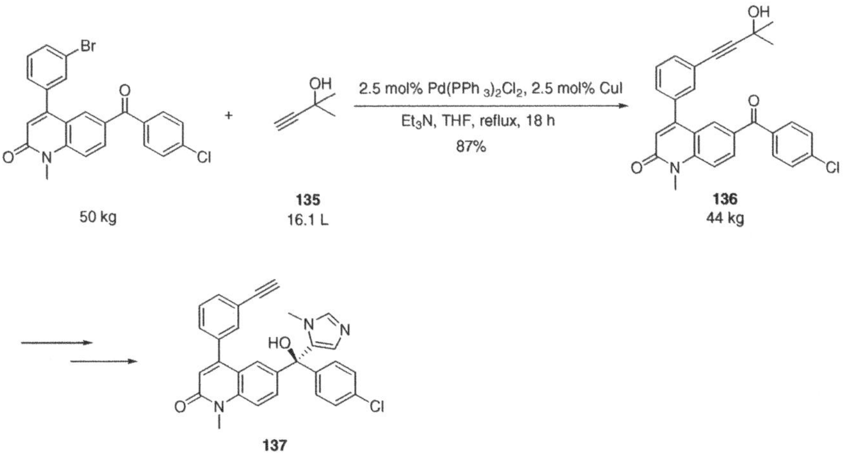

SCHEME 2.49   Pfizer's synthesis of FTase inhibitor via a Sonogashira coupling.

The further addition of water allowed the desired product 133 to precipitate with an excellent yield.

Farnesyltransferase (FTase) inhibitors represent a novel class of oncology agents to target abnormal cellular transformation and malignancy. In the synthesis of FTase inhibitor 137 [131], the Pfizer process chemists conducted a multikilogram-scale Sonogashira coupling to give a very high yield of the alkynyl arene product 136 (Scheme 2.49) [132]. It was noted that the coupling reaction was cleaner with unprotected alkynyl alcohol 135 than with the corresponding trimethylsilylated alcohol. Furthermore, the selection of reaction solvent was critical for the scalability of the reaction. When acetonitrile was used as the reaction solvent, following solvent switch by ethyl acetate caused emulsions. If dichloroethane was used to displace acetonitrile, a large volume was required to remove acetonitrile, and an even larger volume of hexanes was required to precipitate the product. Finally, THF was selected as the reaction solvent as only a small volume of dichloroethane was needed to displace the reaction solvent.

The Abbott Lab developed a convergent multikilogram-scale synthesis of A-79195 (138), a 5-LO inhibitor (Scheme 2.50) [133]. The choice of base was key to the success of the reaction. Interestingly, diisopropylamine (DIPA) enabled the reaction to be completed in 2 h at room temperature, whereas the selection of commonly used triethylamine caused sluggish reaction and formation of by-products.

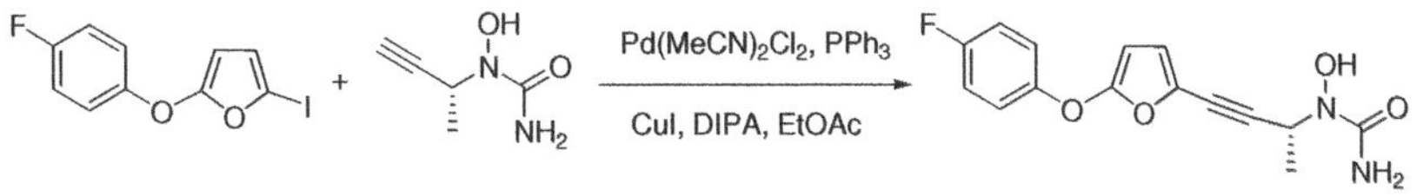

SCHEME 2.50   Abbott's synthesis of A-79195 via a Sonogashira coupling.

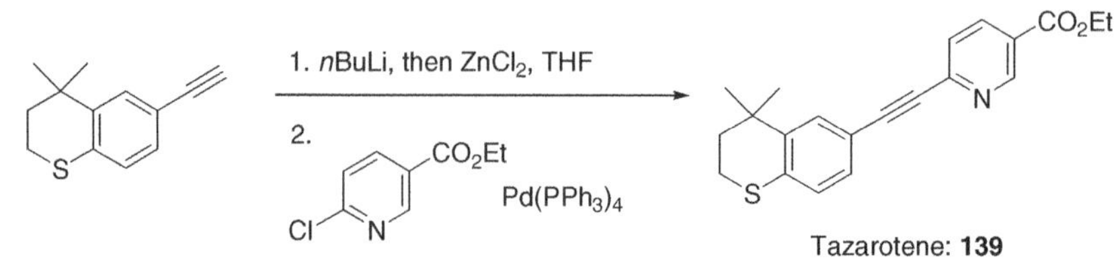

SCHEME 2.51   Allergan's synthesis of tazarotene via a Sonogashira coupling.

Tazarotene (**139**) belongs to the acetylenic retinoid class and it has been commercialized by Allergan for the treatment of psoriasis and acne (Scheme 2.51) [134]. The preparation of this drug involved a variation of a typical Sogonashira coupling, in which a zinc acetylide was generated to couple with heteroaryl halide without the use of a copper cocatalyst.

A Sonogashira coupling/pyrrole formation sequence was applied by a Pfizer group in the synthesis of pyrrolquinolone PHA-529311 (**141**) (Scheme 2.52) [135]. This compound exhibited potent inhibition against HCMV, HSV-1, and VZV polymerases, as well as good selectivity against herpes virus DNA polymerases over human polymerases [136]. In the Pd-catalyzed pyrrole formation, the use of excess acetylene **140** (1.5 equiv) was due to alkyne homocoupling. A 10:1 ratio of CuI:Pd (PPh$_3$)$_2$Cl$_2$ was necessary for the full conversion. It is noteworthy that pyrrole formation that involved a Sonogashira coupling as the first step was quite facile and no uncyclized product was isolated. As screening of metal chelators of Cu and Pd failed to control the levels of both metals, aqueous acid such as HCl was used to extract the compound and the mixture was filtered with Deloxan THP. The filtrate was then kept for crystallization and the solid was obtained with reasonable Pd and Cu levels (20 and 2 ppm, respectively). Alternatively, Si-thiol, a thiol-capped silica gel was used to decrease the Pd level to 17 ppm and Cu level to 1 ppm in the reaction mixture.

To generate ABT-239 (**142**), a histamine receptor H$_3$R antagonist for treatment of cognitive disorders, Abbott developed a facile and scalable route that consisted of a

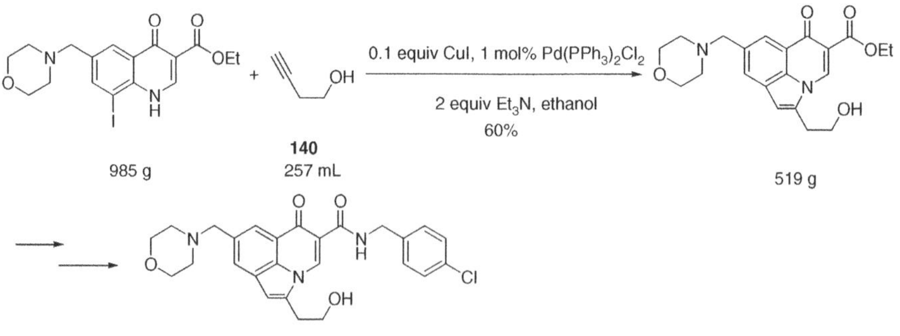

SCHEME 2.52   Pfizer's synthesis of PHA-529311 via a Sonogashira coupling.

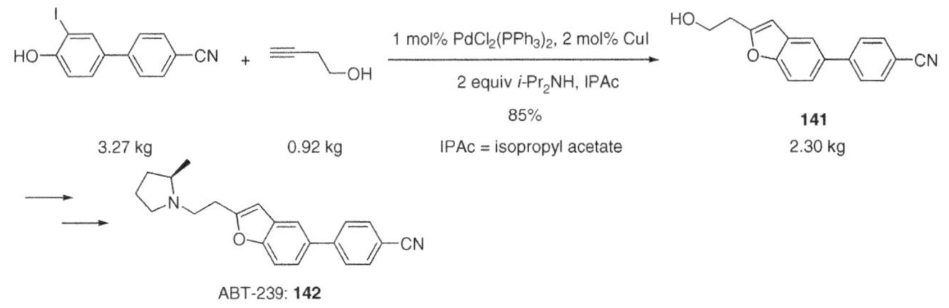

SCHEME 2.53 Abbott's synthesis of ABT-239 via a Sonogashira coupling.

Sonogashira coupling and a subsequent cyclization to assemble benzofuran 141 (Scheme 2.53) [137]. The reaction proceeded smoothly in dichloromethane, but environmentally benign isopropyl acetate was eventually selected as the reaction solvent. An alternative route with the corresponding bromide substrate was abandoned due to the sluggishness of reaction, the necessity of using excess of reagent (3 equiv of 3-butyn-1-ol), higher catalyst loading, and higher reaction temperature for reaction to complete.

2.2.6.2 Vinyl Alkynyl Coupling

In the synthesis of the broad-spectrum antimycotic Lamisil (145) by Sandoz (Novartis), vinyl chloride 143 and alkyne 144 coupled efficiently in the presence of less than 0.05 mol% of the catalyst Pd(PPh$_3$)$_2$Cl$_2$ and CuI cocatalyst (Scheme 2.54) [138].

Eniluracil is a potent dihydropyrimidine dehydrogenase inactivator [139]. The GSK chemistry development group used a Sonogashira coupling to prepare multikilogram of eniluracil (147) (>60 g batch size, 1500 L plant) (Scheme 2.55) [140]. During the process, they overcame the issues of high levels of Cu and Pd, the color of the drug substance, less than optimal yields, and complex workup/isolation procedures. It was found that effective deoxygenation of the reaction mixture was required to lower heavy metal contamination in the product. Lowering both Cu and Pd catalysts to 0.5 mol% still allowed the reaction to complete. The solvent screening established ethyl acetate as the optimal solvent. Although ~600 ppm of Pd and ~100 ppm of Cu were observed after the Sonogashira coupling, an activated charcoal workup generated the crystalline product 146 with minimal Pd (<2 ppm) and Cu (<1 ppm) levels.

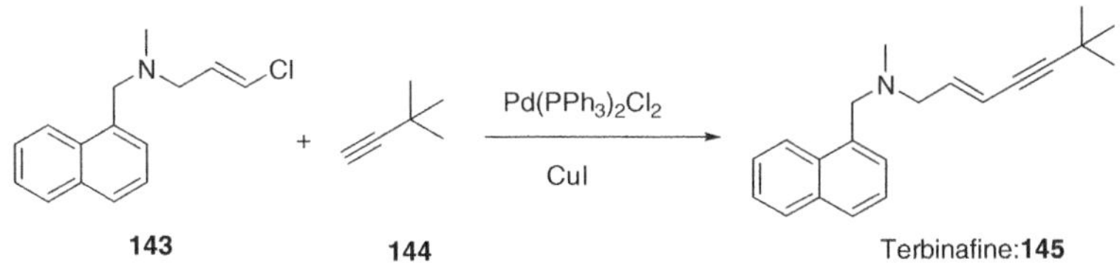

SCHEME 2.54 Sandoz's synthesis of terbinafine via a Sonogashira coupling.

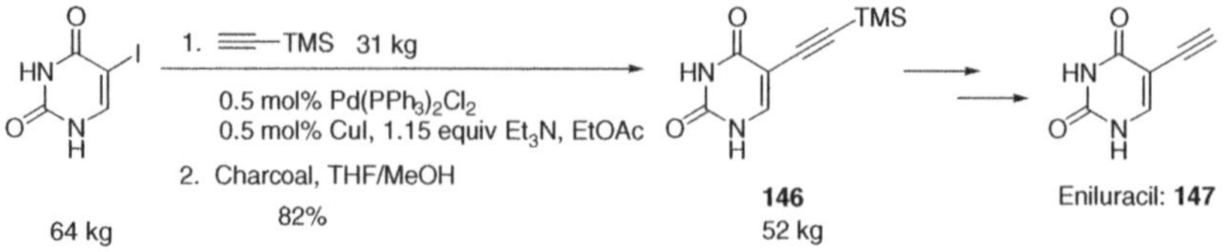

SCHEME 2.55   GSK's synthesis of eniluracil via a Sonogashira coupling.

2.2.7   Trost–Tsuji Coupling

The Trost–Tsuji coupling reaction is a powerful method to create a C–C bond between a carbon nucleophile and an allyl group [141]. This reaction can proceed asymmetrically in the presence of chiral ligands, and it has been extensively used in the pharmaceutical industry to prepare APIs.

For example, in the synthesis of mGluR 2 receptor agonist MGS0028 (**150**), a potential useful drug to treat schizophrenia and anxiety, a Merck process chemistry team employed the Trost asymmetric allylic alkylation (AAA) reaction to form chiral intermediate **149** in the presence of the standard Trost ligand **148** (Scheme 2.56). Interestingly, it was found that decreasing Pd catalyst loading and using bulkier n-Hex$_4$N$^+$ counterion significantly improved the ee to 94–96% range.

The Merck Process Research group developed an efficient and practical asymmetric synthesis of an anti-HIV CCR5 antagonist **154** [142] by using a Mo-catalyzed allylic alkylation, first reported by the Trost group (Scheme 2.57) [143]. The original catalyst (EtCN)$_3$Mo(CO)$_3$ applied by the Trost group was replaced with more readily available, cost-effective, and air-stable Mo(CO)$_6$ precatalyst. This reaction could be performed in the presence of relatively high catalyst and ligand loading on a multikilogram scale to offer the branched product **153** from racemic allylic carbonate **151** when chiral ligand **152** was employed. As a result, excellent yield and enantioselectivity of the chiral product were obtained.

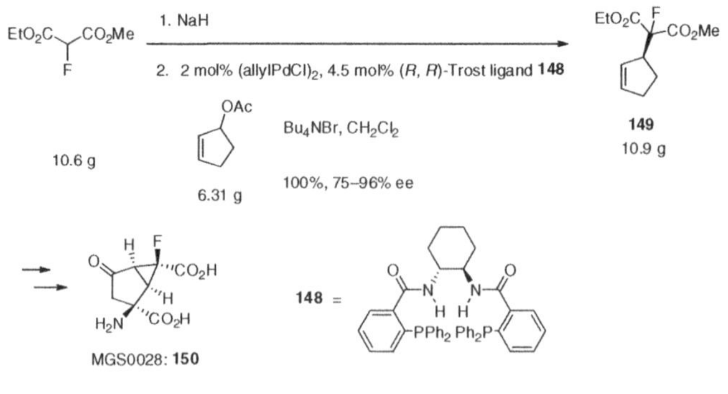

SCHEME 2.56   Merck's synthesis of MGS0028 using a Trost–Tsuji reaction.

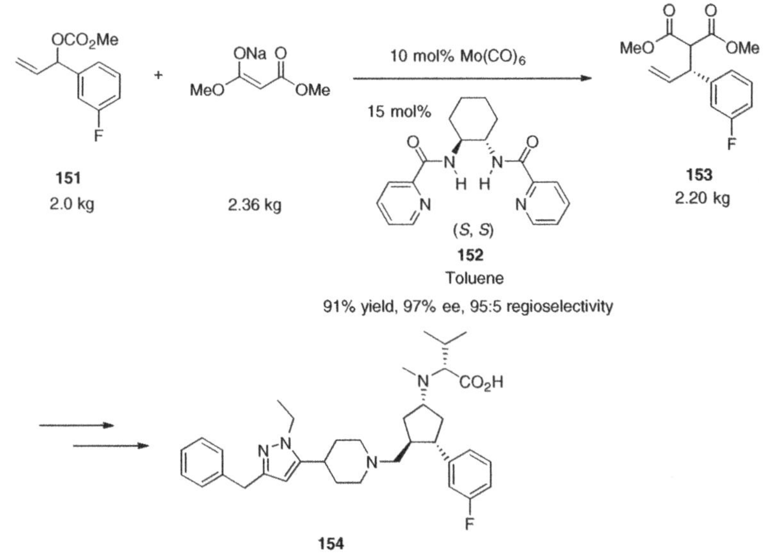

SCHEME 2.57 Merck's synthesis of a chiral cyclopentanone using a Trost–Tsuji reaction.

2.2.8 α-Arylation

The α-aryl carbonyl motif is highly useful as it is present in numerous natural products and pharmaceutical agents [144]. The pioneering work of Buchwald [145] and Hartwig [146] led to the discovery of α-arylation of esters, amides, ketones, aldehydes, and nitriles [147].

To develop a practical synthesis of a chiral analog of FTY720 (**156**), a novel immunomodulator that agonizes S1P receptor, a Novartis team employed an α-arylation reaction of ethyl acetoacetate to form intermediate **155** (Scheme 2.58) [148]. To minimize formation of the des-bromo by-product, the Pd-catalyzed α-arylation reaction needed to be conducted at strictly controlled temperature, namely, 85–90°C.

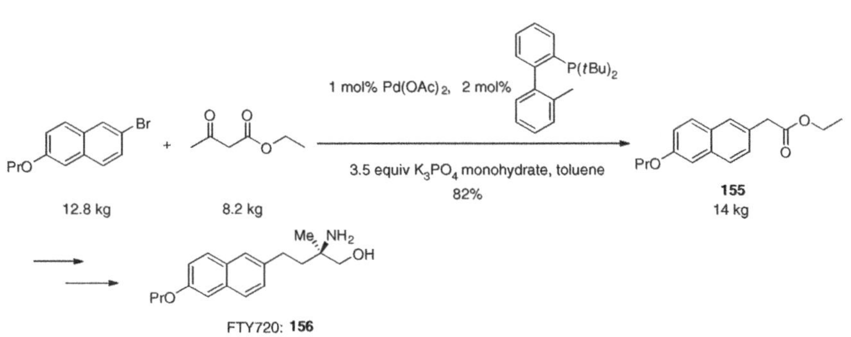

SCHEME 2.58 Novartis's synthesis of FTY720 via the α-arylation.

Technologies.

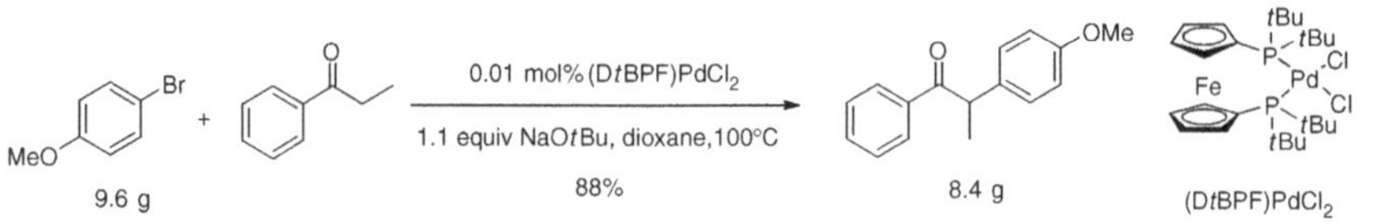

SCHEME 2.59   α-Arylation of ketones developed by Johnson Matthey Catalysis and Chiral Technologies.

After the α-arylation was complete, the reaction temperature was raised to 100–104°C to allow the deacetylation reaction to proceed.

A highly general and practical route for α-arylation of ketones was described by a group in Johnson Matthey Catalysis and Chiral Technologies [149]. This group discovered that a highly reactive, air-stable, and preformed catalyst (D*t*BPF)PdCl$_2$ (D*t*BPF = 1,1′-bis(di-*tert*-butylphosphino)ferrocene) could be used for α-arylation reactions of ketones using aryl bromides or chlorides. In one example (Scheme 2.59), only 0.01 mol% of catalyst was used and a good TOF (2700 h^{-1}) was achieved.

2.2.9   C–H Activation/Direct Arylation

The Pd-catalyzed chelation-directed C–H activation/cross-coupling reactions have emerged as an efficient approach for C–C bond formations [150]. The directing groups, to name a few, can be amides [151], imines [152], amines [153], N-heterocycles [154], esters [155], carboxylic acids [156], and pyridine N-oxides [157].

In addition to the chelation-controlled Pd-catalyzed C–H activation and C–C bond formation, other transition metal catalysts, such as Rh, Ru, Fe, Ir, and Ni catalysts, also enable direct arylation of heteroarenes or arenes by C–H bond cleavage [158].

Recently several Pd-catalyzed C–H activation/cyclization methods have been reported [159]. The Pfizer group used such strategy as a key step in their synthesis of a long-term oxazolidinone antibacterial agent (**158**) (Scheme 2.60) [160]. After the screening of various ligands for the Pd-catalyzed oxindole formation,

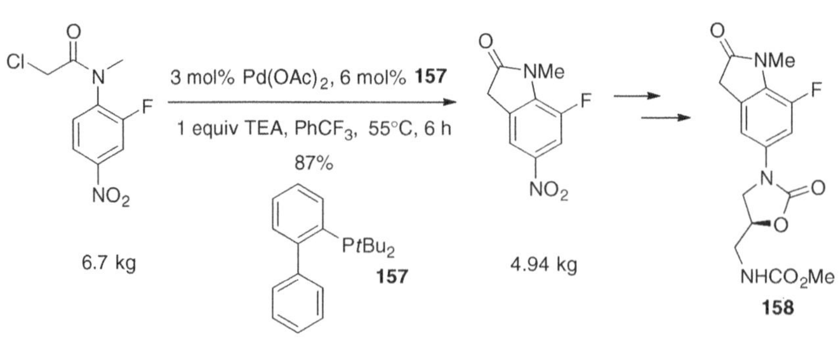

SCHEME 2.60   Pfizer's synthesis of an antibacterial agent via a C–H activation/cyclization.

the di-*tert*-butyl biphenyl phosphine ligand (**157**) appeared to be the only acceptable ligand. Trifluorotoluene was used as the solvent due to its low cost ($26.40/L) and the fact that it was not listed as hazardous air pollutant by the U.S. EPA. This example represents one of the first applications of the Pd-catalyzed C–H activation/cyclization method on large-scale synthesis in the industry.

A process chemistry team in Amgen reported a highly efficient Pd/Cu cocatalytic system for direct arylation of heteroaryls, such as benzothiazoles, benzoxazoles, and benzimidazoles (Scheme 2.61) [161]. Distinct from many direct arylation reactions, this process has low catalyst loading and good functional group compatibility. Furthermore, a multikilogram process based on this system has been run successfully. It is intriguing that neither Pd nor Cu catalysts alone can effectively catalyze the reaction. The authors postulated that the presence of Cu catalyst may lower the pK_a of the proton to be deprotonated, leading to the resulting aryl copper species for the subsequent transmetalation reaction. The unique role of the Cu(Xantphos)I cocatalyst may be potentially suitable for other arylation reactions of nitrogen-containing heteroaromatics.

2.2.10 Sequential Cross-Coupling Reactions

Sequential transition metal-catalyzed cross-coupling reactions provide an efficient approach to rapidly construct complexity. In some cases, the same catalyst can be used to catalyze a cascade of bond formation events in the same reaction vessel.

One early example is the Merck synthesis of rizatriptan benzoate (Maxalt, **159**), a 5-HT$_{1D}$ receptor agonist. This method involves the Larock's indole synthesis via a sequential carbopalladation/*N*-vinylation (Scheme 2.62) [162]. The triethylsilyl group was used to protect 3-butyn-1-ol in order to control the regioselectivity of the reaction and minimize desilylation.

In a more recent example of the Pd-catalyzed indolization using the Larock's protocol to prepare a drug candidate (**160**) [163], the BMS chemists investigated how to remove Pd from the reaction mixture (Scheme 2.63) [164]. The most effective method to reduce the Pd residue from 600–650 to 20–60 ppm in the formed indole intermediate was to apply 2,4,6-trimercapto-*s*-triazine (TMT, **161**). Subsequent recrystallization typically further decreased the Pd level in this active pharmaceutical ingredient to < 1 ppm.

Another example from Merck is the employment of double Suzuki coupling reaction to enable a facile synthesis of etoricoxib (**162**), a COX-2 inhibitor for pain relief (Scheme 2.64) [165]. In this reaction, the lithium triethoxypyridylboronate generated as a crude was directly used for the Suzuki coupling to afford the desired product **162** in good yield. It is worth noting that the second Suzuki coupling reaction

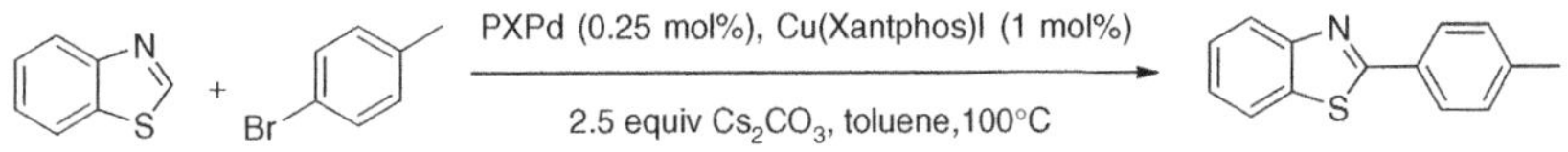

SCHEME 2.61 Amgen's synthesis of a biaryl compound via a direct arylation.

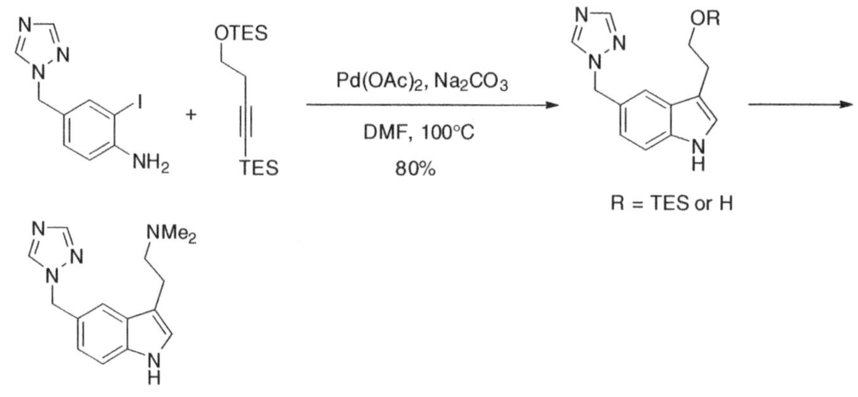

Rizatriptan: **159**

SCHEME 2.62   Merck's synthesis of rizatriptan via Larock's indole synthesis.

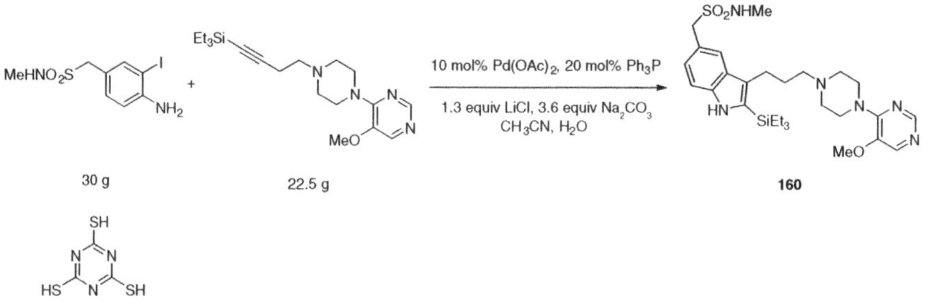

TMT: **161**

SCHEME 2.63   BMS's synthesis of an indole via Larock's indolization.

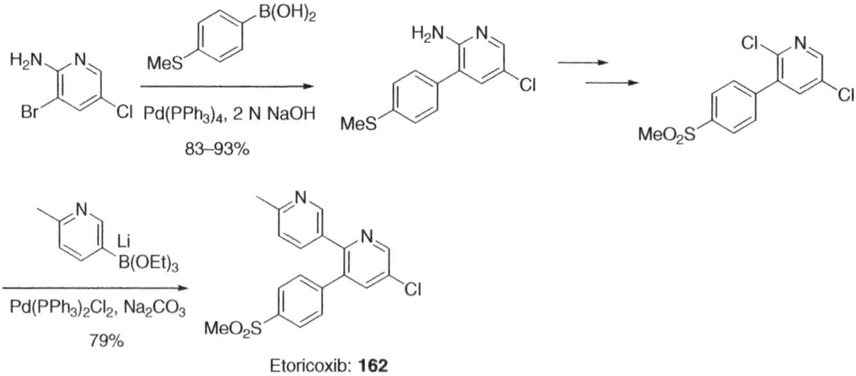

Etoricoxib: **162**

SCHEME 2.64   Merck's synthesis of etoricoxib via a double Suzuki coupling strategy.

was quite chemoselective in that the 2-chloro group reacted preferentially over the 5-chloro group.

A group from deCODE applied sequential intermolecular and intramolecular Heck reactions in one pot to form indole intermediate **163** (Scheme 2.65) [166]. This intermediate was then used to prepare DG-041 (**164**), a potent EP_3 receptor antagonist that may have therapeutic effects on peripheral artery disease (PAD).

A Merck team developed a novel synthesis of 3-methylindoles (**166**) from chlorotriflates (**165**) and allylamine derivatives by harnessing a regioselective Heck reaction, intramolecular *N*-arylation, and isomerization process (Scheme 2.66) [167]. This protocol is highly efficient and general, enabling the rapid access to a variety of 3-methylindoles in large scales.

A Novartis group reported double Sonogashira coupling reactions to construct compound **169** [168], an antimitotic agent that inhibits the proliferation of human epidermal cells (Scheme 2.67) [169]. Catalyst and ligand screening led to the selection of $Pd(PPh_3)_2Cl_2$ as a soluble and air-stable catalyst. It was critical to have a thorough degassing as oxygen promoted the dimerization of the alkyne. The first coupling reaction provided intermediate **167**, which could then be demasked by releasing acetone and exposing the terminal alkyne in intermediate **168**. Without the isolation of **168**, the second Sonogashira coupling proceeded upon the addition of aryl bromide, NaOH, and TBAB to deliver biaryl alkyne **169** in good yield. The last step of the synthesis was the hydrogenation of the alkyne to **170**. As the wide range of Pd and P levels could cause variable exotherms in the hydrogenation reaction, phosphine was removed via its oxidation to phosphine oxide and the subsequent removal using Panther Creek Clay. However, this treatment did not effectively reduce

SCHEME 2.65 deCODE's synthesis of DG-041 via sequential intramolecular and intermolecular Heck coupling reactions.

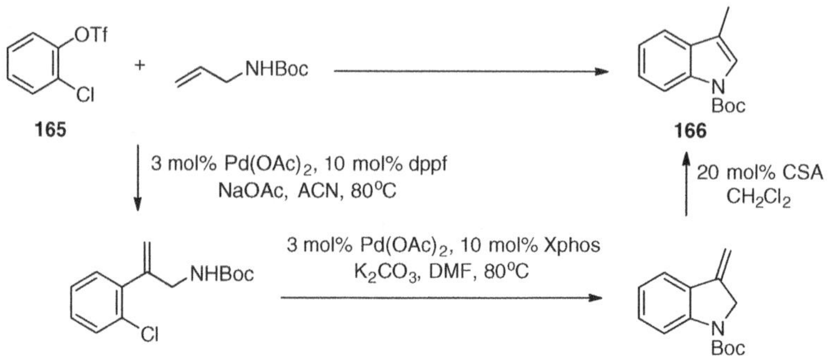

SCHEME 2.66 Merck's synthesis of indole 166 via Heck and *N*-arylation reactions.

the Pd level. Thus, a crystallization of the product from the reaction mixture followed by the treatment of thiourea or *N*-acetylcysteine successfully reduced Pd level to below 20 ppm and P level below 5 ppm. Ultimately, this innovative route was successfully scaled up on a multikilogram scale.

A Merck group used a strategy to prepare boronic acid *in situ* via lithiation of fluorochlorobenzene, followed by quenching with B(O*i*Pr)$_3$ (Scheme 2.68) [170]. The resulting boronic acid nicely coupled with bromide in the presence of Pd(P-*t*Bu$_3$) as the catalyst and KF as a base to form biaryl intermediate **175**. The Heck-type

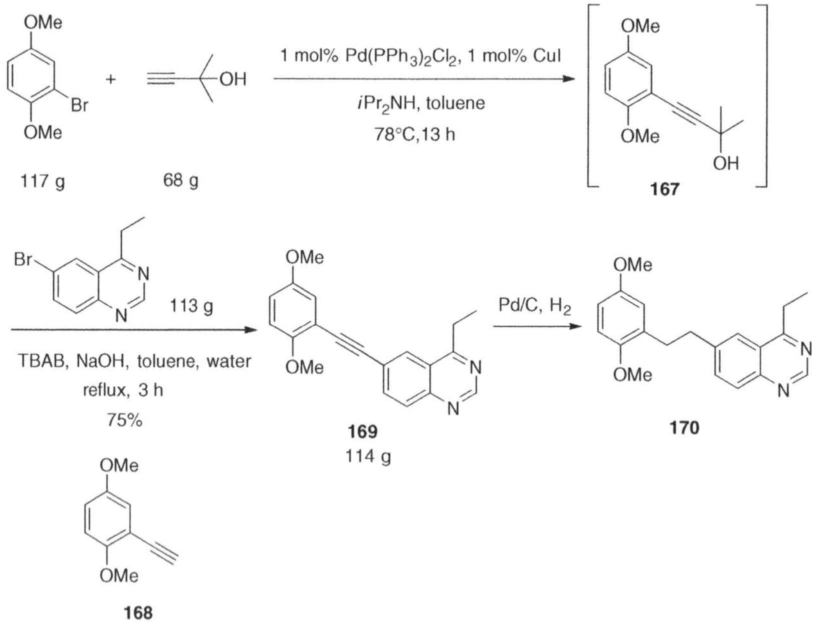

SCHEME 2.67 Novartis's synthesis of an antimitotic agent via double Sonogashira coupling reactions.

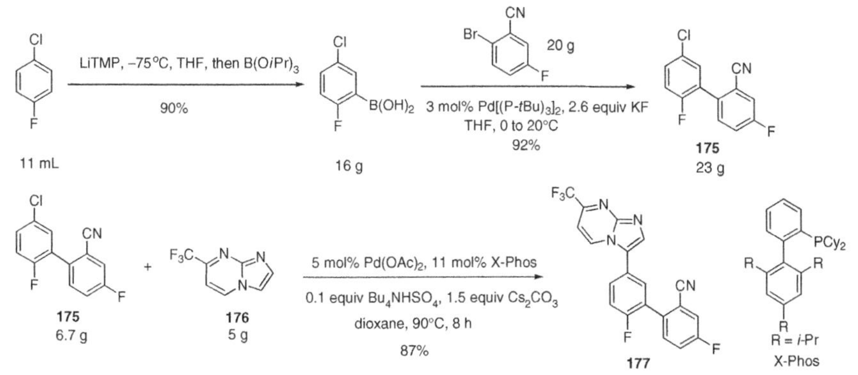

SCHEME 2.68   Merck's synthesis of a GABA $\alpha 2/3$ agonist via a Suzuki coupling and a Heck-type coupling.

coupling between **175** and imidazolopyrimidine **176** was operated under Jeffrey-type conditions [171]. After considerable reaction optimizations, it was found that the use of Bu_4NHSO_4 as the salt and X-Phos as the ligand was critical for the high yield of the reaction. The resulting product **177** is a GABA $\alpha 2/3$ agonist that might demonstrate anxiolytic benefit [172].

To prepare PDE472 (**181**), an inhibitor of type 4D phosphodiesterase presumably for the treatment of asthma, a Novartis group utilized a sequential Kumada/Negishi coupling reaction to install the triaryl moiety (Scheme 2.69) [173]. The Kumada coupling was preferred over the Suzuki coupling to prepare biaryl intermediate **178** due to economical and logistic reasons. Two key procedures were adopted to ensure

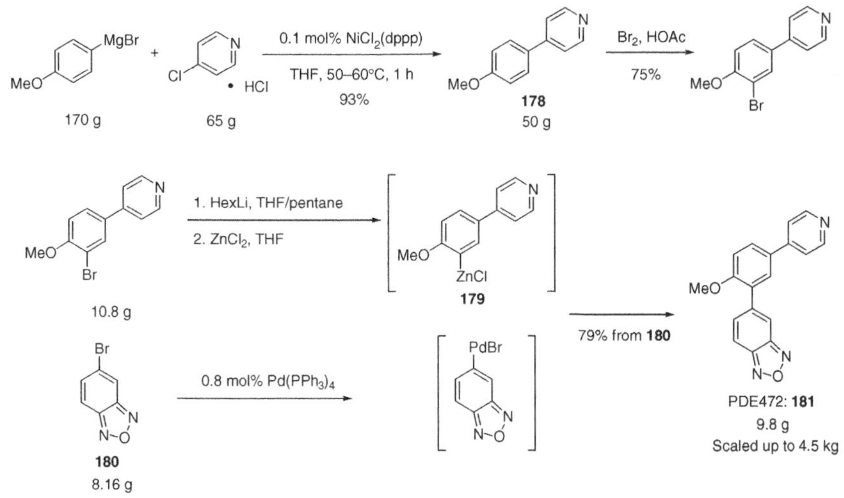

SCHEME 2.69   Novartis's synthesis of PDE472 via sequential Kumada and Negishi coupling reactions.

successful scale-up of the Kumada coupling. The first was to dry the toluene solution of chloropyridine by azeotropic distillation under reflux using a water pump to avoid autopolymerization of the starting material. The second was to use a small aliquot of the Grignard reagent to reduce the nickel catalyst to minimize the reaction acceleration due to reagent accumulation. In addition, the use of a low catalyst loading allowed very low levels of nickel contamination in the biaryl product **178**. Prior to the Negishi coupling was the preparation of zincate **179** from the corresponding lithium intermediate derived from lithium/bromide exchange. The subsequent Negishi coupling with bromide **180** in the presence of $Pd(PPh_3)_4$ produced PDE472, which was successfully scaled up by a factor of ∼500 to yield 4.5 kg of PDE472 (**181**).

A sequential Sonogashira/reductive Heck reaction was used by Eli Lilly chemists in the synthesis of antirhinoviral agent **183** (Scheme 2.70) [174]. This strategy allowed a stereoselective formation of the trisubstituted olefin. The alkynyl amide starting material **182** of the reductive Heck reaction was derived from a Sonogashira coupling in excellent yield. The regioselectivity of the reductive Heck reaction was controlled by using phosphine-free catalysts such as $Pd(dba)_2$ or $Pd(MeCN)_2Cl_2$. In addition, excellent olefin E/Z selectivity was obtained.

In an interesting regioselective C–C bond formation reaction conducted by Shionogi chemists, benzyl zinc intermediate **184** reacted regioselectively with one of the two chlorides of compound **185**, depending on the reaction conditions (Scheme 2.71) [175]. In the presence of $Pd(PPh_3)_4$, **186** was obtained in 53% yield, accompanied with 7% of the 7-benzyl regioisomer. In the absence of the Pd catalyst, nucleophilic substitution proceeded using 2 equiv of LiCl to provide **187** as the predominant product. Each of the resulting chlorides **186** and **187** can react in a subsequent Suzuki coupling leading to two regioisomers **188** and **189**, respectively, which are angiotensin II receptor antagonists for hypertension indications.

Elegant Pd-catalyzed sequential reactions were used by a team at Merck to cast the framework of laropiprant (**193**), a prostaglandin D_2 receptor antagonist and also a component of Tredaptive™ to suppress niacin-induced flushing (Scheme 2.72) [176]. This sequence involves an asymmetric allylic alkylation of racemic allyl acetate **190**, adopting a protocol developed by the Trost group [177], and a Pd-mediated

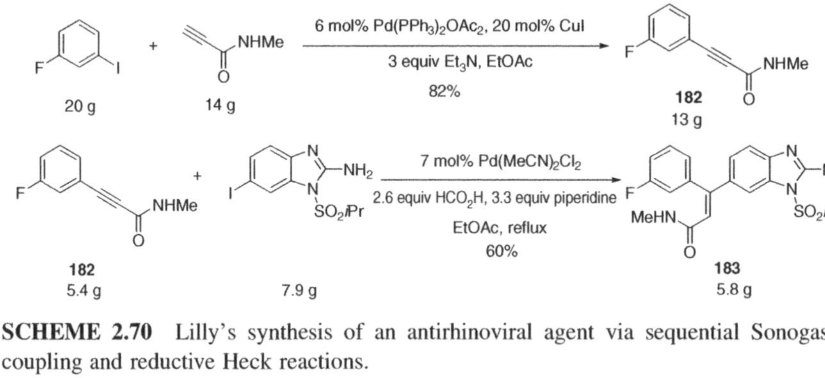

SCHEME 2.70 Lilly's synthesis of an antirhinoviral agent via sequential Sonogashira coupling and reductive Heck reactions.

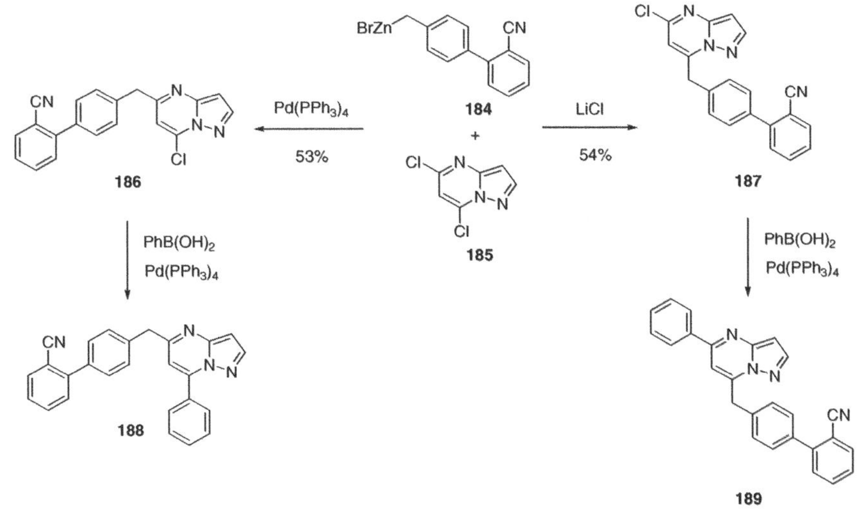

SCHEME 2.71 Shionogi's synthesis of angiotensin II antagonists via Negishi and Suzuki coupling reactions.

N-cyclization to presumably form bicyclic intermediate **191**, followed by an intramolecular Heck reaction to assemble tetracyclic compound **192**. Intermediate **192** was then converted to laropiprant (**193**) over a few more steps.

To prepare compound **195**, a key intermediate leading to selective nuclear hormone receptor modulators, an Eli Lilly group developed a sequence including a Sonogashira coupling and an efficient Pd-catalyzed intramolecular carbometalation, followed by a Suzuki coupling to afford dibenzoxapine **194** containing a tetrasubstituted exocyclic alkene (Scheme 2.73) [178]. This process was successfully

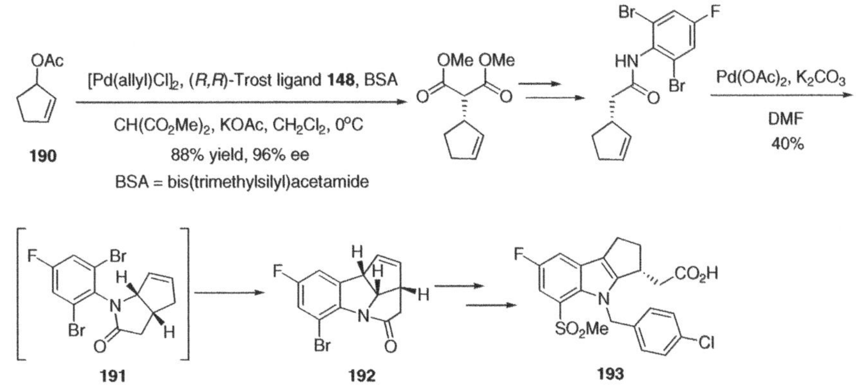

SCHEME 2.72 Merck's synthesis of laropiprant via a Trost AAA reaction, Pd-catalyzed *N*-cyclization, and Heck reactions.

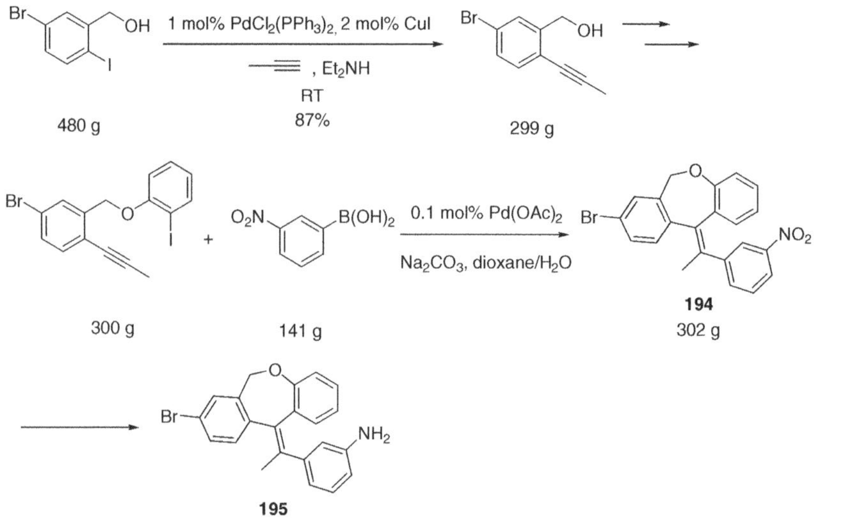

SCHEME 2.73 Lilly's synthesis of a tricyclic intermediate via a Sonogashira/carbopallada-tion/Suzuki coupling sequence.

scaled up to a 12 L scale, providing 2 kg of the target compound. In the carbopallada-tion/Suzuki coupling reaction step, the low loading of catalyst (0.1%) was critical to reduce the formation of polymeric by-products.

More recently, sequential Pd-catalyzed reactions involving Suzuki, Sonogashira, and Negishi coupling reactions were developed to prepare a trisubstituted olefin (**197**) stereoselectively (Scheme 2.74) [179]. The Sonogashira and Negishi coupling reactions were each optimized with respect to the choice of catalyst, base, and solvent. The Sonogashira adduct propargylic alcohol **196** reacted with Red-Al to generate an aluminate *in situ* that could then undergo a transmetalation to the corresponding zincate. The resulting Negishi coupling in the presence of the PEPPSI catalyst afforded the trisubstituted olefin **197** in a higher yield than other Pd catalyst systems. The Pd level in the product was determined to be under 50 ppm.

A SmithKline Beecham team reported a nicely orchestrated sequence of four Pd-catalyzed transformations, including two Sonogashira couplings, one Suzuki coupling, and one indole formation, to assemble their clinical candidate SB-242784 (**198**), a development compound for the treatment of osteoporosis (Scheme 2.75) [180]. This sequence exerted a complete control of double-bond geometry. Remarkably, no protection of the aniline amino group was needed and no isomerization of double bond was observed. After further modifications, SB-242784 was obtained in good overall yield.

2.2.11 Cyanation

Aryl nitriles are frequently present in drug candidates, agrochemicals, and dyes. They can also be important synthetic intermediates leading to useful target molecules.

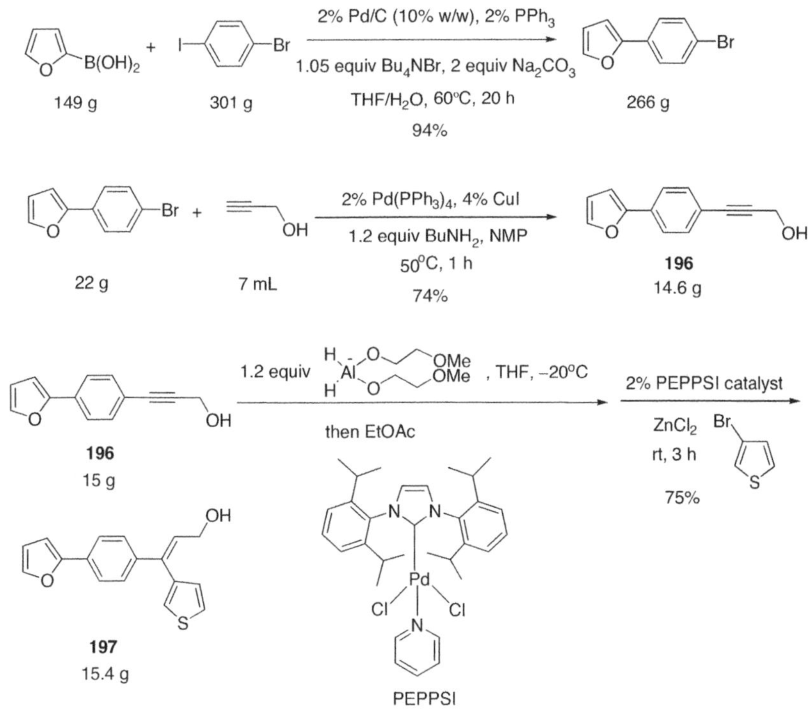

SCHEME 2.74 Johnson & Johnson's synthesis of a trisubstituted olefin via Suzuki, Sonogashira, and Negishi coupling reactions.

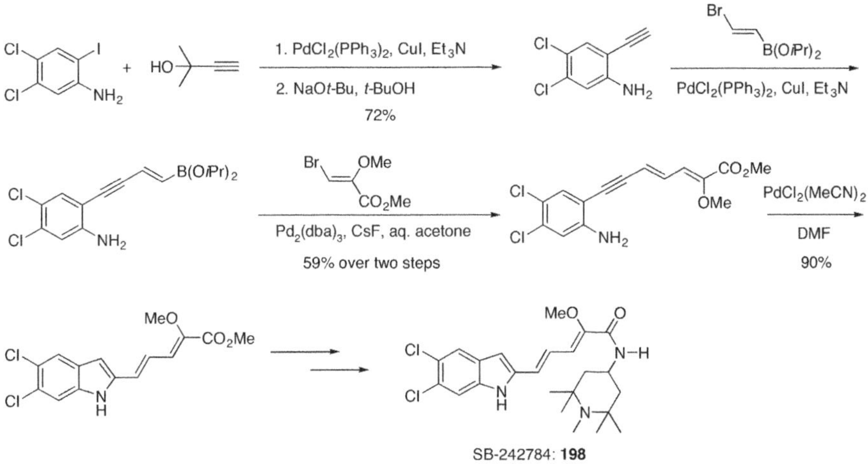

SB-242784: **198**

SCHEME 2.75 SmithKline Beecham's synthesis of SB-242784 via four Pd-catalyzed reactions.

Conventional methods to convert aryl halides to aryl cyanides are the Sandmeyer reaction [181] or the Rosenmund–von Braun reaction [182]. However, these reaction conditions typically require heating of the reaction mixture at $>150°C$ with a stoichiometric amount of copper cyanide. In contrast, the Pd-catalyzed cyanation reaction of aryl halides and even aryl chlorides could be achieved under milder conditions [183]. The challenge of Pd-catalyzed cyanation lies in the cyanide poisoning of the Pd catalyst [184]. To address this problem, solutions such as slow addition of acetone cyanohydrin [185] and a slurry of $Zn(CN)_2$ and aryl halides [186] to control the cyanide concentration in solution have been developed.

During the development of a mild and robust large-scale cyanation reaction to prepare aryl cyanide **200** at AstraZeneca (Scheme 2.76), an initial catalyst screening identified $[BrPdP(t-Bu)_3]_2$ or $Pd(dba)_2/P(t-Bu)_3$ to be optimal for conversion. It was then discovered that the order of reagent addition was crucial for the scalability of this reaction [187]. $Zn(CN)_2$ was needed to be added to the preheated DMF solution of aryl bromide **199**, zinc dust, and Pd catalyst. This protocol was able to drive the reaction to completion within 1 h under moderate heating $(50°C)$ at a small scale. It appeared that the addition of $P(t-Bu)_3$ last led to the conversion problem presumably due to the catalyst poisoning. Finally, $[BrPdP(t-Bu)_3]_2$ was selected as the catalyst for the manufacturing route due to its improved air stability compared to $Pd(dba)_2/P(t-Bu)_3$.

UK-371104 (**202**) is a crystalline nucleoside analog discovered by a Pfizer group as an A_{2a} agonist that could potentially be used to treat COPD [188]. One key step involved in the large-scale synthesis of UK-371104 (**202**) was the Pd-catalyzed cyanation of chloride **201** (Scheme 2.77) [189]. Major impurities of this reaction were the THP-deprotected chloride and cyanide, which could be easily removed by selective solvent extraction or recrystallization. The purpose of the addition of triethylamine was to deactivate the acidic by-product generated in this reaction.

A Merck group developed a catalytic enantioselective process route to taranabant (**203**) [190], a cannabinoid-1 receptor inverse agonist for the treatment of obesity (Scheme 2.78) [191]. The most efficient catalyst, $Pd[P(o-tol)_3]_4$, was formed *in situ*

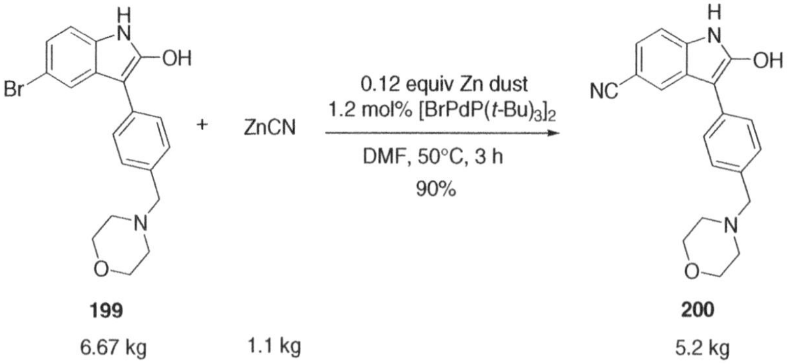

SCHEME 2.76   AstraZeneca's synthesis of cyanide 200 via a cyanation.

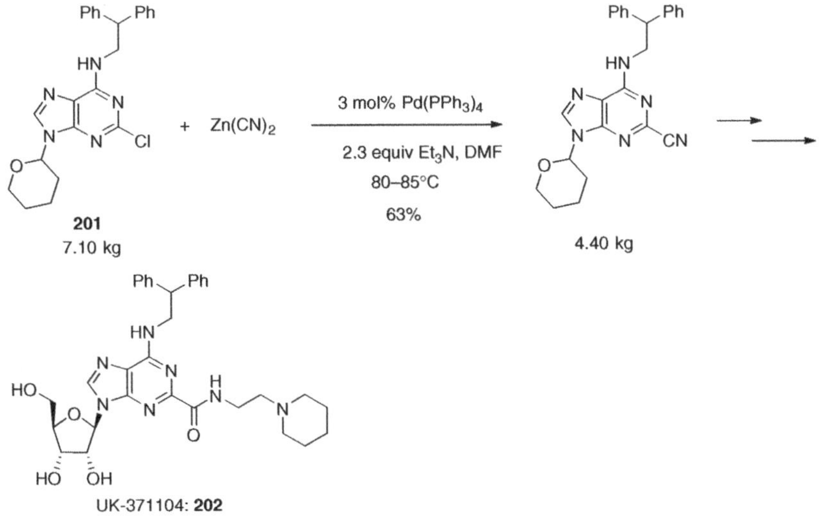

SCHEME 2.77 Pfizer's synthesis of UK-371104 via a cyanation.

by treating $Pd(OAc)_2$ and $P(o\text{-tol})_3$ with diethyl zinc. The efficiency of the reaction was dependent on the quality of the cyanide. As little as 0.6 equiv of $Zn(CN)_2$ was enough for the full conversion, but excessive reagent was needed with certain batches of cyanide. This transformation was highly sensitive to oxygen, thus degassing with nitrogen was necessitated.

Pd-catalyzed cyanations of aryl halides are often air sensitive. However, an open air and robust method for large-scale Pd-catalyzed cyanation of aryl halides was

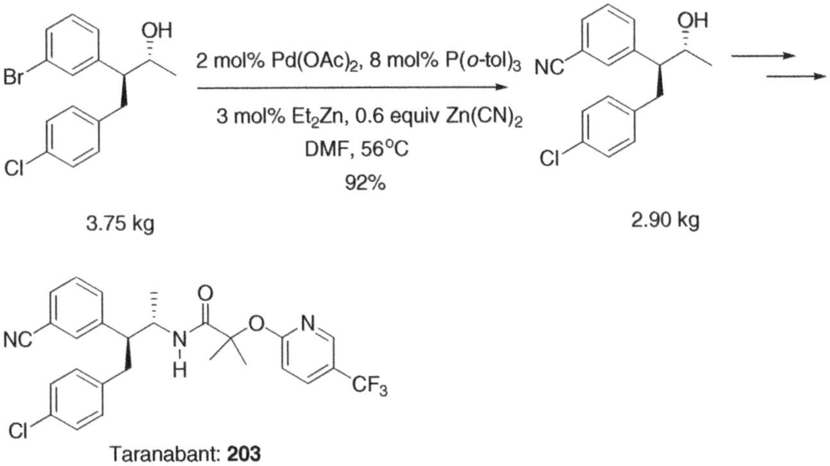

SCHEME 2.78 Merck's synthesis of taranabant via a cyanation.

Technology.

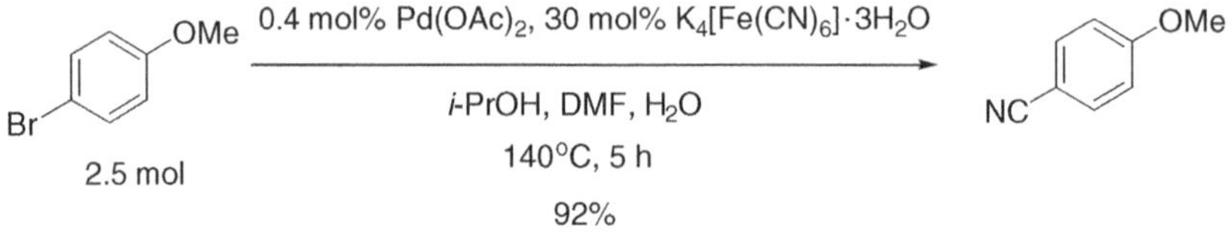

SCHEME 2.79 A robust synthesis of aryl cyanide by Henan University of Science and Technology.

developed using *i*-PrOH to prevent catalyst poisoning by oxygen (Scheme 2.79) [192]. This method used Weissman's ligand-free protocol. The addition of *i*-PrOH improved the yield from 44% to 93% in a model reaction with the reaction mixture being exposed to air. Presumably the inactive Pd(II) species could be reduced to Pd(0) in the presence of *i*-PrOH.

2.2.12 Carbonylation

Pd-catalyzed carbonylation reactions have been increasingly applied in pharmaceutical industry because of their atom-economical nature. By using inexpensive CO and aryl or benzyl halides, products such as aldehydes, acids, esters, and amides can be readily prepared. The scope of this reaction has been explored using various catalysts [193]. Major pharmaceutical products such as ibuprofen are prepared via a carbonylation reaction developed by Hoechst–Celanese [194].

A Pd-catalyzed alkoxycarbonylation of chloropyridine **204** chemoselectively activated the 2-chloro group at 80°C, leading to monoester **205** in almost quantitative yield (Scheme 2.80) [195]. At a higher temperature such as 150°C, double carbonylation could occur to provide the bis-ester in excellent yield.

SB-214857-A (lotrafiban, **208**), a GPIIb/IIIa receptor antagonist, is a clinical candidate developed by GSK to inhibit platelet aggregation [196]. The manufacturing route for lotrafiban used an aminocarbonylation reaction to form amide **207** (Scheme 2.81). An earlier approach to conduct carbonylation with 4,4′-bipiperidine provided low yield of the desired product, accompanied with multiple unidentified by-products. 4,4′-Pyridyl piperidine (**206**) was then applied successfully in the carbonylation reaction. It should be noted that the reaction profile was improved

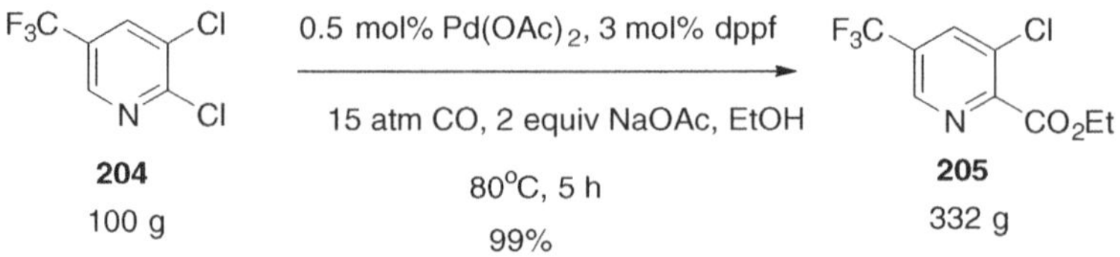

SCHEME 2.80 Lonza's synthesis of a 2-pyridyl ester via an alkoxycarbonylation reaction.

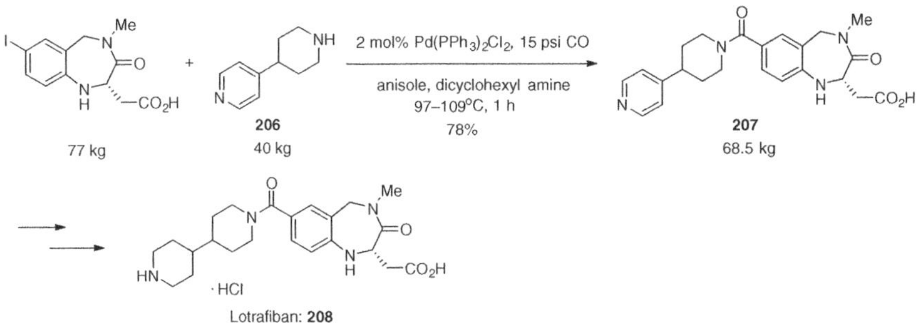

SCHEME 2.81 GSK's synthesis of lotrafiban via an aminocarbonylation.

using a secondary amine such as dicyclohexylamine, and this additive did not undergo competitive aminocarbonylation reaction.

2.2.13 New Technology Enabled C–C Cross-Coupling Reactions

Besides the development of novel ligands, new technologies involving new instruments, automation, environmentally benign reaction medium, and others have introduced significant benefits to transition metal-catalyzed C–C cross-coupling reactions. On numerous occasions, these tools have enabled rapid and efficient process development of large-scale reactions.

2.2.13.1 Microwave-Promoted Cross-Coupling Reactions
In the past decade, a great number of transition metal-catalyzed C–C coupling reactions have been facilitated by microwave technology [197]. Compared to conventional heating, which typically takes hours or days to reach completion, microwave-assisted cross-coupling reactions are typically accomplished in minutes. The microwave method has been widely used in both academia and industry for small-scale reactions. It appears that most transition metal-catalyzed thermal reactions could take advantage of microwave assistance to save time and improve efficiency.

The major challenge in applying microwave conditions to large batch reactors for industrial-scale reactions is the limited penetration depth of the microwave, which is typically on the order of several centimeters depending on the cavity dimension and the medium dielectrical properties. Thus, microwave scale-up beyond 1 L currently seems unfeasible [198]. To address this issue, two strategies have emerged. The first is to introduce multivessel reactor to enhance the throughput of total reaction volume [199]. The second is to apply continuous-flow reactor to take advantage of both microwave heating and flow processing [200]. The major drawback of the latter approach is that it may be difficult to process solids, viscous liquids, or heterogeneous mixtures. To overcome the clogging issue in heterogeneous reaction mixtures, the stop-flow technique was developed. This technology was applied to the pumping of reaction mixture into a microwave reactor, microwave irradiation, reaction vessel evacuation, and the launch of a second batch.

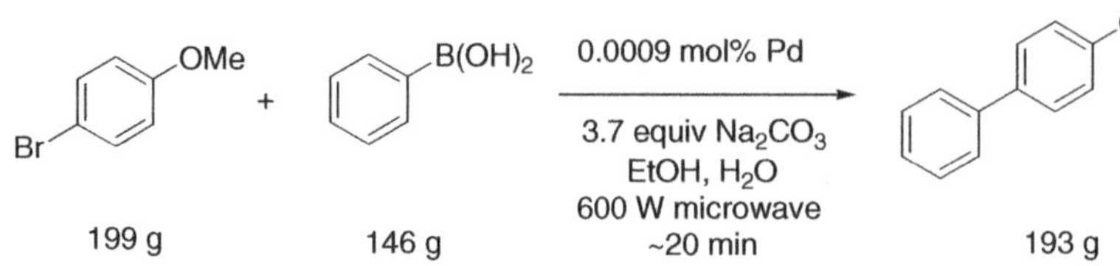

SCHEME 2.82 A microwave-promoted Suzuki coupling with extremely low catalyst loading conducted by University of Connecticut and CEM Microwave Technology.

Remarkably, a large-scale Suzuki coupling was established in an open-vessel microwave using water as the solvent and extremely low loading of the Pd catalyst (1–5 ppm) without a ligand [201]. The reaction was performed in a 3 L flask located in a microwave cavity with an open condenser at the top. The microwave irradiation power was 600 W, which caused the reaction mixture to reflux in 3–4 min. The reaction mixture was then held at this temperature for 20 min to achieve completion (Scheme 2.82). It has been proposed that the microwave might facilitate agglomeration of Pd atoms to form highly active Pd nanoparticles that effectively catalyzed the Suzuki coupling.

Besides Suzuki coupling reactions, Heck reactions have also been conducted under microwave conditions using an automated stop-flow reactor (Scheme 2.83) [202]. In addition, carbonylation reactions have also been demonstrated on large scale using microwave technology [203].

2.2.13.2 Cross-Coupling Reactions Catalyzed by Immobilized Catalysts

Catalysts can be immobilized on a glass tube or silica surface to allow heterogeneous catalysis. The major advantages of the use of immobilized catalysts are the easy separation of products and catalysts and the effective recycling of catalysts [204]. In one example, optimization of Pd-based supported liquid-phase catalyst in the Heck reaction was accomplished in a batch reactor [205]. In this case, sulfonylated triphenylphosphine Pd complexes were held in solution on a polar and hydrophilic film supported by porous glass beads. The reactants and products were restricted to a immiscible solvent phase. In another example of using a practical heterogeneous catalyst anchored onto silica-coated Fe_3O_4, Suzuki, Sonogashira, and Stille coupling reactions all proceeded with less reactive aryl chlorides [206].

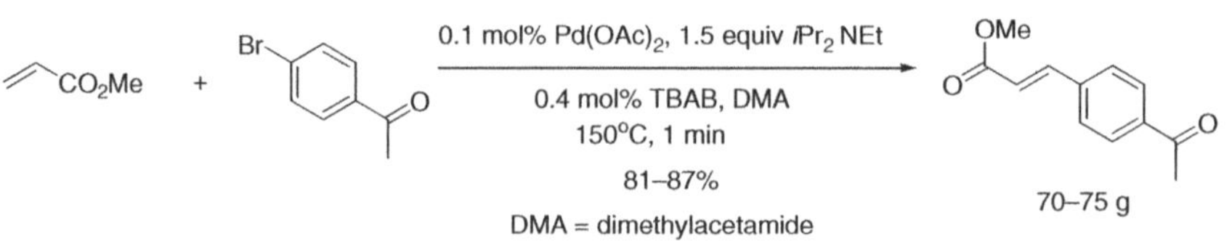

SCHEME 2.83 AstraZeneca's synthesis of a vinyl ester by microwave-promoted Heck reaction.

2.2.13.3 Automated High-Throughput Reaction Condition Screening Inspired by the extensive implementation of combinatorial technology in the 1990s [207], automated synthesis in chemical process research and development represents a powerful strategy to accelerate process optimization and project advancement. As a result, this technology enabled shortened timeline for drug synthesis to supply the need of clinical trials and market sales [208]. For example, an automated system called Anachem SK233 and the REACTarray reaction block have been developed for a Pd-catalyzed Suzuki coupling reaction by GSK [209]. The Design Expert software facilitates rapid screening of reaction concentration, water level, catalyst loading, and boronic acid stoichiometry on the reaction outcome under an inert atmosphere. Consequently, these results have led to robust and reliable scale-up processes. The Otera group at Okayama University of Science also developed an automated synthesizer referred to as MEDLEY, in which a series of reactions can be connected sequentially in one reactor [210]. The utility of MEDLEY was then demonstrated in many reactions, including a Suzuki coupling reaction [211].

The optimization of catalysts, bases, solvents, and additives sometimes can be extremely time-consuming and the best choice is often not obvious. The high-throughput screening technology enables a systematic, comprehensive, and fast optimization that may be particularly useful to identify processes for challenging transformations. As such, several pharmaceutical and chemical companies established a catalysis team to address this need. For example, Merck has a catalysis group to set up reagent platforms/toolkits that could be readily used to optimize multiple catalytic and asymmetric transformations. The power of this approach was demonstrated by a challenging regioselective Suzuki coupling of 1,6-naphthyridone dichloride **209** (Scheme 2.84) [212]. The purpose was to develop a highly regioselective coupling with just one of the four chlorides present in the substrate to yield **210**. A total of approximately 80 commercially available ligands were screened (representative ligands are shown in Scheme 2.84). This ligand collection contains mono- or bidentate, electron rich or electron poor, sterically encumbered or small, and conformationally rigid or flexible phosphines, and heterocyclic carbenes. In addition, four bases (KF, K_3PO_4, K_2CO_3, Cs_2CO_3) and five solvents (DMF, toluene, THF, dioxane, or IPA) were examined under typical literature conditions. Ultimately, $Pd(dba)_3$–$CHCl_3$ complex was selected as the precatalyst, and $(2-MeOPh)_3P$ or IMes HCl was identified as the best ligand to afford over 95% conversion and 92% regioselectivity. The best solvent and base combination was DMF/K_3PO_4. Furthermore, a kinetic study revealed that the first coupling step was 10 times faster than the competing second coupling so that a regioselective Suzuki coupling could be achieved.

2.2.13.4 Cross-Coupling Reactions in Supercritical CO₂ Suzuki coupling has also been achieved using a commercial-scale continuous-flow method in supercritical carbon dioxide ($scCO_2$) [213]. The use of $scCO_2$ offers an environmentally more benign reaction medium because it is nontoxic, cheap, readily separable from organic substances, and potentially recyclable. Immobilized Pd EnCat was employed as the stationary phase.

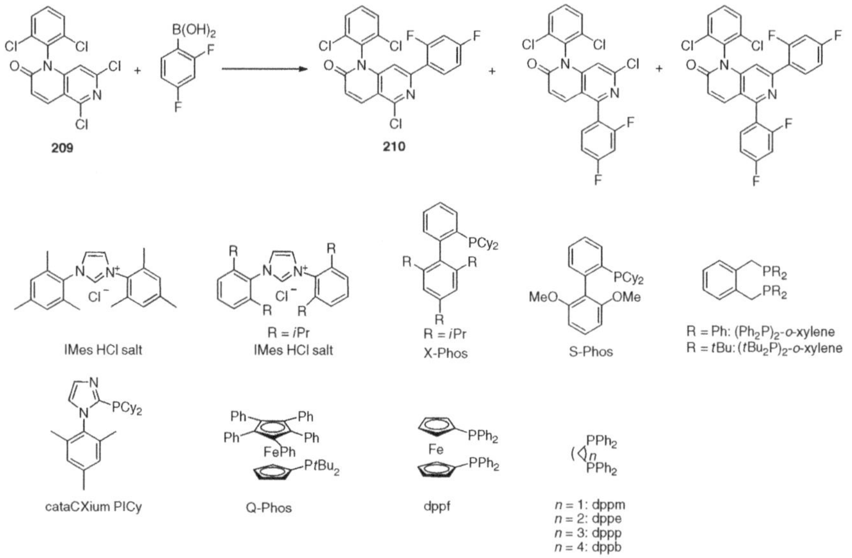

SCHEME 2.84   Merck's high-throughput catalyst screening for a challenging Suzuki coupling.

2.2.13.5   Catalyst Screening Using Calorimetry   A kinetic screening of catalysts using reaction calorimetry provides a multidimensional kinetic and stability profile of a catalyst [214]. In a series of experiments, heat flow signals were captured by adding consecutive pulses of reactants. When reaction heat is significant and rate is fast (such as a model Heck reaction), the analysis of the heat flow profiles allows rapid quantitative measurements of reaction rate for both liquid and multiphase reactions. This method generates a comprehensive picture of catalyst properties, thereby assisting detailed kinetic and mechanistic studies of metal-catalyzed transformations and catalyst optimization.

2.2.13.6   Continuous Microflow Synthesis   Microreactors have several attractive features, such as a high volume-to-area ratio, excellent operational safety, high efficiency of micromixing, heat transfer, and temperature control. Transition metal-catalyzed homogeneous transformations have been achieved with excellent catalyst recycling by using low-viscosity ionic liquids in a continuous microflow system. For example, a Heck coupling reaction has been effectively conducted on over 100 g scale with 80% yield. The Pd catalyst could be continuously recycled with the ionic liquid [215].

2.3   CONCLUSION AND PROSPECTS

The utility and impact of transition metal-catalyzed C–C bond formation reactions have been demonstrated in large-scale processes in the pharmaceutical industry. Thirty years ago, large-scale transition metal-catalyzed reactions were mainly limited to hydrogenation reactions. Now it is fair to say that transition metal-catalyzed reactions have revolutionized the repertoire of the C–C bond formation to access small molecule drug candidates. Of all C–C cross-coupling reactions, the Suzuki coupling is most widely harnessed in both medicinal and process chemistry due to its excellent functional group compatibility, high success rate, and availability of a broad range of organoboron reagents. Heck and Sonogashira coupling reactions are also extensively applied. In numerous scale-up processes for clinical candidates, multiple C–C and C–N cross-coupling reactions have been executed to allow efficient assembly of target molecules.

Future novel applications of large-scale transition metal-catalyzed C–C bond formation reactions may fall into three categories: (1) new reagents including both coupling partners and catalysts, (2) novel methods, and (3) development of conditions that are efficient, cost-effective, convenient, and environmentally benign.

Novel reagents, such as Molander's trifluoroborates [216] and Burke's N-methyliminodiacetic acid (MIDA) boronates [217], provide significant advantages over the corresponding boronic acids/esters. In particular, these reagents are often shelf-stable, easily prepared, and crystalline, and the preparations are normally scalable and inexpensive. In addition, coupling reactions using only stoichiometric amount of reagents are now possible as these compounds are clean monomers as opposed to their corresponding boronic acids. It is particularly worth noting that MIDA boronates have been developed for iterative cross-couplings that enable rapid assembly of drug-like molecules [218]. With all these desirable features, it is envisaged that these reagents will bode well for scalable processes in the near future.

Recently, new electrophiles, such as carbamates [219], ethers [220], carbonates, and sulfamates [221], have been used in Suzuki coupling reactions catalyzed by nickel catalysts. In addition, aryldimethylsilanolates have been used for cross-coupling reactions [222]. Esters [223], tosylates, sulfamates, and phosphates [224] have also been used for iron-catalyzed cross-coupling with Grignard reagents. These reagents may provide advantages in shortening steps and enhancing chemoselectivity in a synthetic sequence. It remains to be seen whether these variations of the corresponding halides can be adopted into process research.

The development of new catalysts, particularly ligands, has rendered less reactive aryl chlorides amenable to C–C bond cross-coupling reactions, which compare favorably to the reactions using the corresponding aryl bromides or iodides that are typically more expensive and less stable (The Aldrich price is $97/2 L for chlorobenzene, $106/2 L for bromobenzene, and $122/500 g for iodobenzene.). Previously less reactive heteroarenes due to their role as ligands for transition metal are also excellent substrates for cross-coupling reactions now as a result of the use of powerful ligands [225].

With respect to new synthetic methods, transition metal-catalyzed direct C–H arylation has emerged as an elegant and efficient method to construct C–C bond [226]. This strategy precludes the conventional use of stoichiometric amounts of organometallic reagents as nucleophiles in cross-coupling reactions. Despite the relatively limited use in the pharmaceutical industry to date, it is anticipated that direct arylation will be applied more routinely in both medicinal and process chemistry in light of the remarkable advance in this field recently.

In addition to C–H activation or cleavage in C–C bond formation, it is noteworthy that Johnson & Johnson chemists recently reported an interesting Pd-catalyzed direct arylation of tautomerizable heterocycles with aryl boronic acids via C–OH bond activation by forming phosphonium salt *in situ* [227].

Transition metal-catalyzed aryl–alkyl coupling or alkyl–alkyl coupling reactions are much less employed in industry [228]. Although elegant chemistry for coupling reactions involving sp^3 hybridized carbons have been discovered [229], the development of scalable and robust protocols for such transformations may facilitate their future applications.

Hiyama coupling has been underutilized in industry, yet, recently, there are several publications in this field taking advantage of the stability and low toxicity of silicon-containing reagents [230]. Metal-catalyzed decarboxylative cross-couplings of aryl carboxylates have emerged as effective tools to form C–C bonds between electrophilic and nucleophilic substrates [231]. This new strategy uses carboxylates to replace sensitive and likely costly organometallic reagents and has been applied to oxidative arylations [232], Heck-type reactions [233], allylations [234], and non-redox cross-coupling reactions [235].

Green chemistry that minimizes the negative impacts on environment has also been actively explored in both academia and industry. Recently, the Lipshutz group used the PTS–H$_2$O system to conduct a variety of C–C cross-coupling reactions in water at room temperature via micellar catalysis [236]. These reactions include Suzuki, Heck, Sonogashira, and Negishi coupling reactions and olefin metathesis. Besides reactivity enhancement compared to reactions conducted in organic solvents, these reactions in water simply minimize the use of organic solvents.

The C–C bond formation via hydrogenation has been applied on large scale (millions of tons) in industry represented by the Fischer–Tropsch process [237] and alkene formylation reactions [238]. Beyond this scope, recently powerful hydrogen-mediated and transition metal-catalyzed C–C coupling reactions developed by the Krische group have received attention as potentially "green" methods in process chemistry [239]. Some of these transformations have been accomplished in a highly regioselective, diastereoselective, and enantioselective fashion. The use of hydrogen instead of stoichiometric amount of organometallic reagents elegantly practices the concept of "atom economy" [240].

In summary, despite the extensive applications of C–C cross-coupling in the pharmaceutical industry today, numerous new catalysts, novel coupling reagents, and efficient methods that have emerged recently may be further developed and incorporated into the arsenal of process and medicinal chemists for large-scale synthesis of drug candidates in the near future.

REFERENCES

[1] (a) Anctil, E. J.-G.; Snieckus, V. *Metal-Catalyzed Cross Coupling Reactions*, 2nd ed.; Wiley: New York, **2004**, p. 761. (b) Whisler, M. C.; MacNeil, S.; Snieckus, V.; Beak, P. *Angew. Chem., Int. Ed.* **2004**, *43*, 2206. (c) Suzuki, A. *Metal-Catalyzed Cross Coupling*

Reactions; Wiley-VCH: New York, **1998**, p. 49. (d) Corbet, J.-P.; Mignani, G. *Chem. Rev.* **2006**, *106*, 2651. (e) Blaser, H.-U.; Indolese, A.; Naud, F.; Nettekoven, U.; Schnyder, A. *Adv. Synth. Catal.* **2004**, *346*, 1583. (f) Zapf, A.; Beller, M. *Top. Catal.* **2002**, *19*, 101. (g) Beller, M.; Bolm, C. *Transition Metals for Organic Synthesis*; Wiley-VCH: Weinheim, 1998.(h) Deiderich, F.; Stang, P. J. *Metal-Catalyzed Cross-Coupling Reactions*; Wiley-VCH: Weinheim, 1998.(i) Cornils, B.; Hermann, W. A. *Applied Homogeneous Catalysis with Organometallic Compounds*; Wiley-VCH: Weinheim, 1996. (j) King, A. O.; Yasuda, N. *Top. Organomet. Chem.* **2004**, *6*, 205. (k) Trost, B. M. *Comprehensive Organic Synthesis*; Pergamon Press: Oxford, **1991**; Vol. 3, Chapter 2. (l) Genet, J. P.; Savignac, M. *J. Organomet. Chem.* **1999**, *576*, 305. (m) Boudier, A.; Bromm, L. O.; Lotz, M.; Knochel, P. *Angew. Chem., Int. Ed.* **2000**, *39*, 4414. (n) Hassan, J.; Sevignon, M.; Gozzi, C.; Schulz, E.; Lemaire, M. *Chem. Rev.* **2002**, *102*, 1359. (o) Calderazzo, F.; Catellani, M.; Chiusoli, G. P. In *Metal Catalysis in Industrial Organic Processes*; Chiusoli, G. P.; Maitlis, P. M., Eds.; RSC Publishing: Cambridge, **2006**. (p) Tucker, C. E.; de Vries, J. G. *Top. Catal.* **2002**, *19*, 111. (q) Carey, J. S.; Laffan, D.; Thomson, C.; Williams, M. T. *Org. Biomol. Chem.* **2006**, *4*, 2337. (r) Beller, M.; Zaft, A. In *Handbook of Organopalladium Chemistry for Organic Synthesis*; Negishi, E.,Ed., Wiley: New York, **2002**; Vol. 1, p. 1209. (s) Tsuji, J. *Organic Synthesis with Palladium Compounds*; Springer: Berlin, 1980. (t) Hegedus, L. S. *Tetrahedron* **1984**, *40*, 2415.

[2] Duncia, J. V.; Carini, D. J.; Chiu, A. T.; Johnson, A. L.; Price, W. A.; Wong, P. C.; Wexler, R. R.; Timmermans, P. B. M. W. M. *Med. Res. Rev.* **1992**, *12*, 149.

[3] Kubo, K.; Kohara, Y.; Imamiya, E.; Sugiura, Y.; Inada, Y.; Furukawa, Y.; Nishikawa, K.; Naka, T. *J. Med. Chem.* **1993**, *36*, 2182.

[4] Pouleur, H. G. *Am. J. Hypertens.* **1997**, *10*, 318.

[5] Hannah, J.; Ruyle, W. V.; Jones, H.; Matzuk, A. R.; Kelly, K. W.; Witzel, B. E.; Holtz, W. J.; Houser, R. W.; Shen, T. Y.; Sarett, L. H. *Br. J. Clin. Pharmacol.* **1977**, *4*, 7S.

[6] (a) Prasit, P.; Wang, Z.; Brideau, C.; Chan, C.-C.; Charleson, S.; Cromlish, W.; Ethier, D.; Evans, J. F.; Ford-Hutchinson, A. W.; Gauthier, J. Y.; Gordon, R.; Guay, J.; Gresser, M.; Kargman, S.; Kennedy, B.; Leblanc, Y.; Léger, S.; Mancini, J.; O'Neill, G. P.; Ouellet, M.; Percival, M. D.; Perrier, H.; Riendeau, D.; Rodger, I.; Tagari, P.; Thérien, M.; Vickers, P.; Wong, E.; Xu, L.-J.; Young, R. N.; Zamboni, R.; Boyce, S.; Rupniak, N.; Forrest, M.; Visco, D.; Patrick, D. *Bioorg. Med. Chem. Lett.* **1999**, *9*, 1773. (b) Desmond, R.; Dolling, U.; Marcune, B.; Tillyer, R.; Tschaen, D. WO9608482 (1996).

[7] Friesen, R. W.; Brideau, C.; Chan, C. C.; Charleson, S.; Deschênes, D.; Dubé, D.; Ethier, D.; Fortin, R.; Gauthier, J. Y.; Girard, Y.; Gordon, R.; Greig, G. M.; Riendeau, D.; Savoie, C.; Wang, Z.; Wong, E.; Visco, D.; Xu, L. J.; Young, R. N. *Bioorg. Med. Chem. Lett.* **1998**, *8*, 2777.

[8] Miyashiro, J. M.; Bertenshaw, S. R.; Carter, J. S. et al. Abstracts of Papers, 222nd ACS National Meeting, Chicago, IL, August 26–30, 2001; ORGN-423.

[9] Piliero, P. J. *Expert Opin. Invest. Drugs* **2002**, *11*, 1295.

[10] (a) Jones, C. D.; Jevikar, M. G.; Pike, A. J.; Peters, M. K.; Black, L. J.; Thompson, A. R.; Falcone, J. F.; Clemens, J. A. *J. Med. Chem.* **1984**, *27*, 1057. (b) Sato, M.; Grese, T. A.; Dodge, J. A.; Bryant, H. U.; Turner, C. H. *J. Med. Chem.* **1999**, *42*, 1.

[11] Druker, B. J.; Lydon, N. B. *J. Clin. Invest.* **2000**, *105*, 3.

[12] (a) Chandrarratna, R. A. S. U.S. Patent 5,089,509 (1992). (b) Chandraratna, R. A. S. U.S. Patent 5,602,130 (1997).

[13] Chen, C. Y.; Larsen, R. D.; Verhoeven, T. R. U.S. Patent, 5,567,824 (1996).

[14] Labelle, M.; Belley, M.; Gareau, Y.; Gauthier, J. Y.; Guay, D.; Gordon, R.; Grossman, S. G.; Jones, T. R.; Leblanc, Y.; McAuliffe, M.; MaFarlane, C. S.; Masson, P.; Metters, K. M.; Quimet, N.; Patrick, D. H.; Piechuta, H.; Rochette, C.; Sawyer, N.; Xiang, Y. B.; Pickett, C. B.; Ford-Hutchinson, A. W.; Zamboni, R. J.; Young, R. N. *Bioorg. Med. Chem. Lett.* **1995**, *5*, 283.

[15] (a) Chockalingam, K.; Rouge, B. U.S. Patent 5,315,029 (1994). (b) Jang, E. J.; Lee, K. H.; Kim, Y. G. *J. Mol. Catal. A* **1999**, *138*, 25.

[16] Beutler, U.; Mazacek, J.; Penn, G.; Schenkel, B.; Wasmuth, D. *Chimia* **1996**, *50*, 154.

[17] (a) Meng, C. Q. *Curr. Opin. Cent. Nerv. Syst. Invest. Drugs* **2002**, *2*, 186. (b) Perkins, J. F. European Patent 1088817 (2001).

[18] The BIG 1-98 Collaborative Group. *N. Engl. J. Med.* **2009**, *361*, 766.

[19] (a) Suzuki, A. *J. Organomet. Chem.* **1999**, *576*, 147. (b) Suzuki, A. *Pure Appl. Chem.* **1991**, *63*, 419. (c) Miyaura, N.; Suzuki, A. *Chem. Commun.* **1979**, 866.

[20] (a) Negishi, E.; King, A.; Okukado, N. *J. Org. Chem.* **1977**, *42*, 1821. (b) Negishi, E. *Acc. Chem. Res.* **1982**, *15*, 340. (c) Knochel, P.; Singer, R. *Chem. Rev.* **1993**, *93*, 2117.

[21] (a) Tamao, K. In *Comprehensive Organic Synthesis*; Trost, B.,Ed.; Permagon: Oxford, 1991; Vol. 3, p. 435. (b) Banno, T.; Hayakawa, Y.; Umeno, M. *J. Organomet. Chem.* **2002**, *653*, 288.

[22] (a) Hiyama, T. In *Metal-Catalyzed Cross-Coupling Reactions*; Diederih, F., Stang, J., Eds.; Wiley-VCH: Weinheim, 1998; Chapter 10, p. 421. (b) Hiyama, T.; Hatanaka, Y. *Pure Appl. Chem.* **1994**, *66*, 1471. (c) Denmark, S. E.; Sweis, R. F. *Acc. Chem. Res.* **2002**, *35*, 835. (d) Spivey, A. C.; Gripton, C. J. G.; Hannah, J. P. *Curr. Org. Synth.* **2004**, *1*, 211.

[23] (a) Stille, J. *Angew. Chem., Int. Ed. Engl.* **1986**, *25*, 508. (b) Milstein, D.; Stille, J. *J. Am. Chem. Soc.* **1978**, *100*, 3636. (c) Mitchell, T. *Synthesis* **1992**, 803. (d) Farina, V.; Krishnamurthy, V.; Scott, W. *J. Org. React.* **1997**, *50*, 1.

[24] (a) Heck, R. F.; Nolley, J. P. Jr., *J. Org. Chem.* **1972**, *37*, 2320. (b) Heck, R. F. *Org. React.* **1982**, *27*, 345. (c) Cabri, W.; Candiani, I. *Acc. Chem. Res.* **1995**, *28*, 2. (d) Alonso, F.; Beletskaya, I.; Yus, M. *Tetrahedron* **2005**, *61*, 11771.

[25] (a) Sonogashira, K. In *Metal-Catalyzed Cross-Coupling Reactions*; Diederich, F.; Stang, J., Eds.; Wiley-VCH: Weinheim, 1998; Chapter 5, p. 203. (b) Kraft, P.; Bajgrowitcz, J.; Denis, C.; Frater, G. *Angew. Chem., Int. Ed.* **2000**, *39*, 2980.

[26] Trost, B. M.; Van Vranken, D. L. *Chem. Rev.* **1996**, *96*, 395.

[27] (a) Hayashi, T.; Yamasaki, K. *Chem. Rev.* **2003**, *103*, 2829. (b) Fagnou, K.; Lautens, M. *Chem. Rev.* **2003**, *103*, 169. (c) Navarre, L.; Darses, S.; Genet, J.-P. *Angew. Chem., Int. Ed.* **2004**, *43*, 719.

[28] (a) Fox, J. M.; Huang, X.; Chieffi, A.; Buchwald, S. L. *J. Am. Chem. Soc.* **2000**, *122*, 1360. (b) Kawatsura, M.; Hartwig, J. F. *J. Am. Chem. Soc.* **1999**, *121*, 1473. (c) Hama, T.; Liu, Z. X.; Culkin, D. A.; Hartwig, J. F. *J. Am. Chem. Soc.* **2003**, *125*, 11176. (d) Moradi, W. A.; Buchwald, S. L. *J. Am. Chem. Soc.* **2001**, *123*, 7996. (e) Hama, T.; Culkin, D. A.; Hartwig, J. F. *J. Am. Chem. Soc.* **2006**, *128*, 4976. (f) Lee, S.; Hartwig, J. F. *J. Org. Chem.* **2001**, *66*, 3402. (g) Ahman, J.; Wolfe, J. P.; Troutman, M. V.; Palucki, M.; Buchwald, S. L. *J. Am. Chem. Soc.* **1998**, *120*, 1918. (h) Martín, R.; Buchwald, S. L. *Angew. Chem., Int. Ed.* **2007**, *46*, 7236.

[29] (a) Ackermann, L.; Vicente, R.; Kapdi, A. R. *Angew. Chem., Int. Ed.* **2009**, *48*, 9792. (b) Ritleng, V.; Sirlin, C.; Pfeffer, M. *Chem. Rev.* **2002**, *102*, 1731.

[30] (a) Tsuji, J. *Transition Metal Reagents and Catalysts: Innovations in Organic Synthesis*; Wiley: Chichester, 2000, pp. 227–297. (b) Colquhoun, H. M.; Thompson, D. J.; Twigg, M. V. *Carbonylation: Direct Synthesis of Carbonyl Compounds*; Plenum Press: New York, 1991. (c) Lin, Y.-S.; Yamamoto, A. *Handbook of Organopalladium Chemistry for Organic Synthesis*; Negishi, E., Ed.; Wiley: New York, 2002; Vol. 2, p. 2389.

[31] (a) Bryndza, H. E.; Tam, W. *Chem. Rev.* **1988**, *88*, 1163. (b) Yang, C.; Williams, J. M. *Org. Chem.* **2004**, *6*, 2837 and references therein.

[32] Dugger, R. W.; Ragan, J. A.; Ripin, D. H. B. *Org. Process Res. Dev.* **2005**, *9*, 253.

[33] (a) Li, C.-J. *Acc. Chem. Res.* **2002**, *35*, 533. (b) Chemler, S. R.; Fuller, P. H. *Chem. Soc. Rev.* **2007**, *36*, 1153. (c) Monnier, F.; Taillefer, M. *Angew. Chem., Int. Ed.* **2008**, *47*, 3096. (d) Plenio, H. *Angew. Chem., Int. Ed.* **2008**, *47*, 6954. (e) Ma, D.; Cai, Q. *Chem. Rev.* **2008**, *41*, 1450.

[34] (a) Montgomery, J. *Acc. Chem. Res.* **2000**, *33*, 467. (b) Ikeda, S. *Acc. Chem. Res.* **2000**, *33*, 511.

[35] (a) Sherry, B. D.; Fürstner, A. *Acc. Chem. Res.* **2008**, *41*, 1500. (b) Correa, A.; Mancheno, O. G.; Bolm, C. *Chem. Soc. Rev.* **2008**, *37*, 1108. (c) For an important paper on the effect of impurities in iron that could be the real catalyst for cross-coupling reactions, see Buchwald, S. L.; Bolm, C. *Angew. Chem., Int. Ed.* **2009**, *48*, 5586.

[36] (a) Frisch, A.; Beller, M. *Angew. Chem., Int. Ed.* **2005**, *44*, 674. (b) Littke, A. F.; Fu, G. C. *Angew. Chem., Int. Ed.* **2002**, *41*, 4176.

[37] (a) Biscoe, M. R.; Fors, B. P.; Buchwald, S. L. *J. Am. Chem. Soc.* **2008**, *130*, 6686. (b) Ogata, T.; Hartwig, J. F. *J. Am. Chem. Soc.* **2008**, *130*, 13848. (c) Gabriele, B.; Mancuso, R.; Salerno, G.; Plastina, P. *J. Org. Chem.* **2008**, *73*, 756. (d) Reddy, C. V.; Kingston, J. V.; Verkade, J. G. *J. Org. Chem.* **2008**, *73*, 3047.

[38] Farina, V. *Adv. Synth. Catal.* **2004**, *364*, 1553.

[39] (a) Nishiyama, H. *Chem. Soc. Rev.* **2008**, *36*, 1133. (b) Surry, D. S.; Buchwald, S. L. *Angew. Chem., Int. Ed.* **2008**, *47*, 6338. (c) Hartwig, J. F. *Acc. Chem. Res.* **2008**, *41*, 1534. (d) Fu, G. C. *Acc. Chem. Res.* **2008**, *41*, 1555. (e) Würtz, S.; Glorius, F. *Acc. Chem. Res.* **2008**, *41*, 1523. (f) Martin, R.; Buchwald, S. L. *Acc. Chem. Res.* **2008**, *41*, 1461. (g) Benito-Garagorri, D.; Krichner, K. *Acc. Chem. Res.* **2008**, *41*, 1555.

[40] (a) Bhanage, B. M.; Arai, M. *Catal. Rev. Sci. Eng.* **2001**, *43*, 315. (b) Cole-Hamilton, D. J.; Tooze, R. P. Eds., *Catalyst Separation, Recovery and Recycling*; Springer: Netherlands, 2006.

[41] (a) Phan, L.; Andretta, J. R.; Horvey, L. K.; Edie, C. F.; Luco, A.-L.; Mirchandani, A.; Darensbourg, D. J.; Jessop, P. G. *J. Org. Chem.* **2008**, *73*, 127. (b) Jessop, P. G.; Heldebrandt, D. J.; Wang, L.; Eckert, C. A.; Liotta, C. L. *Nature* **2005**, *436*, 1102.

[42] (a) Diijkstra, H. P.; van Klink, G. P. M.; van Koten, G. *Acc. Chem. Res.* **2002**, *35*, 798. (b) Wong, H.; Pink, C. J.; Ferreira, F. C.; Livingston, A. G. *Green Chem.* **2006**, *8*, 373.

[43] (a) Behr, A.; Henze, G.; Schomäcker, R. *Adv. Synth. Catal.* **2006**, *348*, 1485. (b) Behr, A.; Fängewisch, C. *Chem. Eng. Technol.* **2002**, *25*, 143.

[44] Larsen, R. D.; King, A. O.; Chen, C. Y.; Corley, E. G.; Foster, B. S.; Roberts, F. E.; Yang, C.; Lieberman, D. R.; Reamer, R. A.; Tschaen, D. M.; Verhoeven, T. R.; Reider, P. J.; Lo, Y. S.; Rossano, L. T.; Brookes, A. S.; Meloni, D.; Moore, J. R.; Arnett, J. F. *J. Org. Chem.* **1994**, *59*, 6391.

[45] Ennis, D. S.; McManus, J.; Wood-Kaczmar, W.; Richarson, J.; Smith, G. E.; Carstairs, A. *Org. Process Res. Dev.* **1999**, *3*, 248.

[46] Magnus, N. A.; Aikins, J. A.; Cronin, J. S.; Diseroad, W. D.; Hargis, A. D.; LeTourneau, M. E.; Parker, B. E.; Reutzel-Edens, S. M.; Schafer, J. P.; Staszak, M. A.; Stephenson, G. A.; Tameze, S. L.; Zollars, L. M. H. *Org. Process Res. Dev.* **2005**, *9*, 621.

[47] Miller, W. D.; Fray, A. H.; Quatroche, J. T.; Sturgill, C. D. *Org. Process Res. Dev.* **2007**, *11*, 359.

[48] Urawa, Y.; Miyazawa, M.; Ozeki, N.; Ogura, K. *Org. Process Res. Dev.* **2003**, *7*, 191.

[49] Robichaud, A. J.; Lee, T.; Deng, W.; Mitchell, I. S. W.; McClung, C. D.; Calvello, E. J. B.; Zawrotny, D. M. WO2000077010 (2000).

[50] Hobson, L. A.; Nugent, W. A.; Anderson, S. R.; Deshmukh, S. S.; Haley, J. J., III; Liu, P.; Magnus, N. A.; Sheeran, P.; Sherbine, J. P.; Stone, B. R. P.; Zhu, J. *Org. Process Res. Dev.* **2007**, *11*, 985.

[51] Gosselin, F.; Lau, S.; Nadeau, C.; Trinh, T.; O'Shea, P. D.; Davies, I. W. *J. Org. Chem.* **2009**, *74*, 7790.

[52] Dantas Ramos, A. L.; da Silva Alves, P.; Aranda, D. A. G.; Schmal, M. *Appl. Catal. A* **2004**, *277*, 71.

[53] (a) Papp, A.; Galbacs, G.; Molnar, A. *Tetrehedron Lett.* **2005**, *46*, 7725. (b) Crudden, C. M.; Sateesh, M.; Lewis, R. *J. Am. Chem. Soc.* **2005**, *127*, 10045.

[54] Okumura, K.; Not, K.; Yoshida, K.; Niwa, M. *J. Catal.* **2005**, *231*, 245.

[55] (a) Corma, A.; Garcia, H.; Leyva, A.; Primo, A. *Appl. Catal. A* **2004**, *257*, 77. (b) Shimizu, K.; Maruyama, R.; Komai, S.; Kitayama, Y. *J. Catal.* **2004**, *227*, 202.

[56] Thiot, C.; Schmutz, M.; Wagner, A.; Mioskowski, C. *Angew. Chem., Int. Ed.* **2006**, *45*, 2868.

[57] Ruiz, J. R.; Jimenez-Sanchidrian, C.; Mora, M. *Tetrahedron* **2006**, *62*, 2922.

[58] Shieh, W.-S.; Shekhar, R.; Blacklock, T.; Tedesco, A. *Synth. Commun.* **2002**, *32*, 1059.

[59] Okamoto, K.; Akiyama, R.; Kobayashi, S. *Org. Lett.* **2004**, *6*, 1987.

[60] (a) Smith, M. D.; Stephan, A. F.; Ramarao, C.; Brennan, P. E.; Ley, S. V. *Chem. Commun.* **2003**, 2652. (b) Andrews, S. P.; Stephan, A. F.; Tanaka, H.; Ley, S. V.; Smith, M. D. *Adv. Synth. Catal.* **2005**, *347*, 647.

[61] Jiang, X.; Sclafani, J.; Prasad, K.; Repic, O.; Blacklock, T. J. *Org. Process Res. Dev.* **2007**, *11*, 769.

[62] Kerkesky, F. A. J.; Leanna, M. R.; Zhang, J.; Li, W.; Lallaman, J. E.; Ji, J.; Morton, H. E. *Org. Process Res. Dev.* **2006**, *10*, 512.

[63] Conlon, D. A.; Drahus-Paone, A.; Ho, G.-J.; Pipik, B.; Helmy, R.; McNamara, J. M.; Shi, Y.-J.; Williams, J. W.; Macdonald, D.; Deschênes, D.; Gallant, M.; Mastracchio, A.; Roy, B.; Scheigetz, J. G. *Org. Process Res. Dev.* **2006**, *10*, 36.

[64] Macdonald, D.; Mastracchio, A.; Perrier, H.; Dubé, D.; Gallant, M.; Lacombe, P.; Deschênes, D.; Scheigetz, J.; Roy, B.; Bateman, K.; Li, C.; Trimble, L. A.; Day, S.; Chauret, V.; Nicoll-Griffith, D. A.; Silva, J. M.; Huang, Z.; Laliberté, F.; Liu, S.; Ethier, D.; Pon, D.; Muise, E.; Boulet, L.; Chan, C. C.; Styhler, A.; Charleson, S.; Mancini, J.; Masson, P.; Claveau, D.; Nicholso, D.; Turner, M.; Young, R. N.; Girard, Y. *Bioorg. Med. Chem. Lett.* **2005**, *15*, 5241.

[65] Bullock, K. M.; Mitchell, M. B.; Toczko, J. F. *Org. Process Res. Dev.* **2008**, *12*, 896.

[66] Fassler, A.; Bold, G.; Capraro, H.; Cozens, R.; Mestan, J.; Poncioni, B.; Rosel, J.; Tintelnot-Blomley, M.; Lang, M. *J. Med. Chem.* **1996**, *39*, 3203.

[67] Xu, Z.; Singh, J.; Schwinden, M. D.; Zheng, B.; Kissick, T. P.; Patel, B.; Humora, M. J.; Quiroz, F.; Dong, L.; Hsieh, D.-M.; Heikes, J. E.; Pudipeddi, M.; Lindrud, M. D.; Srivastava, S. K.; Kronenthal, D. R.; Mueller, R. H. *Org. Process Res. Dev.* **2002**, *6*, 323.

[68] Li, B.; Buzon, R. A.; Zhang, Z. *Org. Process Res. Dev.* **2007**, *11*, 951.

[69] Fray, M. J.; Gillmore, A. T.; Glossop, M. S.; McManus, D. J.; Moses, I. B.; Praquin, C. F. B.; Reeves, K. A.; Thompson, L. R. *Org. Process Res. Dev.* **2010**, *14*, 263.

[70] Wood, M. R., Schirripa K. M.; Kim, J. J; Kuduk, S. D.; Chang, R. K.; DiMarco, C. N.; DiPardo, R. M.; Wan, B.-L.; Murphy, K. L.; Ransom, R. W.; Chang, R. S. L.; Holahan, M. A.; Cook, J. J.; Lemaire, W.; Moser, S. D.; Bednar, R. A.; Tang, C.; Prueksaritanont, T.; Wallace, A. A.; Mei, Q.; Yu, J.; Bohn, D. L.; Clayton, F. C.; Adarayn, E. D.; Sitko, G. R.; Leonard, Y. M.; Freidinger, R. M.; Pettibone, D. J.; Bock, M. G. *Bioorg. Med. Chem. Lett.* **2008**, *18*, 716.

[71] Menzel, K.; Machrouhi, F.; Bodenstein, M.; Alorati, A.; Cowden, C.; Gibson, A. W.; Bishop, B.; Ikemoto, N.; Nelson, T.; Kress, M. H.; Frantz, D. E. *Org. Process Res. Dev.* **2009**, *13*, 519.

[72] Hartung, C. G.; Fecher, A.; Chapell, B.; Snieckus, V. *Org. Lett.* **2003**, *5*, 1899.

[73] Lipton, M. F.; Mauragis, M. A.; Maloney, M. T.; Veley, M. F.; VanderBor, D. W.; Newby, J. J.; Appell, R. B.; Daugs, E. D. *Org. Process Res. Dev.* **2003**, *7*, 385.

[74] Dai, Y.; Hartandi, K.; Ji, Z.; Ahmed, A. A.; Albert, D. H.; Bauch, J. L.; Bouska, J. J.; Bousquet, P. F.; Cunha, G. A.; Glaser, K. B.; Harris, C. M.; Hickman, D.; Guo, J.; Li, J.; Marcotte, P. A.; Marsh, K. C.; Moskey, M. D.; Martin, R. L.; Olson, A. M.; Osterling, D. J.; Pease, L. J.; Soni, N. B.; Stewart, K. D.; Stoll, V. S.; Tapang, P.; Reuter, D. R.; Davidsen, S. K.; Michaelides, M. R. *J. Med. Chem.* **2007**, *50*, 1584.

[75] Kruger, A. W.; Rozema, M. J.; Chu-Kung, A.; Gandarilla, J.; Haight, A. R.; Kotecki, B. J.; Richter, S. M.; Schwartz, A. M.; Wang, Z. *Org. Process Res. Dev.* **2009**, *13*, 1419.

[76] O'Brien, S.; Kipps, T. J.; Faderi, S. *Blood* **2005**, *106*, 135a, abstract number 446.

[77] Daïri, K.; Yao, Y.; Faley, M.; Tripathy, S.; Rious, E.; Billot, X.; Rabouin, D.; Gonzalez, G.; Lavallée, J.-F.; Attardo, G. *Org. Process Res. Dev.* **2007**, *11*, 1051.

[78] Desmond, R.; Dolling, U.; Marcune, B.; Tillyer, R.; Tschaen, D.WO96/08482 (1996).

[79] Yasuda, N.; Huffman, M. A.; Ho, G.-J.; Xavier, L. C.; Yang, C.; Emerson, K. M.; Tsay, F.-R.; Li, Y.; Kress, M. H.; Rieger, D. L.; Karady, S.; Sohar, P.; Abramson, N. L.; DeCamp, A. E.; Mathre, D. J.; Douglass, A. W.; Dolling, U.-H.; Grabowski, E. J. J.; Reider, P. J. *J. Org. Chem.* **1998**, *63*, 5438.

[80] Narukawa, Y.; Hishi, K.; Onoue, H. *Tetrahedron* **1997**, *53*, 539.

[81] Jacks, T. E.; Belmont, D. T.; Briggs, C. A.; Horne, N. M.; Kanter, G. D.; Karrick, G. L.; Krikke, J. J.; McCabe, R. J.; Mustakis, J. G.; Nanninga, T. N.; Risedorph, G. S.; Seamans, R. E.; Skeean, R.; Winkle, D. D.; Zennie, T. M. *Org. Process Res. Dev.* **2004**, *8*, 201.

[82] Doherty, A. M. *J. Med. Chem.* **1992**, *35*, 1493.

[83] Winkle, D. D.; Schaab, K. M. *Org. Process Res. Dev.* **2001**, *5*, 450.

[84] Ager, D. J.; Anderson, K.; Oblinger, E.; Shi, Y.; VanderRoest, J. *Org. Process Res. Dev.* **2007**, *11*, 44.

[85] Mickel, S. J.; Sedelmeier, G. H.; Niederer, D.; Schuerch, F.; Seger, M.; Schreiner, K.; Daeffler, R.; Osmani, A.; Bixel, D.; Loiseleur, O.; Cercus, J.; Stettler, H.; Schaer, K.; Gamboni, R.; Bach, A.; Chen, G.-P.; Chen, W.; Geng, P.; Lee, G. T.; Loeser, E.;

McKenna, J.; Kinder, F. R.; Konigsberger, K.; Prasad, K.; Ramsey, T. M.; Rell, N.; Repič, O.; Rogers, L.; Shieh, W.-C.; Wang, R.-W.; Waykole, L.; Xue, S.; Florence, G.; Paterson, I. *Org. Process Res. Dev.* **2004**. *8*, 113.

[86] Marshall, J. A.; Johns, B. A. *J. Org. Chem.* **1998**, *63*, 7885.

[87] Busacca, C. A.; Cerreta, M.; Dong, Y.; Eriksson, M. C.; Farina, V.; Feng, X.; Kim, J.-Y.; Lorenz, J. C.; Sarvestani, M.; Simpson, R.; Varsolona, R.; Vitous, J.; Campbell, S. J.; Davis, M. S.; Jones, P.-J.; Norwood, D.; Qiu, F.; Beaulieu, P. L.; Duceppe, J.-S.; Haché, B.; Brong, J.; Chiu, F.-T.; Curtis, T.; Kelley, J.; Lo, Y. S.; Powner, T. H. *Org. Process Res. Dev.* **2008**, *12*, 603.

[88] (a) Negishi, E.-I. *Palladium and Nickel Catalyzed Reactions of Organozinc Compounds*; Oxford University Press, Oxford, 1999. (b) Negishi, E.-I.; Valente, L. F.; Kobayashi, M. *J. Am. Chem. Soc.* **1980**, *102*, 3298. (c) Knochel, P.; Singer, R. D. *Chem. Rev.* **1993**, *93*, 2117.

[89] Rieke, R. D.; Hanson, M. V. *Tetrahedron* **1997**, *53*, 1925.

[90] Satow, A.; Maehara, S.; Ise, S.; Hikichi, H.; Fukushima, M.; Suzuki, G.; Kimura, T.; Tanaka, T.; Ito, S.; Kawamoto, H.; Ohta, H. *J. Pharmacol. Exp. Ther.* **2008**, *326*, 577.

[91] Krasiński, A.; Fokin, V. V.; Sharpless, K. B. *Org. Lett.* **2004**, *6*, 1237.

[92] Tsuritani, T.; Mizuno, H.; Nonoyama, N.; Kii, S.; Akao, A.; Sato, K.; Yasuda, N.; Mase, T. *Org. Process Res. Dev.* **2009**, *13*, 1407.

[93] Reeder, M. R.; Gleaves, H. E.; Hoover, S. A.; Imbordino, R. J.; Pangborn, J. J. *Org. Process Res. Dev.* **2003**, *7*, 696.

[94] Scott, R. W.; Neville, S. N.; Urbina, A.; Camp, D.; Stankovic N. *Org. Process Res. Dev.* **2006**, *10*, 296.

[95] Denni-Dischert, D.; Marterer, W.; Bänziger, M.; Yusuff, N.; Batt, D.; Ramsey, T.; Geng, P.; Michael W.; Wnag, R.-M. B.; Taplin, F.; Versace, R.; Cesarz, D.; Perez, L. B. *Org. Process Res. Dev.* **2006**, *10*, 70.

[96] Davies, H.; Bignell, G. R.; Cox, C.; Stephens, P.; Edkins, S.; Clegg, S.; Teague, J.; Woffendin, H.; Garnett, M. J.; Bottomley, W.; Davis, N.; Dicks, E.; Ewing, R.; Floyd, Y.; Gray, K.; Hall, S.; Hawes, R.; Hughes, J.; Kosmidou, V.; Menzies, A.; Mould, C.; Parker, A.; Stevens, C.; Watt, S.; Hopper, S.; Wilson, R.; Jayatilake, H.; Gusterson, B. A.; Copper, C.; Shipley, J.; Hargrave, D.; Pritchard-Jones, K.; Maitland, N.; Chenevix-Trench, G.; Riggins, G. J.; Bigner, D. D.; Palmieri, G.; Cossu, A.; Flanagan, A.; Nicholson, A.; Ho, J. W. C.; Leung, S. Y.; Yuen, S. T.; Weber, B. L.; Seigler, H. F.; Darrow, T. L.; Paterson, H.; Marais, R.; Marshall, C. J.; Wooster, R.; Stratton, M. R.; Futreal, P. A. *Nature* **2002**, *417*, 949.

[97] Pérez-Balado, C.; Willemsens, A.; Ormerod, D.; Aelterman, W.; Mertens, N. *Org. Process Res. Dev.* **2007**, *11*, 237.

[98] (a) Kumada, M. *Pure Appl. Chem.* **1980**, *52*, 669. (b) Corriu, R. J. P.; Masse, J. P. *J. Chem. Soc., Chem. Commun.* **1972**, 144.

[99] Giordano, C.; Coppi, L.; Minsci, F. U.S. Patent 5,312,975 (1994).

[100] Varney, M. D.; Marzoni, G. P.; Palmer, C. L.; Deal, J. D.; Weber, S.; Welsh, K. M.; Bacquet, R. J.; Bartlett, C. A.; Morse, C. A.; Booth, C. L. J.; Herrmann, S. M.; Howland, E. F.; Ward, R. W.; White, J. *J. Med. Chem.* **1992**, *35*, 663.

[101] Marzoni, G.; Varney, M. D. *Org. Process Res. Dev.* **1997**, *1*, 81.

[102] (a) Kosugi, M.; Sasazawa, K.; Shimizu, Y.; Migita, T. *Chem. Lett.* **1977**, 301. (b) Stille, J. K.; Lau, K. S. Y. *Acc. Chem. Res.* **1977**, *10*, 434. (c) Stille, J. K. *Angew. Chem., Int. Ed. Engl.* **1986**, *25*, 508–524.

[103] Yasuda, N. *J. Organomet. Chem.* **2002**, *253*, 279.

[104] (a) Farina, V.; Baker, S. R.; Sapino, C., Jr., *Tetrahedron Lett.* **1988**, *29*, 6043. (b) Nagano, N.; Itahana, H.; Hisamichi, H.; Sakamoto, K.; Hara, R. *Tetrahedron Lett.* **1994**, *35*, 4577. (c) Baker, S. R.; Roth, G. P.; Sapino, C. *Synth. Commun.* **1990**, *20*, 2185. (d) Roth, G. P.; Sapino, C. *Tetrahedron Lett.* **1991**, *32*, 4073. (e) Armitage, M. A.; Lathbury, D. C.; Sweeney, J. B. *Tetrahedron Lett.* **1995**, *36*, 775. (f) Yamamoto, H.; Terasawa, T.; Ohki, A.; Shirai, F.; Kawabata, K.; Sakane, K.; Matsumoto, S.; Matsumoto, Y.; Tawara, S. *Bioorg. Med. Chem. Lett.* **2000**, *8*, 43. (g) Kohrt, J.; Filipski, K. J.; Rapundalo, S. T.; Cody, W. L.; Edmunds, J. J. *Tetrahedron Lett.* **2000**, *41*, 6041.

[105] Ragan, J. A.; Raggon, J. W.; Hill, P. D.; Jones, B. P.; McDermoot, R. E.; Munchhof, M. J.; Marx, M. A.; Casavant, J. M.; Copper, B. A.; Doty, J. L.; Lu, Y. *Org. Process Res. Dev.* **2003**, *7*, 676.

[106] (a) Heck, R. F. *J. Am. Chem. Soc.* **1968**, *90*, 5518. (b) Heck, R. F. *Acc. Chem. Res.* **1979**, *12*, 146. (c) Heck, R. F. *Org. React.* **1982**, *27*, 345.

[107] De Meijere, A.; Meyer, F. E. *Angew. Chem., Int. Ed. Engl.* **1994**, *33*, 2379.

[108] Beletskaya, I. P.; Cheprakov, A. V. *Chem. Rev.* **2000**, *100*, 3009.

[109] Dounay, A. B.; Overman, L. E. *Chem. Rev.* **2003**, *103*, 2945.

[110] Jones, T. R.; Zamboni, R.; Belley, M.; Champion, E.; Charette, L.; Ford-Hutchinson, A. W.; Frenette, R.; Gauthier, J.-Y.; Leger, S.; Masson, P.; McFarlane, C. S.; Piechuta, H.; Rokach, J.; Williams, H.; Young R. N.; DeHaven, R. N.; Pong, S. S. *Can. J. Physiol. Pharmacol.* **1989**, *67*, 17.

[111] (a) King, A. O.; Corley, E. G.; Anderson, R. K.; Larsen, R. D.; Verhoeven, T. R.; Reider, P. J.; Xiang, Y. B.; Belley, M.; Leblanc, Y.; Labelle, M.; Prasit, P.; Zamboni, R. J. *J. Org. Chem.* **1993**, *58*, 3731. (b) Larsen, R. D.; Corley, E. G.; King, A. O.; Carroll, J. D.; Davis, P.; Verhoeven, T. R.; Reider, P. J.; Labelle, M.; Gauthier, J. Y.; Xiang, Y. B.; Zamboni, R. J. *J. Org. Chem.* **1996**, *61*, 3398.

[112] Shinkai, I.; King, A. O.; Larsen, R. D. *Pure Appl. Chem.* **1994**, *66*, 1551.

[113] Macor, J. E.; Wythes, M. J. WO9206973 (1992).

[114] Waite, D. C.; Mason, C. P. *Org. Process Res. Dev.* **1998**, *2*, 116.

[115] Dack, K. N.; Dickinson, R. P.; Long, C. J.; Steele, J. *Bioorg. Med. Chem. Lett.* **1998**, *8*, 2061.

[116] Ainge, D.; Vaz, L.-M. *Org. Process Res. Dev.* **2002**, *6*, 811.

[117] Ripin, D. H. B.; Rourassa, D. E.; Brandt, T.; Castaldi, M. J.; Frost, H. N; Hawkins, J.; Johnson, P. J.; Massett, S. S.; Neumann, K.; Phillips, J.; Raggon, J. W.; Rose, P. R.; Rutherford, J. L.; Sitter, B.; Stewart, A. M.; Vetelino, M. G.; Wei, L. *Org. Process Res. Dev.* **2005**, *9*, 440.

[118] Janssen, P. A. J.; Lewi, P. J.; Arnold, E.; Daeyaert, F.; de Jonge, M.; Heeres, J.; Koymans, L.; Vinkers, M.; Guillemont, J.; Pasquier, E.; Kukla, M.; Ludovici, D.; Andries, K.; de Bethune, M.-P.; Pauwels, R.; Das, K.; Clark, A. D., Jr., ; Frenkel, Y. V.; Hughes, S. H.; Medaer, B.; De Knaep, F.; Bohets, H.; De Clerck, F.; Lampo, A.; Williams, P.; Stoffels, P. *J. Med. Chem.* **2005**, *48*, 1901.

[119] Schils, D.; Stappers, F.; Solberghe, G.; van Heck, R.; Coppens, M.; Van den Heuvel, D.; Van der Donck, P.; Callewaert, T.; Meeussen, F.; De Bie, E.; Eersels, K.; Schouteden, E. *Org. Process Res. Dev.* **2008**, *12*, 530.

[120] Raggon, J. W.; Snyder, W. M. *Org. Process Res. Dev.* **2002**, *6*, 67.

[121] Miller, W. H.; Alberts, D. P.; Bhanager, P. K.; Bondinell, W. E.; Callahan, J. F.; Calvo, R. R.; Cousins, R. D.; Erhard, K. F.; Heerding, D. A.; Keena, R. M.; Kwon, C.; Manley, P. J.; Newlander, K. A.; Ross, S. T.; Samanen, J. M.; Uzinskas, I. N.; Venslavsky, J. W.; Yuan, C. C.-K.; Haltiwanger, R. C.; Gowen, M.; Hwang, S.-M.; James, I. E.; Lark, M. W.; Rieman, D. J.; Stroup, G. B.; Azzarano, L. M.; Salyers, K. L.; Smith, B. R.; Ward, K. W.; Johanson, K. O.; Huffman, W. F. *J. Med. Chem.* **2000**, *43*, 22.

[122] Wallace, M. D.; McGuire, M. A.; Yu, M. S.; Goldfinger, L.; Liu, L.; Dai, W.; Shilcrat, S. *Org. Process Res. Dev.* **2004**, *8*, 738.

[123] Camp, D.; Matthews, C. F.; Neville, S. t.; Rouns, M.; Scott, R. W.; Truong, Y. *Org. Process Res. Dev.* **2006**, *10*, 814.

[124] (a) Di Fabio, R. WO99/64411 (1999). (b) Orlandi, A. WO01/42238 (2001). (c) Chiamulera, C.; Reggiani, A.; Trist, D. G.; Teneggi, V. WO2005/053693 (2005).

[125] Banks, A.; Breen, G. F.; Caine, D.; Carey, J. S.; Drake, C.; Forth, M. A.; Gladwin, A.; Guelfi, S.; Hayes, J. F.; Maragni, P.; Morgan, D. O.; Oxley, P.; Perboni, A.; Popkin, M. E.; Rawlinson, F.; Roux, G. *Org. Process Res. Dev.* **2009**, *13*, 1130.

[126] Stephens, R. D.; Castro, C. E. *J. Org. Chem.* **1963**, *28*, 3313.

[127] (a) Sonogashira, K.; Tohda, Y.; Hagihara, N. *Tetrahedron Lett.* **1975**, 4467. (b) Sonogashira, K. In *Comprehensive Organic Synthesis*; Trost, B. M.; Fleming, I., Eds.; Pergamon Press: New York, **1991**; Vol. 3, p. 521.

[128] Thomas, A. V.; Patel, H. H.; Reif, L. A.; Chemburkar, S. R.; Sawick, D. P.; Shelat, B.; Balmer, M. K.; Patel, R. R. *Org. Process Res. Dev.* **1997**, *1*, 294.

[129] de Dios, A.; Shih, C.; de Uralde, B. L.; Sánchez, C.; del Prado, M.; Cabrejas, L. M. M.; Pleite, S.; Blanco-Urgoiti, J.; Lorite, M. C.; Nevill, R. C., Jr., ; Bonjouklian, R.; York, J.; Vieth, M.; Wang, Y.; Magnus, N. A.; Campbell, R. M.; Anderson, B. D.; MaCann, D. J.; Giera, D. D.; Lee, P. A.; Schultz, R. M.; Li, L. C.; Johnson, L. M.; Wolos, J. A. *J. Med. Chem.* **2005**, *4*, 2270.

[130] Magnus, N. A.; Diseroad, W. D.; Nevill, R.; Wepsiec, J. P. *Org. Process Res. Dev.* **2006**, *10*, 556.

[131] Lyssikatos, J. P.; LaGreca, S. D. U.S. Patent 6,150,337 (2000).

[132] Andresen, B. M.; Couturier, M.; Cronin, B.; D'Occhio, M.; Ewing, M. D.; Guinn, M.; Hawkins, J. M.; Jasys, V. J.; LaGreca, S. D.; Lyssikatos, J. P.; Moraski, G.; Ng, K.; Raggon, J. W.; Stewart, A. M.; Tickner, D. L.; Tucker, J. L.; Urban, F. J.; Vazquez, E.; Wei, L. *Org. Process Res. Dev.* **2004**, *8*, 643.

[133] Dickman, D. A.; Ku, Y.-Y.; Morton, H. E.; Chemburkar, S. R.; Patel, H. H.; Thomas, A.; Plata, D. J.; Sawick, D. P. *Tetrahedron Asymmetry* **1997**, *8*, 1791.

[134] Chandraratna, R. A. S. U.S. Patent 5,089,509 (1992).

[135] Dorow, R. L.; Herrinton, P. M.; Hohler, R. A.; Maloney, M. T.; Mauragis, M. A.; McGhee, W. E.; Moeslein, J. A.; Strohbach, J. W.; Veley, M. F. *Org. Process Res. Dev.* **2006**, *10*, 493.

[136] Vaillancourt, V. A.; Staley, S.; Huang, A.; Nugent, R. A.; Chen, K.; Nair, S. K.; Nieman, J. A.; Strohbach, J. W. WO0202558 (2002).

[137] Pu, Y.-M.; Grieme, T.; Gupta, A.; Plata, D.; Bhatia, A. V.; Cowart, M.; Ku, Y.-Y. *Org. Process Res. Dev.* **2005**, *9*, 45.

[138] (a) Beutler, U.; Mazacek, J.; Penn, G.; Schenkel, B.; Wasmuth, D. *Chimia* **1996**, *50*, 154. (b) Nussbaumer, P.; Leitner, I.; Mraz, K.; Stütz, A. *J. Med. Chem.* **1995**, *38*, 1831.

[139] Porter, D. J. T.; Chestnut, W. G.; Merrill, B. M.; Spector, T. *J. Biol. Chem.* **1992**, *267*, 5236.

[140] Cooke, J. W. B.; Bright, R.; Coleman, M. J.; Jenkins, K. P. *Org. Process Res. Dev.* **2001**, *5*, 383.

[141] Trost, B. M.; Crawley, M. L. *Chem. Rev.* **2003**, *103*, 2921.

[142] (a) Palucki, M.; Um, J. M.; Yasuda, N.; Conlon, D. A.; Tsay, F.-R.; Hartner, F. W.; Hsian, Y.; Marcune, B.; Karady, S.; Hughes, D. L.; Dormer, P. G.; Reider, P. J. *J. Org. Chem.* **2002**, *67*, 5508. (b) Conlon, D. A.; Jensen, M. S.; Palucki, M.; Yasuda, N.; Um, J. M.; Yang, C.; Hartner, F. W.; Tsay, F.-R.; Hsian, Y.; Pye, P.; Rivera, N. R.; Hughes, D. L. *Chirality* **2005**, *17*, S149.

[143] Trost, B. M.; Hachiya, I. *J. Am. Chem. Soc.* **1998**, *120*, 1104.

[144] (a) Shen, T. Y. *Angew. Chem., Int. Ed. Engl.* **1972**, *11*, 460. (b) Venkatesan, H.; Davis, M. C.; Altas, Y.; Snyder, J. P.; Liotta, D. C. *J. Org. Chem.* **2001**, *66*, 3653. (c) Edmondson, S.; Danishefsky, S. J.; Sepp-Lorenzino, L.; Rosen, N. *J. Am. Chem. Soc.* **1999**, *121*, 2147.

[145] (a) Palucki, M.; Buchwald, S. L. *J. Am. Chem. Soc.* **1997**, *119*, 11108. (b) Spielvogel, D. J.; Buchwald, S. L. *J. Am. Chem. Soc.* **2002**, *123*, 3500. (c) Nguyen, H. N.; Huang, X.; Buchwald, S. L. *J. Am. Chem. Soc.* **2003**, *125*, 11818.

[146] (a) Culkin, D. A.; Hartwig, J. F. *Acc. Chem. Res.* **2003**, *36*, 234. (b) Kawatsura, M.; Hartwig, J. F. *J. Am. Chem. Soc.* **1999**, *121*, 1473. (c) Hama, T.; Culkin, D. A.; Hartwig, J. F. *J. Am. Chem. Soc.* **2006**, *128*, 4976.

[147] Johansson, C. C. C.; Colacot, T. J. *Angew. Chem., Int. Ed.* **2010**, *49*, 676.

[148] Jing, X.; Gong, B.; Prasa, K.; Repič, O. *Org. Process Res. Dev.* **2008**, *12*, 1164.

[149] Grasa, G. A.; Colacot, T. J. *Org. Process Res. Dev.* **2008**, *12*, 522.

[150] (a) Dick, A. R.; Sanford, M. S. *Tetrahedron* **2006**, *62*, 2439. (b) Yu, J.-Q.; Giri, R.; Chen, X. *Org. Biomol. Chem.* **2006**, *4*, 4041. (c) Daugulis, O.; Zaitsev, V. G.; Shabashov, D.; Pham, Q.; Lazareva, A. *Synlett* **2006**, 3382.

[151] (a) Zaitsev, V. G.; Daugulis, O. *J. Am. Chem. Soc.* **2005**, *127*, 4156. (b) Li, B.; Tian, S.; Fang, Z.; Shi, Z. *Angew. Chem., Int. Ed.* **2008**, *47*, 1115.

[152] (a) Desai, L. V.; Hull, K. L.; Sanford, M. S. *J. Am. Chem. Soc.* **2004**, *126*, 9542. (b) Thu, H.; Yu, W.; Che, C.-M. *J. Am. Chem. Soc.* **2006**, *128*, 9048.

[153] (a) Cai, G.; Fu, Y.; Li, Y.; Wan, X.; Shi, Z. *J. Am. Chem. Soc.* **2007**, *129*, 7666. (b) Li, J.-J.; Mei, T.-S.; Yu, J.-Q. *Angew. Chem., Int. Ed.* **2008**, *47*, 6452.

[154] (a) Dick, A. R.; Hull, K. L.; Sanford, M. S. *J. Am. Chem. Soc.* **2004**, *126*, 2300. (b) Whitfield, S. R.; Sanford, M. S. *J. Am. Chem. Soc.* **2007**, *129*, 15142. (c) Shi, B.-F.; Maugel, N.; Zhang, Y.-H.; Yu, J.-Q. *Angew. Chem., Int. Ed.* **2008**, *47*, 4882.

[155] Xiao, B.; Fu, Y.; Xu, J.; Gong, T.-J.; Dai, J.-J.; Yi, J.; Liu, L. *J. Am. Chem. Soc.* **2010**, *132*, 468.

[156] (a) Wang, D.-H.; Mei, T.-S.; Yu, J.-Q. *J. Am. Chem. Soc.* **2008**, *130*, 10676. (b) Giri, R.; Yu, J.-Q. *J. Am. Chem. Soc.* **2008**, *130*, 14082. (c) Giri, R.; Maugel, N.; Li, J.-J.; Wang, D.-H.; Breazzano, S. P.; Saunders, L. B.; Yu. J.-Q. *J. Am. Chem. Soc.* **2007**, *129*, 3510.

[157] Campeau, L.-C.; Schipper, D. J.; Fagnou, K. *J. Am. Chem. Soc.* **2008**, *130*, 3266.

[158] (a) Ackermann, L.; Vicente, R.; Kapdi, A. R. *Angew. Chem., Int. Ed.* **2009**, *48*, 9792. (b) Campeau, L.-C.; Stuart, D. R.; Fagnou, K. *Aldrichim. Acta* **2007**, *40*, 35. (c) Liégault, B.; Lapointe, D.; Caron, L.; Vlasosva, A.; Fagnou, K. *J. Org. Chem.* **2009**, *74*, 1826.

[159] (a) Shaughnessy, K. H.; Hamann, B. C.; Hartwig, J. F. *J. Org. Chem.* **1998**, *63*, 6546. (b) Lee, S.; Hartwig, J. F. *J. Org. Chem.* **2001**, *66*, 3402. (c) Hennessy, E. J.; Buchwald, S. L. *J. Am. Chem. Soc.* **2003**, *125*, 12084.

[160] Choy, A.; Colbry, N.; Huber, C.; Pamment, M.; Van Duine, J. *Org. Process Res. Dev.* **2008**, *12*, 884.

[161] Huang, J.; Chan, J.; Chen, Y.; Borths, C. J.; Baucom, K. D.; Larsen, R. D.; Faul, M. M. *J. Am. Chem. Soc.* **2010**, *132*, 3674.

[162] Chen, C.-Y.; Lieberman, D. R.; Larsen, R. D.; Reamer, R. A.; Verhoeven, T. R.; Reider, P. J.; Cottrell, I. F.; Haughton, P. G. *Tetrahedron Lett.* **1994**, *35*, 6981.

[163] Larock, R. C.; Yum, E. K. *J. Am. Chem. Soc.* **1991**, *113*, 6689.

[164] Rosso, V. W.; Lust, D. A.; Bernot, P. J.; Grosso, J. A.; Modi, S. P.; Rusowicz, A.; Sedergran, T. C.; Simpson, J. H.; Srivastava, S. K.; Humora, M.; Anderson, N. G. *Org. Process Res. Dev.* **1997**, *1*, 311.

[165] Friesen, R. W.; Brideau, C.; Chan, C. C.; Charleson, S.; Deschênes, D.; Dubé, D.; Ethier, D.; Fortin, R.; Gauthier, J. Y.; Girard, Y.; Gordon, R.; Greig, G. M.; Riendeau, D.; Savoie, C.; Wang, Z.; Wong, E.; Visco, D.; Xu, L. J.; Young, R. N. *Bioorg. Med. Chem. Lett.* **1998**, *8*, 2777.

[166] Zegar, S.; Tokar, C.; Enache, L. A.; Rajagopol, V.; Zeller, W.; O'Connell, M.; Singh, J.; Muellner, F. W.; Zembower, D. E. *Org. Process Res. Dev.* **2007**, *11*, 747.

[167] Baxter, C. A.; Cleator, E.; Alam, M.; Davies, A. J.; Goodyear, A.; O'Hagan, M. *Org. Lett.* **2010**, *12*, 668.

[168] Königsberger, K.; Chen, G.-P.; Wu, R. R.; Girgis, M. J.; Prasad, K.; Repič, O.; Blacklock, T. J. *Org. Process Res. Dev.* **2003**, *7*, 733.

[169] Nussbaumer, P. WO9628430 (1996).

[170] Cameron, M.; Foster, B. S.; Lynch, J. E.; Shi, Y.-J.; Dolling, U.-H. *Org. Process Res. Dev.* **2006**, *10*, 398.

[171] Jeffrey, T. *J. Chem. Soc., Chem. Commun.* **1984**, 1287.

[172] Chambers, M. S.; Goodacre, S. C.; Haller, D. J.; Jennings, A.; Jones, P.; Lewis, R. T.; Moore, K. W.; Russell, M. G. N.; Street, L. J.; Szekeres, H. J. WO02/074773 A1 (2002).

[173] Manley, P. W.; Acemoglu, M.; Martere, W.; Pachinger, W. *Org. Process Res. Dev.* **2003**, *7*, 436.

[174] Hay, L. A.; Koenig, T. M.; Ginah, F. O; Copp, J. D.; Mitchell, D. *J. Org. Chem.* **1998**, *63*, 5050.

[175] Shiota, T.; Yamamori, T. *J. Org. Chem.* **1999**, *64*, 453.

[176] Campos, K. R.; Journet, M.; Lee, S.; Grabowski, E. J. J.; Tillyer, R. D. *J. Org. Chem.* **2005**, *70*, 268.

[177] (a) Trost, B. M.; Crawley, M. L. *Chem. Rev.* **2003**, *103*, 2921. (b) Trost, B. M.; Bunt, R. C. *J. Am. Chem. Soc.* **1994**, *116*, 4089.

[178] Richey, R. N.; Yu, H. *Org. Process Res. Dev.* **2009**, *13*, 315.

[179] Houpis, I. N.; Shilds, D.; Nettekoven, U.; Schnyder, A.; Bappert, E.; Weerts, K.; Canters, M.; Vermuelen, W. *Org. Process Res. Dev.* **2009**, *13*, 598.

[180] Yu, M. S.; Lopez, L. L.; McGuire, M. A.; Botha, G. *Tetrahedron Lett.* **1998**, *39*, 9347.

[181] Sandmeyer, T. *Ber. Dtsch. Chem. Ges.* **1885**, *18*, 1492.

[182] Ellis, G. P.; Romney-Alexander, T. M. *Chem. Rev.* **1987**, *87*, 779.

[183] Jin, F.; Confalone, P. N. *Tetrahedron Lett.* **2000**, *41*, 3271.

[184] Dobbs, K. D.; Marshall, W. J.; Grushin V. V. *J. Am. Chem. Soc.* **2007**, *129*, 30.

[185] Sundermeier, M.; Zapf, A.; Beller, M. *Angew. Chem., Int. Ed.* **2003**, *42*, 1661.

[186] Marcantonio, K. M.; Frey, L. F.; Liu, Y.; Chen, Y.; Strine, J.; Phenix, B.; Wallace, D. J.; Chen, C.-Y. *Org. Lett.* **2004**, *6*, 3723.

[187] Ryberg, P. *Org. Process Res. Dev.* **2008**, *12*, 540.

[188] Mantell, S. J.; Monaghan, S. M. WO0077018 (2000).

[189] Challenger, S.; Dessi, Y.; Fox, D. E.; Hesmondhalgh, L. C.; Pascal, P.; Pettman, A. J.; Smith, J. D. *Org. Process Res. Dev.* **2008**, *12*, 575.

[190] Chen, C.; Frey, L. F.; Wallace, D. J.; Marcantonio, K.; Payack, J. F.; Vazquez, E.; Springfield, S. A.; Zhou, G.; Liu, P.; Kieczykowski, G. R.; Chen, A. M.; Phenix, B. D.; Singh, U.; Strine, J.; Izzo, B.; Krska, S. W. *Org. Process Res. Dev.* **2007**, *11*, 616.

[191] Lin, L. S.; Lanza, T. J.; Jewell, J. J. P.; Liu, P.; Shah, S. K.; Qi, H.; Tong, X.; Wang, J.; Xu, S. S.; Fong, T. M.; Shen, C.-P.; Lao, J.; Xiao, J. C.; Shearman, L. P.; Sribling, D. S.; Rosko, S.; Strack, A.; Marsh, D. J.; Feng, Y.; Kumar, S.; Samuel, K.; Yin, W.; der Ploeg, L. V.; Mills, S. G.; MacCoss, M.; Goulet, M. T.; Hagmann, W. K. *J. Med. Chem.* **2006**, *49*, 7584.

[192] Ren, Y.; Liu, Z.; He, S.; Zhao, S.; Wang, J.; Niu, R.; Yin, W. *Org. Process Res. Dev.* **2009**, *13*, 764.

[193] Barnard, C. F. *Org. Process Res. Dev.* **2008**, *12*, 566.

[194] Elango, V.; Davenport, K. G.; Murphy, M. A.; Mott, G. N.; Zey, E. G.; Smith, B. L.; Moss, G. L. European Patent Application EP 0 400 892 (1990).

[195] Crettaz, R.; Waser, J.; Bessard, Y. *Org. Process Res. Dev.* **2001**, *5*, 572.

[196] Samanen, J. M.; Ali, F. E.; Barton, L. S.; Bondinell, W. E.; Burgess, J. L.; Callahan, J. F.; Calvo, R. R.; Chen, W.; Chen, L.; Erhard, K.; Feuerstein, G.; Heys, R.; Hwang, S.-M.; Jakas, D. R.; Keenan, R. M.; Ku, T. W.; Kwon, C.; Lee, C.-P.; Miller, W. H.; Newlander, K. A.; Nichols, A.; Parker, M.; Peishoff, C. E.; Rhodes, G.; Ross, S.; Shu, A.; Simpson, R.; Takata, D.; Yellin, T. O.; Uzsinskas, I.; Venslavsky, J. W.; Yuan, C.-K.; Huffman, W. F. *J. Med. Chem.* **1996**, *39*, 4867.

[197] (a) Singh, B. K.; Kaval, N.; Tomar, S.; Van der Eycken, E.; Parmar, V. S. *Org. Process Res. Dev.* **2008**, *12*, 468. (b) Bidström, P.; Tierney, J.; Wathey, B.; Westman, J. *Tetrahedron* **2001**, *57*, 9225. (c) Larhed, M.; Hallberg, A. *Drug Discov. Today* **2001**, *6*, 406.

[198] Moseley, D.; Lenden, P.; Lockwood, M.; Ruda, K.; Sherlock, J. P.; Thomson, A. D.; Gilday, J. P. *Org. Process Res. Dev.* **2008**, *12*, 30.

[199] Glasnov, T. N.; Kappe, C. O. *Macromol. Rapid Commun.* **2007**, *28*, 395.

[200] Cablewski, T.; Faux, A. F.; Strauss, C. R. *J. Org. Chem.* **1994**, *59*, 3408.

[201] Leadbeater, N. E.; Williams, V. A.; Barnard, T. M.; Collins, M. J., Jr., *Org. Process Res. Dev.* **2006**, *10*, 833.

[202] Moseley, J. D.; Woodman, E. K. *Org. Process Res. Dev.* **2008**, *12*, 967.

[203] Iannelli, M.; Bergamelli, F.; Kormos, C. M.; Paravisi, S.; Leadbeater, N. E. *Org. Process Res. Dev.* **2009**, *13*, 634.

[204] (a) Phan, N. T. S.; Sluys, M. V. D.; Jones, C. W. *Adv. Synth. Catal.* **2006**, *348*, 609. (b) Yin, L.; Liebscher, J. *Chem. Rev.* **2007**, *107*, 133.

[205] Mirza, A. R.; Anson, M. S.; Hellgardt, K.; Leese, M. P.; Thompson, D. F.; Tonks, L.; Williams, J. M. J. *Org. Process Res. Dev.* **1998**, *2*, 325.

[206] Jin, M.-J.; Lee, D.-H. *Angew. Chem., Int. Ed.* **2010**, *49*, 1119.

[207] (a) Persidis, A. *Chem. Ind.* **1998**, 782. (b) Früchtel, J. S.; Jung, G. *Angew. Chem., Int. Ed. Engl.* **1996**, *35*, 17.

[208] Harre, M.; Tilstam, U.; Weinmann, H. *Org. Process Res. Dev.* **1999**, *3*, 304.

[209] Armitage, M. A.; Smith, G. E.; Veal, K. T. *Org. Process Res. Dev.* **1999**, *3*, 189.

[210] Orita, A.; Yasui, Y.; Otera, J. *Org. Process Res. Dev.* **2000**, *4*, 333.

[211] Orita, A.; Yasui, Y.; Otera, J. *Org. Process Res. Dev.* **2000**, *4*, 337.

[212] Cai, C.; Chung, J. Y. L.; McWilliams, C.; Sun, Y.; Shultz, S.; Palucki, M. *Org. Process Res. Dev.* **2007**, *11*, 144.

[213] Leeke, G. A.; Santos, R. C. D.; Al-Duri, B.; Seville, J. P. K.; Smith, C. J.; Lee, C. K. Y.; Holmes, A. B.; McConvey, I. F. *Org. Process Res. Dev.* **2007**, *11*, 144.

[214] Balckmond, D. G.; Rosner, T.; Pfaltz, A. *Org. Process Res. Dev.* **1999**, *3*, 275.

[215] Liu, S.; Fukuyama, T.; Sato, T.; Ryu, I. *Org. Process Res. Dev.* **2004**, *8*, 477.

[216] (a) Molander, G. A., Figueroa, R. *Aldrichim. Acta.* **2005**, *38*, 49. (b) Darses, S., Genêt, J.-P. *Eur. J. Org. Chem.* **2003**, 4313. (c) Molander, G. A.; Canturk, B. *Angew. Chem., Int. Ed.* **2009**, *48*, 9240.

[217] (a) Knapp, D. M.; Gillis, E. P.; Burke, M. D. *J. Am. Chem. Soc.* **2009**, *131*, 6961. (b) Uno, B. E.; Gillis, E. P.; Burke, M. D. *Tetrahedron* **2009**, *65*, 3130. (c) Gillis, E. P.; Burke, M. D. *J. Am. Chem. Soc.* **2008**, *130*, 14084.

[218] Gillis, E. P.; Burke, M. D. *J. Am. Chem. Soc.* **2007**, *129*, 6716.

[219] (a) Antoft-Finch, A.; Blackbum, T.; Snieckus, V. *J. Am. Chem. Soc.* **2009**, *131*, 7502. (b) Xu, L.; Li, B.-J.; Wu, Z.-H.; Lu, X.-Y.; Guan, B.-T.; Wang, B.-Q.; Zhao, K.-Q.; Shi, Z.-J. *Org. Lett.* **2010**, *12*, 884.

[220] Shimasaki, T.; Konno, Y.; Tobisu, M.; Chatani, N. *Org. Lett.* **2009**, *11*, 4890.

[221] Quasdorf, K. W.; Riener, M.; Petrova, K. V.; Garg, N. K. *J. Am. Chem. Soc.* **2009**, *131*, 17748.

[222] (a) Denmark, S. E.; Regens, C. S. *Acc. Chem. Res.* **2008**, *41*, 1486. (b) Denmark, S. E.; Sweis, R. F.; Wehrili, D. *J. Am. Chem. Soc.* **2004**, *126*, 4865. (c) Hiyama, T.; Shirakawa, E. *Top. Curr. Chem.* **2002**, *219*, 61.

[223] Li, B.-J.; Wu, Z.-H.; Guan, B.-T.; Sun, C.-L.; Wang, B.-Q.; Shi, Z-J. *J. Am. Chem. Soc.* **2009**, *131*, 14656.

[224] Gøgsig, T. M.; Lindhardt, A. T.; Skrydstrup, T. *Org. Lett.* **2009**, *11*, 4886.

[225] Slagt, V. F.; de Vries, A. H. M.; de Vries, J. G.; Kellogg, R. M. *Org. Process Res. Dev.* **2010**, *14*, 30.

[226] For selected examples and reviews, see (a) Godula, K.; Sames, D. *Science* **2006**, *312*, 67. (b) Proch, S.; Kempe, R. *Angew. Chem., Int.* **2007**, *46*, 3135. (c) Stuart, D. R.; Fagnou, K. *Science* **2007**, *316*, 1172. (d) Turner, G. L.; Morris, J. A.; Greaney, M. F. *Angew. Chem., Int. Ed.* **2007**, *46*, 7996. (e) Do, H.-Q.; Daugulis, O. *J. Am. Chem. Soc.* **2007**, *129*, 12404. (f) Hull, K. L.; Sanford, M. S. *J. Am. Chem. Soc.* **2007**, *129*, 11904. (g) Lewis, L. C.; Bergman, R. G.; Ellman, J. A. *Acc. Chem. Res.* **2008**, *41*, 1013. (h) Alberico, D.; Scott, M. E.; Lautens, M. *Chem. Rev.* **2007**, *107*, 174. (i) Campeau, L.-C.; Fagnou, K.

Chem. Commun. **2006**, 1253. (j) Seregin, I. V.; Gevorgyan, V. *Chem. Soc. Rev.* **2007**, *36*, 1173. (k) Phipps, R. J.; Grimster, N. P.; Gaunt, M. J. *J. Am. Chem. Soc.* **2008**, *130*, 8172. (l) Ueda, S.; Nagasawa, H. *Angew. Chem., Int. Ed.* **2003**, *42*, 5400. (m) Ackermann, L.; Althammer, A.; Fenner, S. *Angew. Chem., Int. Ed.* **2009**, *48*, 201. (n) Wang, X.; Lane, B. S.; Sames, D. *J. Am. Chem. Soc.* **2005**, *127*, 4996. (o) Campeau, L.-C.; Stuart, D. R.; Leclerc, J.-P.; Bertrand-Laperie, M.; Villermure, E.; Sun, H.-Y.; Lasserre, S.; Guimond, N.; Lecavallier, M.; Fangou, K. *J. Am. Chem. Soc.* **2009**, *131*, 3291. (p) Vallée, F.; Mousseau, J. J.; Charrette, A. B. *J. Am. Chem. Soc.* **2010**, *132*, 1514. (q) Liu, W.; Cao, H.; Lei, A. *Angew. Chem., Int. Ed.* **2010**, *49*, 2004. (r) Hachiya, H.; Hirano, K.; Satoh, T.; Miura, M. *Angew. Chem., Int. Ed.* **2010**, *49*, 2202.

[227] Kang, F.-A.; Sui, Z.; Murray, W. V. *J. Am. Chem. Soc.* **2009**, *131*, 4174.

[228] (a) Luh, T.-Y.; Leung, M.; Wong, K.-T. *Chem. Rev.* **2000**, *100*, 3187. (b) Frisch, A. C.; Beller, M. *Angew. Chem., Int. Ed.* **2005**, *44*, 674.

[229] (a) Zhou, J.; Fu, G. C. *J. Am. Chem. Soc.* **2003**, *125*, 14726. (b) González-Bobes, F.; Fu, G. C. *J. Am. Chem. Soc.* **2006**, *128*, 5360. (c) Fisher, C.; Fu, G. C. *J. Am. Chem. Soc.* **2005**, *127*, 4595. (d) Arp, F. O.; Fu, G. C. *J. Am. Chem. Soc.* **2005**, *127*, 10482. (d) Vechorkin, O.; Proust, V.; Hu, X. *J. Am. Chem. Soc.* **2009**, *131*, 9756. (e) Netherton, M. R.; Fu, G. C. *Adv. Synth. Catal.* **2004**, *346*, 1525. (f) Lou, S.; Fu, G. C. *J. Am. Chem. Soc.* **2010**, *132*, 1264.

[230] Zhang, L.; Wu, J. *J. Am. Chem. Soc.* **2008**, *130*, 12250 and references therein.

[231] (a) Gooßen, L. J.; Rodríguez, N.; Gooßen, K. *Angew. Chem., Int. Ed.* **2008**, *47*, 3100. (b) Gooßen, L. J.; Rudolphi, F.; Oppel, C.; Rodríguez, N. *Angew. Chem., Int. Ed.* **2008**, *47*, 3043. (c) Gooßen, L. J.; Zimmermann, B.; Knauber, T. *Angew. Chem., Int. Ed.* **2008**, *47*, 7103. (d) Gooßen, L. J.; Rodríguez, N.; Lange, P. P.; Linder, C. *Angew. Chem., Int. Ed.* **2010**, *49*, 1111.

[232] Wang, C.; Piel, I.; Glorius, F. *J. Am. Chem. Soc.* **2009**, *131*, 4194.

[233] (a) Myers, A. G.; Tanaka, D.; Mannion, M. R. *J. Am. Chem. Soc.* **2002**, *124*, 11250. (b) Forgione, P.; Brochu, M. C.; St-Onge, M.; Thesen, K. H.; Bailey, M. D.; Bilodeau, F. *J. Am. Chem. Soc.* **2006**, *128*, 11350.

[234] Rayabarapu, D. K.; Tunge, J. A. *J. Am. Chem. Soc.* **2005**, *127*, 13510.

[235] (a) Gooßen, L. J.; Deng, G.; Levy, L. M. *Science* **2006**, *313*, 662. (b) Gooßen, L. J.; Rodríguez, N.; Melzer, B.; Linder, C.; Deng, G.; Levy, L. M. *J. Am. Chem. Soc.* **2007**, *129*, 4824.

[236] For a review, see Lipshutz, B. H.; Ghorai, S. *Aldrichim. Acta.* **2008**, *41*, 59.

[237] Fischer, F.; Tropsch, H. *Chem. Ber.* **1923**, *56B*, 2428.

[238] (a) Frohning, C. D.; Kohlpaintner, C. W. In *Applied Homogeneous Catalysis with Organometallic Compounds*; Cornils, B., Herrmann, W. A., Eds.; Wiley-VCH: Weinheim, 1996; Vol. 1, p. 29.(b) van Leeuwen, P. W. N. M. *Homogeneous Catalysis, Understanding the Art*; Kluwer: Dordrecht, 2004.

[239] For a leading review, see Ngai, M.-Y.; Kong, J.-R.; Krische, M. J. *J. Org. Chem.* **2006**, *72*, 1063.

[240] Trost, B. M. *Science* **1991**, *256*, 1471.

3

SELECTED APPLICATIONS OF PD- AND CU-CATALYZED CARBON–HETEROATOM CROSS-COUPLING REACTIONS IN THE PHARMACEUTICAL INDUSTRY

JINGJUN YIN

Applications of Transition Metal Catalysis in Drug Discovery and Development: An Industrial Perspective, First Edition. Edited by Matthew L. Crawley and Barry M. Trost.
© 2012 John Wiley & Sons, Inc. Published 2012 by John Wiley & Sons, Inc.

3.1 INTRODUCTION

In the past 15 years, Pd- [1] and Cu- [2]-catalyzed carbon–heteroatom (N, O, S, B, and P) couplings (Scheme 3.1) have attracted a tremendous amount of interest. These reactions have had a major impact on pharmaceutical synthesis because they provide a very general and efficient access to anilines, aryl ethers, aryl thioethers, and aryl phosphonates that are found in many top-selling drugs, clinical candidates, and numerous bioactive compounds in the early drug discovery space.

This field has been extensively reviewed in recent years [2], so it is not the intent of this chapter to review all the latest method development. Instead, applications in drug discovery and development will be discussed. Selected examples of a few most widely used methods and catalyst systems for each type of carbon–heteroatom coupling will be reviewed in detail to illustrate the impact of this rapid developing research area in the pharmaceutical industry.

In general, continued development of these reactions aims at increasing the substrate scope, allowing milder conditions, and decreasing the catalyst loadings.

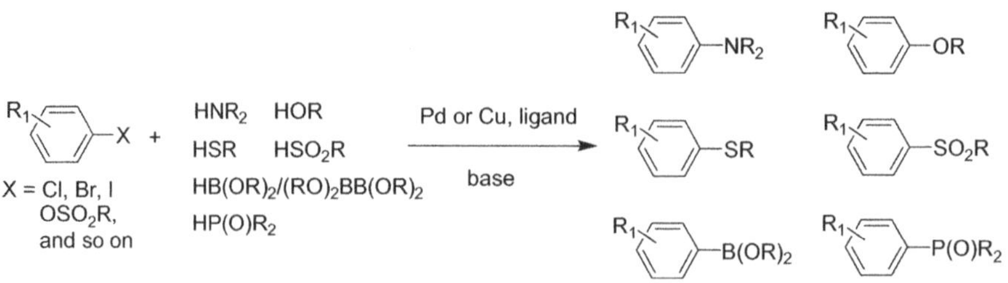

SCHEME 3.1   Pd- or Cu-catalyzed carbon–heteroatom coupling reactions.

All these aspects are important for the synthesis of bioactive molecules. For the early discovery where a large number of molecules need to be synthesized quickly for SAR studies using preferably the same sequence, the generality and functional group tolerance of the catalyst system are very important. High yields and low catalyst loadings are less critical. On the other hand, high efficiency, robustness, and environmentally benign operating conditions are critical for late stage process development where multikilograms of drug candidates are needed.

Like in the development of most transition metal-catalyzed reactions, the supporting ligand plays a critical role. To this end, a large number of ligands have been developed for both Pd- and Cu-catalyzed carbon–heteroatom cross-couplings to enable the more general and efficient access to many new classes of compounds and expand the new chemical space for drug discovery. Phosphine ligands are typically used for Pd-catalyzed reactions that allowed the use of aryl iodides, bromides, chlorides, and various sulfonates to participate in these cross-coupling reactions. Nitrogen- and/or oxygen-based ligands are used for Cu-catalyzed reactions, which so far are mostly limited to aryl iodides, bromides, and few aryl chlorides.

Another key parameter of these reactions is the base that is required to neutralize the acid by-product HX. Inorganic bases are usually used for these reactions. Although strong bases such as NaOtBu, KOtBu, and LiHMDS gave higher reactivities, especially for C–N couplings, they offer relatively narrow functional group tolerance, which is a limitation for the complex molecule synthesis in pharmaceutical research. The use of weaker inorganic bases such as Cs_2CO_3 and K_3PO_4 has significantly expanded the substrate scope for these reactions. Continued development of more active ligands has also allowed these bases to be used with less reactive substrates.

Because of the tight regulations on the residual heavy metal levels (typically < 20 ppm) for active pharmaceutical ingredients (APIs) that are used in clinical studies, the removal of residual Pd or Cu in the APIs presents a challenge for the application of these reactions, especially if they are late in the synthesis. Various methods for metal removal will be discussed in most of the large-scale carbon–heteroatom coupling examples.

3.2 CARBON–HETEROATOM CROSS-COUPLINGS

3.2.1 Pd-Catalyzed C–N Couplings

3.2.1.1 Introduction   The Pd-catalyzed C–N cross-coupling between aryl halides and Bu$_3$Sn-NR$_2$ was first reported by Migita and coworkers in 1983 [3], but it did not generate much interest in the next 12 years, likely due to the use of toxic tin reagent. In 1995, Buchwald and Hartwig independently developed the tin-free cross-couplings of aryl halides with amines in the presence of a strong base [4], and the field has evolved rapidly ever since [1]. With continued development of new conditions and ligands by them and others, the Buchwald–Hartwig amination has become one of the most used reactions for both early discovery and late process development due to the ubiquitous presence of nitrogen atoms in the pharmaceutical industry.

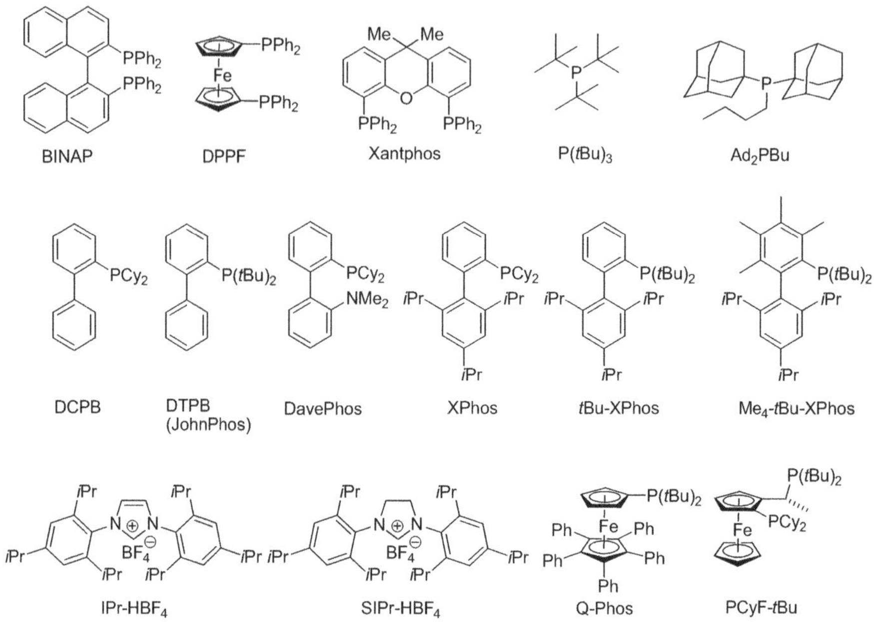

FIGURE 3.1 Commonly used ligands for Pd-catalyzed C–N couplings in pharmaceutical industry.

This reaction can be affected by many reaction parameters such as the Pd source, ligand, Pd/ligand ratio, base, solvent, additive, temperature, order of addition, activation of the catalyst, and so on. However, the most important factor is the choice of the ligand. Many of them have been developed in the past 15 years and a few of the most general and commonly used ligands in the pharmaceutical industry are shown in Figure 3.1.

Both Buchwald and Hartwig initially used P(o-Tol)$_3$ as the ligand for their first papers [4] and then found that chelating ligands BINAP [5] and DPPF [6] improved the substrate scope. Buchwald subsequently developed a series of biphenyl-based, electron-rich, bulky monophosphines such as 2-dicyclohexylphosphino-1,1′-biphenyl (DCPB), 2-di-t-butylphosphino-1,1′-biphenyl (DTPB or JohnPhos) [7], DavePhos [8], and XPhos [9] to significantly expand the substrate scope, in particular, the use of previously inactive aryl chlorides. Hartwig concurrently used P(tBu)$_3$ [10] for general amination reactions of aryl chlorides and others [11]. Most of these early ligands are used for C–N coupling of aryl chlorides, bromides, iodides, and sulfonates with various amines and anilines. Buchwald used 9,9-dimethyl-4-5-bis(diphenylphosphino)xanthene (Xantphos) [12] for the arylation of amides, sulfonamides, and related electron-deficient nitrogen analogs [13], which proves to be very useful in pharmaceutical synthesis. These are currently the most frequently used ligands in drug discovery and development. Buchwald also developed tBu-XPhos [14] for arylation of heteroarylamines and Me$_4$-tBu-XPhos [15] for amidation of aryl chlorides. Hartwig developed Q-Phos [16] and

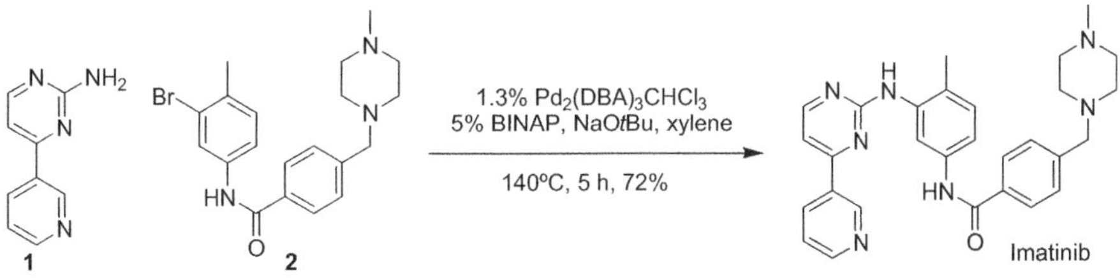

SCHEME 3.2 Pd-catalyzed C–N coupling in the synthesis of Imatinib.

used PCy*t*Bu [17] as more active and general ligands for cross-couplings. Many other ligands were also developed and some examples of their applications will be discussed.

3.2.1.2 Pd-Catalyzed C–N Couplings in the Synthesis of Approved Drugs and Clinical Candidates   Novartis's Gleevec (Imatinib) is one of the top-selling drugs in the world, with $3.9 billion worldwide sales in 2009 [18]. It is a tyrosine kinases inhibitor for treating chronic myelogenous leukemia (CML), gastrointestinal stromal tumors (GISTs), and other cancers. One of the initial processes to access it used a Pd-catalyzed C–N coupling between the aminopyrimidine **1** and the arylbromide **2** (Scheme 3.2) [19]. Tris(dibenzylideneacetone)dipalladium [Pd$_2$(DBA)$_3$] was used as the Pd(0) source and BINAP as the ligand. A strong base, sodium *tert*-butoxide, was used for neutralizing the HBr by-product. A copper-catalyzed process was also developed for its synthesis (see Section 2.2).

Abilify (aripiprazole) is a top-selling antipsychotics drug discovered by Otsuka Pharmaceutical with $3.6 billion worldwide sales in 2008. To synthesize a key metabolite, Otsuka scientists used a Pd-catalyzed amination (Scheme 3.3) [20].

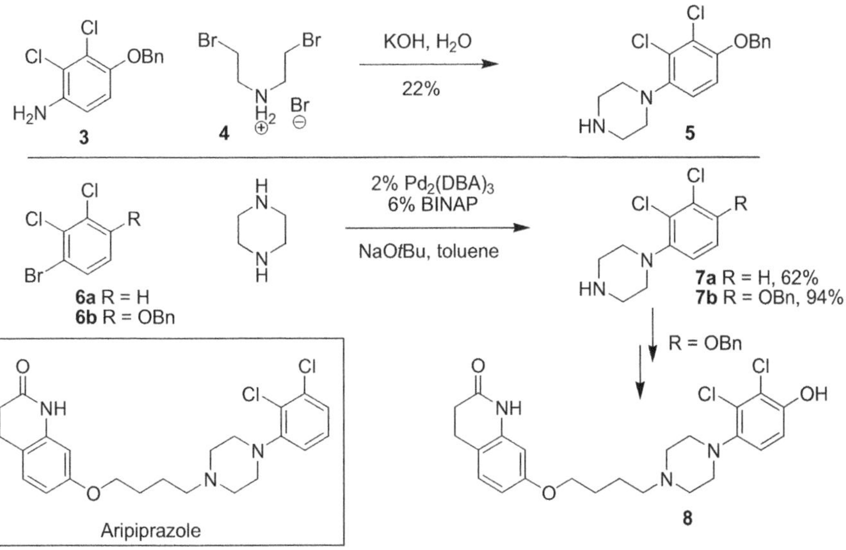

SCHEME 3.3 Pd-catalyzed C–N coupling in the synthesis of aripiprazole.

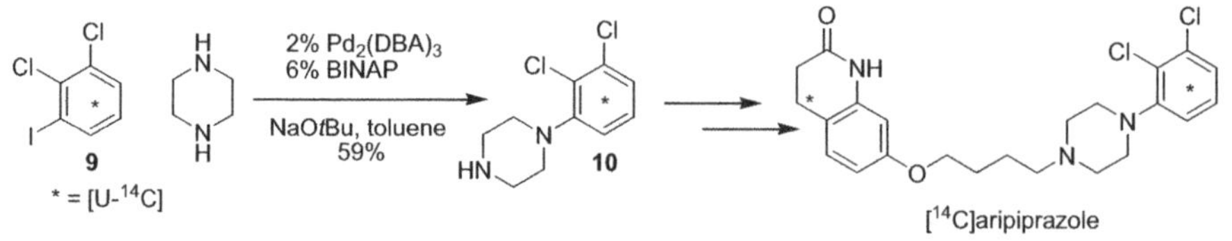

SCHEME 3.4   Pd-catalyzed amination in the synthesis of labeled aripiprazole.

The initial synthesis prepared the piperazine ring by alkylating the aniline **3** with bis(2-bromoethyl)amine hydrobromide (**4**) in only 22% yield in the presence of aqueous potassium hydroxide. Taking advantage of the Pd-catalyzed amination, piperazine was introduced by direct coupling of a large excess of unprotected piperazine (6 equiv) and the aryl bromide **6** catalyzed by the Pd-BINAP catalyst. A moderate yield of 62% was achieved from 1-bromo-2,3-dichlorobenzene (**6a**) that led to aripiprazole. A much better yield of 94% was obtained with the 4-benzyloxy analog **6b**, possibly because the 4-benzyloxy group blocked any potential side reactions of the 3-chloro group. The product **7b** was then converted to the metabolite **8** in two steps.

During the continued development of the drug, a multilabeled [^{14}C]aripiprazole was required for the use in human metabolism and pharmacokinetic studies [21]. The same Pd-catalyzed amination method was used by a BMS group for its synthesis (Scheme 3.4). Aryl iodide **9** was used instead of the bromide to give a 59% yield.

Merck's taranabant is a cannabinoid-1 (CB-1) receptor inverse agonist that was in phase III clinical studies for the treatment of obesity. To address some of the issues in the first-generation process research route such as the use of sodium azide and lack of crystalline intermediates [22], a new asymmetric synthesis was developed based on a Pd-catalyzed amidation reaction and a Rh-catalyzed asymmetric hydrogenation of the enamide (Scheme 3.5) [23]. Initially, enol triflate **11a** was prepared

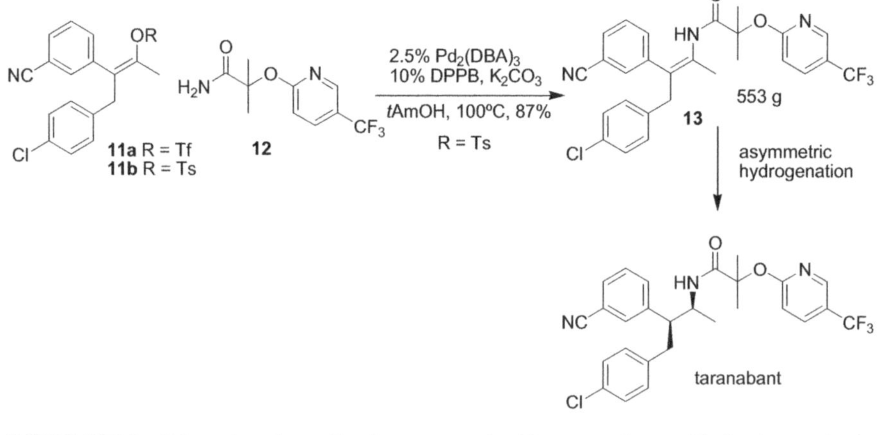

SCHEME 3.5   Pd-catalyzed coupling between enol sulfonates and an amide for the synthesis of taranabant.

and successfully coupled with the amide **12** in 90% yield using the $Pd_2(DBA)_3$-Xantphos catalyst system with Cs_2CO_3 as the base and dioxane as the solvent [24]. Xantphos was found to be the best ligands among some of the commonly used ligands for amination. However, the preparation of the enol triflate substrate proved to be impractical due to the needs of the expensive reagent $NPhTf_2$ and column purification to remove an isomer. Enol tosylate **11b** was prepared efficiently from the ketone precursor as a crystalline solid to address these issues. The Pd-catalyzed coupling of the enol tosylate with the amide **12** was achieved using a cheap commercial ligand 1,4-bisdiphenylphosphinobutane (DPPB) [25]. The use of *tert*-amyl alcohol as the solvent was beneficial for this reaction [26]. Subsequent asymmetric hydrogenation of the tetrasubstituted double bond provided taranabant with excellent ee and in high yield. As a result of this study, the Pd-catalyzed couplings of enol triflates [24] and tosylates [25] with various amides were developed as general methodologies.

Pfizer used a palladium-catalyzed amination reaction for the large-scale synthesis of torcetrapib, a cholesteryl ester transfer protein (CETP) inhibitor for the treatment of cardiovascular disease [27]. The coupling of 4-trifluoromethyl-chlorobenzene (**14**) and the chiral amine **15** was required to provide an advanced intermediate (Scheme 3.6). After extensive screening of the bases, solvents, ligands, and activation protocols, the optimal conditions were achieved with the biphenyl-based Buchwald ligand DavePhos, Cs_2CO_3 as the base, and toluene as the solvent. The Pd(II) precursor $Pd(OAc)_2$ was activated *in situ* by reduction to Pd(0) with $PhB(OH)_2$ [9]. Temperatures above 80°C would lead to a slight ee erosion. The crude product **16** was an oil, so it was hydrolyzed to the primary amide **17** with concentrated sulfuric acid followed by crystallization. On the lab scale, this two-step sequence gave 85% yield. However, when this was scaled up to the kilogram scale, the high density of Cs_2CO_3 caused some agitation problems, leading to slower and incomplete amination reaction for about 77% yield over two steps.

The advantage of the palladium-catalyzed C–N cross-coupling was fully demonstrated in Merck's process development for the synthesis of an orally active KDR

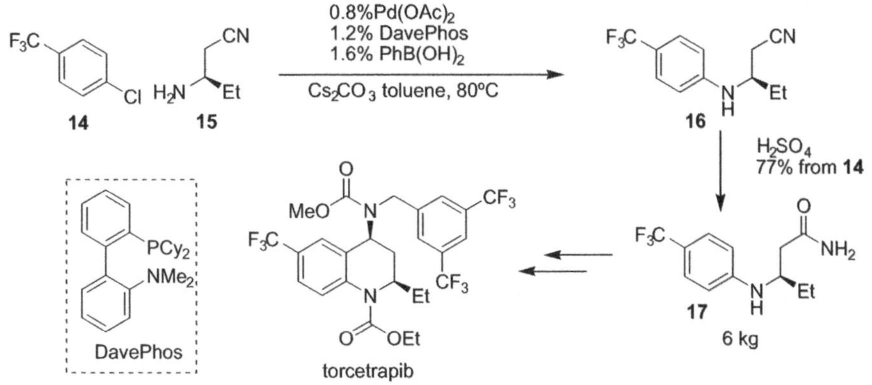

SCHEME 3.6 Amination in the synthesis of torcetrapib.

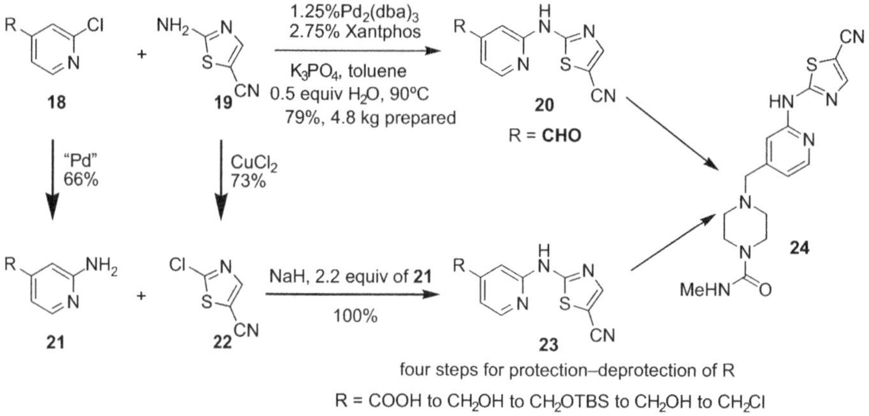

SCHEME 3.7   Pd-catalyzed arylation of an aminoheterocycle for the efficient synthesis of a KDR kinase inhibitor.

kinase inhibitor (Scheme 3.7) [28]. In the original medicinal chemistry approach [29], the chloropyridine **18** was converted to the aminopyridine **21** and the aminothiazole **19** to the chlorothiazole **22**. The aminopyridine **21** was then reacted with the chlorothiazole **22** in an S_NAr manner for the requisite diheteroarylamine **23**; four additional functional group manipulation steps were also required in this sequence. A direct cross-coupling between the aminothiazole **18** and the chloropyridine **19** bearing an aldehyde group would provide the penultimate in one single step, saving a total of six steps. After surveying various reaction parameters, the amination was accomplished in high yield with a Pd-Xantphos catalyst system in toluene. The key to avoid undesired condensation between the amine group and the aldehyde group was the use of toluene solvent that gave a low solubility of the aminothiazole; more polar solvents gave complete decomposition of starting materials. Initially, the reaction gave much lower conversion on scale-up, and careful examination of the bigger scale reactions revealed the loss of water via azeotrope with toluene as the cause for the low yield and poor purity. Addition of a small amount of water reestablished the high yield on scale-up. Even though the exact role of water in the reaction is unclear, it might retard the formation of imine or make K_3PO_4 a more soluble and effective base. The use of powdered K_3PO_4 base also proved critical, as is the case for most of these heterogeneous base-mediated reactions. Final reductive amination provided the API **24** in high yield. The Pd-Xantphos catalyst system proved general for the coupling of (hetero)aryl halides and heteroarylamines including 2-aminopyridines, 2-aminothiazoles, and their analogs to synthesize diheteroarylamines, a very important class of compounds for drug discovery [30].

At Merck, benzopheonone imine (**26**) was used as an ammonia surrogate [31] in the large-scale preparation of **29**, a highly potent, orally active, long acting, selective muscarinic M3 antagonist for the treatment of chronic obstructive pulmonary diseases and urinary incontinence [32]. The coupling between the bromopyridine **25** and benzophenone imine was catalyzed by 0.25 mol% of Pd(OAc)$_2$ and 0.5 mol% of

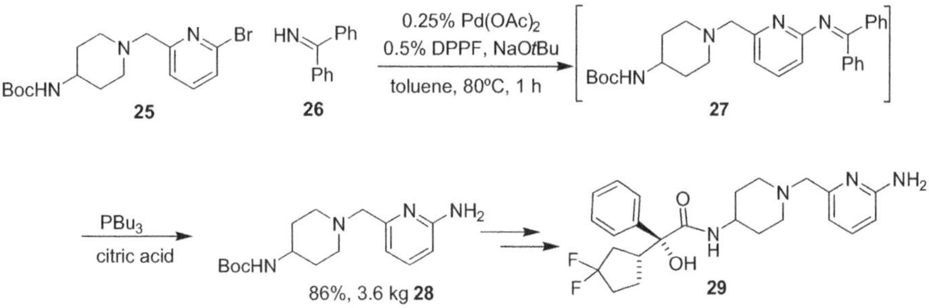

SCHEME 3.8 Benzophenone imine as ammonia surrogate in the synthesis of an M3 antagonist.

DPPF with NaOtBu as the base (Scheme 3.8). To remove palladium after the reaction, (n-Bu)$_3$P was added at the end of the reaction to coordinate to Pd and the toluene mixture was stirred with aqueous citric acid to hydrolyze the imine **27**. The benzophenone by-product and most of the palladium remained in the organic layer. The product **28** was isolated in 86% yield from the aqueous layer after basification, extraction, and crystallization. A copper-catalyzed direct amination process was also developed (see Scheme 2.41).

Novartis also used benzophenone imine on multikilogram scale as an ammonia surrogate for the synthesis of 7-amino-2,1,3-benzothiadiazole **32** (Scheme 3.9) [33]. The initial nitration chemistry to install the nitrogen was not very regioselective and safe for scale-up. For a safe and practical synthesis, the bromide **30** was obtained via selective bromination. Coupling with benzopheonone imine using 0.25% of Pd$_2$(DBA)$_3$ (equivalent to 0.5% Pd) and 0.75% of BINAP catalyst followed by hydrolysis provided the desired product **32** in 86% yield on 14 kg scale in the pilot plant. Interestingly, sodium methoxide could be used as the base despite its potential to act as a reductant via coordination of methoxide to Pd followed by β-hydride elimination.

Abbott developed compound **36** as a potent and selective TRPV1 receptor antagonist that could be used as a non-NSAID, nonopiate treatment for pain. The two synthetic routes for **36** involved Pd-catalyzed C–N couplings (Scheme 3.10) [34]. In the first-generation synthesis, aniline **34** was obtained via the Pd-Xantphos-catalyzed amination using benzophenone imine as an ammonia surrogate. It was then

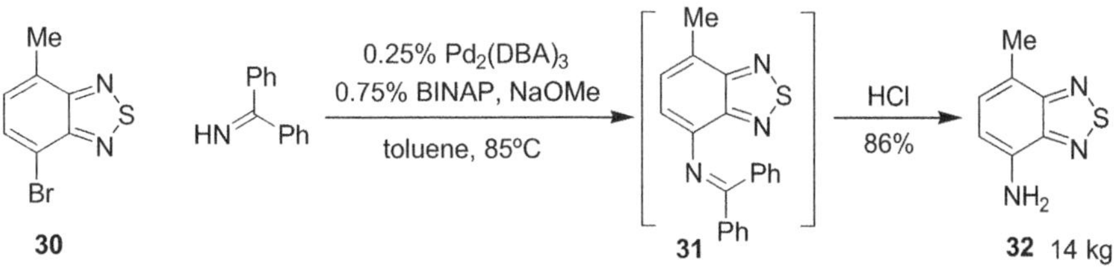

SCHEME 3.9 Use of benzophenone imine in the synthesis of 7-amino-2,1,3-benzothiadiazole **32**.

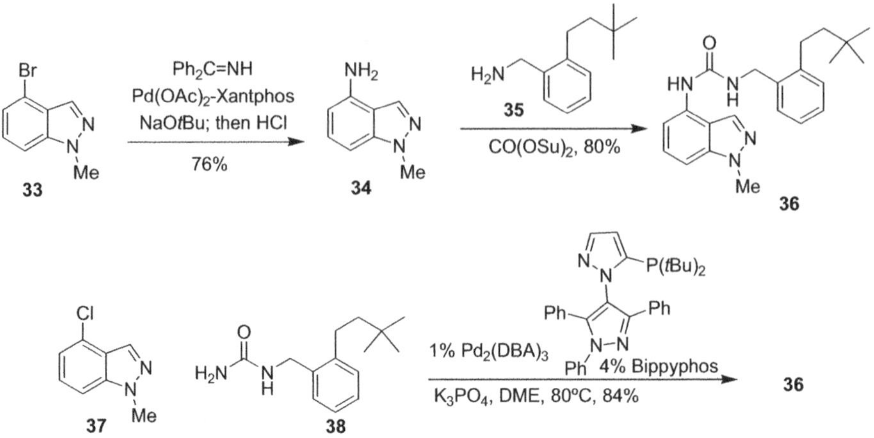

SCHEME 3.10   Amination in the synthesis of a TRPV1 receptor antagonist.

coupled with amine **35** to form the urea product **36**. Due to the instability of the aminoindazole **34** and the difficult purification from benzophenone by-product, the second-generation and more robust synthesis was developed. Based on the Pd-Bippyphos [35]-catalyzed cross-couplings of aryl halides and monosubstituted ureas [36], chloroindazole **37** and the urea **38** were coupled to provide target compound in 84% yield. Screening of ligands showed that only Bippyphos (nonproprietary) and 1,1'-binaphthyl-2-di-*tert*-butylphosphine gave high conversions. Milled K_3PO_4 was the most efficient base for this reaction due to its higher reactive surface. Having a Pd-catalyzed reaction in the final step could be a potential concern for Pd removal in the API, but the authors found that the treatment with thiourea-capped resin and carbon brought the initial Pd levels of > 2000 ppm down to only 35 ppm with $< 1\%$ product loss. The Pd level was further reduced after final crystallization of the API.

AstraZeneca's AR-A2 is a 5-HT$_{1B}$(5-hydroxytryptamine) antagonist for CNS diseases. Its arylpiperazine structure was installed by palladium-catalyzed amination reactions in both the medicinal chemistry approach [37] and the large-scale synthesis (Scheme 3.11) [38]. Pd-BINAP catalyst system was the best. A high ligand/palladium ratio of $\sim$4/1 was important to limit the protodebromination by-product to $< 0.5\%$.

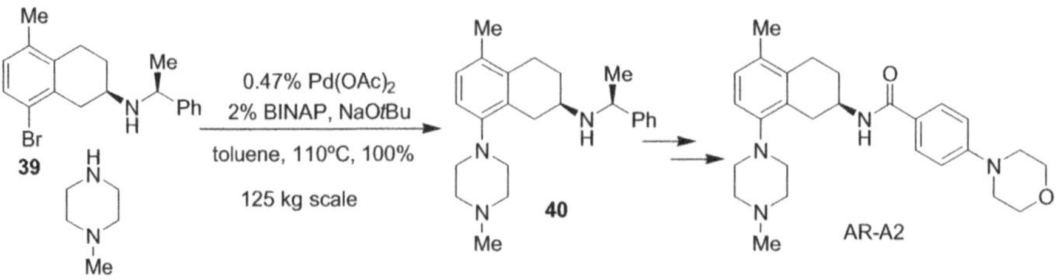

SCHEME 3.11   Piperazine arylation in the synthesis of a 5-HT$_{1B}$ antagonist.

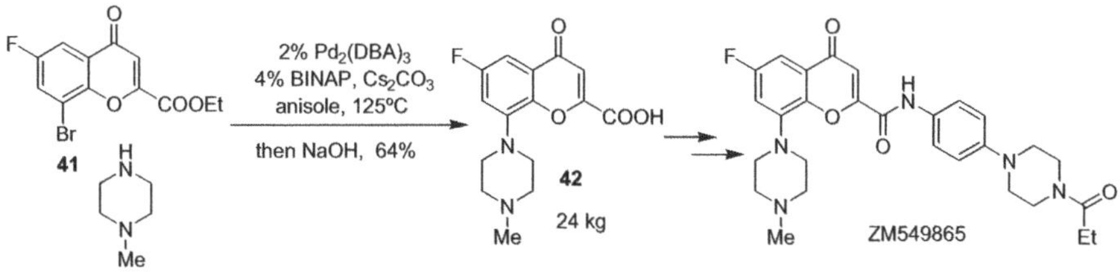

SCHEME 3.12 Amination in the synthesis of ZM549865.

Premixing 0.47% of Pd(OAc)$_2$, 2% of BINAP, and methylpiperazine was critical for a robust and reproducible reaction at as high as 125 kg batch scale. This reaction could be run with as low as 0.065 mol% of Pd, but this was not implemented on the large scale due to the fear of a charging error. This demonstrates a balance between cost and robustness for large-scale catalysis steps to supply critical clinical materials.

In a related AstraZeneca compound ZM549865, a similar amination reaction was achieved using almost the same Pd-BINAP catalyst system (Scheme 3.12) [39]. However, a weaker base, cesium carbonate, had to be used due to the additional functional groups. The yield was moderate due to the lower reactivity of **41** and side reactions such as chromone ring opening by piperazine. The product **42** was isolated as an acid after hydrolysis, and an activated carbon treatment effectively reduced the Pd level to <50 ppm.

At Merck, a palladium-catalyzed C–N coupling between aryl bromide **43** and N-Boc-piperazine was used in the large-scale synthesis of a potent cholecystokinin (CCK) 1R receptor agonist **45** for the potential treatment of obesity (Scheme 3.13) [40]. The same reaction was initially used in the medicinal chemistry synthesis [41], but the yield was low (<50%) and unreliable, possibly due to the use of a strong base, NaOtBu. Switching to a weaker base, K$_3$PO$_4$, and premixing the catalyst (2.5% Pd(DBA)$_2$ and 3% DavePhos) and powdered anhydrous K$_3$PO$_4$ (3.0 equiv) in 2-methyltetrahydrofuran (2-MeTHF) at 78°C proved to give reproducible results.

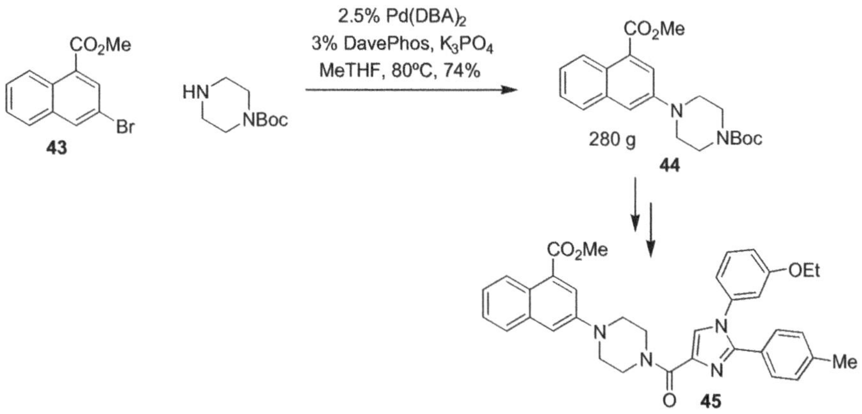

SCHEME 3.13 Amination in the synthesis of a CCK1R antagonist.

However, on larger scale of > 100 g, the reaction required 5 days for completion. This was attributed to the "grinding" effect of the stir bar on the lab scale that was absent or diminished on the bigger scale with overhead stirrer. Regardless, the extended reaction time did not adversely affect the reaction profile and 74% yield was achieved. The treatment with the polystyrene-based resin MP-TMT (trimercaptotriazine) reduced the residual Pd level to 37 ppm and Hg level that was carried over from previous steps to < 3 ppm.

At Abbott, the Pd-catalyzed N-arylation of piperazine was also used in the gram-scale synthesis of **48** [42], the glucuronide metabolite of ABT-724, which is a potent selective D_4 dopamine receptor agonist for the potential treatment of erectile dysfunction (Scheme 3.14). A similar reaction was also used for the initial synthesis of ABT-724 itself [43].

Merck developed a number of kinase insert domain receptor (KDR) inhibitors such as compound **54** that features the indole core structure. A few approaches were evaluated for an efficient synthesis of these compounds [44]. The Fischer indole approach based on Pd-catalyzed amination with benzophenone hydrazone (**50**) [45] was studied (Scheme 3.15). The amination proceeded as expected in almost quantitative yield. Subsequent ketone metathesis took place in 90% yield, but the final indole formation step gave only 45% of the desired product with the deprotection of the N-Ms group being the major by-product (40%). Extensive screening failed to improve isolated yield in the final step.

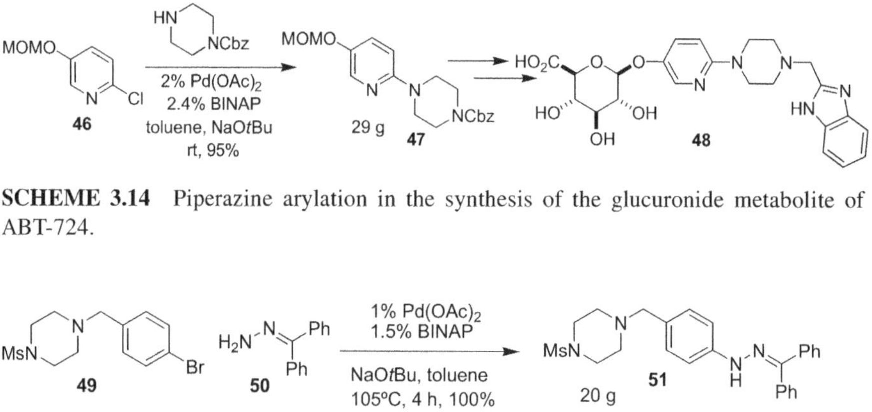

SCHEME 3.14 Piperazine arylation in the synthesis of the glucuronide metabolite of ABT-724.

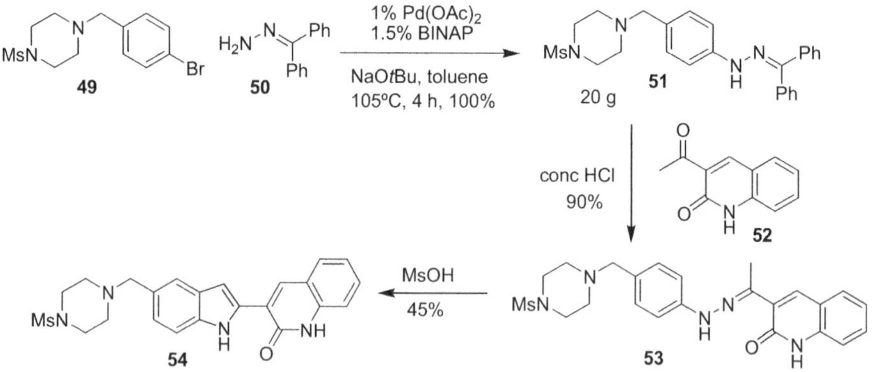

SCHEME 3.15 Arylation of benzophenone hydrazone in the synthesis of a KDR inhibitor.

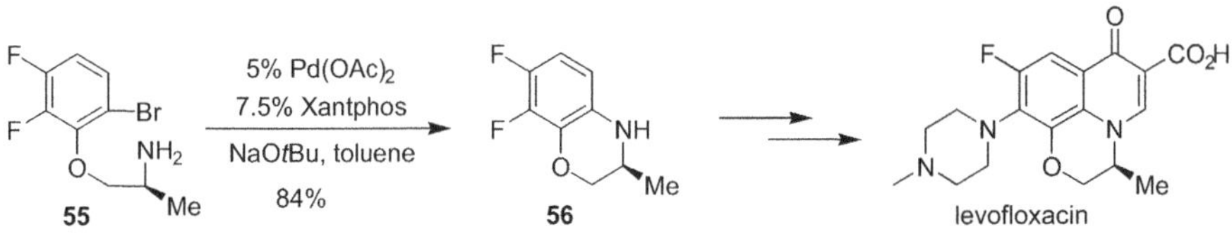

SCHEME 3.16 Pd-catalyzed intramolecular C–N coupling in a synthesis of levofloxacin.

Gallagher developed an asymmetric synthesis of enantiopure 1,4-benzoxazines [46]. The ring-formation step was achieved by a Pd-Xantphos-catalyzed intramolecular amination of **55** (Scheme 3.16). The method was applied to a formal synthesis of levofloxacin (Levaquin, 2008 sales $1.9 billion), Johnson & Johnson's fluoroquinolone drug for treating severe or life-threatening bacterial infections.

In Sepracor's synthesis of Norastemizole, a potent nonsedating histamine H_1-receptor antagonist and an active metabolite of Johnson & Johnson's Astemizole, the primary amine was selectively coupled over a secondary amine to the 2-chloro-1,3-azole **57** with a Pd-BINAP catalyst (Scheme 3.17) [47]. In contrast, the thermal reaction resulted in the selective coupling of the piperidine.

Abbott reported the Pd-BINAP-catalyzed N-arylation of oxazolidinones (Scheme 3.18) [48]. Interestingly, the ketone group was well tolerated despite the possibility of the ketone arylation process under similar reaction conditions [49]. This method was readily applied to the synthesis of an antibacterial agent Dup-721.

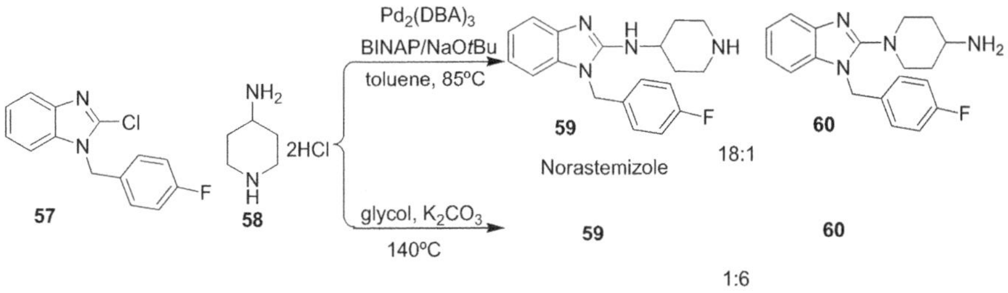

SCHEME 3.17 Pd-catalyzed amination versus thermal reaction in the synthesis of norastemizole.

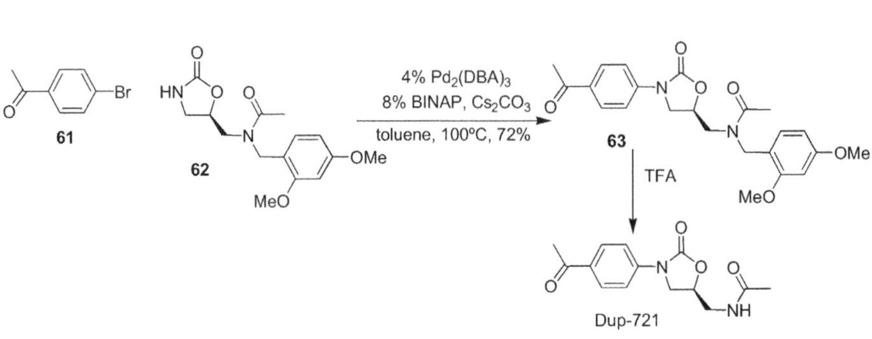

SCHEME 3.18 Arylation of an oxazolidinone in the synthesis of Dup-721.

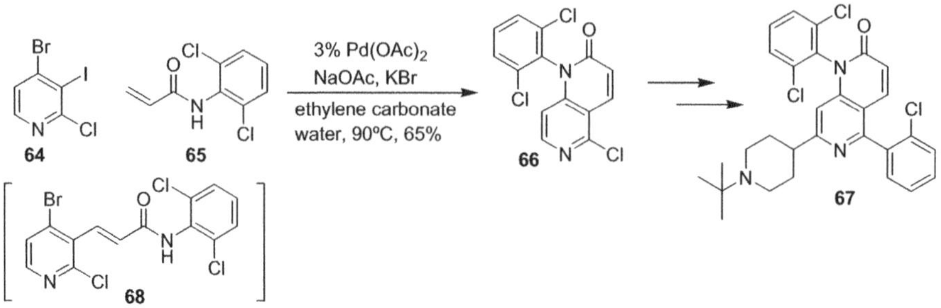

SCHEME 3.19   Intramolecular amination in the synthesis of a MAP kinases inhibitor.

Merck developed a p38 mitogen-activated protein (MAP) kinases inhibitor **67** for the potential treatment of rheumatoid arthritis (RA), Crohn's disease, and psoriasis. An efficient synthesis of the core structure *N*-aryl-1,6-naphthyridone was developed using a unique tandem Heck–lactamization process (Scheme 3.19) [50]. The initial plan was to carry out the Heck reaction between the iodopyridine **64** with the acryl-anilide **65** followed by a separate cyclization step to access the desired product **66**. However, partial cyclization (10–15% yield) was observed under the Heck conditions in DMAC at 130°C. Interestingly, very little Heck intermediate **68** was observed, with the major impurity being the bis-Heck coupling at the iodo and bromo positions of the pyridine. If the Heck intermediate **68** was independently synthesized and subjected to the reaction conditions, only bis-Heck product and decomposition were observed without the desired cyclization. The exact mechanism is unclear, but may be related to a nondisassociated palladium intermediate during the initial Heck reaction. Optimization led to 80% yield of the desired product on the lab scale using ethylene carbonate as solvent and a small amount of water and KBr as additives. The yield dropped to 65% on the kilogram scale.

3.2.1.3   Pd-Catalyzed C–N Couplings in Drug Discovery   Selected examples of Pd-catalyzed C–N coupling reactions in drug discovery will be introduced here with a focus on different substrate classes and a few catalyst systems that are most commonly used by medicinal chemists. Although the scales of these reactions may be small and yields moderate, most of them can likely be optimized on bigger scale in higher yields given the continued development of better catalysts and conditions in this field.

As shown above, many large-scale C–N couplings used the piperazine fragment as the amine partner. *N*-Arylation of piperazines or piperidines has also been widely used in early discovery such as in Wyeth's 5HT program [51], Gmeiner [52] and Boeckler's [53] dopamine receptor ligand synthesis, Otsuka's tuberculosis program [54], Roche's 5HT program [55], Merck's HCV program [56], and GSK's ghrelin program (Scheme 3.20) [57]. Most of these reactions used BINAP as the ligand, Pd(OAc)$_2$ or Pd$_2$(DBA)$_3$ as the Pd source, and a strong base such as NaO*t*Bu or a weaker base such as Cs$_2$CO$_3$, depending on the functional groups.

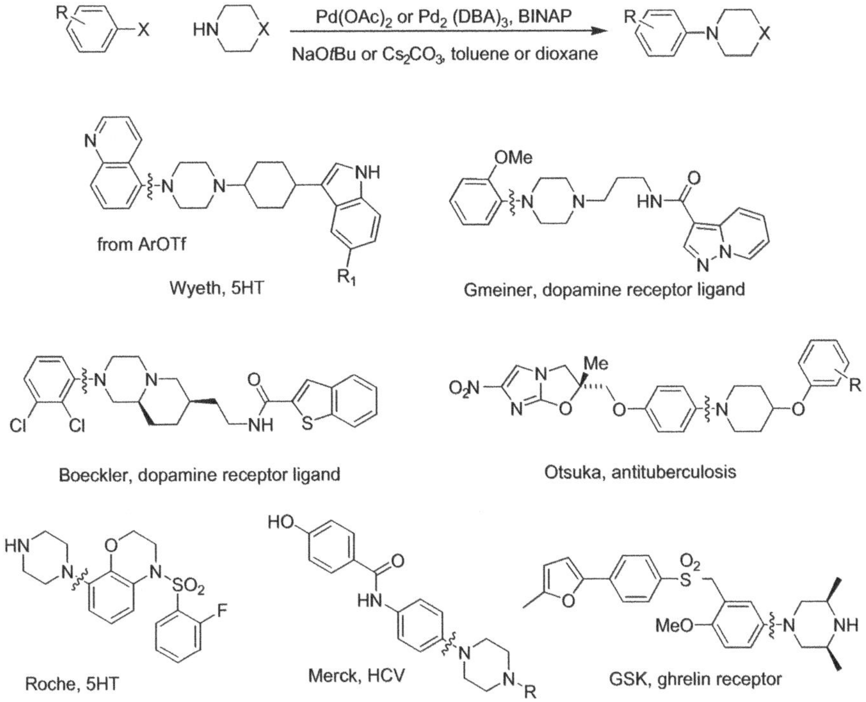

SCHEME 3.20 Pd-catalyzed *N*-arylation of piperazines or piperidines in drug discovery.

Buchwald's biphenyl-based ligands DCPB and DTPB were also used for this reaction in other programs [41,58].

To introduce the cyclopentylamine group to the pyrazolopyridine core to study the antiherpetic activity, GSK chemists used the Pd-BINAP-catalyzed amination in neat cyclopentylamine (Scheme 3.21) [59]. The method was also used to introduce the same group for other analogs [60]. Note a number of heterocycles were tolerated in this reaction. DTBP gave better yields than BINAP in some cases [61].

Oza et al. synthesized and evaluated a series of diclofenac analogs as transthyretin amyloid fibril formation inhibitors [62]. The Pd-BINAP-catalyzed reaction between dichloroanilines and aryl halides/triflates gave moderate to good yields of the diaryl anilines (Scheme 3.22).

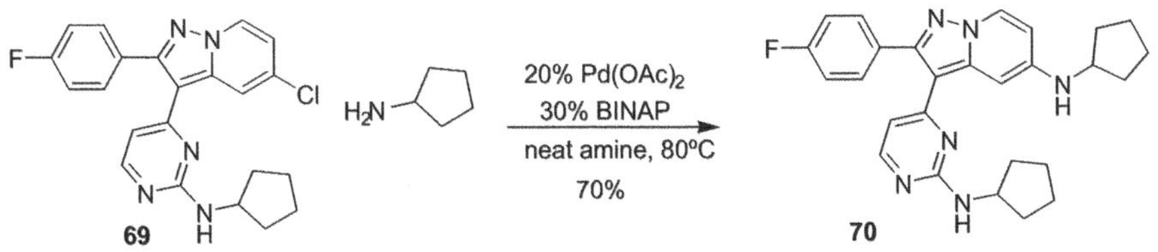

SCHEME 3.21 Pd-BINAP-catalyzed amination in the synthesis of antiherpetic agents.

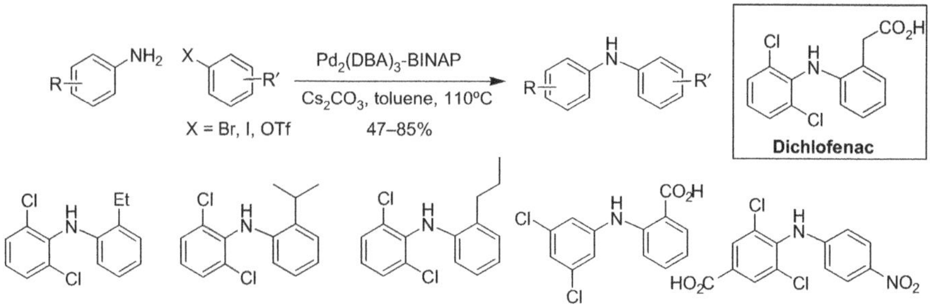

SCHEME 3.22 Amination in the synthesis of diclofenac analogs.

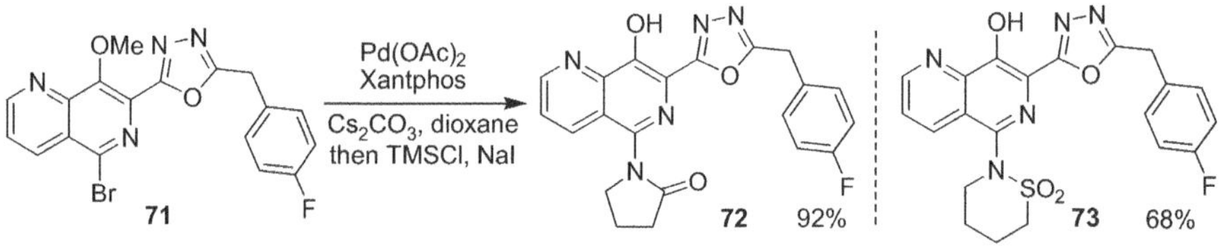

SCHEME 3.23 Pd-Xantphos-catalyzed arylation of lactams and sultams in the synthesis of HIV integrase inhibitors.

The Pd-Xantphos-catalyzed C–N couplings using primary and secondary amides or sulfonamides and related compounds [13] are very useful in early drug discovery. It was applied in the synthesis of a few naphthyridines as HIV-1 integrase inhibitors by GSK (Scheme 3.23) [63]. Instead of previously used Cu_2O-mediated reaction, the Pd catalysis enabled the couplings of the aryl bromide **71** with pyrrolidinone and sultam in good yields. A large number of other analogs were also prepared using the same C–N coupling strategy for SAR.

In addition to benzophenone imine, $BocNH_2$ was also used as an ammonia equivalent in amination reactions. In Johnson & Johnson's synthesis of 5HT reuptake inhibitors, the Pd-Xantphos catalyst was used to couple $BocNH_2$ to aryl bromide **74** in high yield (Scheme 3.24) [64]. TFA deprotection of the crude product **75** gave the aniline **76**, which was further functionalized for SAR. Aniline and substituted anilines were also coupled with the bromide in good yields with BINAP as the ligand.

The N-arylation of heteroarylamines gave access to di(hetero)arylamines that are very useful building blocks in drug discovery. A Merck Process Research group found that the Pd-Xantphos catalyst system was quite general for this reaction [30] and used it in their synthesis of a KDR kinase inhibitor on large scale (Scheme 3.7). More applications of this catalyst in the N-arylation of heteroarylamines in early discovery are given below.

GSK461364 is a potent polo-like kinase (PLK) inhibitor as a clinical candidate for the treatment of cancer. GSK chemists used the Pd-Xantphos-catalyzed arylation of heteroarylamine **77** as a key step for its synthesis (Scheme 3.25) [65]. Other ligands, such as DPPF, XPhos, and BINAP, or copper catalysts were much less effective. The method was also applied to the synthesis of other analogs.

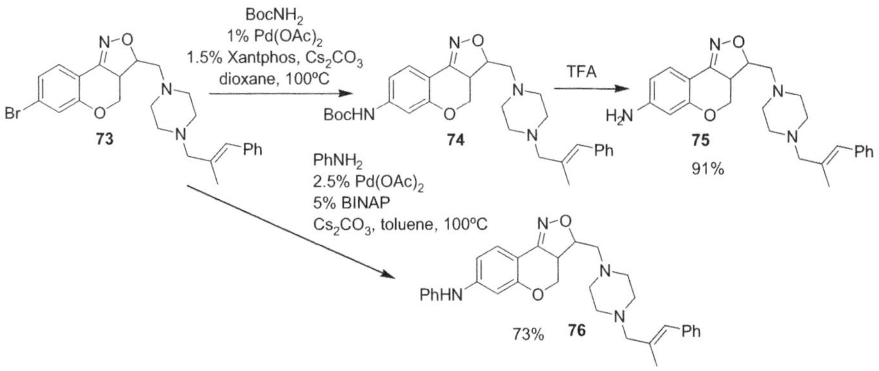

SCHEME 3.24 BocNH$_2$ as ammonia surrogate in the synthesis of 5HT reuptake inhibitors.

The Pd-Xantphos catalyst system was used to couple the chloropyrimidine **80** and various heteroaryl amines for the synthesis of lymphocyte-specific kinase (Lck) inhibitors at Novartis (Scheme 3.26) [66]. In addition to the aminopyridine **81**, aminopyrimidines with different side chains were also coupled in good yields. It is noteworthy that Buchwald also found *t*Bu-XPhos is a general ligand for arylation of heteroarylamines [14].

For Merck's brain-penetrating BACE inhibitors program for Alzheimer's disease [67], a well-defined Pd–ligand complex Pd[P(*t*Bu$_3$)]$_2$ [68] was also used in the macrocyclic amination of **83** in good yields to provide a different series of compounds for SAR (Scheme 3.27).

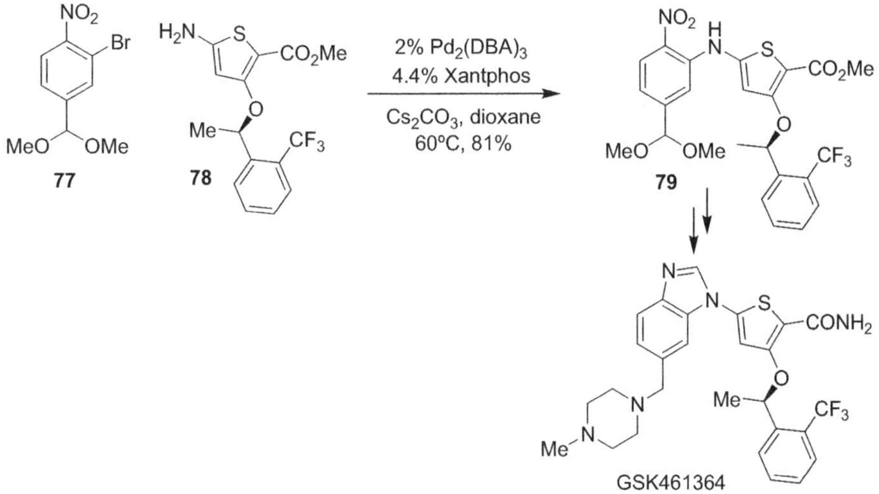

SCHEME 3.25 Pd-Xantphos-catalyzed arylation of a heteroarylamine in the synthesis of GSK461364.

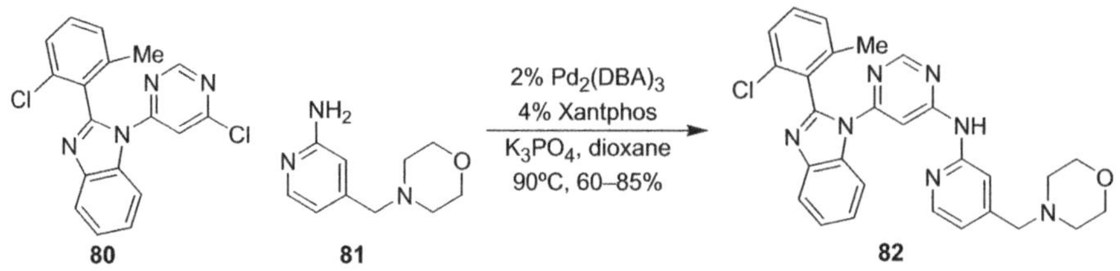

SCHEME 3.26   Arylation of a 2-aminopyridine in the synthesis of a Lck inhibitor.

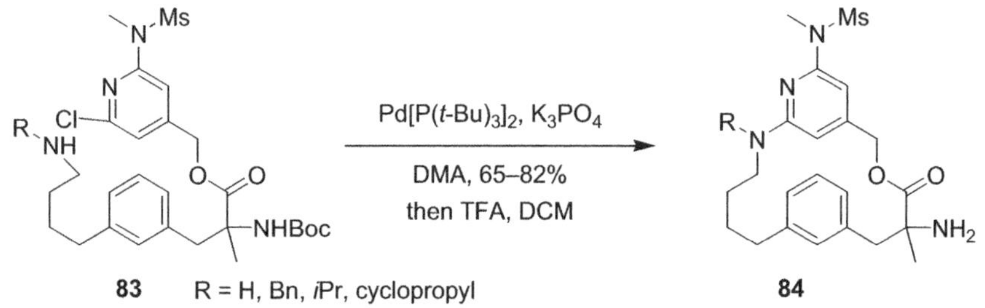

SCHEME 3.27   Macrocyclic C–N coupling in the synthesis of a BACE inhibitor.

Johnson & Johnson also used this Pd[P(tBu$_3$)]$_2$ catalyst for the arylation of ethyl-piperazine (Scheme 3.28) [69]. The reaction was run in a toluene–aqueous KOH biphasic system in the presence of a phase-transfer catalyst. The desired product **86** was isolated in 50% yield, and the des-bromo product from the reduction of the starting material was also observed in 33% yield. The product **86** was then converted to JNJ-39319202, a nonpeptide urotensin-II receptor antagonist.

As shown above, P(tBu)$_3$ is a very versatile ligand for aminations [11]. However, from a practical standpoint, P(tBu)$_3$ is extremely air sensitive and difficult to handle, making it less ideal for early discovery applications. Pd[P(tBu)$_3$]$_2$ is more stable, but still has a limited shelf life [70]. Use of the air-stable phosphonium salts is another possibility [71].

The Buchwald's dialkylphosphino biaryls are very general and versatile ligands for amination of most classes of substrates, including aryl chlorides, bromides, iodides, triflates, tosylates, and mesylates, and primary and secondary amines, anilines,

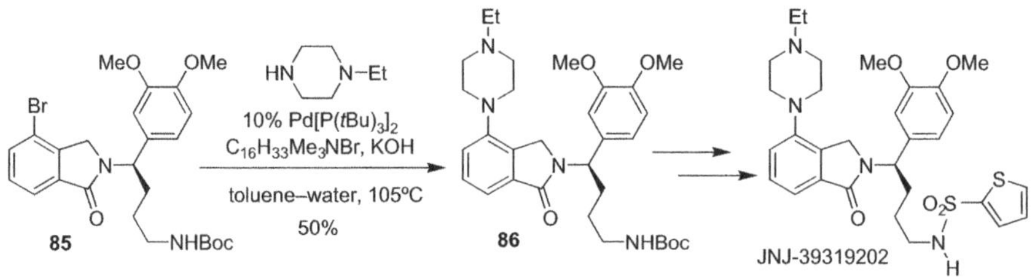

SCHEME 3.28   Pd[P(tBu$_3$)]$_2$-catalyzed amination in the synthesis of JNJ-39319202.

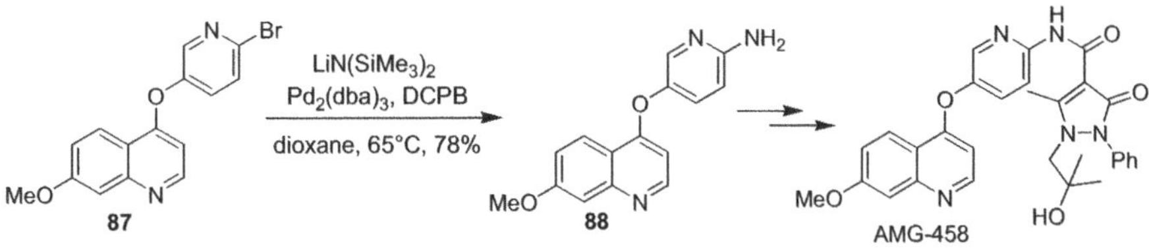

SCHEME 3.29 LiHMDS as ammonia surrogate in the synthesis of AMG-458.

amides, sulfonamides, heteroaryl amines, heterocyclic amines, and so on. These ligands are commercially available and very stable on storage. All these make the diaryl phosphines very attractive for medicinal chemistry applications, especially for many challenging substrates.[1] In a study by Tasler et al., DTPB was identified as a very general ligand for amination reactions to prepare a wide range of bioactive molecules [72].

LiHMDS was used as an ammonia surrogate [73] at Amgen for the c-Met inhibitor AMG-458 for the potential treatment of cancer (Scheme 3.29) [74]. The sterically hindered LiHMDS was coupled with the bromopyridine **87** in good yield using DCPB as the ligand. Upon workup, the free amine was released. Further elaboration of product **88** gave the target molecule AMG-458.

Novartis studied a series of dihydroquinazolinones as p38 kinase inhibitors. In their synthesis, a Pd-catalyzed cyclization of the unprotected ureas **89** onto the chloride was carried out to provide the ring structure **90** (Scheme 3.30) [75]. DCPB was found to be the ligand of choice. It is noteworthy that Merck also used Pd-XPhos-catalyzed cyclization of ureas for the synthesis of benzoimidazolones [76].

In Wyeth's synthesis of a series of inhibitors of mitogen-activated protein kinase-activated protein kinase 2 (MK-2) [77], various heteroaryl amines were coupled to the squarate **91** using XPhos as the ligand (Scheme 3.31). A few amides were also introduced via the Pd-Xantphos-catalyzed amidation reaction.

Heterocyclic amines can be arylated using XPhos [9], *t*Bu-XPhos, or Me$_4$-*t*Bu-XPhos [14]. Kozikowski and coworkers used the Pd-XPhos catalyst for the arylation of the indazole **93** in the synthesis of tetrahydroindazoles as antituberculosis agents (Scheme 3.32) [78].

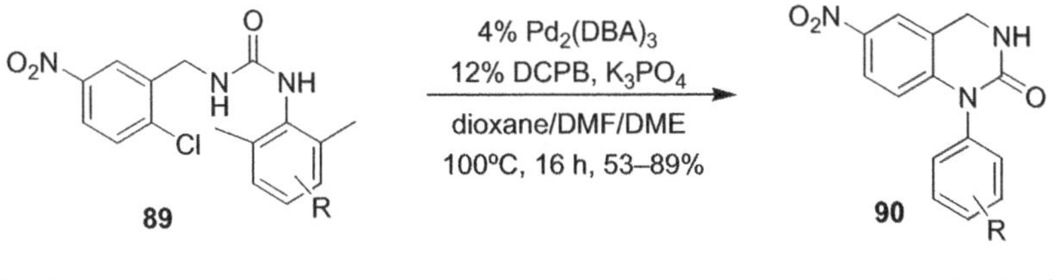

SCHEME 3.30 Intramolecular C–N coupling of unprotected ureas in the synthesis of p38 kinase inhibitors.

---

[1] For additional examples of applications of biaryl phosphine ligands in the pharmaceutical industry, see Ref. 1a.

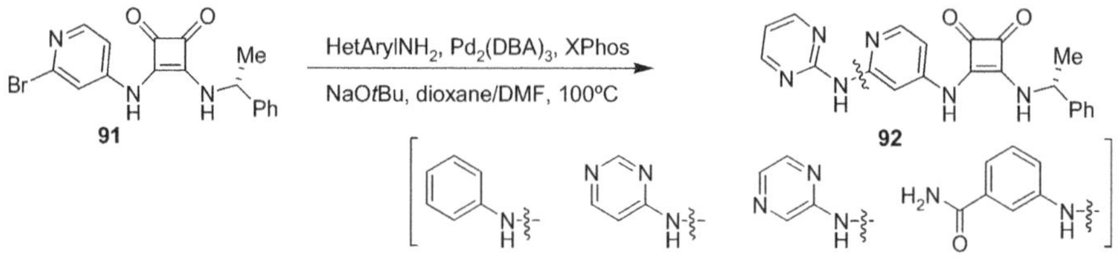

SCHEME 3.31 Pd-XPhos-catalyzed arylation of heteroarylamines in the synthesis of MK-2 inhibitors.

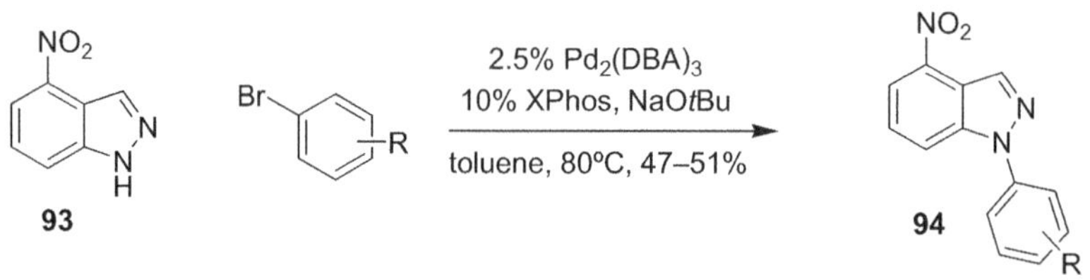

SCHEME 3.32 Arylation of heterocyclic amines in the synthesis of antituberculosis agents.

In addition to the above most commonly used ligands, many others have also been developed for Buchwald–Hartwig amination reactions. Selected examples include Hartwig's Q-Phos [79,80], Beller's n-butyl bis-1-adamantylphosphine (Ad_2PBu) [81,82], and Nolan's $Pd_2(DBA)_3$/IPr-HCl [IPr = 1,3-bis(2,6-diisopropyl-phenyl)imidazol-2-ylidene] system [83,84].

3.2.2 Cu-Catalyzed C–N Couplings

3.2.2.1 Introduction Unlike the Pd-catalyzed C–N coupling reactions, the Cu-catalyzed aminations, Ullmann and Goldberg reactions, have been known for over a century [85,86]. Copper catalysis offers advantages over Pd catalysis such as the lower costs of the metal and possibly the ligand, but it is still mostly limited to aryl and vinyl iodides and bromides. Previously, copper-mediated aminations had been carried out with no ligands under very harsh conditions such as high temperatures and highly polar solvents, limiting their applications in the synthesis of complex bioactive molecules. In the late 1990s, Buchwald and coworkers [87] and Goodbrand and Hu [88] developed Cu-1,10-phenanthroline-catalyzed C–N coupling under milder conditions with heterocyclic amines and anilines, respectively. Buchwald and coworkers then found the use of diamine ligands (Fig. 3.2) [89] to significantly expand the scope of the Cu-catalyzed amination of amides [90] and heterocycle

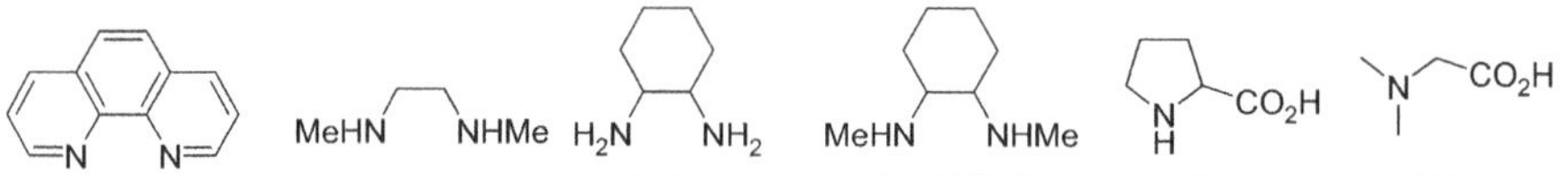

FIGURE 3.2 Commonly used ligands for Cu-catalyzed aminations.

amines [91]. Ma et al. were first to find that the copper-catalyzed C–N coupling with amino acids was dramatically faster than with simple amines, likely due to the amino acids acting as ligands [92]. They then developed proline and *N,N*-dimethyl-glycine (DMG) as ligands for amination using amines and heterocyclic amines [93]. Cu-catalyzed amination sometimes can complement the substrate scope of Pd-catalyzed processes, especially when amides, sulfonamides, and some hetero-cycles are used. Many other ligands have been developed ever since [2], but the inexpensive diamines[2g] and amino acids ligands[2h] are quite general for Cu-catalyzed aminations and seem to be the most used ligands in drug discovery.

3.2.2.2 Cu-Catalyzed C–N Coupling to Make Approved Drugs and Clinical Candidates

As mentioned earlier, a Pd-catalyzed process was initially used for the synthesis of Imatinib (Scheme 3.2). To avoid the use of more expensive Pd metal as the catalyst, Wang and coworkers developed a copper-catalyzed process based on *N,N'*-dimethylethylenediamine (DMEDA) ligand (Scheme 3.33) [94]. The CuI–DMEDA-catalyzed amination between aminopyrimidine **1** and the aryl bromide **95** gave 82% yield of **96**, which was then converted to Imatinib. This coupling may require additional development for larger scale synthesis due to the use of 25% of the catalyst and toxic dioxane solvent. This catalyst system was also successfully used for the *N*-arylation of various heteroarylamines to provide an important class of useful building blocks for bioactive molecules [95].

Temapharm recently used a different disconnection for another approach of Imatinib (Scheme 3.34) [96]. The benzamide **98** was coupled with the aryl bromide **97** in 85% yield as the last step using the same CuI–DMEDA catalyst.

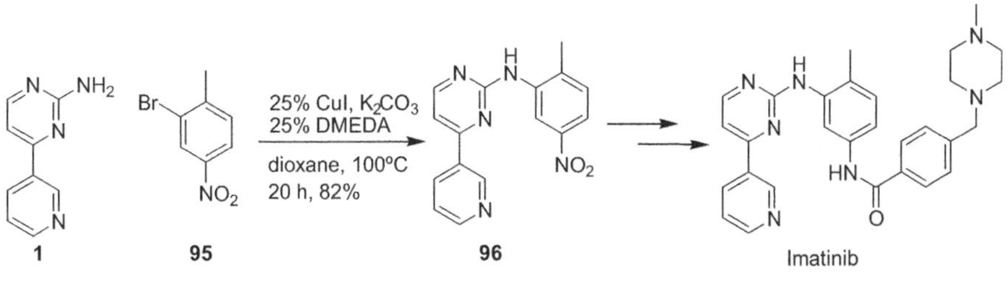

SCHEME 3.33 Cu–DMEDA-catalyzed amination in the synthesis of Imatinib.

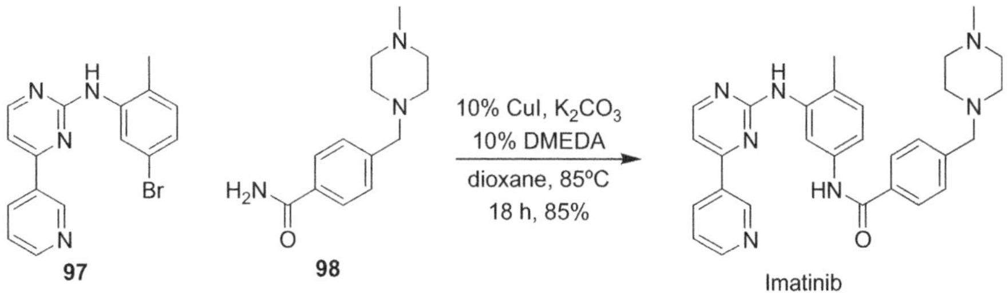

SCHEME 3.34 Cu–DMEDA-catalyzed amination in Temapharm's synthesis of Imatinib.

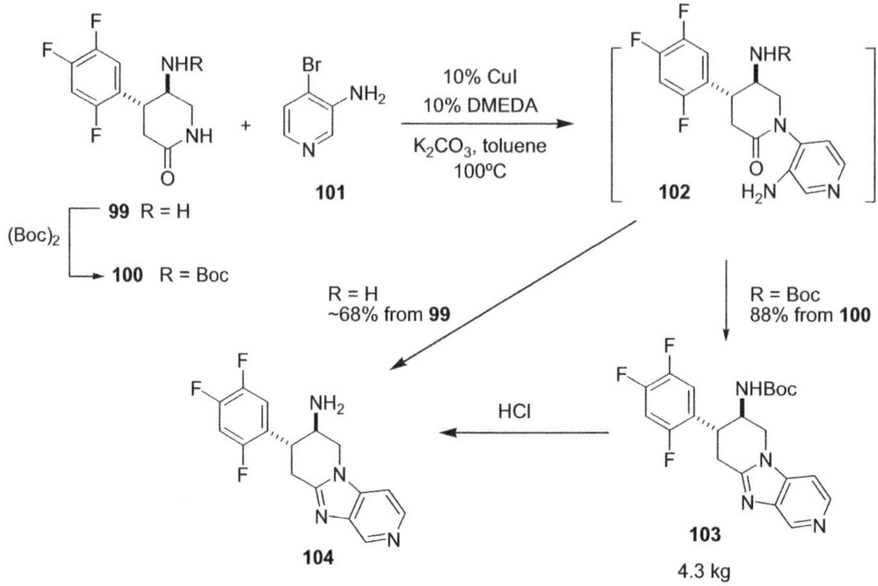

SCHEME 3.35   Cu-catalyzed amination in the synthesis of a DPP4 inhibitor.

A Merck group used the Cu–DMEDA-catalyzed C–N coupling between lactam **100** and the aminobromopyridine **101** (Scheme 3.35). *In situ* cyclization and dehydration gave the piperidine-fused imidazopyridine **103**, which upon deprotection provided **104**, a potent dipeptidyl peptidase IV inhibitor for the treatment of type II diabetes [97]. The free amine group on the pyridine was well tolerated. However, the free amine group on the pyridone moiety **99** resulted in lower yield of the amination–cyclization sequence, so the Boc protection was used to improve the overall yield. The copper salt was removed by filtration through silica gel at the end of cyclization. Deprotection gave the API in high purity.

In a short and scalable synthesis of the κ-opioid receptor agonist CJ-15,161 [98], oxazolidinone **105** was coupled with aryl bromide **106** using a Cu catalyst with 1,2-diaminocyclohexane (DACH) as the ligand (Scheme 3.36). The copper-catalyzed *N*-arylation of oxazolidinones is relatively general and provides an alternative to a Pd catalyst system based on Buchwald's biphenyl-type ligands [99].

Oxindole **112** is UCB pharma's preclinical candidate for the treatment of epilepsy and a practical synthesis was required [100]. A copper-catalyzed C–N coupling

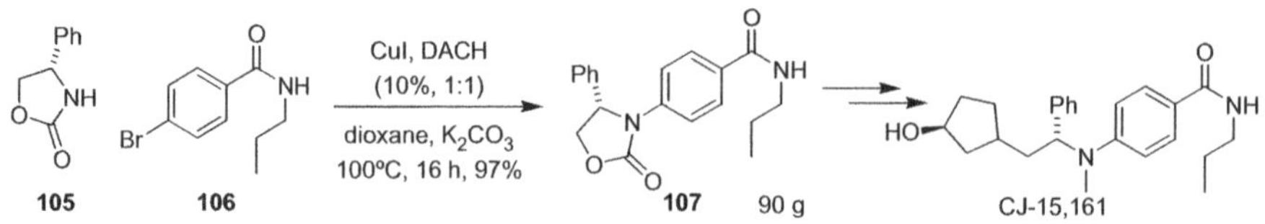

SCHEME 3.36   Cu–DACH-catalyzed arylation of an oxazolidinone in the synthesis of CJ-15,161.

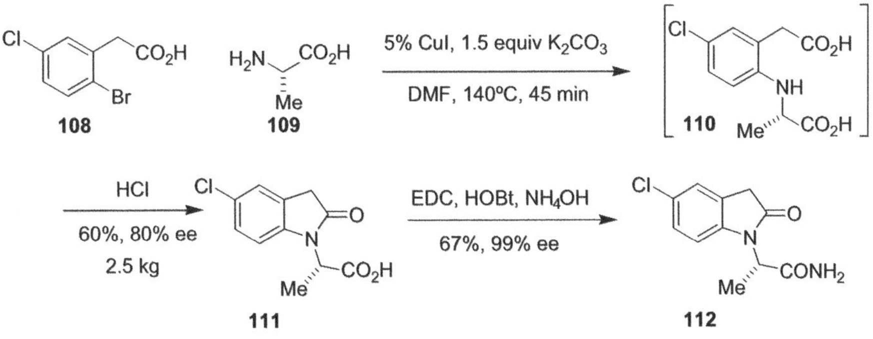

SCHEME 3.37   Cu-catalyzed amination in a practical synthesis of an epilepsy drug candidate.

between the aryl bromide **108** and alanine was the key step for the short and efficient synthesis of **112** (Scheme 3.37). Based on Ma's finding that amino acids functioned as accelerating ligands [92], this reaction did not require an additional ligand. DMF is a much better solvent than toluene, dioxane, *t*BuOH, or water. However, the high temperature that was required for the reaction led to partial ee erosion to give product **111** with 80% ee after acid-mediated cyclization. Subsequent amide formation provided the desired compound **112** in 67% yield and with 99% ee after crystallization to upgrade the ee. Unfortunately, the same reaction with glycine failed to give any desired product and an alternative route had to be developed for a similar preclinical candidate lacking the methyl group.

Abbott's compound **116** has shown potent histamine H_3 receptor antagonism, which could be used for the treatment of cognition and attention-deficit hyperactivity disorder (ADHD). For its kilogram preparation, a C–N coupling of the aryl bromide **113** and pyridazinone **114** was required (Scheme 3.38). On the small scale, this could be achieved with copper powder in refluxing pyridine, but multigram runs resulted in poor reproducibility and lower yields. Despite recent advances in Pd-catalyzed C–N coupling reactions, various Pd-catalyzed conditions gave mostly unreacted starting materials, likely due to the poor nucleophilicity of the pyridazinone ($pK_a = 10.5$). On the other hand, copper-catalyzed reaction proceeded very well to give 85% yield of **115** with no detectable *O*-arylation product. The best ligand for this reaction turned out to be 8-hydroxyquinoline. The methodology was extended to other aryl halides

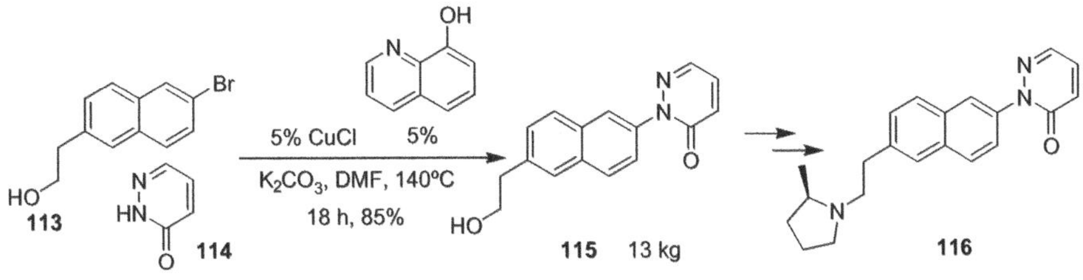

SCHEME 3.38   Cu-catalyzed arylation of a pyridazinone in the synthesis of a H_3 receptor antagonist.

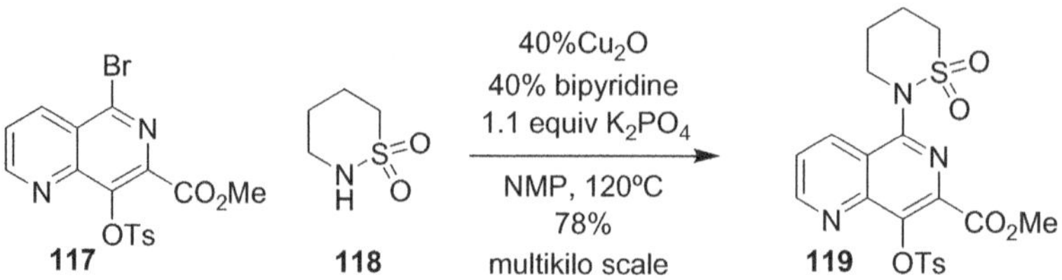

SCHEME 3.39 Cu-catalyzed arylation of sultam in the synthesis of an HIV integrase inhibitor.

for general preparation of aryl pyridazinones, which is a key building block in many bioactive substances [101]. On the large scale, most of the copper was efficiently removed by washing the organic layer with aqueous NH_4OH and Na_2EDTA solution.

In preparing a key advanced intermediate **119** for Merck's HIV integrase inhibitors, coupling of an aryl bromide **117** and sultam **118** was carried out in good yield on multikilogram scale using Cu_2O–bipyridine catalyst system in NMP at 120°C (Scheme 3.39) [102]. The protection of the 8-hydroxy group as the tosylate was important for the high yield. A Pd-Xantphos catalyst system was also found to be general for N-arylation of sultams with most of other aryl halides [103].

Avodart (dutasteride) is a GSK drug with $584 million sales in 2008 for treating prostate diseases such as prostate cancer. An alternative scalable synthesis was recently reported by Reddy and coworkers (Scheme 3.40) [104]. The coupling between the amide **120** and the aryl iodide **121** was achieved with Cu powder at 140–150°C. Most of the copper was removed by filtration after the reaction and the final isolated API had no detectable amounts of Cu. Very high purity of the API was obtained after crystallization to meet all regulatory specifications.

A copper-catalyzed amination of activated aryl halides with ammonia was developed by a Merck group and used in an alternative process for the synthesis of M3 antagonist **124** (Scheme 3.41) [105]. A palladium-catalyzed amination through benzophenone imine had been used on large scale (see Scheme 3.8), but the use of ammonia itself was desired for a one-step amination. During the optimization of the reaction catalyzed by copper, hydroxy or alkoxy pyridine **125** was observed as a major impurity. Ethylene glycol proved to be a great solvent for high selectivity (**124/125** = 94:6). The reaction required a low catalyst loading and modest pressure (50 psi), and Boc group survived under these conditions. Other activated aryl halides were also coupled efficiently with ammonia directly.

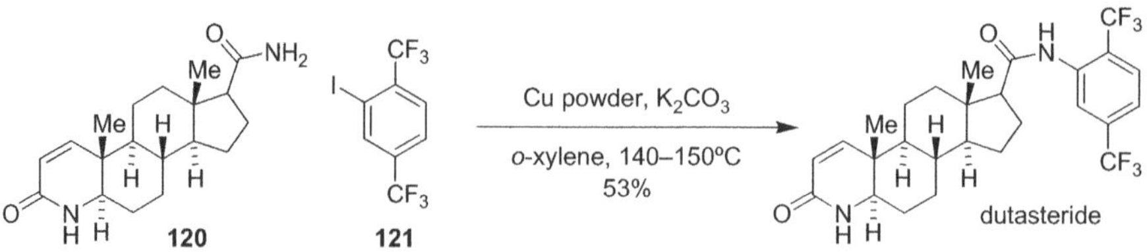

SCHEME 3.40 Cu-catalyzed amidation in a synthesis of dutasteride.

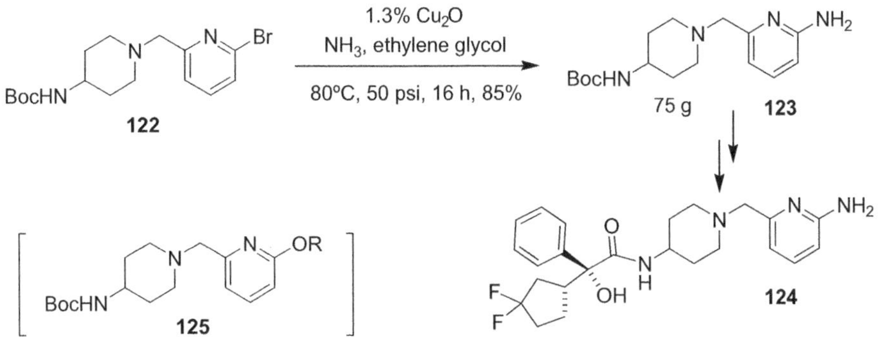

SCHEME 3.41 Cu-catalyzed amination using ammonia in the synthesis of an M3 antagonist.

3.2.2.3 Cu-Catalyzed C–N Coupling in Drug Discovery

In Johnson & Johnson's synthesis of a new melanin-concentrating hormone (MCH) antagonist **128**, the last step is a copper-mediated *N*-arylation of the pyrazinone **126** (Scheme 3.42). Buchwald's CuI–DMEDA catalyst system was used and the microwave irradiation allowed this reaction to complete in 20 min. In fact, the previous two steps to prepare **126** also used microwave for a combined 70 min reaction time over three steps [106].

PF-00277343 is Pfizer's thyroid receptor agonist for the treatment of androgenetic alopecia [107]. The key intermediate **131** was prepared from the CuI–DACH-catalyzed *N*-arylation of 6-azauracil **130** (Scheme 3.43). Subsequent elaboration, including a Cu-catalyzed C–O coupling (Scheme 3.63), gave the desired compound.

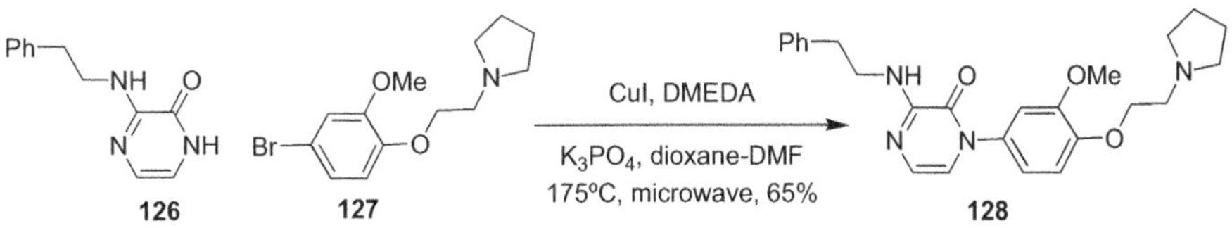

SCHEME 3.42 Cu-mediated *N*-arylation of a pyrazinone in the synthesis of a MCH antagonist.

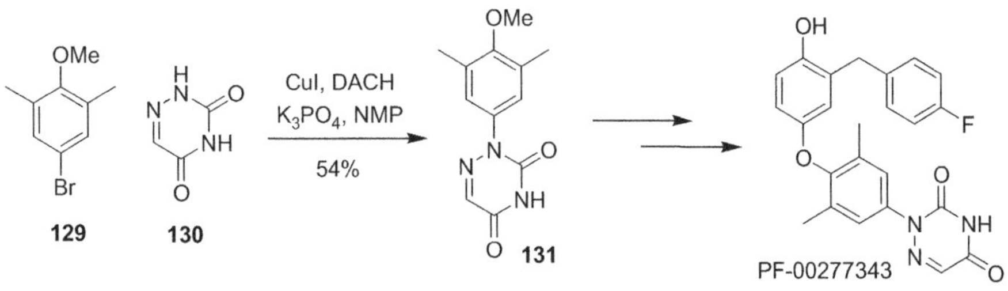

SCHEME 3.43 Cu-catalyzed arylation of a 6-azauracil in the synthesis PF-00277343.

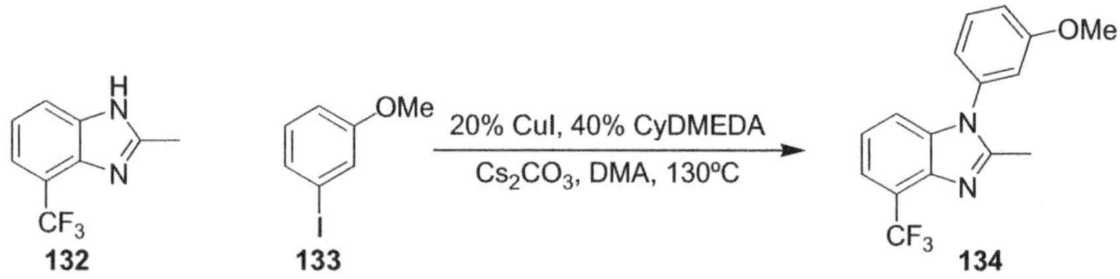

SCHEME 3.44   CuI–CyDMEDA-catalyzed *N*-arylation of benzimidazole in the synthesis of LXR agonists.

The CuI–CyDMEDA-catalyzed *N*-arylation of benzimidazole [91] was applied in Wyeth's LXR program (Scheme 3.44) [108]. A Cu(II)-mediated oxidative coupling using a boronic acid was also used by the same group (see below).

Ma et al. first discovered the significant accelerating effect of amino acids in the Cu-catalyzed C–N coupling reactions [92]. Complexation of the amino acid carboxylate with Cu(I) was proposed for the accelerating effect. This method was applied to the synthesis of benzolactam-V8, a PKC activator (Scheme 3.45). The C–N coupling between the aryl iodide **135** and L-valine (**136**) proceeded in 86% yield in the presence of 10% of CuI with no additional ligand. Further elaboration of product **137** gave benzolactam-V8.

In a similar fashion, Sankyo synthesized arylamine-based cathepsin K inhibitors such as **141** as targets for the treatment of osteoporosis (Scheme 3.46) [109].

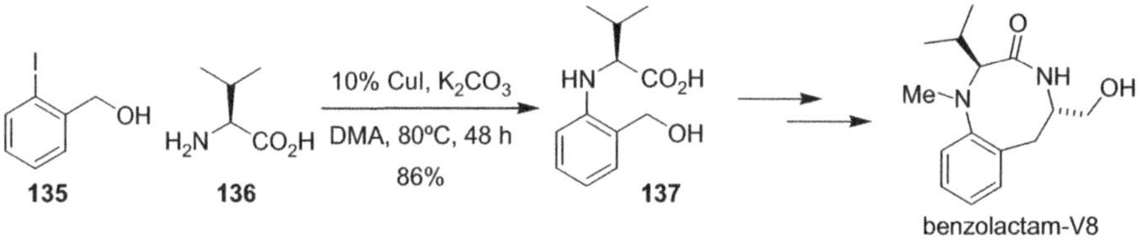

SCHEME 3.45   Cu-catalyzed amination in the synthesis of benzolactam-V8.

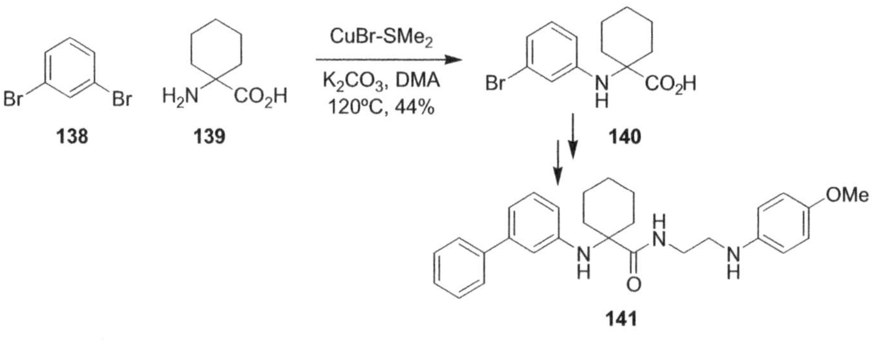

SCHEME 3.46   Cu-catalyzed arylation of an amino acid in the synthesis of cathepsin K inhibitors.

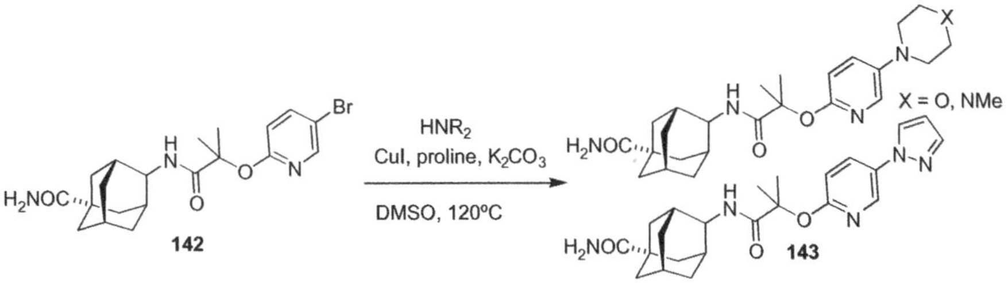

SCHEME 3.47 Cu–proline-catalyzed amination in the synthesis of 11β-HSD1 inhibitors.

Without an external ligand, the amination of 1,3-dibromobenzene and amino acid **139** gave 44% yield of product **140**. Other analogs were also prepared using the same coupling conditions. Merck also had used a similar coupling in the Cat K program [110].

To evaluate a series of admantane-based 11β-HSD1 inhibitors, Abbott used Ma's CuI–proline catalyst system to couple the bromopyridine **142** with morpholine, *N*-methyl-piperazine, and pyrazole (Scheme 3.47) [111]. In another paper by the same group [112], they found that CuI–proline was the best catalyst system for the C–N coupling between morpholine and a similarly functionalized bromopyridine. Other ligands (CyDMEDA and diethylsalicylamide) with Cu or Pd catalysts gave lower conversion and/or much more debromination product.

Buchwald also developed diethylsalicylamide (**145**) and dione ligand **146** for Cu-catalyzed amination using aliphatic amines [113]. Both ligands were found to work very well for Perrio's synthesis of aniline-containing analogs of NMDA NR2B antagonist Ifenprodil [114]. They found that the Pd-catalyzed amination suffered from the oxidation of the benzylic alcohol and debromination, but CuI and the commercial ligands **145** and **146** gave the desired amination product in good yields (Scheme 3.48).

Despite the development of ligands for milder Ullmann and Goldberg reactions, ligandless conditions were sometimes still used in the early discovery, but usually under more forcing conditions. In some cases, the substrate or product could act as a ligand to accelerate the reaction.

A number of heterocyclic amides were coupled with the aryl iodide **148** in good yield without ee erosion in Merck's p38 mitogen-activated protein kinase inhibitors

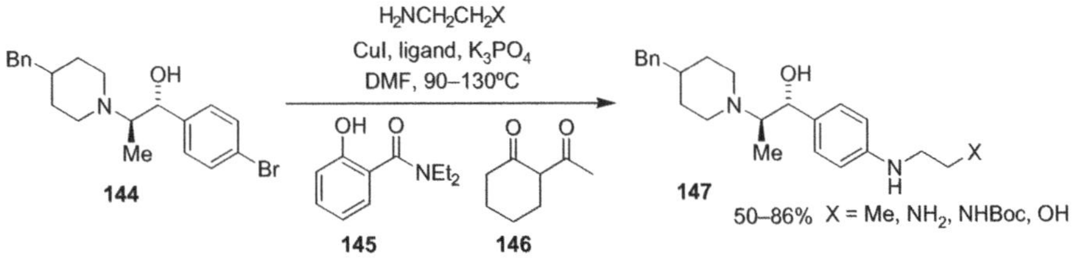

SCHEME 3.48 Cu-catalyzed arylation of primary amines in the synthesis of Ifenprodil analogs.

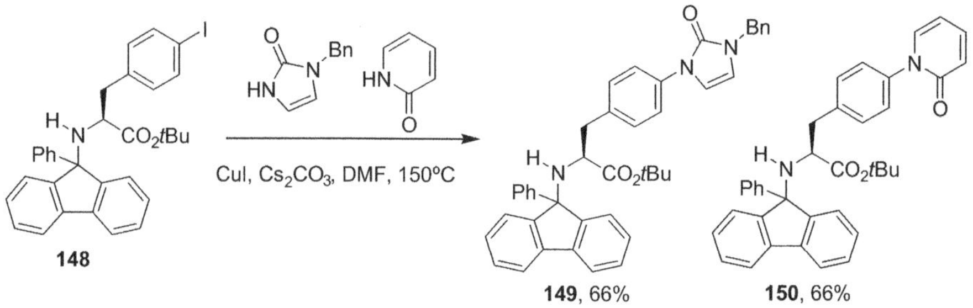

SCHEME 3.49   Ligandless conditions for arylation of heterocycles in the synthesis of MAP kinase inhibitors.

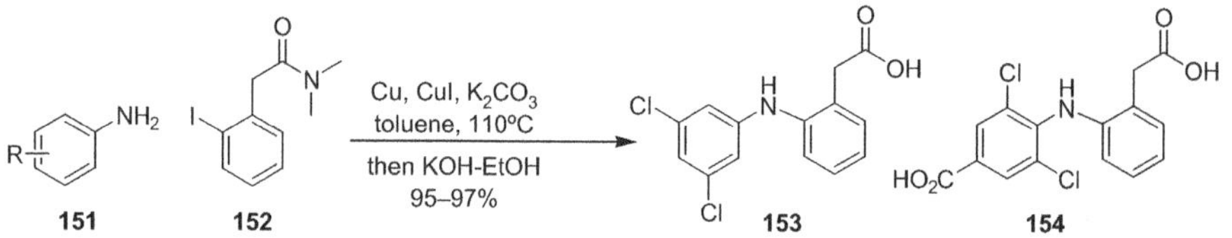

SCHEME 3.50   Cu-catalyzed amination in the synthesis of diclofenac analogs.

program (Scheme 3.49) [115]. No ligand was used with a reaction temperature of 150°C [116].

Oza and Kelly used copper-catalyzed amination in the synthesis of diclofenac analogs (Scheme 3.50) [62]. A mixture of activated copper powder and CuI was used for the coupling between the aniline **151** and aryl iodide **152**.

At Merck, direct conversion of aryl halides to primary aniline with ammonia was achieved in the synthesis of heterobase-modified ribonucleosides as inhibitors of HCV virus RNA replication (Scheme 3.51) [117]. Liquid ammonia was used as the solvent with CuI as catalyst at 130°C to give >38% yield of the monoamino-pyridine **156**. The chloride was then removed by reduction to give a highly active nucleoside **157**.

Müller and coworkers synthesized and evaluated potent inhibitors of *ecto*-5′-nucleotidase based on an anthraquinone scaffold [118]. A catalytic amount of fine copper powder was used in combination with microwave heating to give a rather general amination of the anthraquinonyl bromide **273** with a variety of aliphatic or

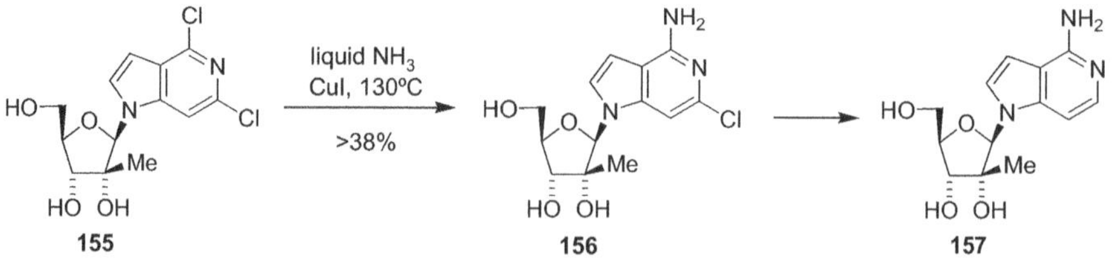

SCHEME 3.51   Cu-catalyzed amination with ammonia in the synthesis of HCV inhibitors.

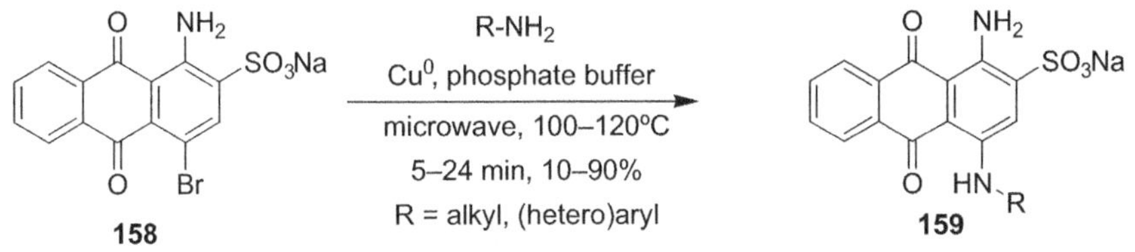

SCHEME 3.52 Cu powder-catalyzed amination in the synthesis of inhibitors of *ecto-5′-nucleotidase*.

(hetero)aromatic amines (Scheme 3.52). Very short reaction times were needed due to the microwave heating. This is also one of the few examples of the Cu-catalyzed amination reactions in water.

In addition to aryl halides, vinyl halides can also be used in the copper-catalyzed C–N couplings. Porco and coworkers developed Cu(I)-catalyzed amidation of vinyl iodides using the tetramethylphenanthroline ligand **162** (Scheme 3.53) [119]. Cu(MeCN)$_4$PF$_6$ gave a slightly better yield than CuI. Rb$_2$CO$_3$ was the optimal base as the weaker base K$_2$CO$_3$ afforded very low conversion and the stronger base Cs$_2$CO$_3$ gave decomposition of the product. This method was applied to the synthesis of antibiotic CJ-15,801 and analogs.

Coleman and Liu also applied the amidation of the vinyl iodide for the synthesis of compound **167**, which could be a useful model substrate as the side chain of antibiotic CJ-12,950 (Scheme 3.54) [120]. CuTc [copper(1)-thiophene-2-carboxylate] and recrystallized CuI were good copper sources for this reaction and CyDMEDA was far superior to 1,10-phenanthroline as the ligand. The base did not have a

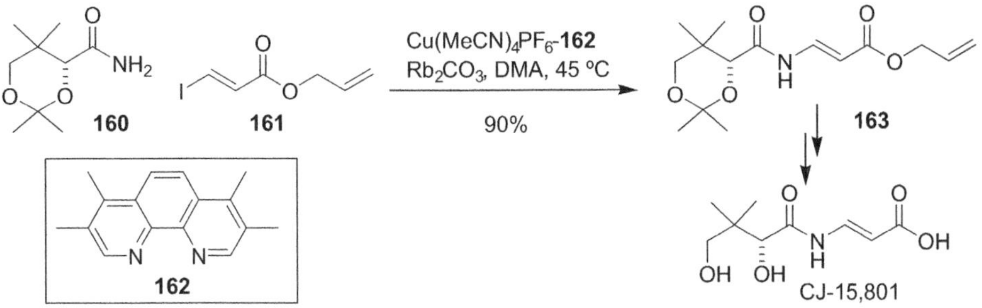

SCHEME 3.53 Amidation of a vinyl iodide in the synthesis of CJ-15,801.

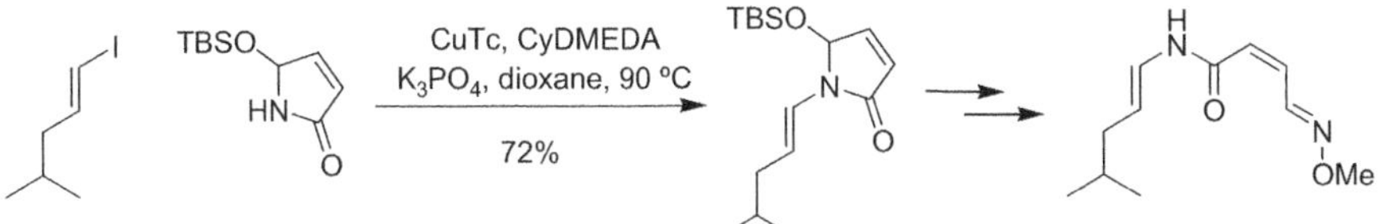

SCHEME 3.54 Cu–CyDMEDA-catalyzed amidation of a vinyl iodide.

significant effect on this reaction. The Pd-catalyzed conditions failed to give any product. Buchwald and coworkers also used the CuI–DMEDA catalyst system for C–N coupling of vinyl halides with lactams or amides [121].

3.2.2.4   Other C–N Couplings

Suzuki et al. first reported CuI-catalyzed azide coupling with aryl iodide in HPMA [122]. Ma and coworkers were able to run the reaction under milder conditions with proline as the ligand [123]. Liang and coworkers found that DMEDA performed better as the ligand in the azide displacement of bromoanilines and other aryl bromides [124].

Aldrich and coworkers used Cu–proline-catalyzed azide displacement of the aryl iodide **168** in his study of 2-triazole-substituted analogs of 5′-O-[N-(salicyl)sulfamoyl]adenosine as antibacterial nucleosides against mycobacterium tuberculosis (Scheme 3.55) [125]. A moderate yield (52%) of the azide **169** was achieved. The azide was then converted to triazoles by coupling with terminal alkynes.

Helquist and coworkers studied N-aryl-3-alkylidenepyrrolinones as potential Niemann–Pick type C disease therapeutics [126]. A Cu–DMEDA-mediated azide displacement was carried out on the aryl bromide **170** (Scheme 3.56), but the resulting azide **171** likely decomposed under the reaction conditions to give the amine **172** directly in 85% overall yield.

Thatcher and coworkers used copper-catalyzed azide displacement of the aryl bromide to synthesize analogs of arzoxifene, a selective estrogen receptor modulator (SERM) that was in phase III clinical studies for postmenopausal osteoporosis (Scheme 3.57) [127]. CuI–proline was used and the resulting azide was converted to the amine under reaction conditions directly, likely due to the azide's instability at 110°C.

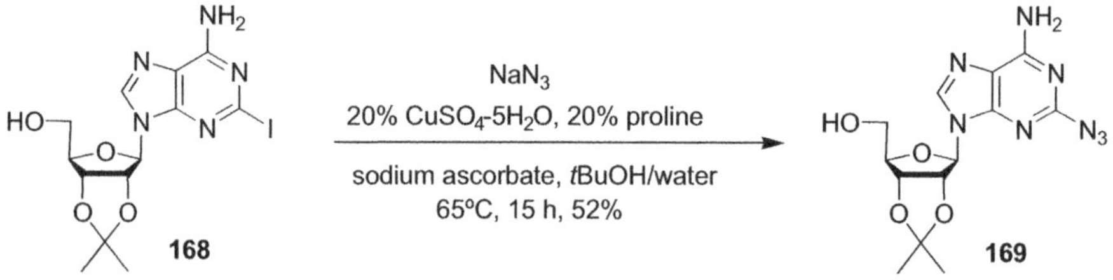

SCHEME 3.55   Cu-catalyzed coupling of an aryl iodide and sodium azide.

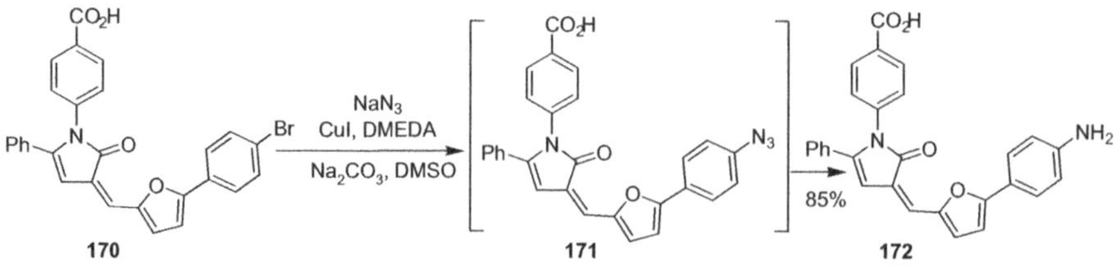

SCHEME 3.56   Cu–DMEDA-mediated azide coupling.

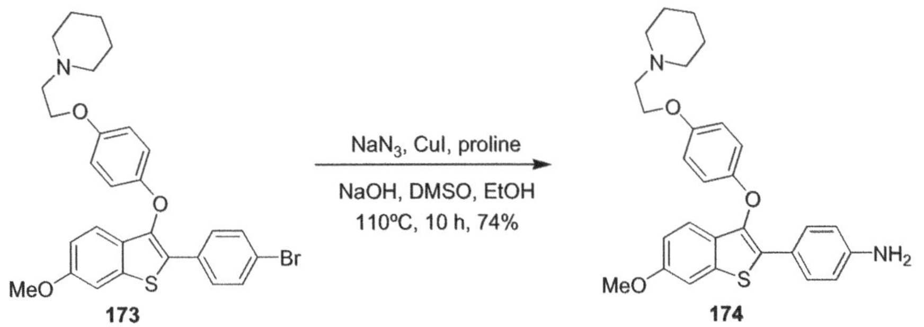

NaN$_3$, CuI, proline
⟶
NaOH, DMSO, EtOH
110°C, 10 h, 74%

173   **174**

SCHEME 3.57   Cu–proline-catalyzed azide coupling.

3.2.3   Pd- or Cu-Catalyzed C–O Coupling

3.2.3.1   Introduction   The early examples of C–O couplings were diaryl ether formation via copper-mediated Ullmann coupling of aryl halides and phenols [128]. In 1997, Buchwald and coworkers found the use of Cs$_2$CO$_3$ and (CuOTf)$_2$PhH-EtOAc–naphthoic acid catalyst system enabled milder coupling conditions [129]. Snieckus and coworkers then introduced Cu(MeCN)$_4$PF$_6$ as a more air-stable catalyst [130]. Gujadhur and Venkataraman used homogeneous copper catalyst Cu(PPh$_3$)$_3$Br for the mild diaryl ether formation [131]. Hauptman and coworkers carried out a 96-well library screening of ligands to identify 8-hydroxyquinoline as a good ligand for diaryl ether formation [132]. Since then, many more catalytic conditions were developed to enable even more general and milder Cu-catalyzed C–O couplings [133]. Most noteworthy contributions that have impacted the drug discovery are Song and coworkers' 2,2,2,6,6,6-hexamethyl-3,5-heptadione (TMHD) ligand [134], Ma and Cai's DMG ligand [135], Buchwald and coworkers' phenanthrolines ligands for arylation of alcohols [136], Beller and coworkers' NMI ligand [137], and Cristau et al.'s Chxn-Py-All ligand (Fig. 3.3) [138].

In 1996, Buchwald and coworkers [139] and Mann and Hartwig [140] reported Pd-catalyzed C–O couplings. Continued development led to more general arylation of phenols, alcohols, and even water [141] using ligands such as DTPB [142], *t*Bu-XPhos and Me$_4$-*t*Bu-XPhos [143], P(*t*Bu)$_3$ [144], and Q-Phos [16]. Unfortunately, some of the other useful ligands developed by Buchwald are not readily available due to

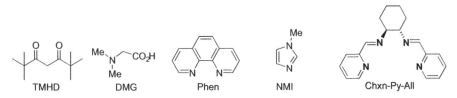

FIGURE 3.3   Commonly used ligands for Cu-catalyzed C–O couplings.

difficult syntheses. Beller also developed 2-di*t*Bu- or 2-diadmantylphosphino-*N*-arylpyrrole and 2-diadmantylphosphino-*N*-arylindoles as ligands for Pd-catalyzed C–O couplings [145].

3.2.3.2 C–O Coupling in the Synthesis of Approved Drugs and Clinical Candidates At Pfizer, tetrasubstituted pyridine **181** was identified as a corticotropin-releasing factor (CRF) antagonist for the potential treatment of anxiety and depression. For its large-scale preparation [146], a copper-mediated C–O coupling for diaryl ether synthesis was developed to avoid the use of the *N*-oxide of **175** that was used in the discovery route, but has significant process safety concerns (Scheme 3.58). The ethyl ester analog **176** was used as a model substrate to study the coupling with the phenol. Without the copper catalyst, direct S_NAr reaction between the 2,4-dichloropyridine **176** and the phenol **177** with Cs_2CO_3 in NMP at 95°C resulted in the predominate 4-substitution with an 8/1 ratio of **180/179**. In the presence of CuBr•Me$_2$S, only desired 2-substitution was observed, albeit in a poor 29% yield of **179**. Similar reactivities were seen with the reaction of 2,4-dichloro-3,6-dimethyl-pyridine and mesitol [147]. The low yield was traced to the instability of the phenol **177** under the reaction conditions, so the mesylate **178** was used as a masked phenol, which provided a higher yield of the desired product. On the lab scale, the copper salts were removed by filtration after treating the crude reaction mixture with aqueous sodium thiosulfate. However, on the large scale, the filtration became too lengthy and difficult that resulted in some hydrolysis of the ester group. Instead, the reaction was quenched with an acidic resin, Amberlite 15, followed by filtration on a silica gel pad to rapidly remove the copper salts. Final displacement of 4-chloro group with the chiral amine gave the target compound **181**.

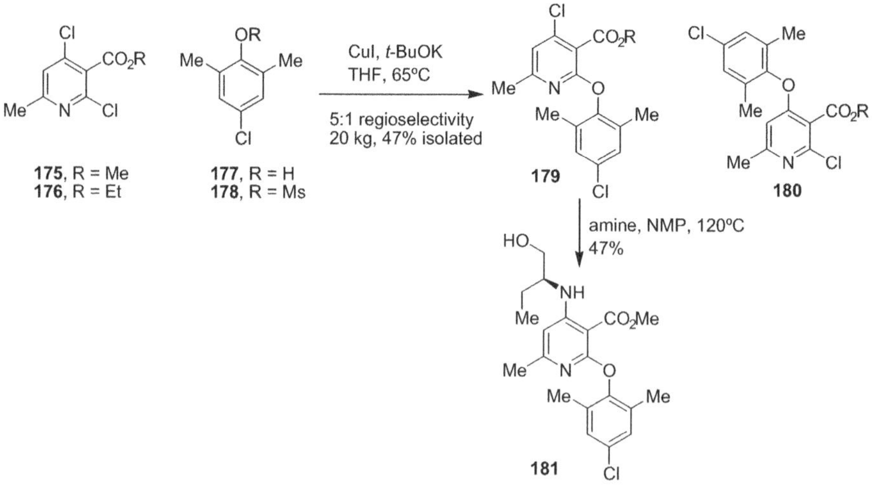

SCHEME 3.58 Cu-catalyzed diaryl ether formation in the synthesis of a CRF antagonist.

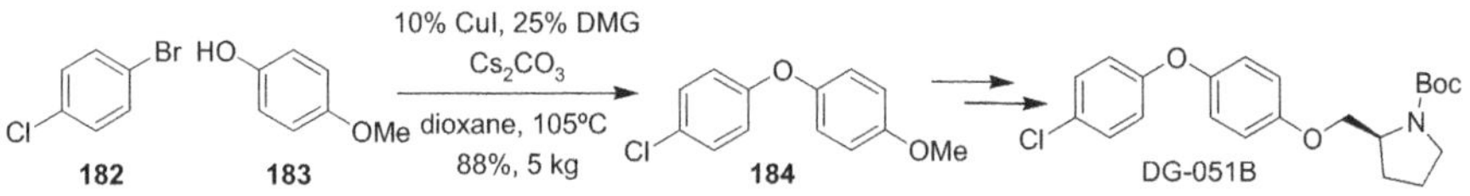

SCHEME 3.59   Cu–DMG-catalyzed diarylation formation in the synthesis of DG-051B.

DG-051B is a first-in-class small-molecule leukotriene A4 hydrolase (LTA4H) inhibitor that is in phase II clinical studies for the treatment of myocardial infarction (MI) and stroke. A scalable synthesis was developed by deCODE Chemistry (Scheme 3.59) [148]. The advanced intermediate diaryl ether **184** was prepared by the CuI–DMG [135]-catalyzed coupling between the 4-chloro-bromobenzene and 4-methoxyphenol in high yield. The same method was also used in the discovery of the molecule [149].

GSK376501A is a selective peroxisome proliferator-activated receptor gamma modulator (SPPARMγ) for the treatment of type 2 diabetes mellitus [150]. Two different C–O bond-forming strategies were considered (Scheme 3.60). With the difluoro compound **185**, direct S_NAr reaction gave the desired product in 83% yield. If the dibromo compound **186** was used, a copper-catalyzed C–O coupling was required. Extensive optimization led to the use of 0.1 equiv of CuI in diglyme–DMF mixed solvents; in this particular solvents system, it was critical to use sodium *tert*-pentoxide instead of sodium *tert*-butoxide to suppress the dehalogenation impurity. In the reaction workup, >95% of the copper was removed via filtration, and crystallization of the product gave <1 ppm of copper. However, some variability concerns of the copper-catalyzed process led to the selection of the direct S_NAr with difluoro compound as their long-term route.

Beller and coworkers developed a copper-catalyzed synthesis of the core structure of insecticide diafenthiuron using *N*-methylimidazole (NMI) as the ligand (Scheme 3.61) [137]. The coupling between the bromoaniline **187** and phenol using just 5% CuCl and 50% of NMI gave the desired diaryl ether **188** in almost quantitative yield on 30 g scale. No protection of the primary aniline was needed. The method was also applicable to a number of other phenols, even including hindered ones such as 2,4-di-*tert*-butylphenol (80% yield) and 2,6-dimethylphenol (55% yield).

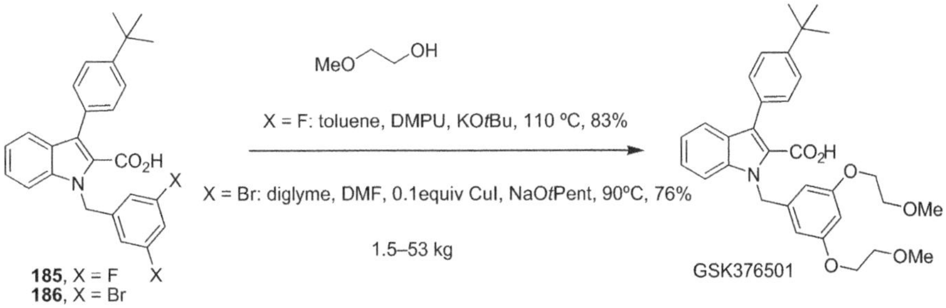

SCHEME 3.60   Cu-catalyzed aryl–alkyl ether formation in the synthesis of GSK376501.

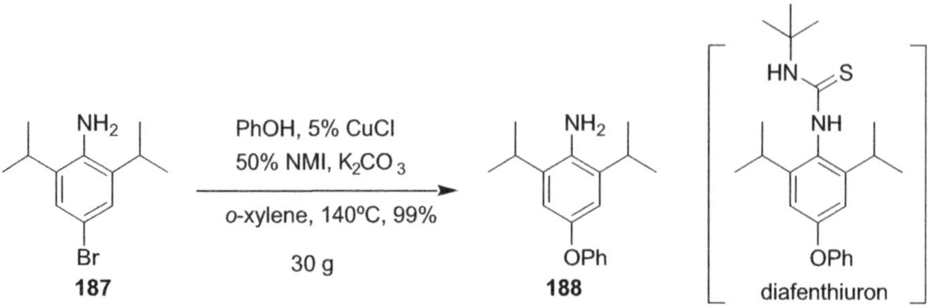

SCHEME 3.61   Cu–NMI-catalyzed diaryl ether formation in the synthesis of diafenthiuron.

Thatcher and coworkers used copper-catalyzed methoxylation of the aryl bromide **189** to synthesize arzoxifene (Scheme 3.62) [127]. A 85% yield was achieved in MeOH–DMF without an external ligand.

3.2.3.3   *C–O Coupling in Drug Discovery*   To synthesize the diaryl ether core of PF-00277343, a thyroid receptor agonist for the treatment of androgenetic alopecia [107], Pfizer chemists used a CuI-TMHD [134]-catalyzed C–O coupling between the phenol **190** and aryl bromide **191** in high yield (Scheme 3.63). A copper-catalyzed C–N coupling was also used in the preparation of the key intermediate **190** (Scheme 3.43).

In Wyeth's LXR program [108], the phenol **193** was coupled with the aryl halide **194** to synthesize **195** for SAR (Scheme 3.64). When the halide was fluoride or

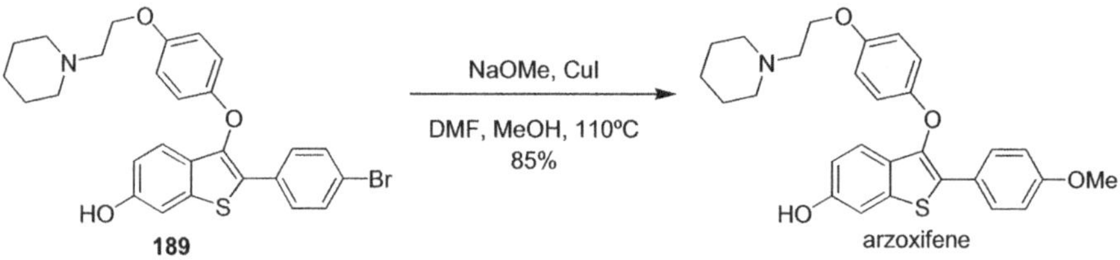

SCHEME 3.62   Cu-catalyzed arylmethyl ether in a synthesis of arzoxifene.

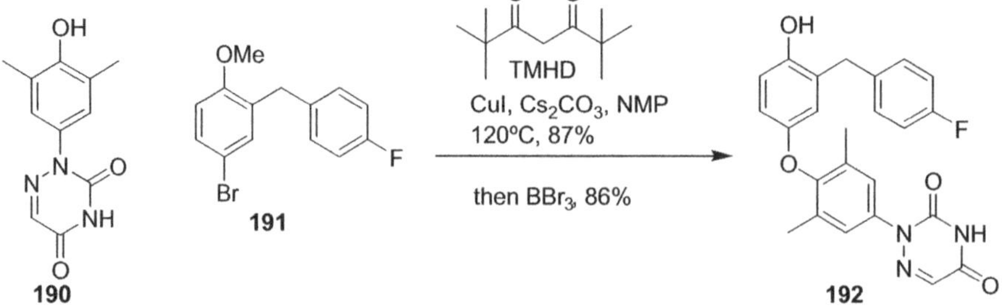

SCHEME 3.63   Cu–TMHD-catalyzed diaryl ether formation in the synthesis of PF-00277343.

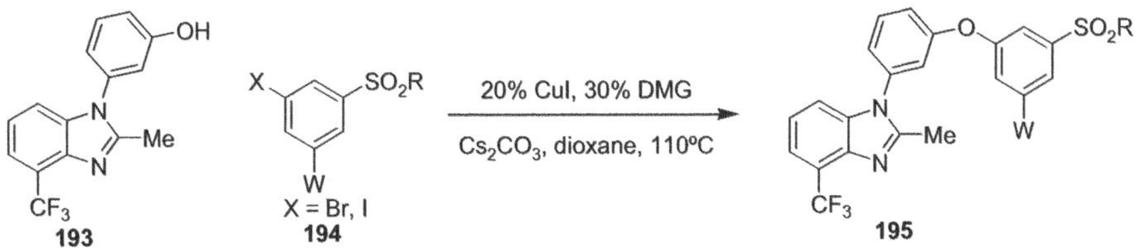

SCHEME 3.64 Cu–DMG-catalyzed diaryl ether formation in the synthesis of LXR agonists.

chloride, direct S_NAr reaction was sufficient. However, when the halide was bromide or iodide, a CuI–DMG [135]-catalyzed C–O coupling was required for the diaryl ether formation.

At Kalypsys, a series of primary alcohols were coupled with the aryl iodide **196** for the SAR study of mercaptoketone-based histone deacetylase inhibitors (Scheme 3.65). Reasonable yields were obtained using the CuI-1,10-phenanthroline system [136]a and 20 equiv of the alcohol [151]. This led to the discovery of KD5170 as a clinical candidate. It is noteworthy that Buchwald also developed CuI-3,4,7,8-tetramethyl-1,10-phenanthroline (Me_4-Phen) catalyst system to allow the use of only a slight excess of the alcohol reagent [136]b.

For the VEGF receptor inhibitors program at Amgen [152], a Pd-catalyzed C–O coupling was used for the synthesis of **200** (Scheme 3.66). Without the *ortho*-fluoride on the chloroquinoline, most of the compounds were synthesized by S_NAr reaction using Cs_2CO_3 in DMSO at 140°C, but the fluoride on **198** presented a problem with

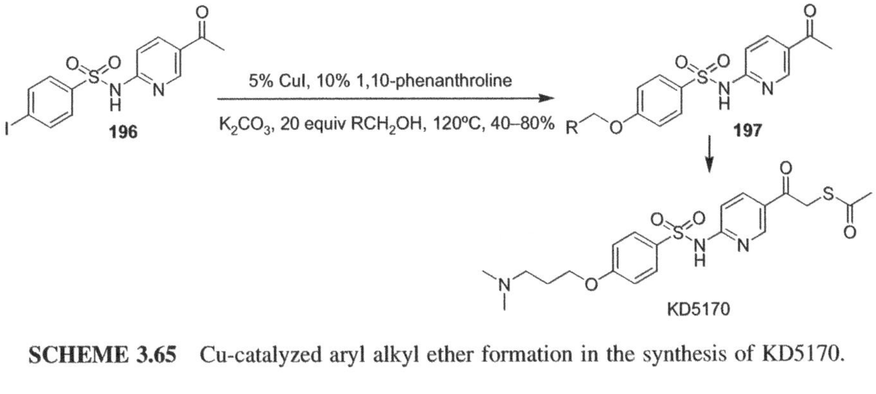

SCHEME 3.65 Cu-catalyzed aryl alkyl ether formation in the synthesis of KD5170.

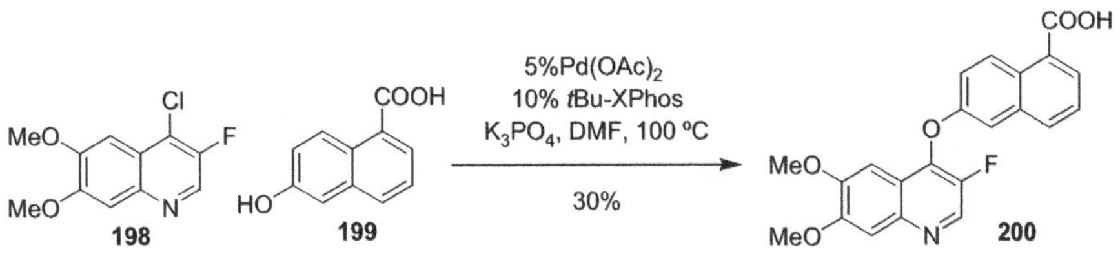

SCHEME 3.66 Pd-catalyzed C–O coupling in the synthesis of a VEGF receptor inhibitor.

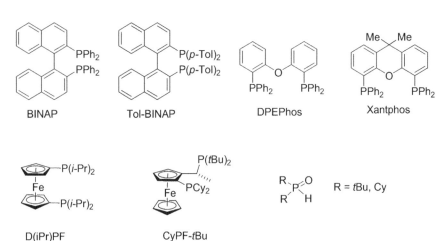

SCHEME 3.67   Pd-catalyzed aryl alkyl ether formation in the synthesis of a PPAR agonist.

this approach due to the competing displacement of the fluoride. The Pd-*t*Bu-XPhos-catalyzed diaryl ether formation under milder conditions [143] was applied to couple the chloride in the presence of the fluoride to access the desired compound [153].

A Pfizer group used the Pd-catalyzed C–O coupling for the synthesis of compound **303**, a peroxisome proliferator-activated receptor (PPAR) agonist (Scheme 3.67) [154]. Racemic 2-(di-*tert*-butylphosphino)-1,1′-binaphthyl [155] was used as the ligand for this coupling in high yield. Use of a weak base Cs_2CO_3 allowed the tolerance of functional group COOMe in the presence of a primary alcohol. The same conditions had been previously used for analog synthesis [156].

3.2.4   Pd- or Cu-Catalyzed C–S Coupling

3.2.4.1   Introduction   First reported by Migita and coworkers using PPh_3 as the ligand [157], the Pd-catalyzed carbon–sulfur bond formation has been revisited since 1998 due to the discovery of various more reactive ligands for carbon–heteroatom cross-couplings [1]. Due to the strong binding of sulfur atom to the Pd, most of the newer catalyst systems consist of a chelating ligand such as BINAP or Tol-BINAP used by Merck [158], DPEPhos by Schopfer [159], Xantphos by Merck [160], D(*i*Pr)PF by Murata and Buchwald [161], and CyPF-*t*Bu by Hartwig and coworkers [162]. Highly active dialkylphosphine oxide ligands were also used by Li et al. (Fig. 3.4) [163].

FIGURE 3.4   Commonly used ligands for Pd-catalyzed C–S couplings.

The Cu-catalyzed C–S coupling was also improved recently due to the use of supporting ligands to allow broader substrate scope.[2c–e] The milder conditions include the use of phosphazene as base by Palomo et al. [164], neocuproine (2,9-dimethyl-1,10-phenanthroline) as ligand by Venkataraman and coworkers [165], and ethylene glycol as an additive by Kwong and Buchwald [166].

3.2.4.2 C–S Coupling in the Synthesis of Approved Drugs and Clinical Candidates

Thymitaq (Nolatrexed) is a noncompetitive, high-affinity antifolate thymidylate synthetase inhibitor for the treatment of inoperable primary liver cancer. For the large-scale manufacturing of this compound at Dupont, a copper-mediated C–S coupling was used in the final step (Scheme 3.68) [167]. This did provide the API in high yield, but also presented a serious problem for heavy metal removal due to the use of stoichiometric amounts of copper and high affinity of the API itself to copper. Copper(I) oxide disproportionated to copper(0) and copper(II) when treated with HCl. Most of the copper(0) and copper (I) could be removed by filtration. Treating the product in the filtrate with ammonium hydroxide precipitated out the product, leaving most of the Cu(II) in solution. However, repeating this sequence still left traces of copper in the API. To achieve very low levels of copper in the API, 2,4,6-trimercapto-s-triazine (TMT) was introduced to successfully remove the residual copper.

Very recently, Xue and coworkers developed another synthesis of thymitaq featuring the same end game C–S coupling (Scheme 3.69) [168]. The charge of the copper catalyst was significantly reduced to only 2.5% of CuBr and 2.5% of Cu_2O, and a weaker base K_2CO_3 replaced NaOH. A much higher temperature was required. Addition of saturated aqueous Na_2S to the crude product in EtOH at pH 6–7 was very efficient to remove the copper salts. The heavy metals including copper after isolating the thymitaq dihydrochloride salt were less than 20 ppm.

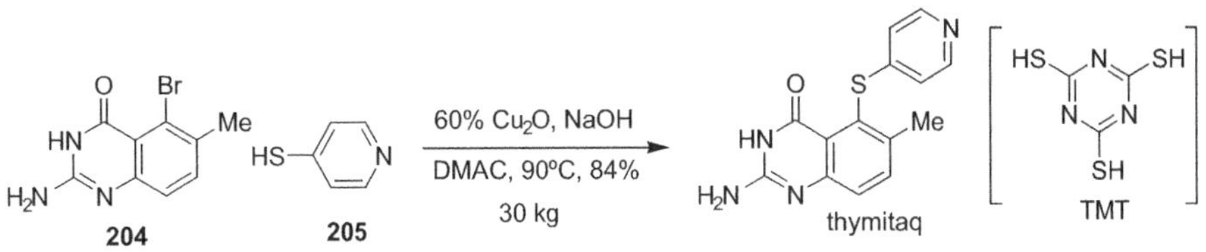

SCHEME 3.68 Cu-catalyzed C–S coupling in the synthesis of thymitaq.

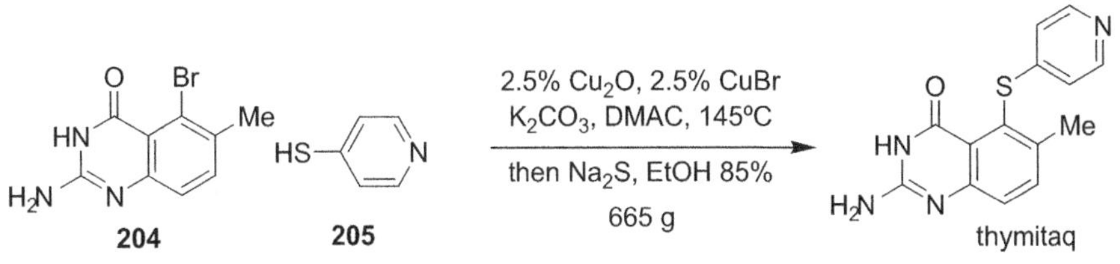

SCHEME 3.69 Another Cu-catalyzed C–S coupling in the synthesis of thymitaq.

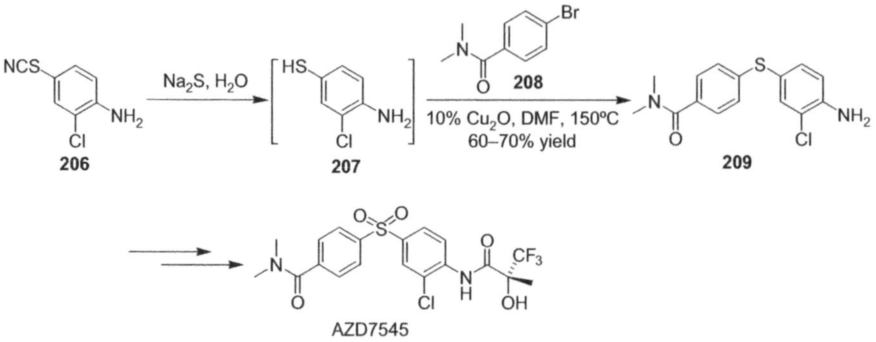

SCHEME 3.70   Cu-catalyzed C–S coupling in the synthesis of AZD7545.

AZD7545 was a pyruvate dehydrogenase kinase (PDK) inhibitor for an oral treatment of type II diabetes. To evaluate many potential routes as the long-term manufacturing route of this compound, the AstraZeneca group used the Kepner–Tregoe decision analysis (KTDA) tool [169] as an aid. Many of the promising routes featured the copper-catalyzed C–S coupling reaction [170]. The thiocyanate **206** was reduced *in situ* with Na$_2$S followed by coupling with the bromide **208** with 0.1 equiv of Cu$_2$O as the catalyst without a supporting ligand (Scheme 3.70). The thioether **209** was obtained in moderate yield and then readily converted to AZD7545 in a few more steps.

AstraZeneca's ZD4407 is a 5-lipoxygenase inhibitor for the potential treatment of a variety of inflammatory conditions [171]. To develop a practical route for the early deliveries, the Pd-catalyzed C–S coupling between bromooxindole **210** and thiol **211** was used in the final step (Scheme 3.71). The original Migita and coworkers' Pd(PPh$_3$)$_4$ catalyst [157] was used and only 41% yield was achieved after optimization of the reaction time, temperature, and molar ratio. The acidic proton on the oxindole **210** was believed to cause the low yield. A major by-product is the dimerization of the aryl bromide via homocoupling. When the Pd catalyst was replaced with a Ni catalyst, a similar yield was obtained with the homocoupling of the thiol as the major by-product. The use of trialkylstannyl thioether gave a good yield, but removal of toxic tin was too much of a problem.

At Pfizer, a practical and efficient synthesis of the antiasthma candidate **215** was required to provide clinical material for phase II studies [172]. The carbon–sulfur bond was installed by Pd-catalyzed coupling between aryl bromide **212** and thiol **213**

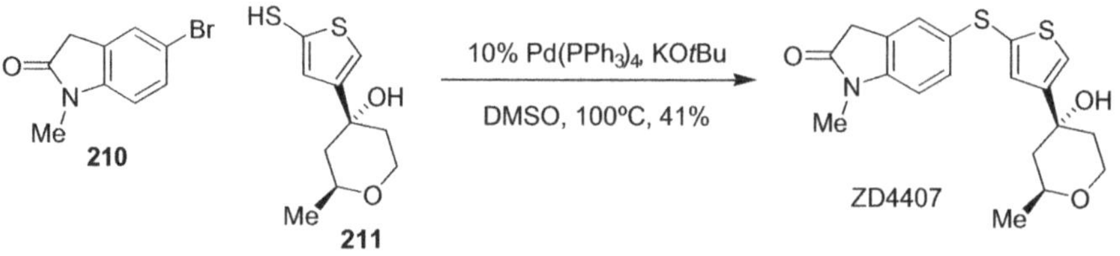

SCHEME 3.71   Pd-catalyzed C–S coupling in the synthesis of ZD4407.

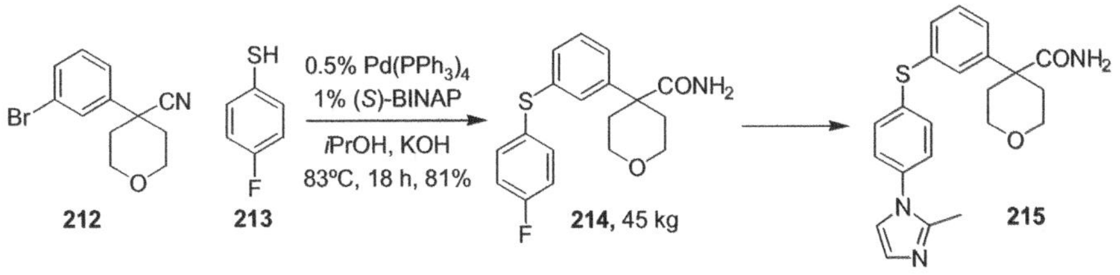

SCHEME 3.72 Pd-catalyzed C–S coupling in the synthesis of an antiasthma candidate.

with *in situ* hydrolysis of the nitrile group to the primary amide **214** (Scheme 3.72). The Pd(PPh$_3$)$_4$ alone could also catalyze the reaction, but the yield was significantly improved with the additional (*S*)-BINAP. The racemic BINAP did not work as well, possibly due to its lower solubility in the reaction system.

Axitinib (AG13736) is a vascular endothelial growth factor (VEGF) inhibitor that is being studied by Pfizer as an oncology drug. In the initial synthetic route, a Pd-catalyzed C–S bond formation between the aryl iodide **216** and the thiol **217** was carried out to give the penultimate **218**, which in turn gave the API upon deprotection (Scheme 3.73). DPPF was used as the ligand to give a moderate 62% yield. In a new manufacturing route for phase II, the total steps were cut down to three, two of which

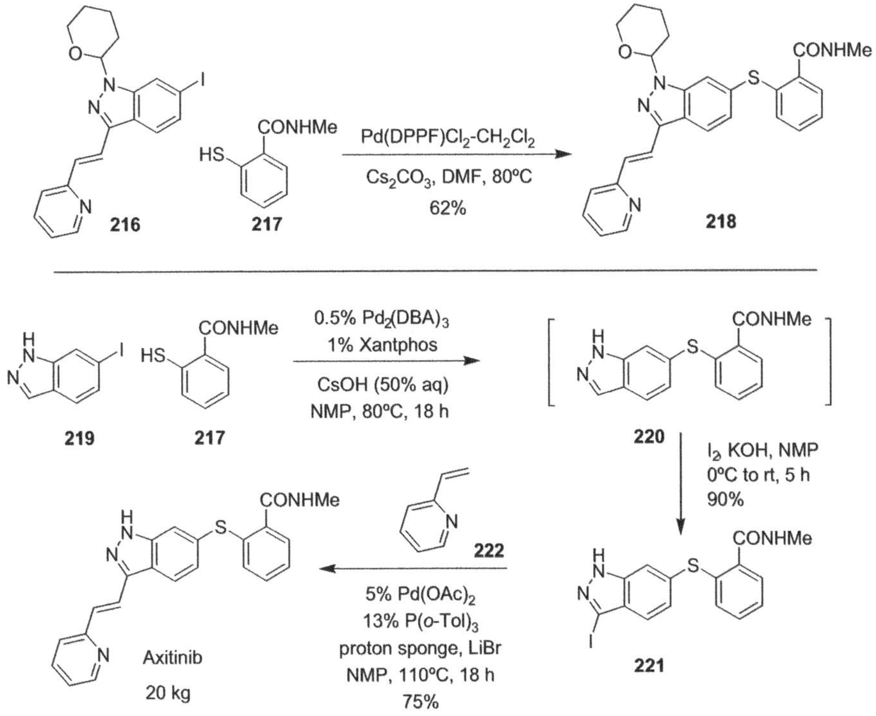

SCHEME 3.73 Pd-catalyzed C–S coupling in the synthesis of axitinib.

were catalyzed by Pd [173]. The C–S coupling between **219** and **217** was catalyzed by 1% of Pd and 1% of Xantphos. Aqueous CsOH was used as the base to give greater than 90% yield of the intermediate **220**, which was carried directly to iodination to give **221** in a one-pot process. The final Heck reaction with 2-vinylpyridine (**222**) gave the crude API in 75% yield. Given the two Pd-catalyzed steps in a three-step synthesis, the high Pd loading (5%) in the final step, and the API's tendency to chelate to Pd, the Pd removal presented a great challenge. After extensive screening of Pd scavengers, the authors found the combination of 1,2-diaminopropane and DPPE helped to reduce the Pd level to <20 ppm. They reasoned that the diamine helped to remove cationic Pd^{2+} species and the diphosphine the neutral Pd^0 species. However, only $\sim70\%$ recovery of the API was achieved in the Pd removal–recrystallization process.

3.2.4.3   *C–S Coupling in Drug Discovery*   A Merck group also used the copper–ethylene glycol catalyst system [166] for the diaryl sulfide synthesis in their 5-HT$_{2A}$ program (Scheme 3.74) [174]. Interestingly, to selectively functionalize the 2-chloride in the presence of 5-iodide of the pyridine **223**, direct S$_N$Ar reaction was carried out and the 5-iodide of the same pyridine could be selectively coupled under the copper-catalyzed conditions. This is yet another example where a catalyst was used to switch the selectivity of various halides on the same aromatic ring.

Johnson & Johnson applied both Pd- and Cu-catalyzed C–S couplings in their synthesis of pyrazole-based thioethers as cathepsin S inhibitors (Scheme 3.75) [175]. In most cases, Pd-DPPF catalyst was used for the coupling, but CuI–neocuproine was used for the specific case of thioalcohol.

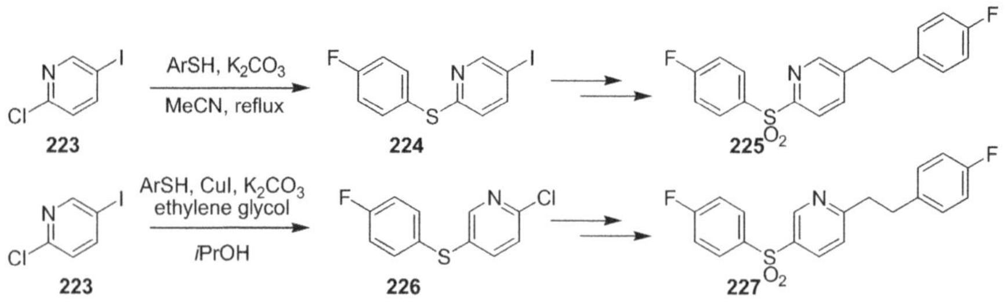

SCHEME 3.74   Cu-catalyzed versus thermal C–S coupling on a dihalopyridine.

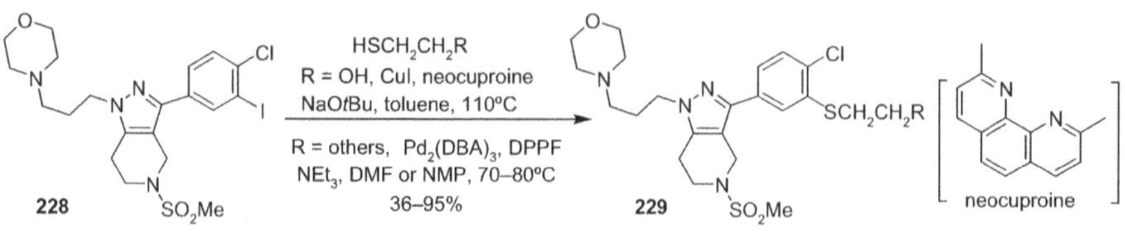

SCHEME 3.75   Pd- and Cu-catalyzed C–S couplings in the synthesis of cathepsin S inhibitors.

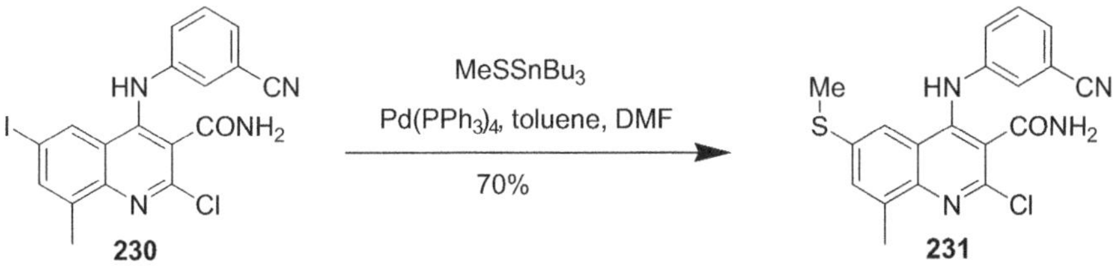

SCHEME 3.76 Pd-catalyzed C–S coupling using (methylthio)tributylstannane.

Nonnucleophilic (methylthio)tributylstannane reagent [176] has also been used for the C–S coupling. For selective introduction at the iodo-position of **230** in the presence of a highly activated 2-chloropyridine moiety, this protocol was used to provide product **231** in good yield (Scheme 3.76). Subsequent oxidation and hydrolysis of the chloroquinoline provided an oral PDE4 inhibitor by GSK [177]. Other alkylthiols such as ethyl and *t*-butyl were also used.

3.2.5 Pd- or Cu-Catalyzed C–SO₂R Coupling

Instead of C–S coupling followed by the oxidation to sulfones as described in many examples above, Pd- [178] or Cu-catalyzed [179] coupling reaction of aryl halides with sulfinic acid salts can give direct access to sulfones that are widely present in the bioactive molecules.

MK-0524 (laropiprant) is a potent and selective prostaglandin D_2 (PGD_2) receptor antagonist that reduces flushing induced by niacin, a cholesterol-lowering drug. Merck is studying the combination of MK-0524 and niacin to reduce LDL. The methylsulfone group was installed via the Cu-catalyzed coupling of sodium methylsulfinate and the aryl bromide **232** in good yield (Scheme 3.77). Hydrolysis gave the API as the acid [180]. This was also used in the additional analog synthesis [181]. The Merck Process group developed an early asymmetric route that used this coupling on a similar substrate [182].

BMS used the Cu–DMEDA system [179]b for the sulfone formation in the BACE program (Scheme 3.78) [183]. The diamine ligand was used in combination with a more soluble copper source, (CuOTf)₂PhH.

Thatcher and coworkers used the copper-catalyzed sulfonylation of the aryl bromide to synthesize arzoxifene, a selective estrogen receptor modulator that was in phase III clinical studies for postmenopausal osteoporosis (Scheme 3.79) [127].

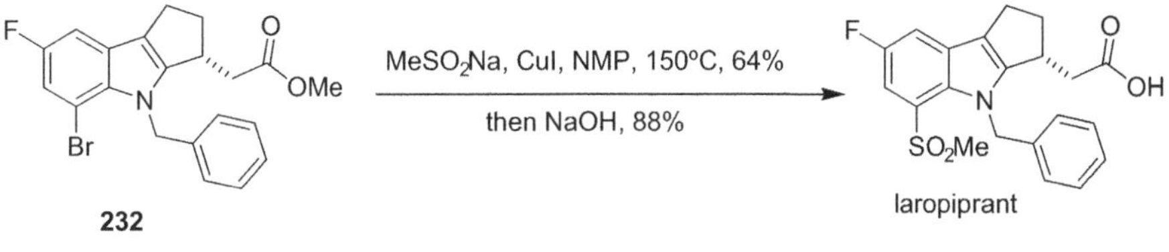

SCHEME 3.77 Cu-catalyzed C-SO₂Me coupling in the synthesis of laropiprant.

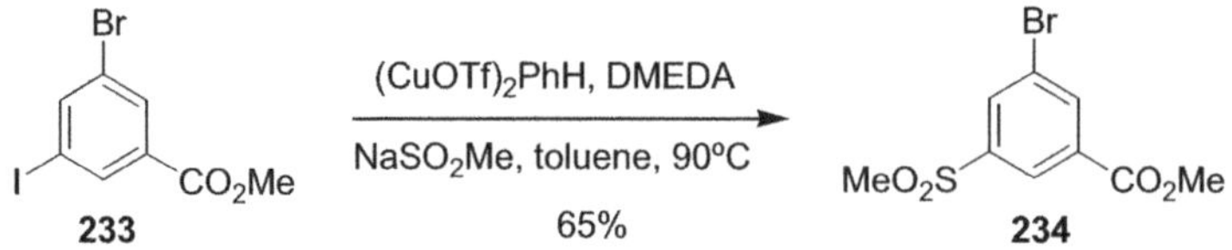

233 234

SCHEME 3.78 Cu–DMEDA-catalyzed sulfone formation in the synthesis of BACE inhibitors.

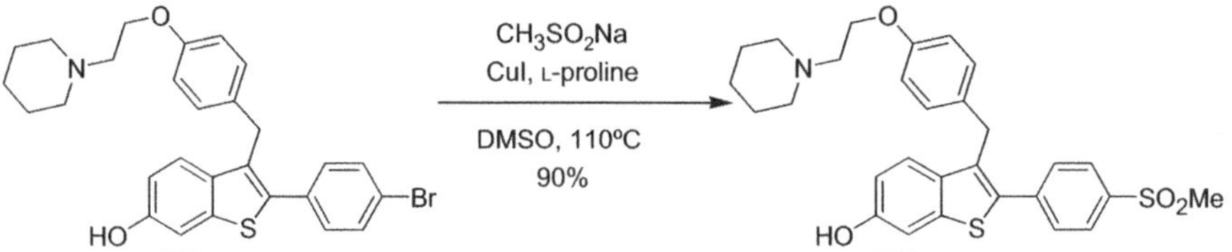

235 236

SCHEME 3.79 Cu–proline-catalyzed sulfone formation in the synthesis of arzoxifene.

Ma's CuI–proline catalyst system [179]c was used to couple bromide **235** with sodium methylsufinate in the presence of the free phenol moiety in 90% yield.

In GSK's PDE4 inhibitor synthesis [177], the CuI–DACH catalyst was used for the introduction of the methylsufone group directly as an alternative approach (Scheme 3.80) to C–S coupling followed by oxidation (Scheme 3.76).

Reynolds and coworkers used the Pd-Xantphos-catalyzed C–SO_2R coupling [178] in an FabH inhibitor analog synthesis (Scheme 3.81) [184].

3.2.6 Cu-Mediated Oxidative Coupling of Boronic Acids and Heteroatoms

Chan, Evans, and Lam first reported the $Cu(OAc)_2$-mediated oxidative couplings between amines and phenols with aryl boronic acids to form C–N and C–O bonds as

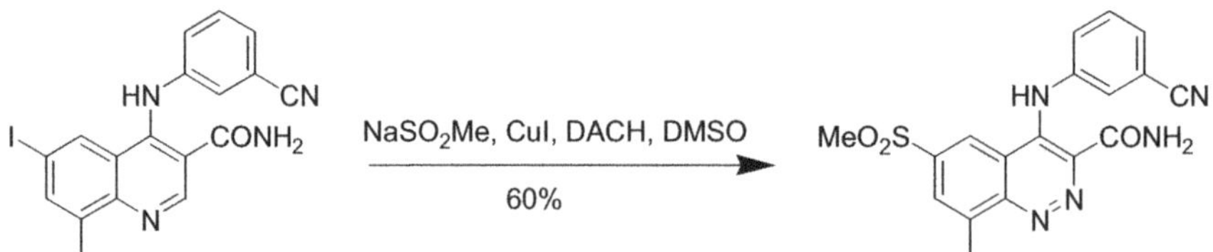

237 238

SCHEME 3.80 Cu–DACH-catalyzed sulfone formation in the synthesis of PDE4 inhibitors.

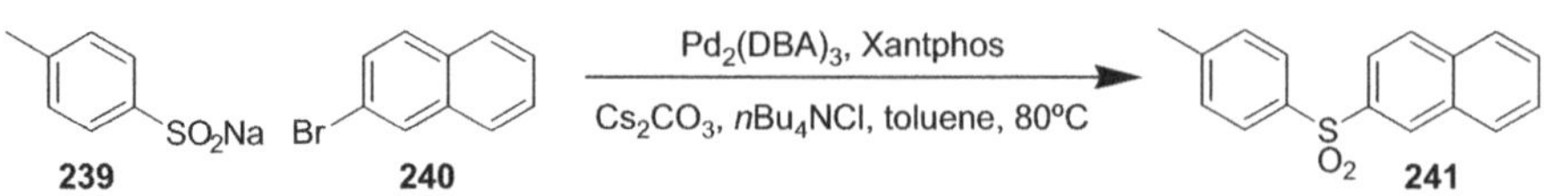

239 240 241

SCHEME 3.81 Pd-Xantphos-catalyzed sulfone formation.

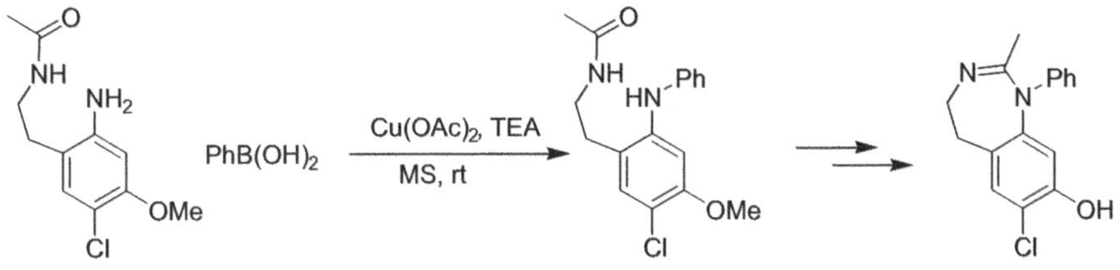

SCHEME 3.82 Cu(II)-mediated oxidative coupling in the synthesis of dopamine antagonists.

an alternative to the Pd- or Cu-catalyzed reactions using aryl halides/sulfonates [185]. Catalytic versions were also reported since then [186,187]. These oxidative couplings are not ideal for large-scale syntheses, but the relatively mild and simple conditions without the need for external ligands and reasonable generality, especially for hindered substrates, made them very attractive for early drug discovery applications.

At Shering Plough, the Cu(II)-mediated oxidative coupling was used to make aniline derivatives for SAR in the dopamine antagonist program (Scheme 3.82) [188]. The primary aniline **242** was coupled with phenylboronic acid in the presence of an amide and an aryl chloride functional group.

In Wyeth's synthesis of ADAMTS-5 inhibitors such as **247**, the copper-mediated oxidative coupling was used to introduce the aryl group to the highly electron-deficient nitrogen of **245** (Scheme 3.83) [189]. This coupling would have been quite challenging with an aryl halide.

The Cu(II)-catalyzed oxidative coupling with catalytic amount of myristic acid was developed by Antilla and Buchwald [186]d. BMS chemists used this method for the arylation of the pyridine **248** bearing an iodide group, which would have made the typical C–N coupling with aryl halide much more difficult (Scheme 3.84).

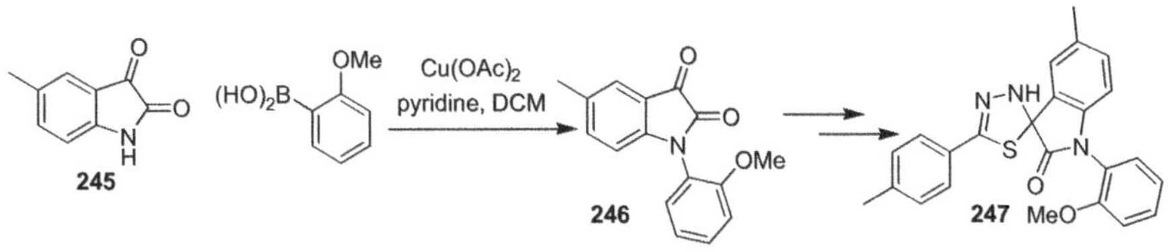

SCHEME 3.83 Oxidative coupling in the synthesis of ADAMTS-5 inhibitors.

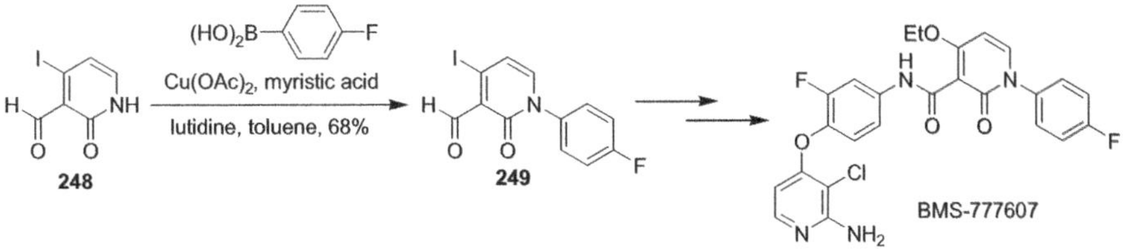

SCHEME 3.84 Oxidative coupling with a pyridone in the synthesis of BMS-777607.

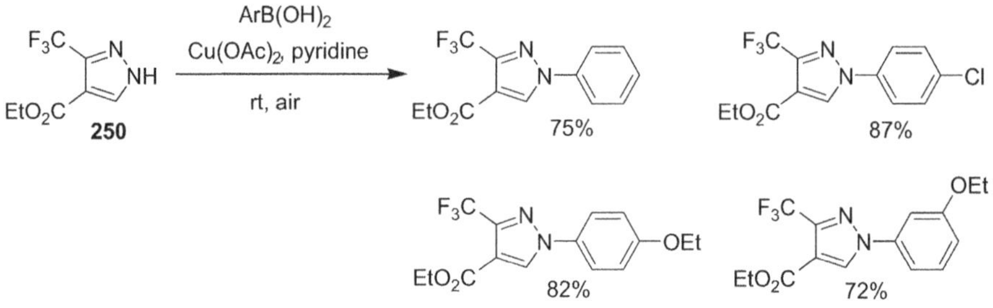

SCHEME 3.85   Oxidative coupling with a pyrazole in the synthesis of DGAT inhibitors.

The aldehyde group also survived the oxidative conditions. The product **249** was converted to BMS-777607, a selective Met Kinase inhibitor that has been in phase 1 clinical trials [190].

In Takeda's synthesis of diacylglycerol acyltransferase (DGAT) inhibitors [191], the Cu(II)-mediated C–N coupling was used (Scheme 3.85). Very good yields were obtained at room temperature. In this case, the CF_3 group was possibly directing the regioselectivity.

BMS synthesized various azetidinone acids as dual PPARα/γ agonists [192]. Aryl groups were attached to the β-lactam via the boronic acids in almost quantitative yields (Scheme 3.86). Interestingly, the corresponding dimethylacetal (in place of methyl ester group) analog gave poor yield in the same reaction.

One of the advantages of the oxidative coupling is the ability to tolerate hindered substrates. For example, the 2,6-diiodo-phenol **254** was coupled with the boronic acid **253** in good yield under mild conditions (Scheme 3.87). Phenols thus hindered have not been used in the metal-catalyzed C–O coupling of aryl halides. The oxidative coupling is favored over the homocoupling between the phenol and the iodide. Product **255** was used in Metabasis Therapeutics' synthesis of thyroid hormone receptor agonists [193].

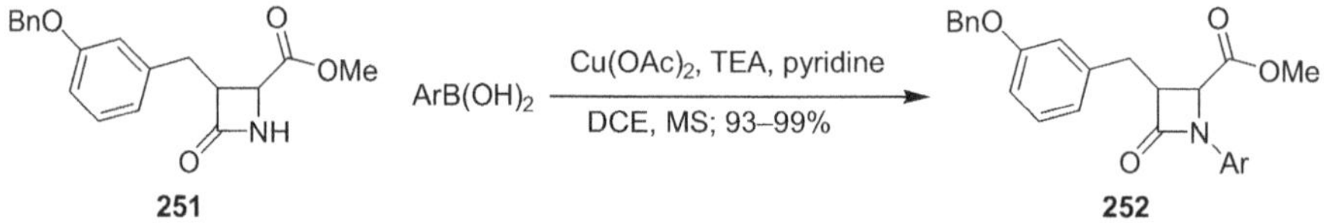

SCHEME 3.86   Oxidative coupling with a lactam in the synthesis of PPAR agonists.

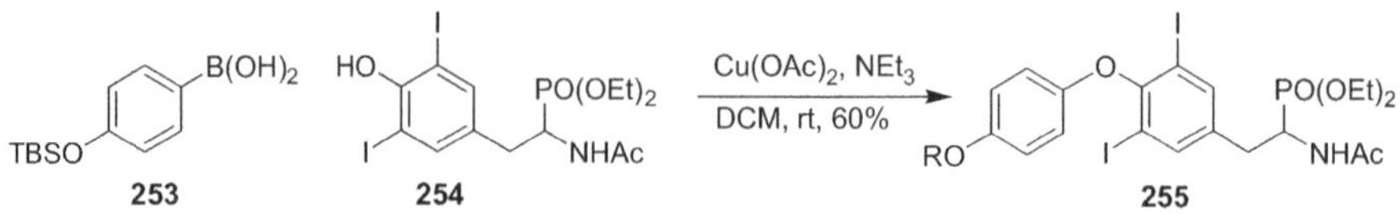

SCHEME 3.87   Oxidative coupling of a 2,6-diiodophenol.

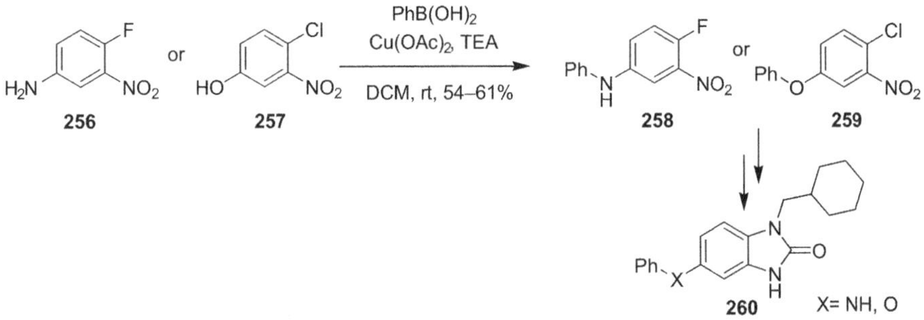

SCHEME 3.88 Oxidative couplings in the synthesis of MAP kinase inhibitors.

Boehringer Ingelheim synthesized benzimidazolones **260** as p38 MAP kinase inhibitors [194]. The Cu(OAc)$_2$-mediated oxidative couplings of both the aniline **256** and the phenol **257** with phenyl boronic acid were carried out under the same conditions in good yields (Scheme 3.88), demonstrating the versatility and generality of this protocol, especially for early discovery chemistry.

Most of the above C–O oxidative couplings used phenols and boronic acids. Quach and Batey found that potassium aryl trifluoroborates and aliphatic alcohols can be coupled with a Cu(II) catalyst and air as the oxidant (Scheme 3.89) [195]. Alkenyl trifluoroborates and phenols also worked.

In addition to *N*- and *O*-arylations mediated by Cu(II), the C–S coupling between thiols and aryl boronic acids promoted by Cu(OAc)$_2$-1,10-phenanthroline was first reported by Guy and coworkers [196]. Using a modified procedure [197], Botta and coworkers synthesized a series of *S*-aryl-*S*-DABO's **264** as HIV-1 inhibitors (Scheme 3.90) [198]. The use of microwave heating allowed very short reaction times; the reactions did work at room temperature albeit with longer times and in lower yields.

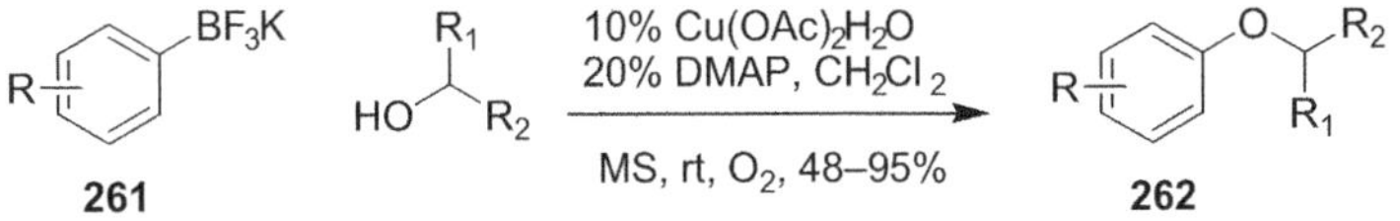

SCHEME 3.89 Oxidative coupling with aryl trifluoroborates.

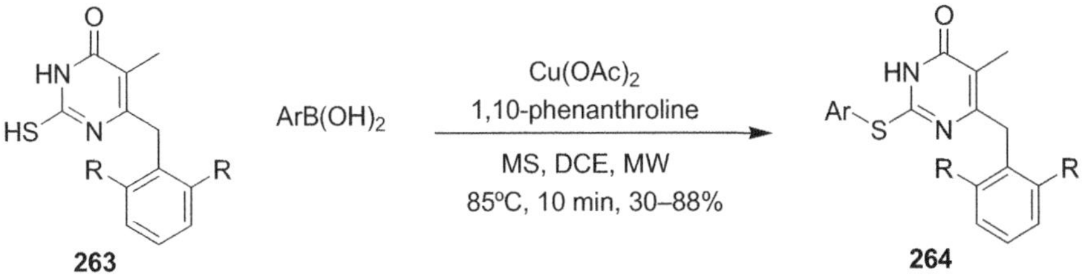

SCHEME 3.90 Oxidative C–S coupling.

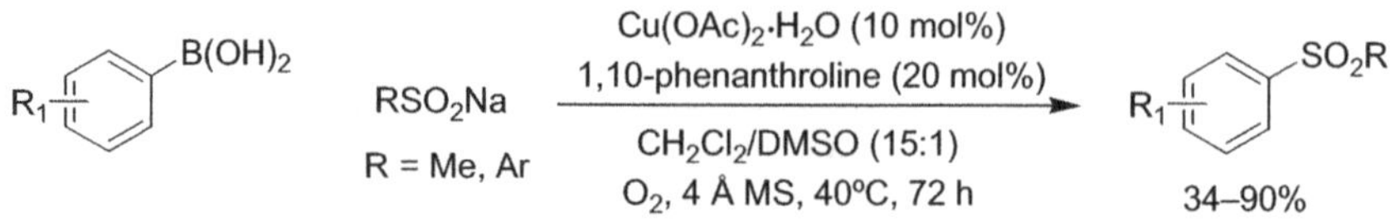

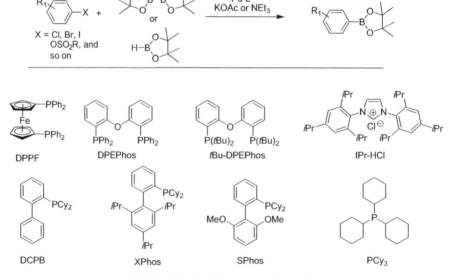

SCHEME 3.91 Oxidative coupling with sulfinates.

Cu(OAc)$_2$–phenanthroline-catalyzed oxidative couplings between aryl boronic acids and sufinates were developed by Huang and Batey (Scheme 3.91) [199]. This could be a very useful method for sulfone formation.

3.2.7 Pd- or Cu-Catalyzed C–B Coupling

3.2.7.1 Introduction The Pd-catalyzed borylation of aryl halides with bis (pinacolato)diboron or pinacolborane is a very useful method to synthesize aryl boronates, which have been widely used for Suzuki couplings in drug discovery and development [200]. First reported by Ishiyama et al., aryl bromides and iodides could be converted to boronate with PdCl$_2$(DPPF) as a catalyst and KOAc as base in DMSO [201]. Many ligands were developed for this reaction, including for borylation with bis(pinacolato)diboron, DPPF [202], PCy$_3$ [203], carbene ligand (IPr-HCl) [204], and XPhos or SPhos [205] and for borylation with pinacolborane, DPPF [206], DPEPhos [207], bis(2-di-*tert*-butyl-phosphinophenyl)ether [208], DCPB [209], and XPhos or SPhos (Scheme 3.92) [210]. Most of the time, KOAc or NEt$_3$ was used as the base, which is important to prevent the product from reacting

SCHEME 3.92 Pd-catalyzed borylation.

with the starting aryl halides/triflate via Suzuki couplings. Many of the reactions can be carried forward to a subsequent *in situ* Suzuki coupling by adding another aryl halide/triflate and a stronger base such as aqueous K_2CO_3. Borylation with pinacolborane sometimes suffers from the dehalogenation side reaction. Zhu and Ma also reported a CuI–proline-catalyzed borylation of aryl iodides by pinacolborane at room temperature [211].

3.2.7.2 C–B Coupling in the Synthesis of Approved Drugs and Clinical Candidates
The Merck process research group applied the Pd-catalyzed borylation in the large-scale preparation of **269**, a potent bradykinin 1 antagonist for the potential treatment of inflammatory pain (Scheme 3.93) [212]. Because of the presence of the NHAc group in the bromopyridine **265**, the typical metalation–borylation protocol would have been difficult. After extensive optimization, a one-pot process was established by borylation with $PdCl_2(DPPF)$ and KOAc in toluene followed by *in situ* Suzuki coupling with bromide **267** with K_3PO_4 as the base. Converting the bromopyridine **265** to trifluoroboronate [213] for Suzuki coupling was also considered, but abandoned due to the liberation of HF that causes the corrosion of the stainless steel and glassware.

The Eli Lilly chemists used the cheaper pinacol borane for the borylation of the iodide **270** (Scheme 3.94) [214]. Subsequent hydrolysis and Suzuki coupling with another aryl iodide **273** provided LY451395, an AMPA potentiator developed for Alzheimer's disease. A key issue of the synthesis is the formation of the boronic acid homocoupling impurity **274**, which differs from the desired API in only one methyl group, so the very tight control of the homocoupling became critical. They authors found the homocoupling was most likely formed when boronic acid **272** was exposed to Pd(II) species either on storage or during the Suzuki reaction. To limit the residual

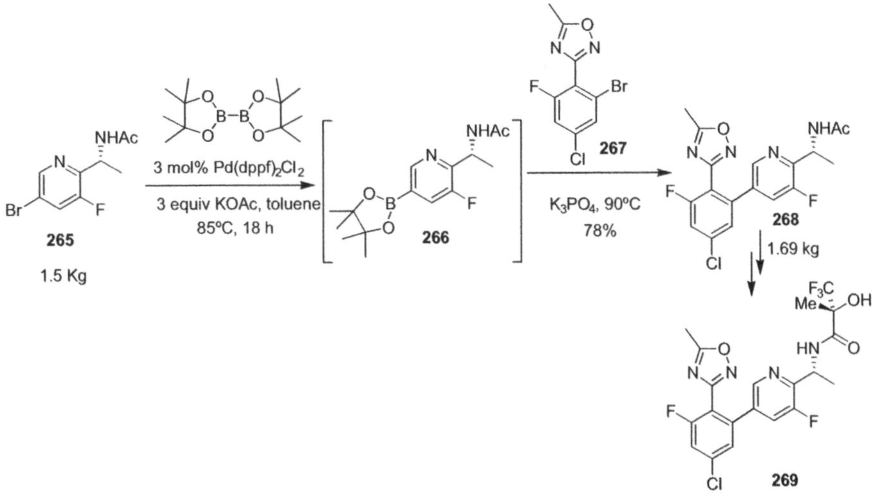

SCHEME 3.93 Borylation in the synthesis of a bradykinin 1 antagonist.

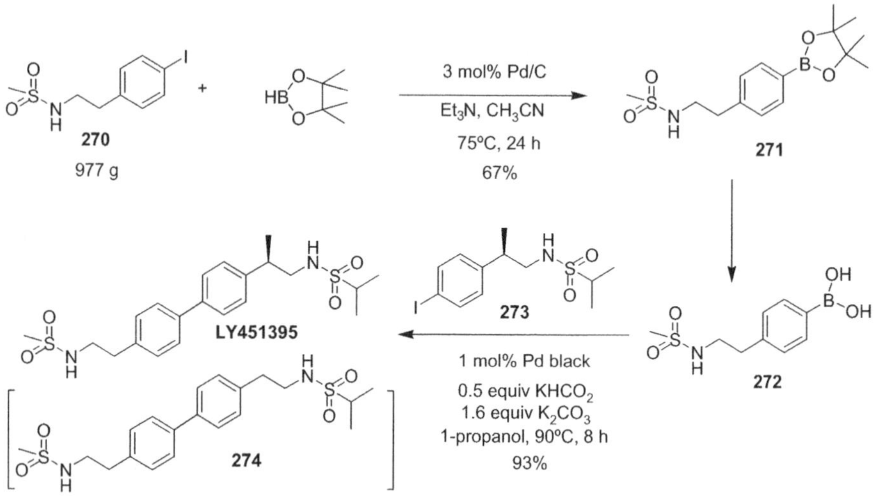

SCHEME 3.94   Borylation in the synthesis of LY451395.

Pd in the boronic acid from the borylation step, heterogeneous Pd on carbon was used for the reaction, which was almost completely removed by a simple filtration through Celite. Rigorous deoxygenation by a subsurface sparge of nitrogen and addition of potassium formate helped suppress the dimer formation during the Suzuki reaction. This was scaled up successfully at kilogram scale.

Lexicon Pharmaceuticals used the borylation of the triflate **275** in a practical synthesis of protected 4-boronophenylalanine (BPA), a very useful building block in drug discovery (Scheme 3.95) [215]. This had been achieved by the C–B coupling using 8% of Pd and the expensive reagent bis(pinacolato)diboron. The authors found that the cheaper pinacolborane and a very low loading of 0.5 mol% of $PdCl_2(DPPF)$ were sufficient for the borylation. The use of a weak base such as N-methylmorpholine (NMM), triethylamine, or Hünig's base was critical to suppress the reduction process. They proposed that a stronger base (DABCO and DMAP) binds to the borane more tightly to facilitate the hydride transfer instead of boron transfer. Commonly used inorganic base, KOAc, and others such as K_2CO_3 and Cs_2CO_3 gave mostly reduction product. Other ligands such as DPPP, DPPB, DPEPhos,

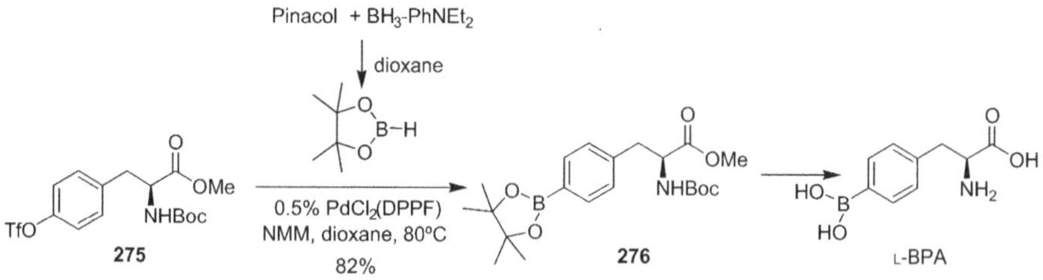

SCHEME 3.95   Borylation in the synthesis of L-BPA.

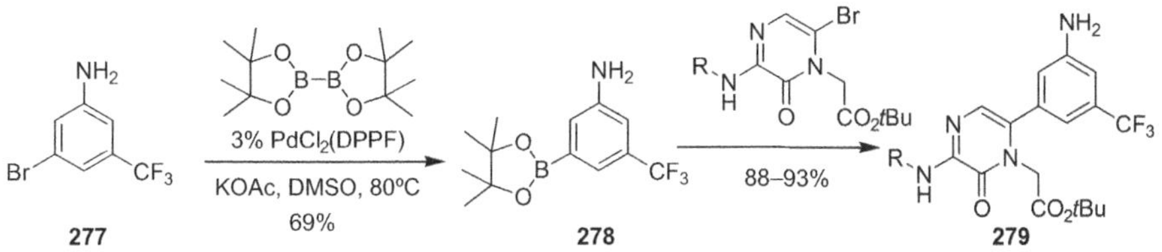

SCHEME 3.96 Borylation in the synthesis of tissue factor–factor VIIa inhibitors.

BINAP, and PCy$_3$ were less effective. To further reduce the cost, the pinacolborane was prepared *in situ* from pinacol and borane–PhNEt$_2$ complex.

3.2.7.3 C–B Coupling in Drug Discovery At Pfizer, the boronate **278** was prepared from the bromoaniline **277** in the tissue factor–factor VIIa program (Scheme 3.96) [216]. A primary aniline was tolerated and the product was used directly in the Suzuki coupling to provide the target **279** for SAR.

PdCl$_2$(DPPF) has been one of the most used Pd–ligand complexes for this reaction due to its generality and ease of use. A few more examples from Pfizer [217], Amgen [218,219], Daiichi Asubio [220], Merck [221], Ligand [222], BMS [223], and Johnson & Johnson [224] have been summarized below (Scheme 3.97). Aryl bromides, iodides, and triflates were all viable substrates.

However, one problem of PdCl$_2$(DPPF) is that it can not catalyze the borylation of unactivated aryl chlorides. In Merck KGaA's synthesis of 2,3-diaryl-5-cyano-4-azaindoles **282** as c-Met inhibitors [225], the aryl chloride **280** was borylated using Buchwald's procedure with XPhos as the ligand [205] in very good yields (Scheme 3.98). Again, the primary aminopyridine was tolerated.

Researchers at NIH used the borylation of halopyridine **283** to prepare heterobicyclic templates for metabotropic glutamate receptor subtype 5 antagonists (Scheme 3.99) [226]. SPhos ligand was used to activate both the aryl chloride and the bromide.

A Merck group converted the unactivated aryl chloride **285** to phenol **287** via the boronate **286** in their synthesis of a cannabinoid-1 receptor (CB1R) positron emission tomography tracer (Scheme 3.100) [227]. The aryl chloride **285** was borylated using PCy$_3$ as the ligand [203] and subsequent oxone oxidation provided the phenol **287** in 95% yield over two steps.

GSK chemists used a slightly different borylating reagent for their synthesis of AKT inhibitors (Scheme 3.101) [228]. The authors found that the borylation of the bromochloropyridine **288** using the most commonly used bis(pinacolato)diboron failed. Borylation using 5,5,5′,5′-tetramethyl-2,2′-bi-1,3,2-dioxaborinane did provide the desired reaction at the bromo position. Further functionalization via two Suzuki reactions followed by deprotection provided the target compound **290**.

3.2.7.4 C–H Activation for C–B Coupling Instead of using an aryl halide for the C–B coupling, C–H activation of an arene followed by C–B bond formation can provide a more direct access to boronic esters [229].

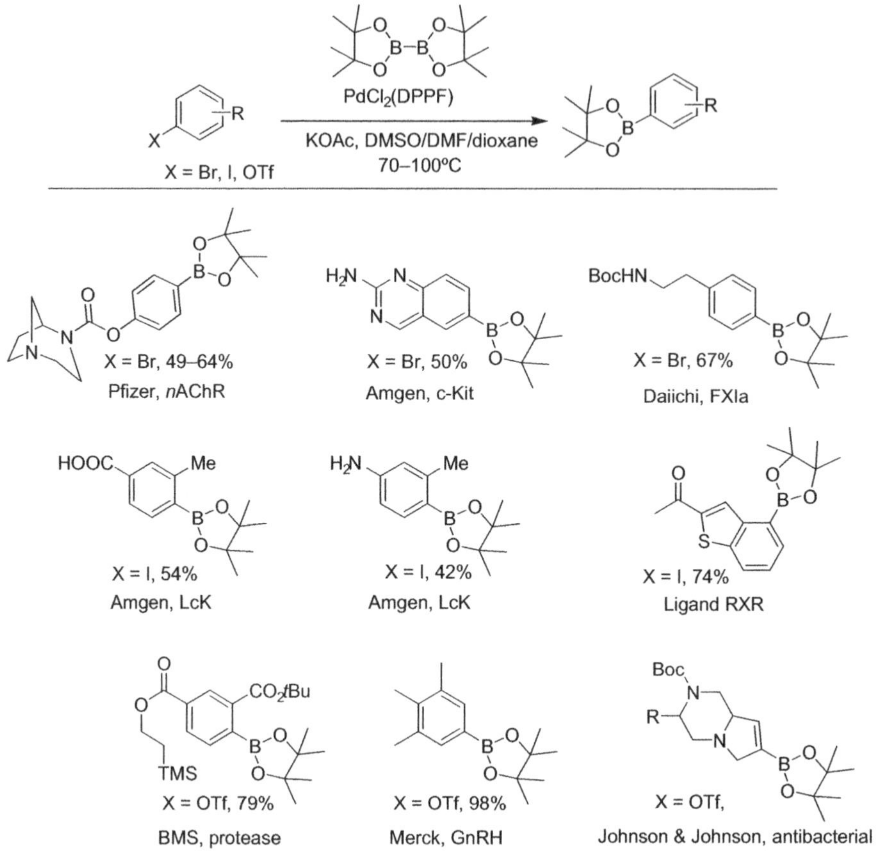

SCHEME 3.97   PdCl$_2$(DPPF)-catalyzed borylation in drug discovery.

This has not yet seen many applications in drug discovery, but its power was demonstrated in an elegant total synthesis of (+)-complanadine A by Fischer and Sarpong (Scheme 3.102) [230]. The 2-pyridyl triflate **291** was reduced to **292** using the Pd-catalyzed transfer hydrogenation. The boronate at the 3-position in **293** was then installed in good yield via direct C–H functionalization catalyzed by Ir(I) [231].

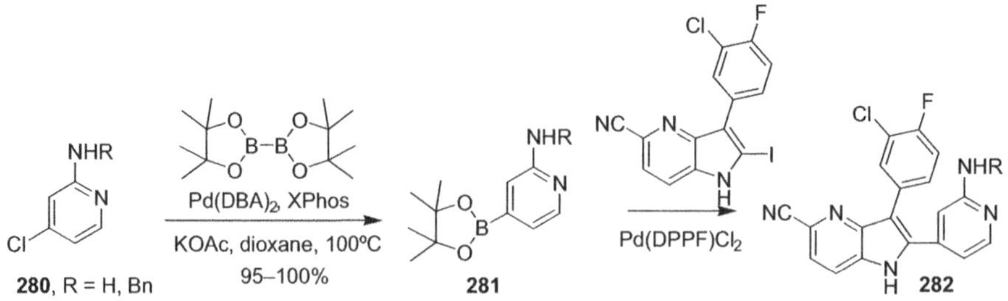

SCHEME 3.98   Pd-XPhos-catalyzed borylation of aryl chlorides.

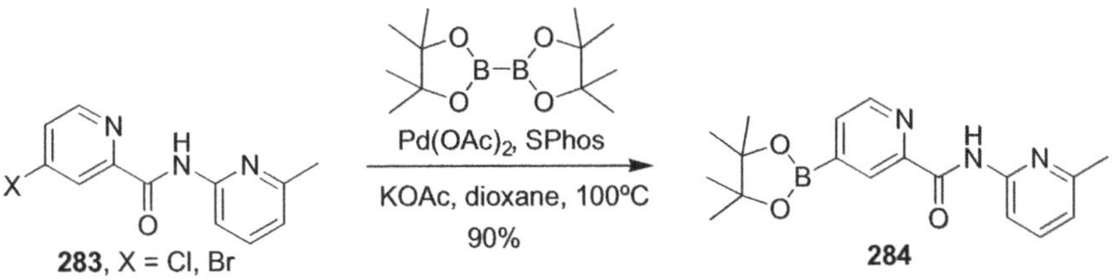

SCHEME 3.99 Pd-SPhos-catalyzed borylation of aryl halides.

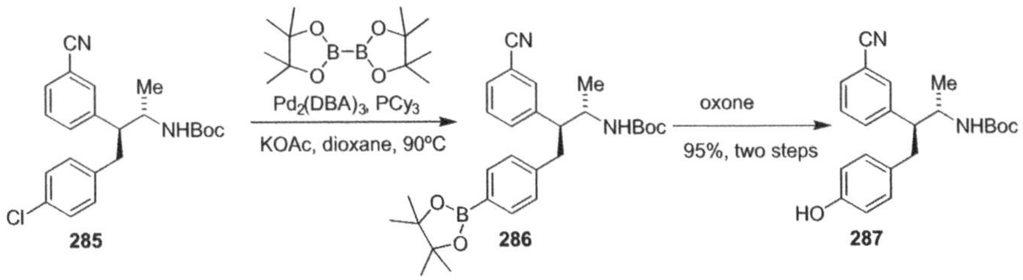

SCHEME 3.100 Borylation followed by oxidation to form the phenol.

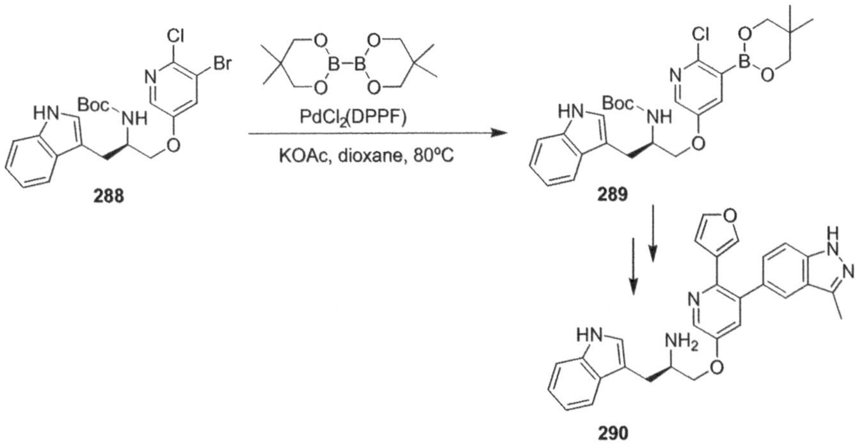

SCHEME 3.101 Borylation in the synthesis of an AKT inhibitor.

The selectivity was mostly controlled by steric factors. Subsequent Suzuki coupling of **293** with the SM **291** followed by deprotection provide (+)-complanadine A, an acetylcholinesterase (AChE) inhibitor.

3.2.8 Pd- or Cu-Catalyzed C–P Coupling

First reported by Hirao, the Pd(PPh$_3$)$_4$-catalyzed coupling between aryl and vinyl bromides with diethylphosphite provides the aryl and vinyl phosphonates in one step [232]. Subsequently, other ligands such as DPPP, DPPB, and DPPF have been

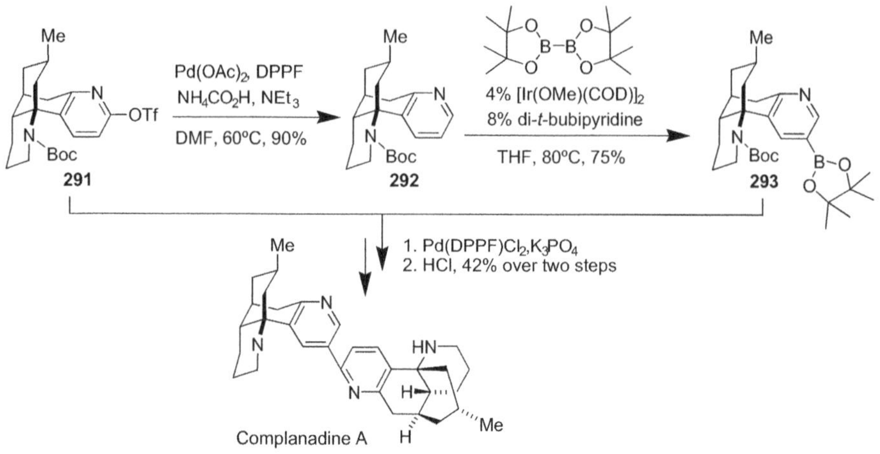

SCHEME 3.102   C–H borylation in the synthesis of (+)-complanadine A.

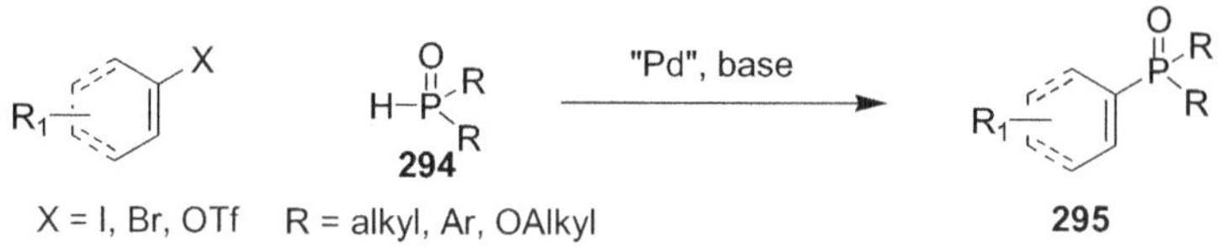

X = I, Br, OTf   R = alkyl, Ar, OAlkyl

SCHEME 3.103   Pd-catalyzed coupling between aryl and vinyl bromides with dialkyl- and diarylphosphite.

used for the phosphonylation of aryl/vinyl iodides, bromides, and triflates (Scheme 3.103) [233]. Cu-catalyzed processes have also been reported [234]. This is one of the most used methods to introduce phosphorous atoms for phosphine ligand synthesis with both R groups of **294** being alkyl or aryl groups. The carbon–phosphorus bond forming reactions have not seen as wide applications in drug discovery as the C–N, C–O, C–S, and C–B counterparts. In a few selected examples shown below, at least one of the R groups is an alkoxy group.

Researchers at LEO Pharma used the Pd-catalyzed C–P coupling in the synthesis of cyclic phosphoamides as potent matrix metalloproteinase inhibitors with antitumor activity (Scheme 3.104) [235]. Various diethyl arylphosphonates **296** were

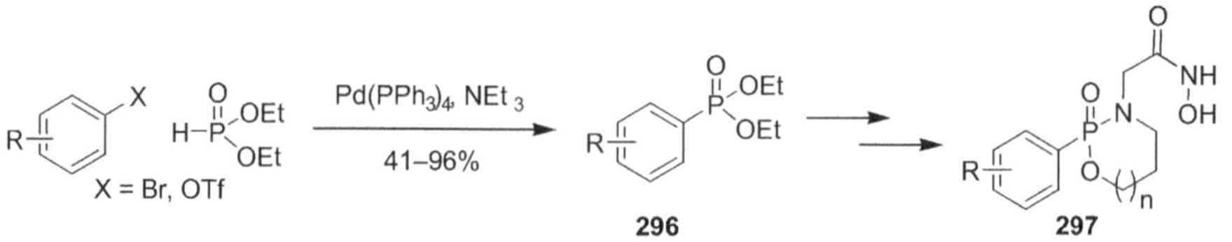

SCHEME 3.104   Diethyl arylphosphonates formation in the synthesis of matrix metalloproteinase inhibitors.

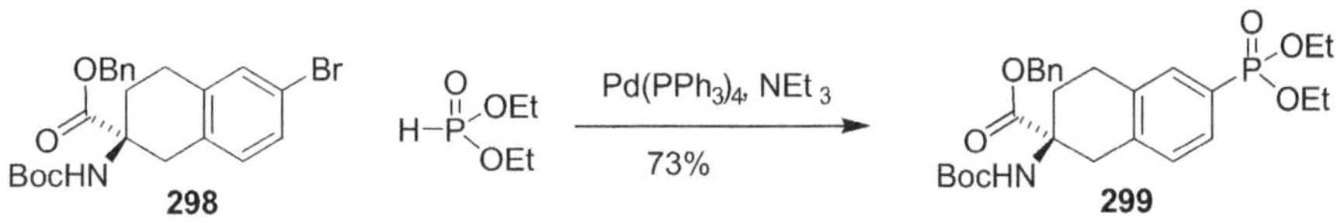

SCHEME 3.105 Arylphosphonates in the study of cellular signal transduction processes.

prepared in good yields from the corresponding aryl bromides or triflates and then subsequently converted to the target compounds **462** for SAR. Chemists at Nippon Organon applied the same method to prepare diethyl arylphosphonates for their metalloproteinase inhibitor synthesis [236].

At NIH, the same method was used to synthesize arylphosphonates such as **299** as conformationally constrained phosphotyrosyl mimetics in the study of cellular signal transduction processes (Scheme 3.105) [237].

At Johnson & Johnson, a number of phosphorus-containing 11β-aryl-substituted steroids **301** were synthesized as new progesterone receptor antagonists [238]. The dialkylphosphites or dialkylphosphine oxides were coupled with the aryl triflate **300** using DPPP as the ligand and microwave heating (Scheme 3.106). PPh₃ ligand gave much lower conversions under the same conditions.

Kim et al. synthesized new sildenafil analogs such as **304** to study their phosphodiesterase 5 inhibitory activities (Scheme 3.107) [239]. The alkenyl triflate **302** was coupled with diethyl phosphate to provide the vinyl phosphonate **303**, which was then converted to the target **304**.

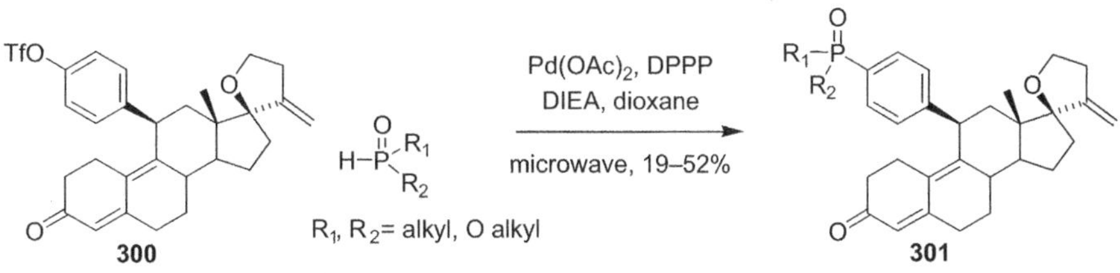

SCHEME 3.106 Synthesis of dialkylphosphine oxides and dialkylphosphites.

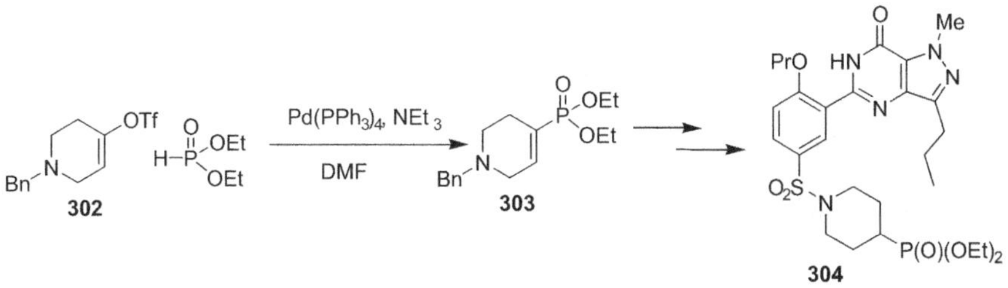

SCHEME 3.107 Synthesis of a vinyl phosphonate.

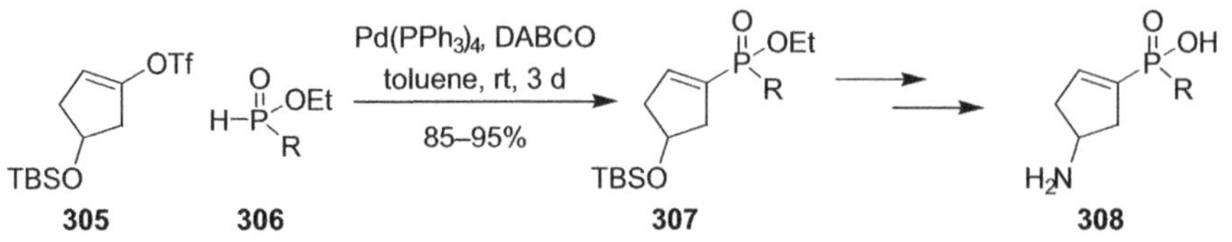

SCHEME 3.108   Pd-catalyzed C–P bond formation between a cyclopentenyl triflate and hypophosphites.

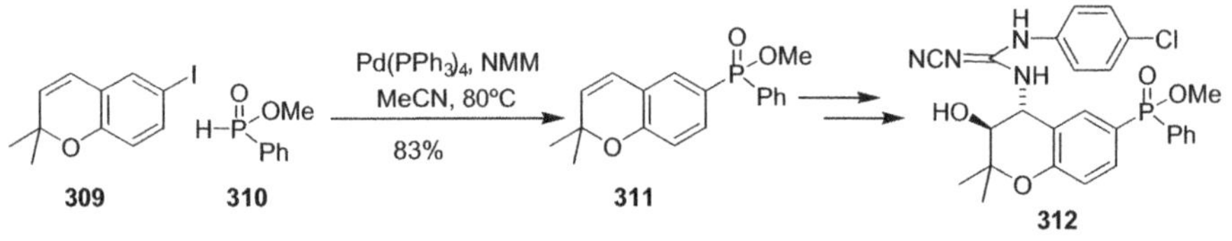

SCHEME 3.109   C–P coupling with phenylphosphinic acid methyl ester.

Hanrahan and coworkers synthesized γ-aminobutyric acid (GABA) analogs based on 4-amino-cyclopent-1-enyl phosphinic acid as potential selective homomeric F1 GABAC receptor antagonists (Scheme 3.108) [240]. The phosphinic acid moiety was introduced by the Pd-catalyzed C–P bond formation [241] between cyclopentenyl triflate **305** and hypophosphites **306** in high yields at room temperature.

At BMS, the C–P coupling was used in the synthesis of one of the analogs of BMS-180448 to study its potency and selectivity as a cardioprotective K_{ATP} opener (Scheme 3.109) [242]. Following literature conditions [243], phenylphosphinic acid methyl ester (**310**) was coupled directly with the aryl iodide **309** in 83% yield at 2 g scale using a weak base, *N*-methylmorpholine.

3.3   CONCLUSIONS AND OUTLOOK

The Pd- and Cu-catalyzed carbon–heteroatom couplings have had a major impact in the pharmaceutical industry. This is still a very active research field with continued development to further expand the substrate scope, which is important for the drug discovery community to efficiently access a diverse array of complex compounds. Other transition metals such as nickel [244] and iron/copper cocatalyst [245] have also been used for these reactions [246]. New reactions have surfaced such as the carbon–fluoride coupling [247] and direct azole amination via C–H functionalization [248]. Challenges still remain for this very important field such as developing more general catalyst systems that tolerate many functional groups for rapid SAR without fine-tuning the reaction conditions. The catalyst cost and heavy metal removal are concerns for the application of these reactions in the manufacturing settings, which can be addressed by developing more active and readily accessible catalysts with higher stability and turnover numbers.

REFERENCES

[1] (a) Surry, D.; Buchwald, S. L. *Angew. Chem., Int. Ed.* **2008**, *47*, 6338–6361. (b) Hartwig, J. F. *Acc. Chem. Res.* **2008**, *41*, 1534–1544. (c) Hartwig, J. F. *Nature* **2008**, *455*, 314–322. (d) Janey, J. M. Buchwald–Hartwig amination. In *Name Reactions for Functional Group Transformations*; Li, J. J., Corey, E. J., Eds.; John Wiley & Sons, Inc.: Hoboken, NJ, 2007; pp. 564–609. (e) Buchwald, S. L.; Mauger, C.; Mignani, G.; Scholz, U. *Adv. Synth. Catal.* **2006**, *348*, 23–39. (f) Jiang, L.; Buchwald, S. L. In *Metal-Catalyzed Cross-Coupling Reactions*, 2nd ed.; de Meijere, A.; Diederich, F., Eds.; Wiley-VCH: Weinheim, Germany, 2004. (g) Hartwig, J. F. *Synlett* **2006**, 1283–1294. (h) Schlummer, B.; Scholz, U. *Adv. Synth. Catal.* **2004**, *346*, 1599–1626. (i) Prim, D.; Campagne, J.-M.; Joseph, D.; Andrioletti, B. *Tetrahedron* **2002**, *58*, 2041–2075. (j) Muci, A. R.; Buchwald, S. L. *Top. Curr. Chem.* **2002**, *219*, 131. (k) Hartwig, J. F. In *Handbook of Organopalladium Chemistry for Organic Synthesis*; Negishi, E., Ed.; Wiley-Interscience: New York, **2002**; p. 1051. (l) Hartwig, J. F. In *Comprehensive Coordination Chemistry II*; McCleverty, J. A.; Meyer, T. J., Eds.; Elsevier Ltd.: Oxford, 2004; Vol. *9*, pp. 369–398.

[2] (a) Monnier, F.; Taillefer, M. *Angew. Chem., Int. Ed.* **2009**, *48*, 6954–6971. (b) Evano, G.; Blanchard, N.; Toumi, M. *Chem. Rev.* **2008**, *108*, 3054. (c) Beletskaya, I. P.; Cheprakov, A. V. *Coord. Chem. Rev.* **2004**, *248*, 2337–2364. (d) Kunz, K.; Scholz, U.; Ganzer, D. *Synlett* **2003**, *15*, 2428–2439. (e) Ley, S. V.; Thomas, A. W. *Angew. Chem., Int. Ed.* **2003**, *42*, 5400–5449. (f) Thomas, A. W.; Ley, S. V. In *Modern Arylation Methods*; Ackerman, L., Ed.; Wiley-VCH: Weinheim, Germany, **2009**; pp. 121–154. (g) Surry, D. S.; Buchwald, S. L. *Chem. Sci.* **2010**, *1*, 13–31. (h) Ma, D.; Cai, Q. *Acc. Chem. Res.* **2008**, *41*, 1450–1460.

[3] Kosugi, M.; Kameyama, M.; Migita, T. *Chem. Lett.* **1983**, 927.

[4] (a) Guram, A. S.; Rennels, R. A.; Buchwald, S. L. *Angew. Chem., Int. Ed. Engl.* **1995**, *34*, 1348. (b) Louie, J.; Hartwig, J. F. *Tetrahedron Lett.* **1995**, *36*, 3609.

[5] (a) Wolfe, J. P.; Wagaw, S.; Buchwald, S. L. *J. Am. Chem. Soc.* **1996**, *118*, 7215. (b) Wolfe, J. P.; Buchwald, S. L. *J. Org. Chem.* **2000**, *65*, 1144.

[6] Driver, M. S.; Hartwig, J. F. *J. Am. Chem. Soc.* **1996**, *118*, 7217.

[7] Wolfe, J. P.; Tomori, H.; Sadighi, J. P.; Yin, J.; Buchwald, S. L. *J. Org. Chem.* **2000**, *65*, 1158.

[8] Old, D. W.; Wolfe, J. P.; Buchwald, S. L. *J. Am. Chem. Soc.* **1998**, *120*, 9722.

[9] Huang, X.; Anderson, K. W.; Zim, D.; Jiang, L.; Klapars, A.; Buchwald, S. L. *J. Am. Chem. Soc.* **2003**, *125*, 6653.

[10] (a) First used by Nishiyama for amination: Nishiyama, M.; Yamamoto, T.; Koie, Y. *Tetrahedron Lett.* **1998**, *39*, 617. (b) Yamamoto, T.; Nishiyama, M.; Koie, Y. *Tetrahedron Lett.* **1998**, *39*, 2367.

[11] Hartwig, J. F.; Kawatsura, M.; Hauck, S. I.; Shaughnessy, K. H.; Alcazar-Roman, L. M. *J. Org. Chem.* **1999**, *64*, 5575.

[12] (a) First developed by van Leeuwen: Kranenburg, M.; van der Burgt, Y. E. M.; Kamer, P. C. J.; van Leeuwen, P. W. N. M. *Organometallics* **1995**, *14*, 3081. (b) Guari, Y.; van Es, D. S.; Reek, J. N. H.; Kamer, P. C. J.; van Leeuwen, P. W. N. M. *Tetrahedron Lett.* **1999**, *40*, 3789.

[13] (a) Yin, J.; Buchwald, S. L. *Org. Lett.* **2000**, *2*, 1101. (b) Yin, J.; Buchwald, S. L. *J. Am. Chem. Soc.* **2002**, *124*, 6043.

[14] Anderson, K. W.; Tundel, R. E.; Ikawa, T.; Altman, R. A.; Buchwald, S. L. *Angew. Chem., Int. Ed.* **2006**, *45*, 6523.

[15] Ikawa, T.; Barder, T. E.; Biscoe, M. R.; Buchwald, S. L. *J. Am. Chem. Soc.* **2007**, *129*, 13001.

[16] Kataoka, N.; Shelby, Q.; Stambuli, J. P.; Hartwig, J. F. *J. Org. Chem.* **2002**, *67*, 5553.

[17] Shen, Q.; Ogata, T.; Hartwig, J. F. *J. Am. Chem. Soc.* **2008**, *130*, 6586.

[18] Novartis 2009 Annual Report: Healthcare Portfolio. Available at http://www.novartis. com/downloads/investors/sales-results/Q4 -2009-media-release_EN.pdf

[19] Loiseleur, O.; Kaufmann, D.; Abel, S.; Buerger, H. M.; Meisenbach, M.; Schmitz, B.; Sedelmeier, G. WO 03/066613 (2003).

[20] Morita, S.; Kitano, K.; Matsubara, J.; Ohtani, T.; Kawano, Y.; Otsubo, K.; Uchida, M. *Tetrahedron* **1998**, *54*, 4811.

[21] Bonacorsi, S. J., Jr.; Waller, S. C.; Rinehart, J. K. *J. Labelled Comp. Radiopharm.* **2006**, *49*, 1.

[22] Chen, C.-y.; Frey, L. F.; Shultz, S.; Wallace, D. J.; Marcantonio, K.; Payack, J. F.; Vazquez, E.; Springfield, S. A.; Zhou, G.; Liu, P.; Kieczykowski, G. R.; Chen, A. M.; Phenix, B. D.; Singh, U.; Strine, J.; Izzo, B.; Krska, S. *Org. Process Res. Dev.* **2007**, *11*, 616.

[23] Wallace, D. J.; Campos, K. R.; Shultz, C. S.; Klapars, A.; Zewge, D.; Crump, B. R.; Phenix, B. D.; McWilliams, J. C.; Krska, S.; Sun, Y.; Chen, C.-y.; Spindler, F. *Org. Process Res. Dev.* **2009**, *13*, 84.

[24] Wallace, D. J.; Klauber, D. J.; Chen, C.-y.; Volante, R. P. *Org. Lett.* **2003**, *5*, 4749.

[25] Klapars, A.; Campos, K. R.; Chen, C.-y.; Volante, R. P. *Org. Lett.* **2005**, *7*, 1185.

[26] For beneficial effects of *tert*-amyl alcohol in palladium-catalyzed processes, see: (a) Huang, X.; Anderson, K. W.; Zim, D.; Jiang, L.; Klapars, A.; Buchwald, S. L. *J. Am. Chem. Soc.* **2003**, *125*, 6653. (b) Bagdanoff, J. T.; Ferreira, E. M.; Stolz, B. M. *Org. Lett.* **2003**, *5*, 835.

[27] Damon, D. B.; Dugger, R. W.; Hubbs, S. E.; Scott, J. M.; Scott, R. W. *Org. Process Res. Dev.* **2006**, *10*, 472.

[28] Zhao, M.; Yin, J.; Huffman, M. A., McNamara, J. M. *Tetrahedron* **2006**, *62*, 1110.

[29] Bilodeau, M. T.; Balitza, A. E.; Koester, T. J.; Manley, P. J.; Rodman, L. D.; Buser-Doepner, C.; Coll, K. E.; Fernandes, C.; Gibbs, J. B.; Heimbrook, D. C.; Huckle, W. R.; Kohl, N.; Lynch, J. J.; Mao, X.; McFall, R. C.; McLoughlin, D.; Miller-Stein, C. M.; Rickert, K. W.; Sepp-Lorenzino, L.; Shipman, J. M.; Subramanian, R.; Thomas, K. A.; Wong, B. K.; Yu, S.; Hartman, G. D. *J. Med. Chem.* **2004**, *47*, 6363.

[30] Yin, J.; Zhao, M. M.; Huffman, M. A.; McNamara, J. M. *Org. Lett.* **2002**, *4*, 3481–3484.

[31] Wolfe, J. P.; Åhman, J.; Sadighi, J. P.; Singer, R. A.; Buchwald, S. L. *Tetrahedron Lett.* **1997**, *38*, 6367.

[32] Mase, T.; Houpis, I. N.; Akao, A.; Dorziotis, I.; Emerson, K.; Hoang, T.; Iida, T.; Itoh, T.; Kamei, K.; Kato, S.; Kato, Y.; Kawasaki, M.; Lang, F.; Lee, J.; Lynch, J.; Maligres, P.; Molina, A.; Nemoto, T.; Okada, S.; Reamer, R.; Song, J. Z.; Tschaen, D.; Wada, T.; Zewge, D.; Volante, R. P.; Reider, P. J.; Tomimoto, K. *J. Org. Chem.* **2001**, *66*, 6775.

[33] Liu, Y.; Prashar, M.; Repič, O.; Blacklock, T. J. *J. Heterocycl. Chem.* **2003**, *40*, 713.

[34] Yu, S.; Haight, A.; Kotecki, B.; Wang, L.; Lukin, K.; Hill, D. R. *J. Org. Chem.* **2009**, *74*, 9539.

[35] Singer, R. A.; Dore, M.; Sieser, J. E.; Berliner, M. *Tetrahedron Lett.* **2006**. *47*, 3727.

[36] Kotecki, B.; Fernando, D.; Haight, T.; Lukin, K. *Org. Lett.* **2009**, *11*, 947.

[37] Berg, S.; Linderberg, M.; Ross, S.; Thorberg, S.-O.; Ulff, B.PCT Int. Appl. WO 9,905,134 (1999).

[38] (a) Federsel, H.-J.; Hedberg, M.; Qvarnström, F. R.; Sjögren, M. P. T.; Tian, W. *Acc. Chem. Res.* **2007**, *40*, 1377. (b) Federsel, H.-J.; Hedberg, M.; Qvarnström, F. R.; Tian, W. *Org. Process Res. Dev.* **2008**, *12*, 512.

[39] Robinson, G. E.; Cunningham, O. R.; Dekhane, M.; McManus, J. C.; O'Kearney-McMullan, A.; Mirajkar, A. M.; Mishra, V.; Norton, A. K.; Venugopalan, B.; Wiliams, E. G. *Org. Process Res. Dev.* **2004**, *8*, 925.

[40] Kuethe, J. T.; Childers, K. G.; Humphrey, G. R.; Journet, M.; Peng, Z. *Org. Process Res. Dev.* **2008**, *12*, 1201.

[41] (a) Berger, R.; Zhu, C.; Hansen, A. R.; Harper, B.; Chen, Z.; Holt, T. G.; Hubert, J. A.; Lee, S. J.; Pan, J.; Qian, S.; Reitman, M. L.; Strack, A. M.; Weingarth, D. T.; Wolff, M. S.; MacNeil, D. J.; Weber, A. W.; Edmondson, S. D. *Bioorg. Med. Chem. Lett.* **2008**, *18*, 4833. (b) Zhu, C.; Hansen, A. R.; Bateman, T.; Chen, Z.; Holt, T. G.; Hubert, J. A.; Karanam, B. V.; Lee, S. J.; Pan, J.; Qian, S.; Reddy, V. B. G.; Reitman, M. L.; Strack, A. M.; Tong, V.; Weingarth, D. T.; Wolff, M. S.; MacNeil, D. J.; Weber, A. E.; Duffy, J. L.; Edmondson, S. D. *Bioorg. Med. Chem. Lett.* **2008**, *18*, 4393.

[42] Engstrom, K. M.; Daanen, J. F.; Wagaw, S.; Stewart, A. O. *J. Org. Chem.* **2006**, *71*, 8378.

[43] Cowart, M.; Latshaw, S. P.; Bhatia, P.; Daanen, J. F.; Rohde, J.; Nelson, S. L.; Patel, M.; Kolasa, T.; Nakane, M.; Uchic, M. E.; Miller, L. N.; Terranova, M. A.; Chang, R.; Donnelly-Roberts, D. L.; Namovic, M. T.; Hollingsworth, P. R.; Martino, B. R.; Lynch, J. J., III; Sullivan, J. P.; Hsieh, G. C.; Moreland, R. B.; Brioni, J. D.; Stewart, A. O. *J. Med. Chem.* **2004**, *47*, 3853–3864.

[44] Kuether, J. T.; Wong, A.; Qu, C.; Smitrovich, J.; Davies, I. W.; Hughes, D. L. *J. Org. Chem.* **2005**, *70*, 2555.

[45] (a) Wagaw, S.; Yang, B. H.; Buchwald, S. L. *J. Am. Chem. Soc.* **1998**, *120*, 6621. (b) Wagaw, S.; Yang, B. H.; Buchwald, S. L. *J. Am. Chem. Soc.* **1999**, *121*, 10251.

[46] Bower, J. F.; Szeto, P.; Gallagher, T. *Org. Lett.* **2007**, *17*, 3283.

[47] Hong, Y.; Tanoury, G. J.; Wilkinson, H. S.; Bakale, R. P.; Wald, S. A.; Senanayake, C. H. *Tetrahedron Lett.* **1997**, *38*, 5607.

[48] Madar, D. J.; Kopecka, H.; Pireh, D.; Pease, J.; Pliushchev, M.; Sciotti, R. J.; Wiedeman, P. E.; Djuric, S. W. *Tetrahedron Lett.* **2001**, *42*, 3681.

[49] Fox, J. M.; Huang, X.; Chieffi, A.; Buchwald, S. L. *J. Am. Chem. Soc.* **2000**, *122*, 1360.

[50] (a) Cvetovich, R. J.; Reamer, R. A.; DiMichele, L.; Chung, J. Y. L.; Chilenski, J. *J. Org. Chem.* **2006**, *71*, 8610. (b) Chung, J. Y. L.; Cvetovich, R. J.; McLaughlin, M.; Amato, J.; Tsay, F.-R.; Jensen, M.; Weissman, S.; Zewge, D. *J. Org. Chem.* **2006**, *71*, 8602.

[51] Zhou, D.; Zhou, P.; Evrard, D. A.; Meagher, K.; Webb, M.; Harrison, B. L.; Huryn, D. M.; Colembieski, J.; Hornby, G. A.; Schechter, L. E.; Smith, D. L.; Andree, T. H.; Mewshaw, R. E. *Bioorg. Med. Chem.* **2008**, *16*, 6707.

[52] Bettinetti, L.; Löber, S.; Hübner, H.; Gmeiner, P. *J. Comb. Chem.* **2005**, *7*, 309.

[53] Salama, I.; Schlotter, K.; Utz, W.; Hübner, H.; Gmeiner, P.; Boeckler, F. *Bioorg. Med. Chem.* **2006**, *14*, 5898.

[54] Sasaki, H.; Haraguchi, Y.; Itotani, M.; Kuroda, H.; Hashizume, H.; Tomishige, T.; Kawasaki, M.; Matsumoto, M.; Komatsu, M.; Tsubouchi, H. *J. Med. Chem.* **2006**, *49*, 7854.

[55] Zhao, S.-H.; Berger, J.; Clark, R. D.; Sethofer, S. G.; Krauss, N. E.; Brothers, J. M.; Martin, R. S.; Misner, D. L.; Schwab, D.; Alexandrova, L. *Bioog. Med. Chem. Lett.* **2007**, *17*, 3504.

[56] Conte, I.; Giuliano, C.; Ercolani, C.; Narjes, F.; Koch, U.; Rowley, M.; Altamura, S.; De Franesco, R.; Neddermann, P.; Migliaccio, G.; Stansfield, I. *Bioorg. Med. Chem. Lett.* **2009**, *19*, 1779.

[57] Witherington, J.; Abberley, L.; Briggs, M. A.; Collis, K.; Dean, D. K.; Gaiba, A.; King, N. P.; Kraus, H.; Shuder, N.; Steadman, J. G. A.; Takle, A. K.; Sanger, G.; Wadsworth, G.; Butler, S.; McKay, F.; Muir, A.; Winborn, K.; Heightman, T. D. *Bioorg. Med. Chem. Lett.* **2008**, *18*, 2203.

[58] Schlotter, K.; Boeckler, F.; Hübner, H.; Gmeiner, P. *J. Med. Chem.* **2005**, *48*, 3696.

[59] Gudmundsson, K. S.; Johns, B. A.; Allen, S. H. *Bioorg. Med. Chem. Lett.* **2008**, *18*, 1157.

[60] (a) Johns, B. A.; Gudmundsson, K. S.; Allen, S. *Bioorg. Med. Chem. Lett.* **2007**, *17*, 2858. (b) Gudmundsson, K. S.; Johns, B. A. *Bioorg. Med. Chem. Lett.* **2007**, *17*, 2735. (c) Gudmundsson, K. S.; Johns, B. A.; Wang, Z.; Turner, E. M.; Allen, S. H.; Freemen, G. A.; Boyd, F. L., Jr., ; Sexton, C. J.; Selleseth, D. W.; Moniri, K. R.; Creech, K. L. *Bioorg. Med. Chem.* **2005**, *13*, 5346.

[61] Gudmundsson, K. S.; Johns, B. A. *Org. Lett.* **2003**, *5*, 1369.

[62] Oza, V. B.; Smith, C.; Raman, P.; Koepf, E. K.; Lashuel, H. A.; Petrassi, H. M.; Chiang, K. P.; Powers, E. T.; Sachettinni, J.; Kelly, J. W. *J. Med. Chem.* **2002**, *45*, 321.

[63] Johns, B. A.; Weatherhead, J. G.; Allen, S. H.; Thompson, J. B.; Garvey, E. P.; Foster, S. A.; Jeffrey, J. L.; Miller, W. H. *Bioorg. Med. Chem. Lett.* **2009**, *19*, 1807.

[64] Andrés, J. I.; Alcázar, J.; Alonso, J. M.; De Lucas, A. I.; Iturrino, L.; Biesmans, I.; Megens, A. A. *Bioorg. Med. Chem.* **2006**, *14*, 4361.

[65] Hornberger, K. R.; Badiang, J. G.; Slaovich, J. M.; Kuntz, K. W.; Emmitte, K. A.; Cheung, M. *Tetrahedron Lett.* **2008**, *49*, 6348.

[66] Zhang, G.; Ren, P.; Gray, N. S.; Sim, T.; Liu, Y.; Wang, X.; Che, J.; Tian, S.-S.; Sandberg, M. L.; Spadking, T. A.; Romeo, R.; Iskandar, M.; Chow, D.; Seidel, H. M.; Karanewsky, D. S.; He, Y. *Bioorg. Med. Chem. Lett.* **2008**, *18*, 5618.

[67] Moore, K. P.; Zhu, H.; Rajapakse, H. A.; McGaughey, G. B.; Colussi, D.; Price, E. A.; Sankaranarayanan, S.; Simon, A. J.; Pudvah, N. T.; Hochman, J. H.; Allison, T.; Munshi, S. K.; Graham, S. L.; Vacca, J. P.; Nantermet, P. G. *Bioorg. Med. Chem. Lett.* **2007**, *17*, 5831.

[68] (a) Otsuka, S.; Yoshida, T.; Matsumoto, M.; Nakatsu, K. *J. Am. Chem. Soc.* **1976**, *98*, 5850. (b) Yoshida, T.; Otsuka, S. *Inorg. Synth.* **1990**, *28*, 113.

[69] Lawson, E. C.; Luci, D. K.; Ghosh, S.; Kinney, W. A.; Reynolds, C. H.; Qi, J.; Smith, C. E.; Wang, Y.; Minor, L. K.; Haretlein, B. J.; Parry, T. J.; Damiano, B. P.; Maryanoff, B. E. *J. Med. Chem.* **2009**, *52*, 7432.

[70] Dai, C.; Fu, G. C. *J. Am. Chem. Soc.* **2001**, *123*, 2719.

[71] (a) Netherton, M. R.; Fu, G. C. *Org. Lett.* **2001**, *3*, 4295. (b) Tewari, A.; Hein, M.; Zapf, A.; Beller, M. *Tetrahedron* **2005**, *61*, 9705.

[72] Tasler, S.; Mies, J.; Lang, M. *Adv. Synth. Catal.* **2007**, *349*, 2286.

[73] (a) Huang, X.; Buchwald, S. L. *Org. Lett.* **2001**, *3*, 3417. (b) Lee, S.; Jørgensen, M.; Hartwig, J. F. *Org. Lett.* **2001**, *3*, 2729.

[74] Liu, L.; Siegmund, A.; Xi, N.; Kaplan-Lefko, P.; Rex, K.; Chen, A.; Lin, J.; Moriguchi, J.; Berry, L.; Huang, L.; Teffera, Y.; Yang, Y.; Zhang, Y.; Bellon, S. F.; Lee, M.; Shimanovich, R.; Bak, A.; Dominguez, C.; Norman, M. H.; Harmange, J.-C.; Dussault, I.; Kim, T.-S. *J. Med. Chem.* **2008**, *51*, 3688.

[75] Schlapbach, A.; Heng, R.; Di Padova, F. *Bioorg. Med. Chem. Lett.* **2004**, *14*, 357.

[76] McLaughlin, M.; Palucki, M.; Davies, I. W. *Org. Lett.* **2006**, *8*, 3311.

[77] Lovering, F.; Kirincich, S.; Wang, W.; Combs, K.; Resnick, L.; Sabalski, J. E.; Butera, J.; Liu, J.; Parris, K.; Telliez, J. B. *Bioorg. Med. Chem. Lett.* **2009**, *19*, 3342.

[78] Guo, S.; Song, Y.; Huang, Q.; Yuan, H.; Wan, B.; Wang, Y.; He, R.; Beconi, M. G.; Franzblau, S. G.; Kozikowski, A. P. *J. Med. Chem.* **2010**, *53*, 649.

[79] Kataoka, N.; Shelby, Q.; Stambuli, J. P.; Hartwig, J. F. *J. Org. Chem.* **2002**, *67*, 5553.

[80] Boyd, M. J.; Crane, S. N.; Robichaud, J.; Scheigetz, J.; Black, W. C.; Chauret, N.; Wang, Q.; Masse, F.; Oballa, R. M. *Bioorg. Med. Chem. Lett.* **2009**, *19*, 675.

[81] Rataboul, F.; Zapf, A.; Jackstell, J.; Harkal, S.; Riermeier, T.; Monsees, A.; Dingerdissen, U.; Beller, M. *Chem. Eur. J.* **2004**, *10*, 2983.

[82] Knight, R. L.; Allen, D. R.; Birch, H. L.; Chapman, G. A.; Galvin, F. C.; Jopling, L. A.; Lock, C. J.; Meissner, J. W. G.; Owen, D. A.; Raphy, G.; Watson, R. J.; Williams, S. C. *Bioorg. Med. Chem. Lett.* **2008**, *18*, 629.

[83] Grasa, G. A.; Miciu, M. S.; Huang, J.; Nolan, S. P. *J. Org. Chem.* **2001**, *66*, 7729.

[84] Hanan, E. J.; Fucini, R. V.; Romanowski, M. J.; Elling, R. A.; Lew, W.; Purkey, H. E.; VanderPorten, E. C.; Yang, W. *Bioorg. Med. Chem. Lett.* **2008**, *18*, 5186.

[85] (a) Ullmann, F. *Ber. Dtsch. Chem. Ges.* **1903**, *36*, 2382. (b) Goldberg, I. *Ber. Dtsch. Chem. Ges.* **1906**, *39*, 1691.

[86] For an early review, see: Lindley, J. *Tetrahedron* **1984**, *40*, 1433–1456.

[87] Kiyomori, A.; Marcoux, J.-F.; Buchwald, S. L. *Tetrahedron Lett.* **1999**, *40*, 2657.

[88] Goodbrand, H. B.; Hu, N. -X . *J. Org. Chem.* **1999**, *64*, 670.

[89] Klapars, A.; Antila, J. C.; Huang, X.; Buchwald, S. L. *J. Am. Chem. Soc.* **2001**, *123*, 7727.

[90] Klapars, A.; Huang, X.; Buchwald, S. L. *J. Am. Chem. Soc.* **2002**, *124*, 7421.

[91] Antilla, J. C.; Baskin, J. M.; Barder, T. E.; Buchwald, S. L. *J. Org. Chem.* **2004**, *69*, 5578.

[92] (a) Ma, D.; Zhang, Y.; Yao, J.; Wu, S.; Tao, F. *J. Am. Chem. Soc.* **1998**, *120*, 12459. (b) Ma, D.; Xia, C. *Org. Lett.* **2001**, *3*, 2583.

[93] (a) Ma, D.; Cai, Q.; Zhang, H. *Org. Lett.* **2003**, *5*, 2453. (b) Zhang, H.; Cai, Q.; Ma, D. *J. Org. Chem.* **2005**, *70*, 5164.

[94] Liu, Y.-F.; Wang, C.-L.; Bai, J.-J.; Han, N.; Jiao, J.-P.; Qi, X.-L. *Org. Process Res. Dev.* **2008**, *12*, 490.

[95] Liu, Y.-F.; Bai, Y. J.; Zhang, J.; Li, Y. Y.; Jiao, J. P.; Qi, X. L. *Eur. J. Org. Chem.* **2007**, *36*, 6084.

[96] Gruza, H.; Mirek, S.; Jezewski, A.; Wrzosek, A. *WO 2010/014022* (**2010**).

[97] Xu, F.; Corley, E.; Zacuto, M.; Conlon, D. A.; Pipik, B.; Humphrey, G.; Murry, J.; Tschaen, D. *J. Org. Chem.* **2010**, *75*, 1343.

[98] Ghosh, A.; Sieser, J. E.; Caron, S.; Couturier, M.; DuPont-Gaudet, K.; Girardin, M. *J. Org. Chem.* **2006**, *71*, 1258.

[99] Ghosh, A.; Sieser, J. E.; Riou, M.; Cai, W.; Rivera-Ruiz, L. *Org. Lett.* **2003**, *5*, 2207.

[100] Broeders, F.; Defrère, L.; Deltent, M.-F.; Driessens, F.; Gilson, F.; Grooters, L.; Ikonomakos, X.; Limauge, F.; Sergeef, E.; Verstraeten, N. *Org. Process Res. Dev.* **2009**, *13*, 442.

[101] Pu, Y.; Ku, Y.; Grieme, T.; Bhatia, A. V. *Tetrahedron Lett.* **2006**, *47*, 149.

[102] Maligres, P. E.; Askin, D. *WO 2003/016309* 2003.

[103] Steinhuebel, D.; Palucki, M.; Askin, D.; Dolling, U. *Tetrahedron. Lett.* **2004**, *45*, 3305.

[104] Satyanarayana, K.; Srinivas, K.; Himabindu, V.; Reddy, G. M. *Org. Process Res. Dev.* **2007**, *11*, 842.

[105] Lang, F.; Zewge, D.; Houpis, I. N.; Volante, R. P. *Tetrahedron Lett.* **2001**, *42*, 3251.

[106] Oyarzabal, J.; Howe, T.; Alcazar, J.; Ignacio Andrés, J.; Alvarez, R. M.; Dautzenberg, F.; Iturrino, L.; Martínez, S.; Van der Linden, I. *J. Med. Chem.* **2009**, *52*, 2076.

[107] Li, J. J.; Mitchell, L. H.; Dow, R. L. *Bioorg. Med. Chem. Lett.* **2010**, *20*, 306.

[108] Travins, J. M.; Bernotas, R. C.; Kaufman, D. H.; Quinet, E.; Nambi, P.; Feingold, I.; Huselton, C.; Wilhelmsson, A.; Goos-Nilsson, A.; Wrobel, J. *Bioorg. Med. Chem. Lett.* **2010**, *20*, 526.

[109] Shinozuka, T.; Shimada, K.; Matsui, S.; Yamane, T.; Ama, M.; Fukuda, T.; Taki, M.; Takeda, Y.; Otsuka, E.; Yamato, M. Naito, S. *Bioorg. Med. Chem.* **2006**, *14*, 6807.

[110] Robichaud, J.; Bayly, C.; Oballa, R.; Prasit, P.; Mellon, C.; Falgueyret, J. P.; Percival, M. D.; Wesolowski, G.; Rodan, S. B. *Bioorg. Med. Chem. Lett.* **2004**, *14*, 4291.

[111] Yeh, V. S. C.; Patel, J. R.; Yong, H.; Kurukulasuriya, R.; Fung, S.; Monzon, K.; Chiou, W.; Wang, J.; Stolarik, D.; Imade, H.; Beno, D.; Brune, M.; Jacobson, P.; Sham, H.; Link, J. T. *Bioorg. Med. Chem. Lett.* **2006**, *16*, 5414.

[112] Yeh, V. S. C.; Wiedeman, P. E. *Tetrahedron Lett.* **2006**, *47*, 6011.

[113] (a) Kwong, F. Y.; Buchwald, S. L. *Org. Lett.*, **2003**, *5*, 793. (b) Shafir, A.; Buchwald, S. L. *J. Am. Chem. Soc.* **2006**, *128*, 8742.

[114] Bouteiller, C.; Becerril-Ortega, J.; Marchand, P.; Nicole, O.; Barré, L.; Buisson, A.; Perrio, C. *Org. Biomol. Chem.*, **2010**, *8*, 1111.

[115] Yang, G. X.; Chang, L. L.; Truong, Q.; Doherty, G. A.; Magriotis, P. A.; de Laszlo, S. E.; Li, B.; MacCoss, M.; Kidambi, U.; Egger, L. A.; McCauley, E.; Van Riper, G.; Mumbford, R. A.; Schmidt, J. A.; Hagmann, W. K. *Bioorg. Med. Chem. Lett.* **2002**, *12*, 1497.

[116] Sugahara, M.; Ukita, T. *Chem. Pharm. Bull.* **1997**, *45*, 719.

[117] Eldrup, A. B.; Prhavc, M.; Brooks, J.; Bhat, B.; Prakash, T. P.; Song, Q.; Bera, S.; Bhat, N.; Dande, P.; Cook, P. D.; Bennett, C. F.; Carroll, S. S.; Ball, R. G.; Bosserman, M.; Burlein, C.; Colwell, L. F.; Fay, J. F.; Flores, O. S.; Getty, K.; LaFemina, R. L.; Leone, J.; MacCoss, M.; McMasters, D.; Tomassini, J. E.; Von Langen, D.; Wolanski, B.; Olsen, D. B. *J. Med. Chem.* **2004**, *47*, 5284.

[118] Baqi, Y.; Lee, S.-Y.; Iqbal, J.; Ripphausen, P.; Lehr, A.; Scheiff, A. B.; Zimmermann, H.; Bajorath, J.; Müller, C. E. *J. Med. Chem.* **2010**, *53*, 2076.

[119] Han, C.; Shen, R.; Su, S.; Porco, J. A., Jr., *Org. Lett.* **2004**, *6*, 27.

[120] Coleman, R. S.; Liu, P.-H. *Org. Lett.* **2004**, *6*, 577.

[121] Jiang, L.; Job, G. E.; Klapars, A.; Buchwald, S. L. *Org. Lett.* **2003**, *5*, 3667.

[122] Suzuki, H.; Miyoshi, K.; Shinoda, M. *Bull. Chem. Soc. Jpn.* **1980**, *53*, 1765.

[123] (a) Zhu, W.; Ma, D. *Chem. Commun.* **2004**, 888. (b) Cai, Q.; Zhu, W.; Zhang, H.; Zhang, Y.; Ma, D. *Synthesis* **2005**, 496.

[124] Andersen, J.; Madsen, U.; Björkling, F.; Liang, X. *Synlett* **2005**, 2209.

[125] Gupte, A.; Boshoff, H. I.; Wilson, D. J.; Neres, J.; Labello, N. P.; Somu, R. V.; Xing, C.; Barry, C. E.; Aldrich, C. C. *J. Med. Chem.* **2008**, *51*, 7495.

[126] Cosner, C. C.; Markiewicz, J. T.; Bourbon, P.; Mariani, C. J.; Wiest, O.; Rujoi, M.; Rosenbaum, A. Y.; Huang, A. I.; Maxfield, F. R.; Helquist, P. *J. Med. Chem.* **2009**, *52*, 6494.

[127] Qin, Z.; Kastrati, I.; Chandrasena, E. P.; Liu, H.; Yao, P.; Petukhov, P. A.; Bolton, J. L.; Thatcher, G. R. J. *J. Med. Chem.* **2007**, *50*, 2682.

[128] Ullmann, F. *Chem. Ber.* **1904**, *37*, 853.

[129] Marcoux, J.-F.; Doye, S.; Buchwald, S. L. *J. Am. Chem. Soc.* **1997**, *119*, 10539.

[130] Kalinin, A. V.; Bower, J. F.; Riebel, P.; Snieckus, V. *J. Org. Chem.* **1999**, *64*, 2986.

[131] Gujadhur, R. K.; Venkataraman, D. *Synth. Commun.* **2001**, *31*, 2865.

[132] Fagan, P. J.; Hauptman, E.; Shapiro, R.; Casalnuovo, A. *J. Am. Chem. Soc.* **2000**, *122*, 5043.

[133] (a) Frlan, R.; Kikelj, D. *Synthesis* **2006**, 2271. (b) Ouali, A.; Spindler, J.-F.; Jutand, A.; Taillefer, M. *Adv. Synth. Catal.* **2007**, *349*, 1906.

[134] Buck, E.; Song, Z. J.; Tschaen, D.; Dormer, P. G.; Volante, R. P.; Reider, P. J. *Org. Lett.* **2002**, *4*, 1623.

[135] Ma, D.; Cai, Q. *Org. Lett.* **2003**, *5*, 3799.

[136] (a) Wolter, M.; Nordmann, G.; Job, G. E.; Buchwald, S. L. *Org. Lett.* **2002**, *4*, 973. (b) Altman, R. A.; Shafir, A.; Choi, A.; Lichtor, P. A.; Buchwald, S. L. *J. Org. Chem.* **2008**, *73*, 284.

[137] Schareina, T.; Zapt, A.; Cotté, A.; Müller, N.; Beller, M. *Org. Process Res. Dev.* **2008**, *12*, 537.

[138] Cristau, H.-J.; Cellier, P. P.; Hamada, S.; Spindler, J.-F.; Taillefer, M. *Org. Lett.* **2004**, *6*, 913.

[139] Palucki, M.; Wolfe, J. P.; Buchwald, S. L. *J. Am. Chem. Soc.* **1996**, *118*, 10333.

[140] Mann, G.; Hartwig, J. F. *J. Am. Chem. Soc.* **1996**, *118*, 13109.

[141] Anderson, K. W.; Ikawa, T.; Tundel, R. E.; Buchwald, S. L. *J. Am. Chem. Soc.* **2006**, *128*, 10694.

[142] (a) Aranyos, A.; Old, D. W.; Kiyomori, A.; Wolfe, J. P.; Sadighi, J. P.; Buchwald, S. L. *J. Am. Chem. Soc.* **1999**, *121*, 4369. (b) Torraca, K. E.; Kuwabe, S.; Buchwald, S. L. *J. Am. Chem. Soc.* **2000**, *122*, 12907. (c) Parrish, P. A.; Buchwald, S. L. *J. Org. Chem.* **2001**, *66*, 2498.

[143] (a) Burgos, C. H.; Barder, T. E.; Huang, X.; Buchwald, S. L. *Angew. Chem., Int. Ed.* **2006**, *45*, 4321. (b) Vorogushin, A. V.; Huang, X.; Buchwald, S. L. *J. Am. Chem. Soc.* **2005**, *127*, 8146.

[144] Mann, G.; Incarvito, C.; Rheingold, A. L.; Hartwig, J. F. *J. Am. Chem. Soc.* **1999**, *121*, 3224.

[145] Harkal, S.; Kumar, K.; Michalik, D.; Zapf, A.; Jackstell, R.; Rataboul, F.; Riermeier, T.; Monsees, A.; Beller, M. *Tetrahedron Lett.* **2005**, *46*, 3237.

[146] Caron, S.; Do, N. M.; Sieser, J. E.; Whritenour, D. C.; Hill, P. D. *Org. Process Res. Dev.* **2009**, *13*, 324.

[147] Ruggeri, S. G.; Vanderplas, B. C.; Anderson, B. G.; Breitenbach, R.; Urban, F. J.; Stewart, A. M., III; Young, G. R. *Org. Process Res. Dev.* **2008**, *12*, 411.

[148] Enache, L. A.; Kennedy, I.; Sullins, D. W.; Chen, W.; Ristic, D.; Stahl, G. L.; Dzekhtser, S.; Erickson, R. A.; Yan, C.; Muellner, F. W.; Krohn, M. D.; Winger, J.; Sandanayaka, V.; Singh, J.; Zembower, D. E.; Kiselyov, A. S. *Org. Process Res. Dev.* **2009**, *13*, 1177.

[149] Sandanayaka, V.; Mamat, B.; Mishra, R. K.; Winger, J.; Krohn, M.; Zhou, L.-M.; Keyvan, M.; Enache, L.; Sullins, D.; Onua, E.; Zhang, J.; Halldorsdottir, G.; Sigthorsdottir, H.; Thorlaksdttir, A.; Sigthorsson, G.; Thorsteinnsdottir, M.; Davies, D. R.; Stewart, L. L.; Zembower, D. E.; Andresson, T.; Kiselyov, A. S.; Singh, J.; Gurney, M. E. *J. Med. Chem.* **2010**, *53*, 573.

[150] Bullock, K. M.; Burton, D.; Corona, J.; Diederich, A.; Glover, B.; Harvey, K.; Michell, M. B.; Trone, M. D., Yule, R.; Zhang, Y.; Toczko, J. F. *Org. Process Res. Dev.* **2009**, *13*, 303.

[151] Payne, J. E.; Bonnefous, C.; Hassig, C. A.; Symons, K. T.; Guo, X.; Nguyen, P.-M.; Annable, T.; Wash, P. L.; Hoffman, T. Z.; Rao, T. S.; Shiau, A. K.; Malecha, J. W.; Noble, S. A.; Hager, J. H.; Smith, N. D. *Bioorg. Med. Chem. Lett.* **2008**, *18*, 6093.

[152] Weiss, M. M.; Harmange, J.-C.; Polverino, A. J.; Bauer, D.; Berry, L.; Berry, V.; Borg, G.; Bready, J.; Chen, D.; Choquette, D.; Coxon, A.; DeMelfi, T.; Doerr, N.; Estrada, J.; Flynn, J.; Graceffa, R. F.; Harriman, S. P.; Kaufman, S.; La, D. S.; Long, A.; Neervannan, S.; Patel, V. F.; Potashman, M.; Regal, K.; Roveto, P. M.; Schrag, M. L.; Starnes, C.; Tasker, A.; Teffera, Y.; Whittington, D. A.; Zanon, R. *J. Med. Chem.* **2008**, *51*, 1668.

[153] Li, X.; Vince, R. *Bioorg. Med. Chem.* **2006**, *14*, 5742.

[154] Humphries, P. S.; Do, Q.-Q. T.; Wilhite, D. M. *Tetrahedron Lett.* **2009**, *50*, 1765.

[155] Torraca, K. E.; Huang, X.; Parrish, C. A.; Buchwald, S. L. *J. Am. Chem. Soc.* **2001**, *123*, 10770.

[156] Humphries, P. S.; Bailey, S.; Do, Q.-Q. T.; Kellum, J. H.; McClellan, G. A.; Wilhite, D. M. *Tetrahedron Lett.* **2006**, *47*, 5333.

[157] (a) Kosugi, M.; Shimizu, T.; Migita, T. *Chem. Lett.* **1978**, *1*, 3–14. (b) Migita, T; Shimizu, T.; Asami, Y.; Shiobara, J.; Kato, Y.; Kosugi, M. *Bull. Chem. Soc. Jpn.* **1980**, *53*, 1385–1389.

[158] Zheng, N.; McWilliams, J. C.; Fleitz, F. J.; Armstrong, J. D., III; Volante, R. P. *J. Org. Chem.* **1998**, *63*, 9606.

[159] Schopfer, U.; Schlapbach, A. *Tetrahedron* **2001**, *57*, 3069.

[160] (a) Itoh, T.; Mase, T. *Org. Lett.* **2004**, *6*, 4587. (b) Mispelaere-Canivet, C.; Spindler, J.-F.; Perrio, S.; Beslin, P. *Tetrahedron* **2005**, *61*, 5253.

[161] Murata, M.; Buchwald, S. L. *Tetrahedron* **2004**, *60*, 7397.

[162] (a) Fernández-Rodríguez, M. A.; Shen, Q.; Hartwig, J. F. *J. Am. Chem. Soc.* **2006**, *128*, 2180. (b) Fernández-Rodríguez, M. A.; Shen, Q.; Hartwig, J. F. *Chem. Eur. J.* **2006**, *12*, 7782. (c) Fernández-Rodríguez, M. A.; Hartwig, J. F. *J. Org. Chem.* **2009**, *74*, 1663.

[163]   (a) Li, G. Y.; Zheng, G.; Noonan, A. F. *J. Org. Chem.* **2001**, *66*, 8677. (b) Li, G. Y. *Angew. Chem., Int. Ed.* **2001**, *40*, 1513.

[164]   Palomo, C.; Oiarbide, M.; López, R.; Gómez-Bengoa, E. *Tetrahedron Lett.* **2000**, *41*, 1283.

[165]   Bates, C. G.; Gujadhur, R. K.; Venkataraman, D. *Org. Lett.* **2002**, *4*, 2803.

[166]   Kwong, F. Y.; Buchwald, S. L. *Org. Lett.* **2002**, *4*, 3517.

[167]   Malmgren, H.; Bäckström, B.; Sølver, E.; Wennerberg, J. *Org. Process Res. Dev.* **2008**, *12*, 1195.

[168]   Zhao, X.; Li, F.; Zhuang, W.; Xue, X.; Lian, Y.; Fan, J.; Fang, D. *Org. Process Res. Dev.* **2010**, *14*, 346.

[169]   Parker, J. S.; Moseley, J. D. *Org. Process Res. Dev.* **2008**, *12*, 1041–1043.

[170]   Moseley, J. D.; Brown, D.; Firkin, C. R.; Jenkin, S. L.; Patel, B.; Snape, E. W. *Org. Process Res. Dev.* **2008**, *12*, 1044.

[171]   Hutton, J.; Jones, A. D.; Lee, S. A.; Martin, D. M. G.; Meyrick, B. R.; Patel, I.; Peardon, R. F.; Powell, L. *Org. Process Res. Dev.* **1997**, *1*, 61.

[172]   Norris, T.; Leeman, K. *Org. Process Res. Dev.* **2008**, *12*, 869.

[173]   Flahive, E. J.; Ewanicki, B. L.; Sach, N. W.; O'Neill-Slawecki, S. A.; Stankovic, N. S.; Yu, S.; Guinness, S. M.; Dunn, J. *Org. Process Res. Dev.* **2008**, *12*, 637.

[174]   Wilson, K. J.; van Niel, M. B.; Cooper, L.; Bloomfield, D.; O'Connor, D.; Fish, L. R.; MacLeod, A. M. *Bioorg. Med. Chem. Lett.* **2007**, *17*, 2643.

[175]   Lee-Dutra, A.; Wiener, D. K.; Arienti, K. L.; Liu, J.; Mani, N.; Ameriks, M. K.; Axe, F. U.; Gebauer, D.; Desai, P. J.; Nguyen, S.; Randal, M.; Thurmond, R. L.; Sun, S.; Karlsson, L.; Edwards, J. P.; Jones, T. K.; Grice, C. A. *Bioorg. Med. Chem. Lett.* **2010**, *20*, 2370.

[176]   Dickens, M. J.; Gilday, J. P.; Mowlem, T. J.; Widdowson, D. A. *Tetrahedron* **1991**, *47*, 8621.

[177]   Lunniss, C.; Eldred, C.; Aston, N.; Craven, A.; Gohil, K.; Judkins, B.; Keeling, S.; Ranshaw, L.; Robinson, E.; Shipley, T.; Trivedi, N. *Bioorg. Med. Chem. Lett.* **2010**, *20*, 137.

[178]   (a) Cacchi, S.; Fabrizi, G.; Goggiamani, A.; Parisi, L. M. *Org. Lett.* **2002**, *4*, 4719. (b) Cacchi, S.; Fabrizi, G.; Goggiamani, A.; Parisi, L. M.; Bernini, R. *J. Org. Chem.* **2004**, *69*, 5608.

[179]   (a) Suzuki, H.; Abe, H. *Tetrahedron Lett.* **1995**, *36*, 6239. (b) Baskin, J. M.; Wang, Z. *Org. Lett.* **2002**, *4*, 4423. (c) Zhu, W.; Ma, D. *J. Org. Chem.* **2005**, *70*, 2696.

[180]   Sturino, C. F.; O'Neill, G.; Lachance, N.; Boyd, M.; Berthelette, C.; Labelle, M.; Li, L.; Roy, B.; Scheigetz, J.; Tsou, N.; Aubin, Y.; Bateman, K. P.; Chauret, N.; Day, S. H.; Lévesque, J.-F.; Seto, C.; Silva, J. H.; Trimble, L. A.; Carriere, M.-C.; Denis, D.; Greig, G.; Kargman, S.; Lamontagne, S.; Mathieu, M.-C.; Sawyer, N.; Slipetz, D.; Abraham, W. M.; Jones, T.; McAuliffe, M.; Piechutta, H.; Nicoll-Griffith, D. A.; Wang, Z.; Zamboni, R.; Young, R. N.; Metters, K. M. *J. Med. Chem.* **2007**, *50*, 794.

[181]   Beaulieu, C.; Guay, D.; Wang, Z.; Leblanc, Y.; Roy, P.; Dufresne, C.; Zamboni, R.; Berthelette, C.; Day, S.; Tsou, N.; Denis, D.; Greig, G.; Mathieu, M.-C.; O'Neill, G. *Bioorg. Med. Chem. Lett.* **2008**, *18*, 2696.

[182]   Campos, K. R.; Journet, M.; Lee, S.; Grabowski, E. J. J.; Tillyer, R. D. *J. Org. Chem.* **2005**, *70*, 268.

[183] Wu, Y.-J.; Zhang, Y.; Good, A. C.; Burton, C. R.; Toyn, J. H.; Albright, C. F.; Macor, J. E.; Thompson, L. A. *Bioorg. Med. Chem. Lett.* **2009**, *19*, 2654.

[184] Alhamadsheh, M. M.; Waters, N. C.; Sachdeva, S.; Lee, P.; Reynolds, K. A. *Bioorg. Med. Chem. Lett.* **2008**, *18*, 6402.

[185] (a) Chan, D. M. T.; Monaco, K. L.; Wang, R.-P.; Winters, M. P. *Tetrahedron Lett.* **1998**, *39*, 2933. (b) Evans, D. A.; Katz, J. L.; West, T. R. *Tetrahedron Lett.* **1998**, *39*, 2937. (c) Lam, P. Y. S.; Clark, C. G.; Saubern, S.; Adams, J.; Winters, M. P.; Chan, D. M. T.; Combs, A. *Tetrahedron Lett.* **1998**, *39*, 2941.

[186] (a) Collman, J. P.; Zhong, M. *Org. Lett.* **2000**, *9*, 1233. (b) Collman, J. P.; Zhong, M.; Zeng, L.; Costanzo, S. *J. Org. Chem.* **2001**, *66*, 1528. (c) Lam, Y. S. P.; Vincent, G.; Clark, C. G.; Deudon, S.; Jadhav, P. K. *Tetrahedron Lett.* **2001**, *42*, 3415. (d) Antilla, J. C.; Buchwald, S. L. *Org. Lett.* **2001**, *3*, 2077.

[187] For a review, see: Chan, D. M. T.; Lam, P. Y. S. In *Boronic Acids*; Hall, D. G., Ed.; Wiley-VCH: Weinheim, 2005; pp. 205–240. See also Ref. 2e.

[188] Zhu, Z.; Sun, Z.-Y.; Ye, Y.; McKittrick, B.; Greenlee, W.; Czarniecki, M.; Fawzi, A.; Zhang, H.; Lachowicz, J. E. *Bioorg. Med. Chem. Lett.* **2009**, *19*, 5218.

[189] Bursavich, M. G.; Gilbert, A. M.; Lombardi, S.; Georgiadis, K. E.; Reifenberg, E.; Flannery, C. R.; Morris, E. A. *Bioorg. Med. Chem. Lett.* **2007**, *17*, 5630.

[190] Schroeder, G. M.; An, Y.; Cai, Z.-W.; Chen, X.-T.; Clark, C.; Cornelius, L. A. M.; Dai, J.; Gullo-Brown, J.; Gupta, A.; Henley, B.; Hunt, J. T.; Jeyaseelan, R.; Kamath, A.; Kim, K.; Lippy, J.; Lombardo, L. J.; Manne, V.; Oppenheimer, S.; Sack, J. S.; Schmidt, R. J.; Shen, G.; Stefanski, K.; Tokarski, J. S.; Trainor, G. L.; Wautlet, B. S.; Wei, D.; Williams, D. K.; Zhang, Y.; Zhang, Y.; Fargnoli, J.; Borzilleri, R. M. *J. Med. Chem.* **2009**, *52*, 1251.

[191] Nakada, Y.; Aicher, T. D.; Le Huerou, Y.; Turner, T.; Pratt, S. A.; Gonzales, S. S.; Boyd, S. A.; Miki, H.; Yamamoto, T.; Yamaguchi, H.; Kato, K.; Kitamura, S. *Bioorg. Med. Chem.* **2010**, *18*, 2785.

[192] Wang, W.; Devasthale, Farrelly, D.; Gu, L.; Harrity, T.; Cap, M.; Chu, C.; Kunselman, L.; Morgan, N.; Ponticiello, R.; Zebo, R.; Zhang, L.; Locke, K.; Lippy, J.; O'Malley, K.; Hosagrahara, V.; Zhang, L.; Kdiyala, P.; Chang, C.; Muckelbauer, J.; Doweyko, A. M.; Zahler, R.; Ryono, D.; Hariharan, N.; Cheng, P. T. W. *Bioorg. Med. Chem. Lett.* **2008**, *18*, 1939.

[193] Boyer, S. H.; Jiang, H.; Jacintho, J. D.; Reddy, M. V.; Li, H.; Li, W.; Godwin, J. L.; Schulz, W. G.; Cable, E. E.; Hou, J.; Wu, R.; Fujitaki, J. M.; Hecker, S. J.; Erion, M. D. *J. Med. Chem.* **2008**, *51*, 7075.

[194] Hammach, A.; Barbosa, A.; Gaenzler, F. C.; Fadra, T.; Goldberg, D.; Hao, M.-H.; Kroe, R. R.; Liu, P.; Qian, K. C.; Ralph, M.; Sarko, C.; Soleymanzadeh, F.; Moss, N. *Bioorg. Med. Chem. Lett.* **2006**, *16*, 6316.

[195] Quach, T. D.; Batey, R. A. *Org. Lett.* **2003**, *5*, 1381.

[196] Herradura, P. S.; Pendola, K. A.; Guy, R. K. *Org. Lett.* **2000**, *2*, 2019.

[197] Lengar, A.; Kappe, C. O. *Org. Lett.* **2004**, *6*, 771.

[198] Mugnaini, C.; Manetti, F.; Esté, J. A.; Clotet-Codina, I.; Maga, G.; Cancio, R.; Botta, M.; Corelli, F. *Bioorg. Med. Chem. Lett.* **2006**, *16*, 3541.

[199] Huang, F.; Batey, R. A. *Tetrahedron* **2007**, *63*, 7667.

[200] For a review on transition metal-catalyzed carbon–boron bond formation, see: Ishiyama, T.; Miyaura, N. *Chem. Rec.* **2004**, *3*, 271–280.

[201] Ishiyama, T.; Murata, M.; Miyaura, N. *J. Org. Chem.* **1995**, *60*, 7508.

[202] (a) Ishiyama, T.; Itoh, Y.; Kitano, Y.; Miyaura, N. *Tetrahedron Lett.* **1997**, *38*, 3447. (b) Giroux, A.; Han, Y.; Prasit, P. *Tetrahedron Lett.* **1997**, *38*, 3841.

[203] Ishiyama, T.; Ishida, K.; Miyaura, N. *Tetrahedron* **2001**, *57*, 9813.

[204] Fürstner, A.; Seidel, G. *Org. Lett.* **2002**, *4*, 541.

[205] Billingsley, K.; Barder, T. E.; Buchwald, S. L. *Angew. Chem., Int. Ed.* **2007**, *46*, 5359.

[206] (a) Murata, M.; Watanabe, S.; Masuda, Y. *J. Org. Chem.* **1997**, *62*, 6458. (b) Murata, M.; Oyama, T.; Watanabe, S.; Masuda, Y. *J. Org. Chem.* **2000**, *65*, 164.

[207] Broutin, P.-E.; Cerna, I.; Campaniello, M.; Leroux, F.; Colobert, F. *Org. Lett.* **2004**, *6*, 4419.

[208] Murata, M.; Sambommatsu, T.; Watanabe, S.; Masuda, Y. *Synlett* **2006**, 1867.

[209] Baudoin, O.; Guénard, D.; Guéritte, F. *J. Org. Chem.* **2000**, *65*, 9268.

[210] Billingsley, K.; Buchwald, S. L. *J. Org. Chem.* **2008**, *73*, 5589.

[211] Zhu, W.; Ma, D. *Org. Lett.* **2006**, *8*, 261.

[212] Menzel, K.; Machrouhi, F.; Bodenstein, M.; Alorati, A.; Cowden, C.; Gibson, A. W.; Bishop, B.; Ikemoto, N.; Nelson, T.; Kress, M. H.; Frantz, D. E. *Org. Process Res. Dev.* **2009**, *13*, 519.

[213] Molander, G. A.; Biolatto, B. *J. Org. Chem.* **2003**, 4302.

[214] Miller, W. D.; Fray, A. H.; Quatroche, J. T.; Sturgill, C. D. *Org. Process Res. Dev.* **2007**, *11*, 359.

[215] Iimura, S.; Wu, W. *Tetrahedron Lett.* **2010**, *51*, 1353.

[216] Jones, D. E.; South, M. S. *Tetrahedron* **2010**, *66*, 2570.

[217] O'Donnell, C. J.; Peng, L.; O'Neill, B. T.; Arnold, E. P.; Mather, R. J.; Sands, S. B.; Shrikhande, A.; Lebel, L. A.; Spracklin, D. K.; Nedza, F. M. *Bioorg. Med. Chem. Lett.* **2009**, *19*, 4747.

[218] Hu, E.; Tasker, A.; White, R. D.; Kunz, R. K.; Human, J.; Chen, N.; Bürli, R.; Hungate, R.; Novak, P.; Itano, A.; Zhang, X.; Yu, V.; Nguyen, Y.; Tudor, Y.; Plant, M.; Flynn, S.; Xu, Y.; Meagher, K. L.; Whittington, D. A.; Ng, G. Y. *J. Med. Chem.* **2008**, *51*, 3065.

[219] DiMauro, E. F.; Newcomb, J.; Nunes, J. J.; Bemis, J. E.; Boucher, C.; Buchanan, J. L.; Buckner, W. H.; Cee, V. J.; Chai, L.; Deak, H. L.; Epstein, L. F.; Faust, T.; Gallant, P.; Geuns-Meyer, S. D.; Gore, A.; Gu, Y.; Henkle, B.; Hodous, B. L.; Hsieh, F.; Huang, X.; Kim, J. L.; Lee, J. H.; Martin, M. W.; Masse, C. E.; McGowan, D. C.; Metz. D.; Mohn, D.; Morgenstern, K. A.; Oliveira-dos-Santos, A.; Patel, V. F.; Powers, D.; Rose, P. E.; Schneider, S.; Tomlinson, S. A.; Tudor, Y.-Y.; Turci, S.; Welcher, A. A.; White, R. D.; Zhao, H.; Zhu, L.; Zhu, X. *J. Med. Chem.* **2006**, *49*, 5671.

[220] Lazarova, T. I.; Jin, L.; Rynkiewicz, M.; Gorga, J. C.; Bibbins, F.; Meyers, H. V.; Babine, R.; Strickler, J. *Bioorg. Med. Chem. Lett.* **2006**, *16*, 5022.

[221] Walsh, T. F.; Toupence, R. B.; Ujjainwalla, F.; Young, J. R.; Goulet, M. T. *Tetrahedron* **2001**, *57*, 5233.

[222] Michellys, P.-Y.; D'Arrigo, J.; Grese, T. A.; Karanewsky, D. S.; Leibowitz, M. D.; Mais, D. A.; Mapes, C. M.; Reifel-Miller, A.; Runta, D.; Boehm, M. F. *Bioorg. Med. Chem. Lett.* **2004**, *14*, 1593.

[223] Shi, S.; Zhu, S.; Gerritz, S. W.; Rachwal, B.; Ruan, Z.; Hutchins, R.; Kakarla, R.; Sofia, M. J. Sutton, J.; Cheney, D. *Bioorg. Med. Chem. Lett.* **2009**, *19*, 6477.

[224] Zhu, B.; Marinelli, B. A.; Goldschmidt, R.; Foleno, B.; Hilliard, J. J.; Bush, K.; Macielag, M. J. *Bioog. Med. Chem. Lett.* **2009**, *19*, 4933.

[225] Koolman, H.; Heinrich, T.; Böttcher, H.; Rautenberg, W.; Reggelin, M. *Bioorg. Med. Chem. Lett.* **2009**, *19*, 1879.

[226] (a) Kulkarni, S. S.; Newman, A. H. *Bioorg. Med. Chem. Lett.* **2007**, *17*, 2074. (b) Kulkarni, S. S.; Newman, A. H. *Bioorg. Med. Chem. Lett.* **2007**, *17*, 2987.

[227] Liu, P.; Lin, L. S.; Hamill, T. G.; Jewell, J. P.; Lanza, T. J., Jr., ; Gibson, R. E.; Krause, S. M.; Ryan, C.; Eng, W.; Sanabria, S.; Tong, X.; Wang, J.; Levorse, D. A.; Owens, K. A.; Fong, T. M.; Shen, C.-P.; Lao, J.; Kumar, S.; Yin, W.; Payack, J.; Springfield, S.; Hargreaves, R.; Burns, H. D.; Goulet, M. T.; Hagmann, W. K. *J. Med. Chem.* **2007**, *50*, 3427.

[228] Lin, H.; Yamashita, D. S.; Zeng, J.; Xie, R.; Verma, S.; Luengo, J. I.; Rhodes, N.; Zhang, S.; Robell, K. A.; Choudhry, A. E.; Lai, Z.; Kumar, R.; Minthorn, E. A.; Brown, K. K.; Heerding, D. A. *Bioorg. Med. Chem. Lett.* **2010**, *20*, 679.

[229] Mkhalid, I. A. I.; Barnard, J. H.; Marder, T. B.; Murphy, J. M.; Hartwig, J. F. *Chem. Rev.* **2010**, *110*, 890–931.

[230] Fischer, D. F.; Sarpong, R. *J. Am. Chem. Soc.* **2010**, *132*, 5926.

[231] Ishiyama, T.; Takagi, J.; Ishida, K.; Miyaura, N.; Anastasi, N. R.; Hartwig, J. F. *J. Am. Chem. Soc.* **2002**, *124*, 390.

[232] (a) Hirao, T.; Masunaga, T.; Ohshiro, Y.; Agawa, T. *Synthesis* **1981**, 56. (b) Hirao, T.; Masunaga, T.; Yamada, N.; Ohshiro, Y.; Agawa, T. *Bull. Chem. Soc. Jpn.* **1982**, *55*, 909.

[233] (a) For reviews, see: Beletskaya, I. P.; Kazankova, M. A. *Russ. J. Org. Chem.* **2002**, *38*, 1391–1430. (b) Schwan, A. L. *Chem. Soc. Rev.* **2004**, *33*, 218–224. (c) For a review on phosphonylation of heterocycles, see: Van der Jeught, S.; Stevens, C. V. *Chem. Rev.* **2009**, *109*, 2672–2702.

[234] (a) Gelman, D.; Jiang, L.; Buchwald, S. L. *Org. Lett.* **2003**, *5*, 2315. (b) Huang, C.; Tang, X.; Fu, H. A.; Jiang, Y. Y.; Zhao, Y. F. *J. Org. Chem.* **2006**, *71*, 5020.

[235] Sørensen, M. D.; Blæhr, L. K. A.; Christensen, M. K.; Høyer, T.; Latini, S.; Hjarnaac, P.-J. V. Björkling, F. *Bioorg. Med. Chem.* **2003**, *11*, 5461.

[236] Sawa, M.; Kiyoi, T.; Kurokawa, K.; Kumihara, H.; Yamamoto, M.; Miyasaka, T.; Ito, Y.; Hirayama, R.; Inoue, T.; Kirii, Y.; Nishiwaki, E.; Ohmoto, H.; Maeda, Y.; Ishibushi, E.; Inoue, Y.; Yoshino, K.; Kondo, H. *J. Med. Chem.* **2002**, *45*, 919.

[237] Oishi, S.; Kang, S.-U.; Liu, H.; Zhang, M.; Yang, D.; Deschamps, J. R.; Burke, T. R. *Tetrahedron* **2004** *60*, 2971.

[238] Jiang, W.; Allan, G.; Fiordeliso, J. J.; Linton, O.; Tannenbaum, P.; Xu, J.; Zhu, P.; Gunnet, J.; Demarest, K.; Lundeen, S.; Sui, Z. *Bioorg. Med. Chem.* **2006**, *14*, 6726.

[239] Kim, D.-K.; Lee, J. Y.; Park, H.-J.; Thai, K. M. *Bioorg. Med. Chem. Lett.* **2004**, *14*, 2099.

[240] Kumar, R. J.; Chebib, M.; Hibbs, D. E.; Kim, H.-L.; Johnston, G. A. R.; Salam, N. K.; Hanrahan, J. R. *J. Med. Chem.* **2008**, *51*, 3825.

[241] Dumond, Y. R.; Montchamp, J.-L. *J. Organomet. Chem.* **2002**, *653*, 252.

[242] Ding, C. Z.; Rovnyak, G. C.; Misra, R. N.; Grover, G. J.; Miller, A. V.; Ahmed, S. Z.; Kelly, Y.; Normandin, D. E.; Sleph, P. G.; Atwal, K. S. *J. Med. Chem.* **1999**, *42*, 3711.

[243] Lei, H.; Stoakes, M. S.; Schwabacher, A. W. *Synthesis* **1992**, 1255.

[244] For Ni-catalyzed amination of aryl pivalates, see: Shimasaki, T.; Tobisu, M.; Chatani, N. *Angew. Chem., Int. Ed.* **2010**, *49*, 2929. Also see references cited therein for other examples of Ni-catalyzed amination of aryl halides.

[245] Buchwald, S. L.; Bolm, C. *Angew Chem., Int. Ed.* **2009**, *48*, 5586 and references cited therein.

[246] For a review, see: Serdyuk, O. V.; Abaev, V. T. *Russ. Chem. Rev.* **2009**, *78*, 1031–1045.

[247] (a) Watson, D. A.; Su, M.; Teverovskiy, G.; Zhang, Y.; García-Fortanet, J.; Kinzel, T.; Buchwald, S. L. *Science* **2009**, *321*, 1661. (b) Wang, X.; Mei, T.-S.; Yu, J. Q. *J. Am. Chem. Soc.* **2009**, *131*, 7520. (c) Furuya, T.; Kaiser, H. M.; Ritter, T. *Angew. Chem., Int. Ed.* **2008**, *47*, 5993. (d) Furuya, T.; Ritter, T. *J. Am. Chem. Soc.* **2008**, *130*, 10060. (e) Grushin, V. V. *Acc. Chem. Res.* **2010**, *43*, 160. (f) Yandulov, D. V.; Tran, N. T. *J. Am. Chem. Soc.* **2007**, *129*, 1342. (g) Hull, K. L.; Anani, W. Q.; Sanford, M. S. *J. Am. Chem. Soc.* **2006**, *128*, 7134. (h) Kaspi, A. W.; Yahav-Levi, A.; Goldberg, I.; Vigalok, A. *Inorg. Chem.* **2008**, *47*, 5. (i) Vigalok, A. *Chem. Eur. J.* **2008**, *14*, 5102.

[248] For a highlight, see: Armstrong, A.; Collins, J. C. *Angew. Chem., Int. Ed.* **2010**, *49*, 2.

4

ASYMMETRIC CROSS-COUPLING REACTIONS

Vince Yeh and William A. Szabo

4.1 INTRODUCTION

In this chapter, we examine recent advances in transition metal catalysis for the synthesis of enantiomerically enriched small molecules with an emphasis on molecules of therapeutic interest [1]. This is an appropriate connection since much of the innovation in asymmetric catalysis has been driven by efforts in drug development. The role that chirality plays in small-molecule therapeutics cannot be overstated. For example, it is estimated that in 2009, 40% of the drugs on the market in the United States were chiral and nonracemic [2a], and they accounted for over $200 billion in annual sales [2b] (note that sales in 2009 included $11.4 billion for the world's best-selling drug, Pfizer's Lipitor® [2c]). The prevalence of chiral drugs has created a need for APIs and their stereoisomers having high enantiomeric purity. Most importantly, there is a *functional* necessity because stereoisomers may possess different pharmacological properties. Second, there is a *regulatory* necessity. Current FDA guidance [3] mandates that all enantiomers of a chiral API in development be screened separately for pharmacological activity, pharmacokinetic properties, and toxicity. As a result, there is a need for enabling technologies for the synthesis and production of single enantiomers having high degrees of isomeric purity.

The introduction of chirality into a small molecule is generally dependent on the stage of drug development. In the *discovery* phase, the emphasis is on rapid, small-scale synthesis, with little attention paid to the economics of the process. Racemic molecules are often produced by conventional methods (e.g., using chiral-pool building blocks and classical resolution) for biological testing. In the *development* stage, a single isomer is prepared for pharmacokinetic, toxicological, and proof-of-concept studies. Cost becomes a significant issue, and different synthetic routes and methods, including asymmetric synthesis and the use of chiral auxiliaries, are considered. It has been suggested by Pfizer's Hawkins [4] that the chances of using an asymmetric catalytic process on a commercial scale are maximized if that process is used (1) early in designing a synthetic route, and (2) in a late synthetic step. Finally, once a promising chiral drug candidate has been identified, the same considerations apply for larger-scale *production* as they do for any other small molecule, namely scalability (including technical feasibility and robustness), economics (including the availability of the raw materials and the costs of the catalyst metal, ligand, and any royalties relating to freedom-to-operate issues), and environmental concerns.

Asymmetric transition metal catalyzed cross-coupling reactions can provide advantages over more-conventional synthetic methods in several respects. They can offer synthetic efficiency since the one-step creation of a stereogenic center and bond formation is usually more efficient than a corresponding multistep process. They can provide high stereoselectivity, which often rivals that of enzymatic processes. Finally, transition metal catalysts are often available in enantiomeric forms, thereby providing access to *both* enantiomers of a chiral drug for screening.

The examples in this chapter serve to illustrate the proven and potential uses of asymmetric transition metal catalysis in small-molecule drug development.

4.2 CARBON–CARBON COUPLING REACTIONS

4.2.1 Addition to C=O Bonds

Nucleophilic addition to carbonyls groups is one of the most fundamental and reliable processes for generating carbon–carbon bonds. This reaction has the potential to add complexity to a molecule in a single step by forming the carbon–carbon bond while generating a stereogenic center at the same time. Despite the tremendous innovations that have been made by the academic labs in the catalytic asymmetric addition of organometallic species to carbonyl compounds, which include a variety of reaction types such as the aldol reaction, organozinc addition, and addition of cyanides, industrial application of these new technologies has been slow. This is perhaps due to the large amount of research and investment in the area of asymmetric hydrogenation, which in many cases can lead to the same secondary alcohol products. However, several useful applications of asymmetric addition reactions to carbonyl groups have started to appear in the literature by both academic and industrial groups. The area of organozinc addition reactions has been studied extensively since 1984 [5] and hundreds of ligands have been published [6]. However, most of the papers are very limited in scope, describing only diethylzinc as a nucleophile. Phenyl, alkenyl, and alkynyl zinc species can also undergo enantioselective additions to aldehydes yielding more functionalized products.

Such reactions have been utilized by Eli Lilly chemists to synthesize the mGlu2 potentiator **1** for the potential treatment of migraine headache (Scheme 4.1) [7]. The stereogenic center in **2** was generated by the asymmetric aryl transfer of a mixed organozinc species **3** to aldehyde **4**. The methodology was originally developed by Bolm in which an excess of aryl boronic acid (>2 equivalents versus the aldehyde electrophile) was transmetallated to a mixed organozinc species using an excess of diethylzinc (>7 equivalents) at 60°C for 20 h [8]. Over the course of optimization studies by the Lilly chemists in which they monitored the formation of the mixed zinc species **3** by solution IR and real-time gas analyzer and mass spectrometry for the evolution of ethane gas, they found that the transmetallation could be carried out using boroxine **5** instead of boronic acid, and the transmetallation reaction took place over 2 h at 60°C. They also found that the amount of nucleophile could be reduced to nearly half of that of the original method. Under the optimized condition 0.45 equivalent of boroxine (35 mol% excess of aryl equivalent versus the original 130 mol% excess) was treated with 2.8 equivalents of Et$_2$Zn to generate *in situ* the mixed organozinc species which added to cyanoaldehyde **4** with chiral ligand **6** to produce product **2** in 88% yield and $>94\%$ee. The reaction was carried out on 12 g of the aldehyde and the chiral ligand was recovered during workup.

A modification of the above asymmetric arylation reaction has been developed by the Walsh group in which they generated the mixed organozinc species *in situ* from aryl bromides, obviating the need to isolate the boronic acid intermediate, thereby significantly enhancing the step economy of this reaction [9]. An aryl bromide such as **7** (Scheme 4.2) was first transmetallated to an aryllithium with *n*-BuLi. Zinc chloride followed by *n*-BuLi were subsequently added to the reaction mixture to provide the

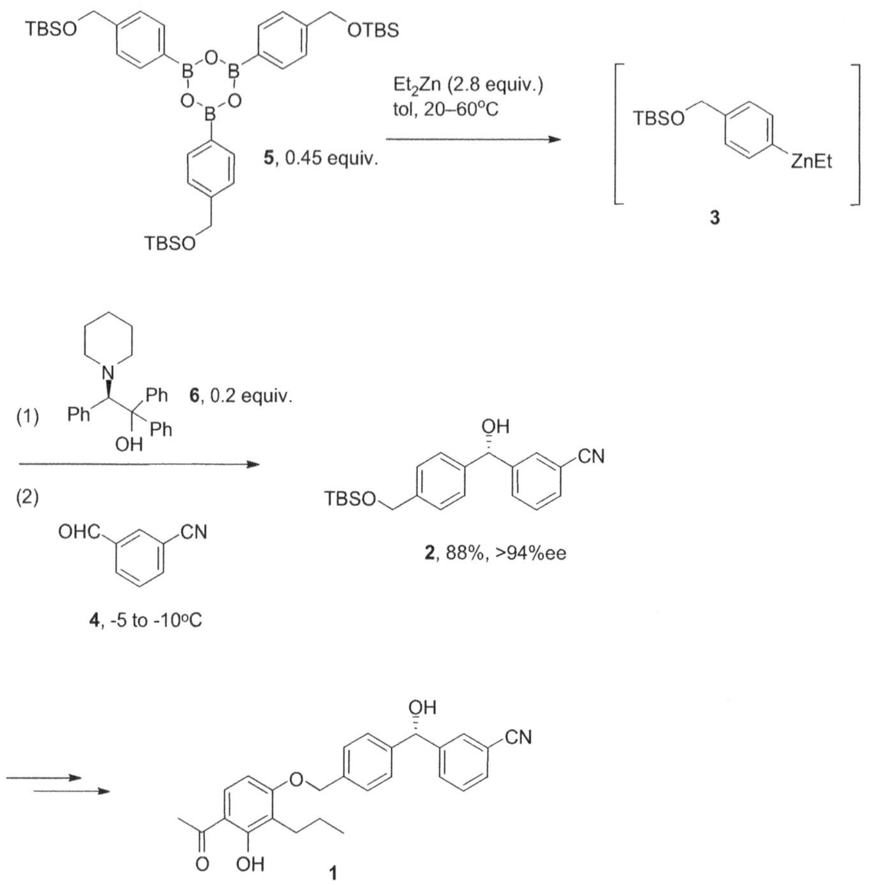

SCHEME 4.1   Synthesis of mGlu2 potentiator **1**.

requisite mixed organozinc species. It was found that the addition of a lithium chelator such as tetraethylethylenediamine (TEEDA) was crucial in reducing lithium salt-promoted background nonenantioselective addition of zinc nucleophiles to the aldehyde. The asymmetric addition was promoted by a MIB ligand (Scheme 4.2), which gave products in 80–90% yields and 85–97%ee. One of the drawbacks of this procedure was the need to use of 2 equivalents of aryl bromide versus the aldehyde coupling partner. Further optimization in the stoichiometry and the choice of ligands might be necessary if the reaction were used in an industrial scale. The utility of this procedure was demonstrated in an asymmetric synthesis of a retinoic acid receptor agonist (S)-BMS184394 (**8**).

3-Aryl-3-hydroxy-2-oxindoles are privileged structures in medicinal chemistry. For example, SM-130686 (**9**, Scheme 4.3) is a highly potent nonpeptidic growth hormone secretagogue [10] and compounds such as **10** show high affinity for the

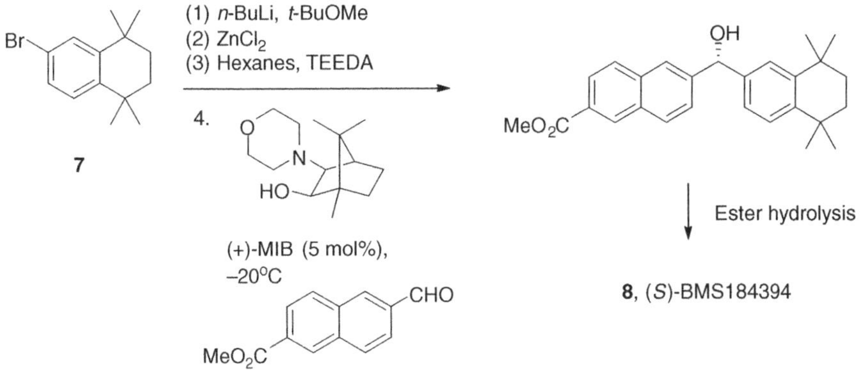

SCHEME 4.2   Synthesis of (S)-BMS184394 (**8**).

arginine–vasopressin (AVP) V1b receptor and the oxytocin receptors [11]. Such structures have inspired a number of academic groups to develop asymmetric additions of aryl nucleophiles to isatins. The Hayashi group has shown that aryl and alkenylboronic nucleophiles can be added asymmetrically to various isatins using a Rh/MeO–mop complex to give arylated products in good yields and high ee's [12]. For example, addition of phenyl boronic acid to isatin **11** using 5 mol% of rhodium and 10 mol% of ligand (R)-MeO-mop (**12**) in a basic aqueous THF solution at 50°C gave adduct **13** in 92% yield and 90%ee, as shown in Scheme 4.3. The published procedure utilized an excess (2 equivalents) of boronic acid per equivalent of isatin, a condition that is not ideal for large-scale synthesis. It is common to observe a large amount of proteo-deboronated product under the current procedure, hence the need for an excess of boronic acid to drive the reaction to completion. However, it is conceivable that further optimization can reduce the amount of nucleophile needed (see Section 2.3.1 on Michael reactions). An alternative route to 3-aryl-3-hydroxy-2-oxindoles was disclosed by the Shibasaki group using inter-molecular aryl silane additions to isatins catalyzed by a CuF/chiral phosphine ligand [13]. The reaction worked well on isatin and 5- or 6-substituted isatins. However, the intermolecular reaction failed to provide product of good yield and ee when it was applied to a 4-trifluoromethyl-substituted isatin required for the synthesis of SM-130686 (**9**).

The challenge was attributed to the steric hindrance by the trifluoromethyl group in the vicinity of the reaction center. To overcome lack of reactivity, an intramolecular version of the arylation was developed. Arylboronates were used instead of arylsi-lanes due to the robustness of arylboronates under various reaction and purification conditions needed to reach the cyclization substrate **14**. The cyclization was catalyzed by a relatively high loading of CuF (10 mol%) and bisphosphine ligand **15** (14 mol%), and the product was obtained in useful 90% yields and 87% enantiomeric excesses. The catalytic asymmetric synthesis of structures with con-gested tetrasubstituted centers such as SM-130686 (**9**) proved to be challenging even with the most state-of-the-art method given here. These examples certainly highlight

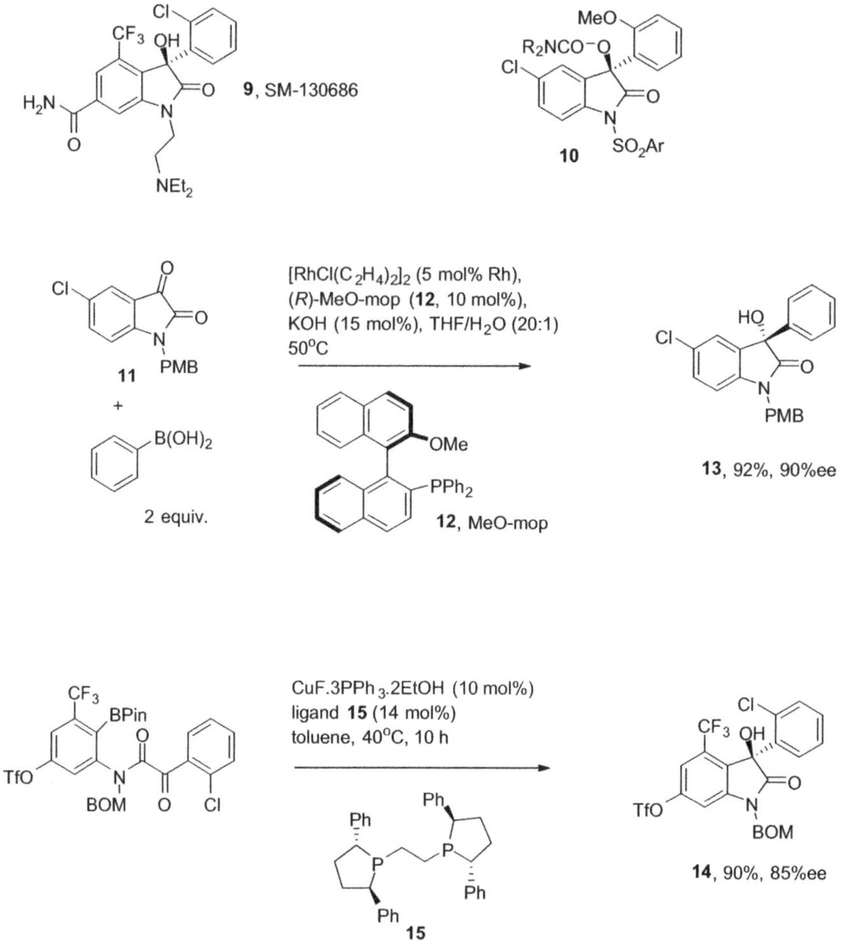

SCHEME 4.3 Syntheses of oxindoles.

the need for continual developments of new C–C bond forming to address these challenges.

Cyanide anion is a powerful and versatile one-carbon nucleophile that synthetic chemists have in their arsenal. Addition of cyanide to carbonyls or imines generates chiral products that can be easily elaborated to acids, amines, aldehydes, or ketones. Asymmetric additions of cyanide to aldehydes and imines (Strecker reactions) have been studied extensively over the years. In addition to a number of metal catalysts, this type of reaction can also be promoted by several highly efficient enzymes and organocatalysts [14]. Unlike asymmetric additions to aldehydes, asymmetric additions to ketones present further challenges. The energy for the transition state of ketone addition is higher than that of aldehyde addition due to a more sterically congested environment as the nucleophile approaches the ketone substrate.

Furthermore, differentiation of the two substituents on a ketone substrate is more challenging than for an aldehyde. Hence, catalysts for such a reaction must have high reactivity and exquisite differentiation between the prochiral faces of the ketone substrates.

Shibasaki's group has developed a class of catalysts for the cyanosilylation of ketones based on carbohydrate-derived ligands such as **16** in Scheme 4.4 [15]. The design was based on the idea that the Lewis acid and Lewis base functionalities are built within the catalyst structure in which the Lewis acid activates the electrophilic substrate and the Lewis base activates the nucleophile. A number of reviews have been published in which Shibasaki illustrates these ideas in depth with a number of versatile catalysts from his group [16]. It was shown that cyanide addition products of opposite absolute stereochemistry can be obtained from a ligand of the same stereochemistry by simply changing the central metal species and metal versus ligand stoichiometry. Hence, a titanium-based catalyst leads to an *R* product **17** and a gadolinium-based catalyst leads to the corresponding *S* product, **18**.

So far there have not been publications from industrial groups utilizing such catalysts in large-scale syntheses. Shibasaki's group has showcased these catalysts in the enantioselective synthesis of (20*S*)-camptothecin (**19**), an important drug in cancer chemotherapy [17]. Addition of TMSCN was effected by a catalyst generated from Sm(O-*i*-Pr)₃ and ligand **16** in a 1:1.8 ratio in cold EtCN, as shown in Scheme 4.5. The resulting adduct **20** was obtained in nearly quantitative yield and 84%ee. It was then improved to >99%ee by recrystallization at a later stage in the synthetic sequence.

Oxybutynin (**21**) is an anticholinergic medication used to relieve urinary and bladder difficulties, including frequent urination and inability to control urination [18].

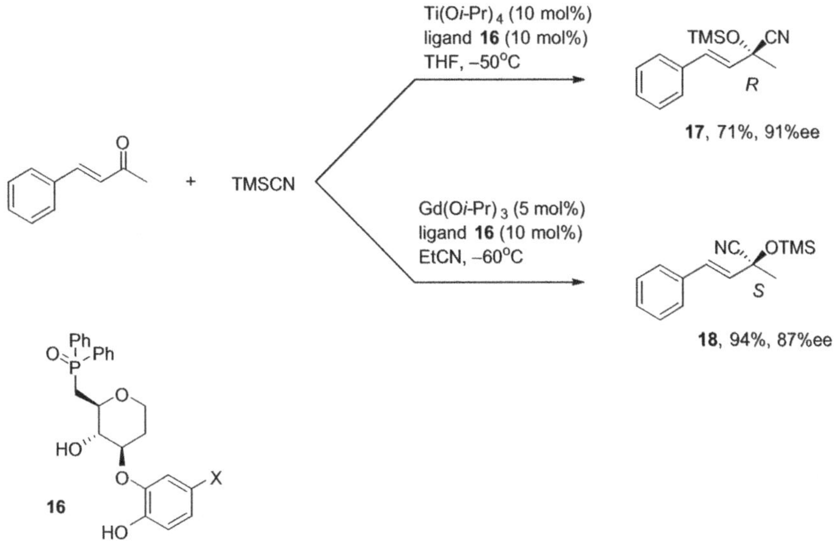

SCHEME 4.4   Cyanosilylation of ketones.

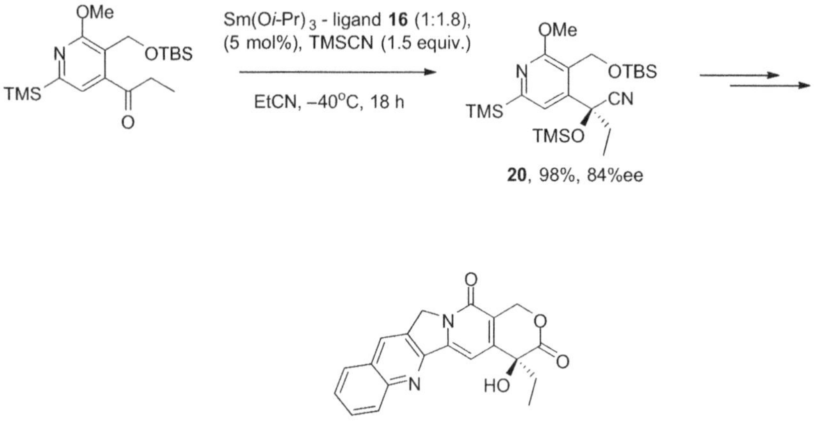

SCHEME 4.5 Synthesis of (20*S*)-camptothecin (**19**).

The enantioselective synthesis was carried out using the Gd/ligand-**16** catalyst on a 100 g scale [19]. Despite the seemingly small steric difference between the phenyl and the cyclohexyl groups, the cyanosilylation product **22** was obtained in quantitative yield and 94%ee. A 100 g demonstration reaction was carried out by the group using only 1 mol% of the gadolinium catalyst. The reaction was completed in 40 h in a −40°C medium. After a standard aqueous workup, 98% of the ligand could be recovered. This example demonstrated the robustness of the gadolinium catalyst in a drug synthesis setting.

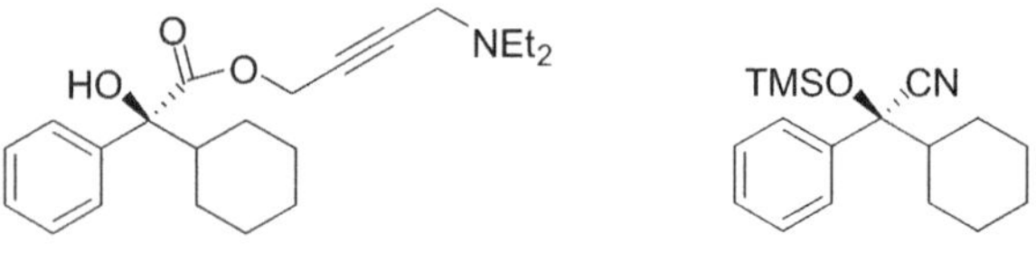

21, Oxybutynin

22, 100%, 94%ee,
>100 g scale

The asymmetric nitro aldol (Henry) reaction has been extensively studied in academia and a number of efficient catalysts have been developed for this process. The industrial applications of these asymmetric Henry reaction catalysts have been rare despite numerous demonstrations of API syntheses [20]. A Novartis process group reported a catalyst-promoted diastereoselective synthesis of a key building block for the antitumor marine natural product, bengamide [21]. The catalyst that the group utilized was a lanthanide-based bis-metallic-BINOL complex developed by Shibasaki's group [22]. Incidentally, this class of catalysts is one of the most studied in Shibasaki's group and variations of the catalyst system have been used in a range of reactions including direct aldol reactions and Michael reactions. As shown in Scheme 4.6, the newly generated stereogenic center of the aldol product was controlled by the catalyst, and the preexisting stereogenic center had minimal effect

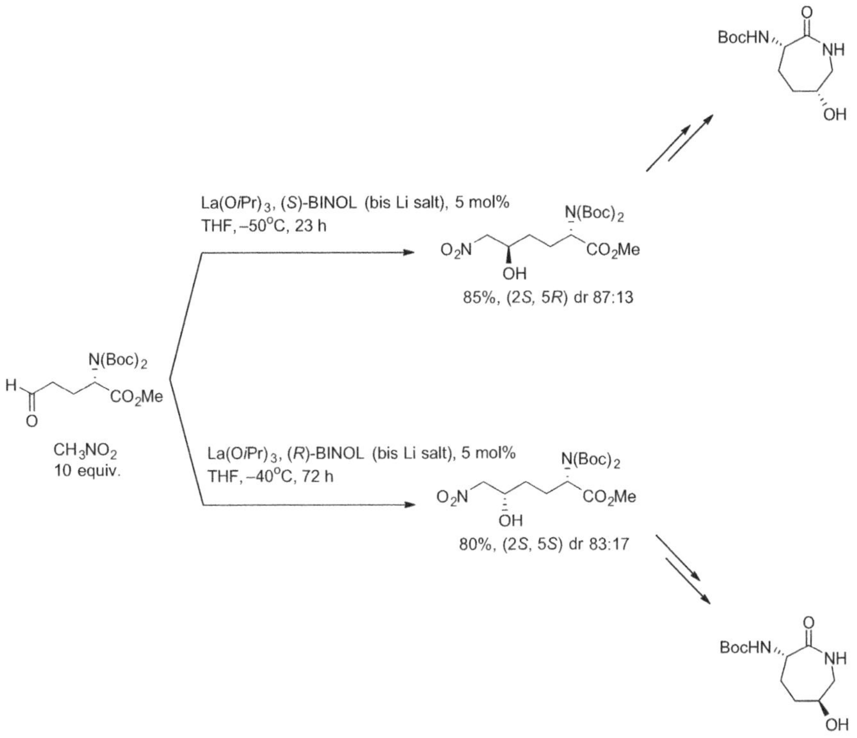

SCHEME 4.6 Asymmetric nitro aldol (Henry) reaction.

on the reaction outcome. The diastereomers could be synthesized in good yields and selectivity using stereo-complimentary catalysts. The stereoselectivity of the reaction can probably be further optimized, but the authors have shown that the desired products could be enriched to near homogeneity by recrystallization.

4.2.2 Addition to C=N Bonds

The importance and prevalence of chiral amines in APIs cannot be overstated. Asymmetric addition of a carbon nucleophile to an imine electrophile is clearly one of the most efficient ways of synthesizing a chiral amine product since the carbon–carbon bond formation is accompanied by the generation of a stereogenic center. A number of chiral auxiliary-based imine addition reactions have been invented. In particular, the chiral sulfinamide chemistry popularized by Ellman's group has found extensive application in API syntheses by industrial groups [23]. Recently, advances have been made in transition metal-based enantioselective catalysis variants of this important reaction. Although a number of organometallic nucleophiles such as organozinc [24], titanium [25], and boron have been examined in their addition to imines, organoboron reagents are the most attractive due to the large number and diversity of commercially available derivatives and their relative stability in ambient conditions versus the other organometallic species. The first example of arylboronic acid addition to an imine was published by Miyaura's group [26]. Since then a number of enantioselective variants such as diene **23** and phosphoramidite **24** have been reported using a variety of chiral ligands. One of the most validated procedures came from the Ellman group using commercially available [RhCl(cod)]₂ and deguPhos (**25**) as catalyst precursors, a significant advantage for the reaction to be applied on an industrial scale [27]. As shown in Scheme 4.7, α-carbamoyl sulfones such as **26** were used as ambiently stable and storable precursors of *N*-Boc imines. The imine functionality was revealed under the basic reaction conditions via elimination of the sulfonate anion. The active catalyst was generated by the preincubation of [RhCl(cod)]₂ and deguPhos before addition to the reaction mixture to ensure high enantioselectivity. The authors commented on the need to recrystallize the starting boronic acid to minimize the unreactive boroxine species to provide consistently high product yields. Both aromatic and aliphatic imines are viable substrates in this reaction. So far only aromatic boronic acids have been used as nucleophiles. Because of its good yields and selectivity, wide substrate scope, and

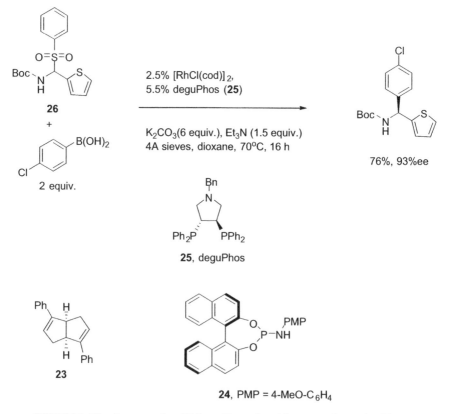

SCHEME 4.7 Asymmetric addition of boronic acids to α-carbamoyl sulfones.

its simplicity of reaction operation, this reaction has the potential to find application on an industrial scale.

An asymmetric synthesis of (+)-dihydrotetrabenazine (**27**, DTBZ) was published by a General Electric group [28] utilizing the Pd(II)-BINAP-catalyzed addition of malonates to dihydroisoquinolines technology first published by the Sodeoka group (Scheme 4.8) [28b]. DTBZ is a synthetic analog of the alkaloid emetine which possess potent inhibitory activity toward vesicular monoamine transporter 2 (VMAT-2). ^{11}C-Labeled (+)-DTBZ can be used as a brain imaging agent for VMAT-2.

Following Sodeoka's original procedure, the GE group performed the reaction on a 7 g scale in 0.66 M CH$_2$Cl$_2$ solution at ~2.5°C using only 0.8 mol% of catalyst. Although the scale was not large, the low catalyst loading, high concentration, and mild reaction temperature of the reaction should make the reaction amenable to scale-up. The malonate adduct **28** was isolated in 94% yield and 97%ee. The synthesis of (+)-DTBZ was completed in eight additional steps in 16% overall yield.

The Reissert reaction is the addition of cyanide ion to quinolines or isoquinolines in the presence of acid chlorides to generate highly useful chiral products, which are considered to be privileged structures in terms of biological activity. Despite the usefulness of the reaction, most of the asymmetric Reissert-type reactions rely on stoichiometric amounts of chiral acylating reagents as chiral auxiliaries [29].

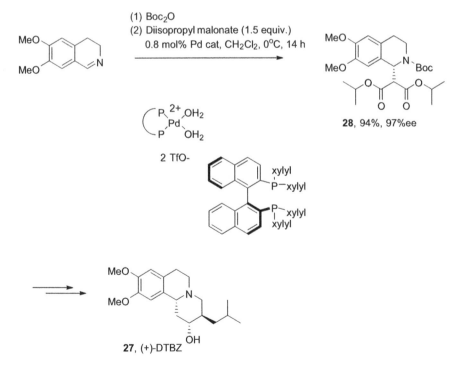

SCHEME 4.8   Asymmetric synthesis of (+)-DTBZ (**27**).

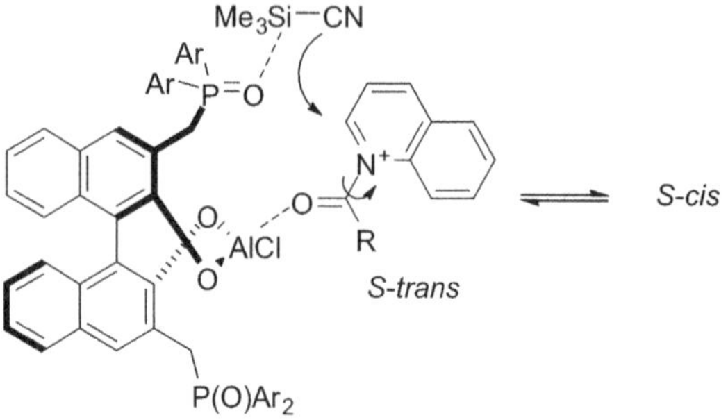

FIGURE 4.1   Aluminum-BINOL-based bifunctional catalyst.

Shibasaki's group was the first to report asymmetric catalytic Reissert reactions using an Al-BINOL-based bifunctional catalyst (**29**) [30]. The catalyst differentiates between the (S)-*trans* and (S)-*cis* isomers of the activated amide intermediate (Fig. 4.1) and selectively delivers the cyanide nucleophile using one of the phosphine oxides to a single enantiotopic face of the activated quinoline.

The group demonstrated the utility of the catalyst in a synthesis of N-methyl-D-aspartate receptor antagonist L-689,560 (**30**, Scheme 4.9). The reaction was carried out on a 1 g scale using 1 mol% of catalyst, 2 equivalents of TMSCN, and activating reagent 2-furoyl chloride. Sodium cyanoborohydride was added in the same reaction vessel at the end of the Reissert reaction to reduce the enamine functionality to provide **31** in 91% yield and 93%ee. On a larger scale, the reaction can probably be further optimized by identifying a less-reactive activating reagent than the 2-furoyl chloride. Such a reagent might have the benefit of minimizing undesired background noncatalyzed cyanide addition, at the same time raising the reaction temperature from −40°C to a more manageable ambient temperature, on scale-up. Variants of this catalyst (e.g., **32**) were also developed by the Shibasaki group for Reissert-type reactions of isoquinolines and pyridine substrates [31].

The enantioselective Strecker reaction is one of the most studied reactions in the catalysis field [14]. The chiral cyanide adduct from the Strecker reaction can be converted to amino acids, which are important building blocks in a plethora of biologically active molecules. Although a large variety of catalysts can promote the enantioselective Strecker reaction of aldimines, cyanide additions to ketimines are far more challenging for the reasons outlined in Section 2.1. Shibasaki's group has discovered a Gd-carbohydrate-based catalyst (**33**) that can efficiently carry out Strecker reactions on ketimine substrates [32]. The catalyst was used in the enantioselective synthesis of sorbinil (**34**), a therapeutic agent developed by Pfizer for the treatment of diabetic neuropathy, as shown in Scheme 4.10 [33]. It was found that the addition of a proton source such as 2,6-dimethylphenol (**35**, DMP) significantly enhanced the catalyst turnover number and turnover frequency allowing the reaction to be carried out with 1 mol% of catalyst and 1.1 equivalents of TMSCN.

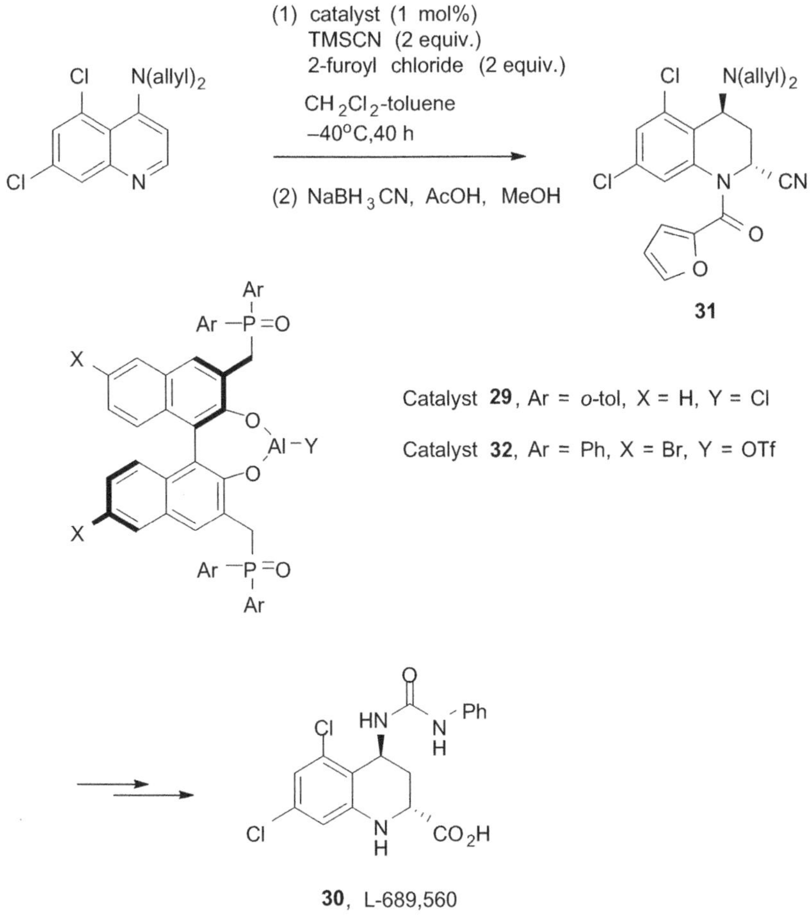

SCHEME 4.9 Asymmetric synthesis of L-689,560 (**30**).

In a detailed study of the catalyst species using ESI-MS, the group found that in the presence of a proton source the catalyst was transformed from a bis-silylated complex **36** to a more reactive *O*-protonated species **37**, as shown in Scheme 4.11. Based on this observation, the authors replaced the phenol proton source with HCN, which facilitated the same catalyst transformation. Under the streamlined reaction condition, the Strecker reaction could be carried out using 0.1 mol% of catalyst using 2.5 mol% of TMSCN and HCN as the stoichiometric cyanide source [34]. Since HCN is a gas, the excess reagent could simply be evaporated, thereby obviating the decomposition of TMSCN and removal of silylated phenol under previous conditions. This particular example illustrates how the detailed characterization of the catalyst species led to insights and reaction condition improvements, which

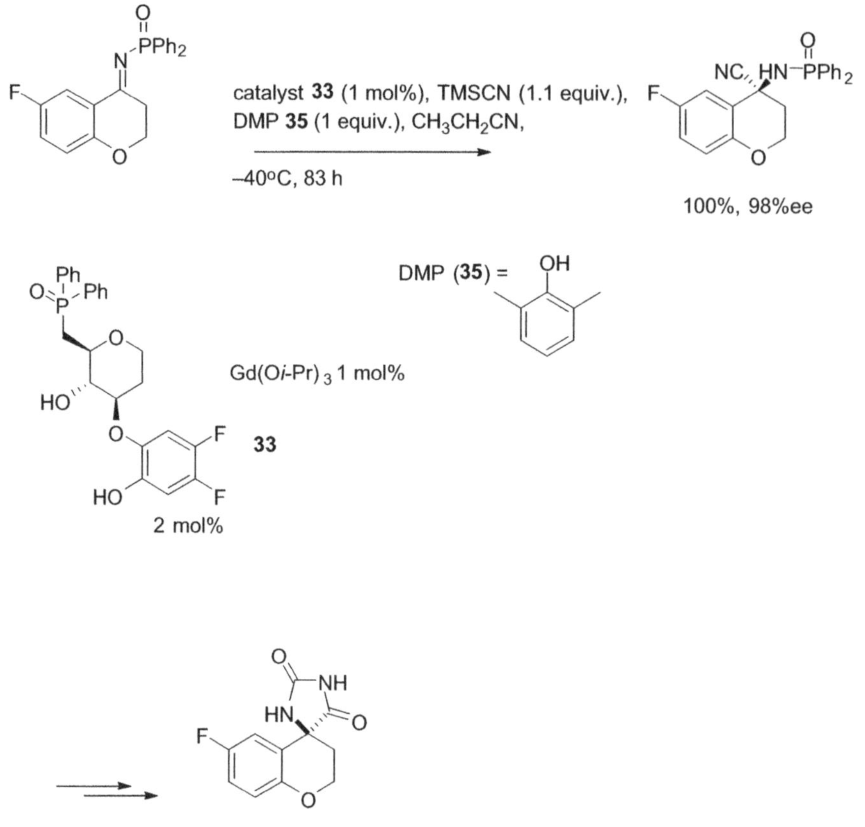

SCHEME 4.10   Asymmetric synthesis of sorbinil (**34**).

significantly reduced the catalyst loading and purification of the reaction product. Having such insights greatly improves the ease of transitioning the catalyst technology on an industrial scale.

4.2.3   Addition to C=C and CHC=C Bonds

4.2.3.1   Conjugate Addition   Conjugate addition of a carbon nucleophile to an electron poor olefin (the Michael reaction) is a reliable yet versatile C–C bond-forming method that has been employed in countless syntheses [35]. The first examples of catalytic enantioselective addition of 1,3-dicarbonyl compounds to nitro alkenes was developed by an Abbott process group lead by Barnes to tackle the large-scale synthesis of ABT-546 (**38**, Scheme 4.12), a highly potent endothelin-1A receptor antagonist [35b]. As with many process syntheses of APIs, the classical resolution of a racemic mixture via the tartaric acid salt first attempted by the Abbott team was not successful. The resolution process required multiple crystallization

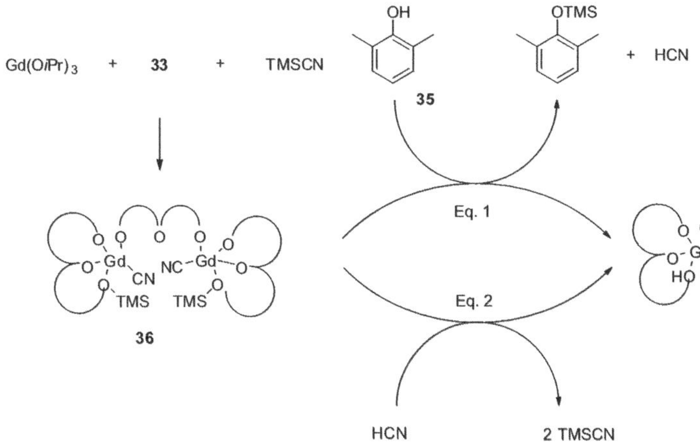

$Gd(OiPr)_3$ + **33** + TMSCN

35

+ HCN

Eq. 1

Eq. 2

36

37

HCN

2 TMSCN

SCHEME 4.11 Proposed mechanism of Gd catalyst.

steps and the desired salt was only isolated in 20% yield. The asymmetric process was then invented out of necessity as the Michael route was deemed the most convergent and required the least number of operational steps. The Michael reaction was promoted by the chiral bis-oxazoline–Mg complex generated by ligand **39** and $Mg(OTf)_2$. N-Methylmorpholine was employed in catalytic amounts to promote the soft enolization of the β-ketoester **40**, and the addition most likely occurred via a catalyst-bound β-ketoester enolate to the nitro alkene **41**. The reaction was carefully optimized with respect to base, solvent, temperature, and additives. It was found that the reaction was sensitive to water and molecular sieves had to be used. Under optimal conditions, product **42** was obtained in 82% yield and 88% selectivity using 4 mol% of catalyst. The synthesis of ABT-546 (**38**) was accomplished in 39% yield in 11 steps from dimethyl acrylate. The impressive overall yield was achieved via the utilization of a highly convergent and selective C–C bond forming Michael reaction step, obviating the need for classical resolution. The reaction was employed in the multikilogram synthesis of ABT-546. The Abbott chemists also demonstrated the generality of the reaction through a number of synthetic examples including an enantioselective synthesis of the anti-inflammatory drug, (R)-rolipram, **43** (Scheme 4.13).

Evans' group has also developed a chiral nickel diamine-based catalyst for the Michael addition of 1,3-dicarbonyl compounds to nitroalkenes [36]. There are some advantages of the Ni-based catalyst compared to the magnesium catalyst discussed above: (1) the nickel complex **44** (Scheme 4.14) is both isolable and bench stable which simplifies the reaction operation; (2) the reaction is not sensitive toward water and can be carried out in a variety of process chemistry friendly solvents such as THF or toluene; and (3) one of the bis-amine ligands dissociates from the complex during reaction which serves as the catalytic base, thereby obviating the need for added base. Furthermore, the catalyst was shown to promote addition of β-keto acids to nitro olefins followed by in situ decarboxylation leading to Michael adducts **45** (Scheme 4.15), which can serve as useful building blocks. Typically, Michael

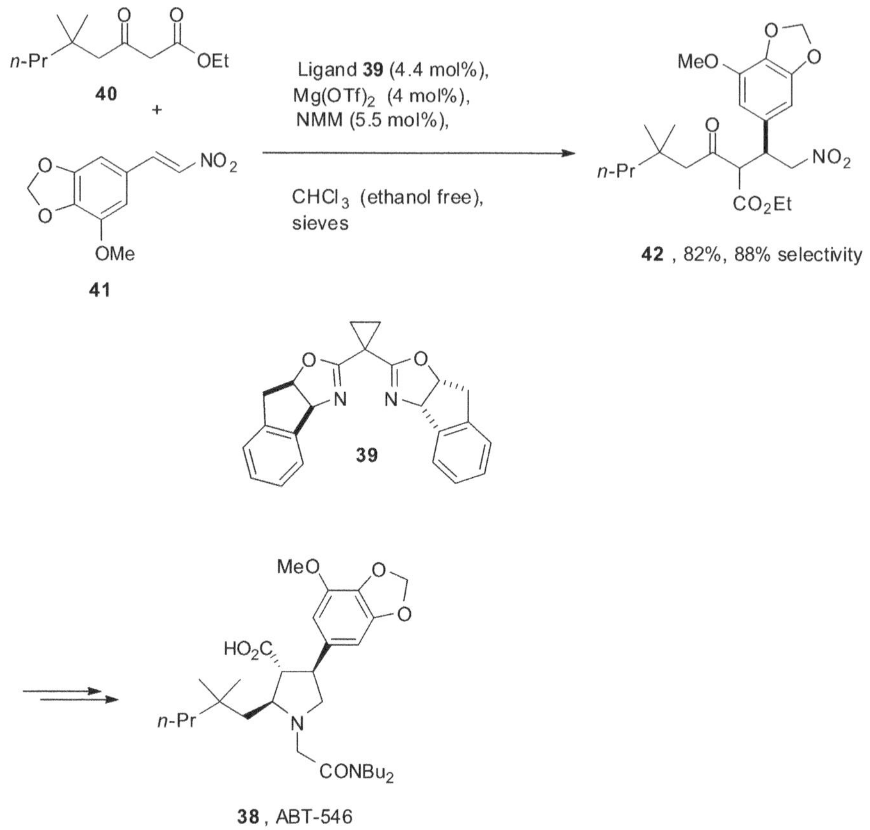

SCHEME 4.12   Asymmetric synthesis of ABT-546 (**38**).

additions catalyzed by **44** yield adducts in $>95\%$ yield and $>90\%$ee. Although there has been no report of the catalyst being employed in an industrial process, its ease of use and broad substrate scope deserve the attention of industrial chemists.

Exploiting the mechanistic concept of cooperative actions of both Lewis acid and Lewis base functionalities within the same catalyst structure, Shibasaki's group has devised a number of highly efficient heterobimetallic catalysts for a variety of transformations. One of the most notable is the aluminum lithium bis(binaphthoxide) (ALB)-catalyzed Michael reaction of malonates to cyclic enones [37]. In an enantioselective synthesis of $(-)$-strychnine, this reaction was road tested on kilogram scales. As shown in Scheme 4.16, Michael addition of dimethyl malonate (**46**) to cyclohexenone **47** has proven to be highly efficient and operationally simple to provide Michael adduct **48** in excellent yield and enantioselectivity. The ALB complex **49** was prepared from $LiAlH_4$ and BINOL and introduced to the reaction mixture as a THF solution. The amount of the catalyst could be as low as 0.05 mol% to provide product in remarkably high yields and selectivity (98%ee). However, on a

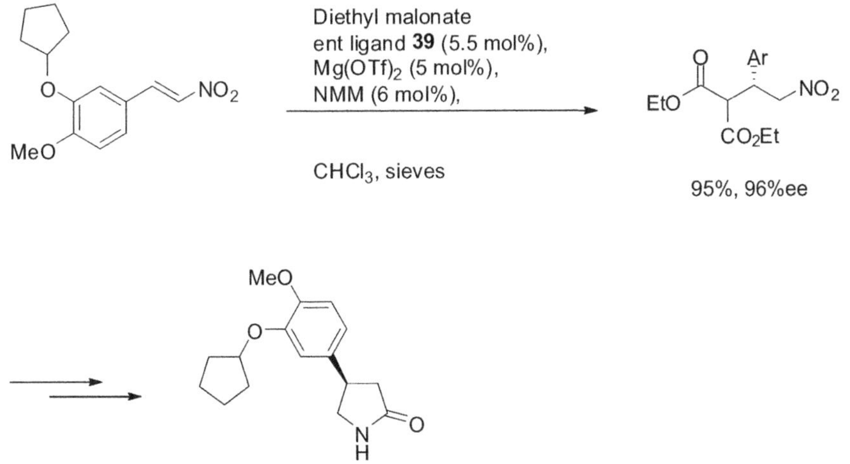

43, (*R*)-rolipram

SCHEME 4.13 Synthesis of (*R*)-rolipram (**43**).

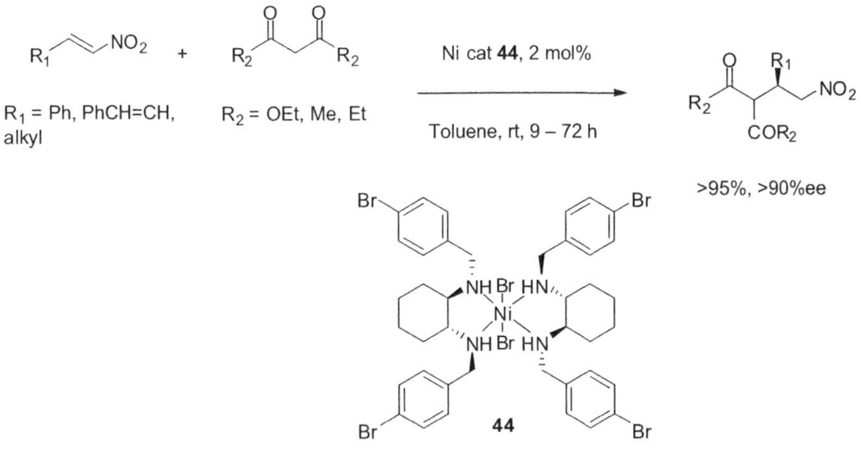

SCHEME 4.14 Asymmetric Michael addition using Ni complex **44**.

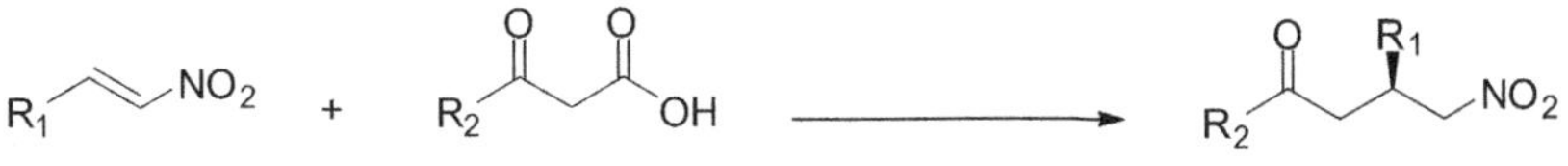

45

SCHEME 4.15 Asymmetric addition of β-keto acids to nitro olefins.

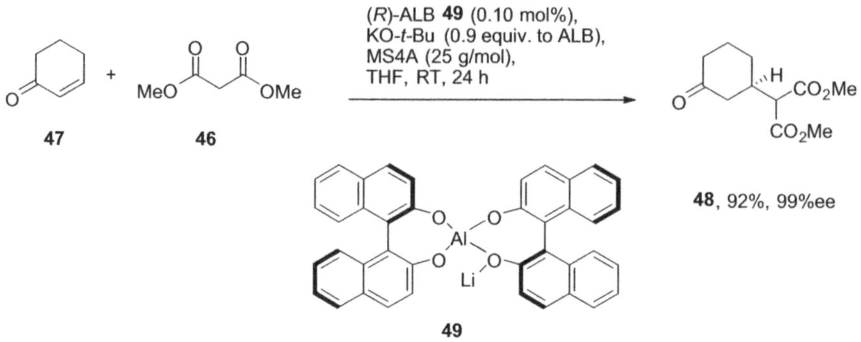

SCHEME 4.16 Asymmetric Michael addition promoted by ALB catalyst.

kilogram scale 0.5 mol% of catalyst was employed to complete the reaction in 24 h. The reaction was run in high concentration (21 mL of THF per mole of substrate), which significantly shortened the reaction time. The group also devised a simple workup procedure that involved filtration to remove the molecular sieves followed by direct crystallization of the product by addition of hexane to the filtrate. In addition, the most expensive ingredient of the reaction, BINOL, was recovered to the extent of 79% from the mother liquor, which greatly improved the economics of this reaction. As the result of the ALB catalyst discovery, Michael adducts such as **48** can be considered easily accessible chiral starting materials for API synthesis, and the procedures described by Shibasaki and coworkers clearly have demonstrated potential for larger industrial scales.

Complementary to the methods described above, rhodium-catalyzed additions of organoboronic acids to α,β-unsaturated carbonyl compounds can provide access to β-substituted aryl, alkenyl, and alkynyl enones or esters. Miyaura and Hayashi first described the reaction in 1997. Since then Hayashi and others have significantly improved the substrate scope and the efficiency of the reaction, providing various Michael adducts in high yield and enantioselectivity [38].

In a synthetic campaign to provide all four stereoisomers of S1P1 receptor agonist VPC01091 (**50**), Abbott chemists developed a practical procedure that significantly improved the utility of this reaction for large-scale applications [39]. The standard reaction protocol published by Hayashi required the use of rhodium complexes that are commercially unavailable or use specialized equipment (e.g., a glove box). Moreover, the metal catalysts were used in relatively high loadings (typically 3 mol%) and the requirement of a large excess of arylboronic acids (1.5–3 equivalents) to compensate for the proteodeboronation during the reaction reduced the cost efficiency of the process. During their optimization studies, the Abbott chemists found that the active catalyst could be generated by premixing the commercially available bis(norbonadiene)rhodium tetrafluoroborate (**51**) with the desired enantiomer of the BINAP ligand in dioxane in the presence of boronic acid. Following 2 h of complex formation, the enone and triethylamine base and water cosolvent were then introduced and the reaction proceeded at ambient temperatures. Under the

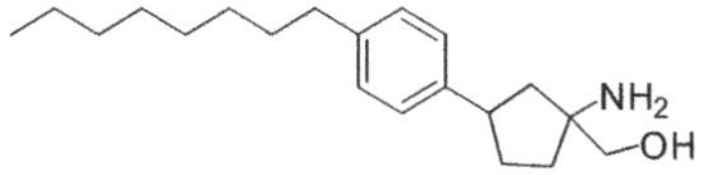

50, VPC01091

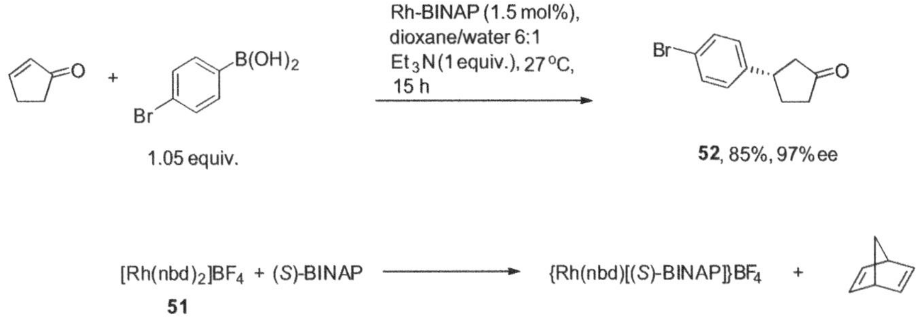

Rh-BINAP (1.5 mol%),
dioxane/water 6:1
Et$_3$N (1 equiv.), 27 °C,
15 h

1.05 equiv.

52, 85%, 97% ee

[Rh(nbd)$_2$]BF$_4$ + (S)-BINAP $\longrightarrow$ {Rh(nbd)[(S)-BINAP]}BF$_4$ +

51

SCHEME 4.17 Rh-catalyzed asymmetric Michael addition of boronic acids.

optimized new reaction conditions only 1.05 equivalents of boronic acid and 1.5 mol % loading of the catalyst were needed to obtain the Michael adduct **52** in high yields (85%) and excellent selectivity (97%ee), as shown in Scheme 4.17. Due to the reduction in boronic acid equivalents, the reaction no longer required silica gel purification to remove the proteodeboronation by-product, hence significantly improving the operational efficiency. This example illustrates that an improvement of the catalyst generation procedure can sometimes take a lab scale enantioselective reaction to an optimized production scale process.

One of the latest developments in the rhodium-catalyzed reactions was the discovery of a number of chiral dienes as ligands [40]. Both the Hayashi [41] and Carreira [42] groups independently reported chiral dienes with bridged carbobi-cyclic frameworks such as [2.2.2]bicyclooctadienes or [3.3.2]bicyclodecadienes. Dienes based on other frameworks such as nonbridged [3.3.0] bicyclics have also appeared in the literature [43]. Chiral 1,4-diene-rhodium complexes can sometimes display higher reactivities and enantioselectivities than chiral phosphine complexes. Although many of them are either not commercially available or have not been tested in an industrial setting, they offer structural alternatives to chiral phosphines and perhaps expanded substrate scopes during optimization studies. For example, the Hayashi group found that diene **53** (Scheme 4.18) provided Michael adducts of arylmethylene cyanoacetates **54** in much higher yields and enantio-selectivities than the standard BINAP ligand. The utility of the ligand was demonstrated in a lab-scale synthesis of the antimuscarinic drug, (R)-tolterodine (**55**, Pfizer's Detrol®) [44].

4.2.3.2 Cycloaddition This section describes the asymmetric synthesis of a variety of three-, five-, and six-membered rings using chiral transition metal catalysts.

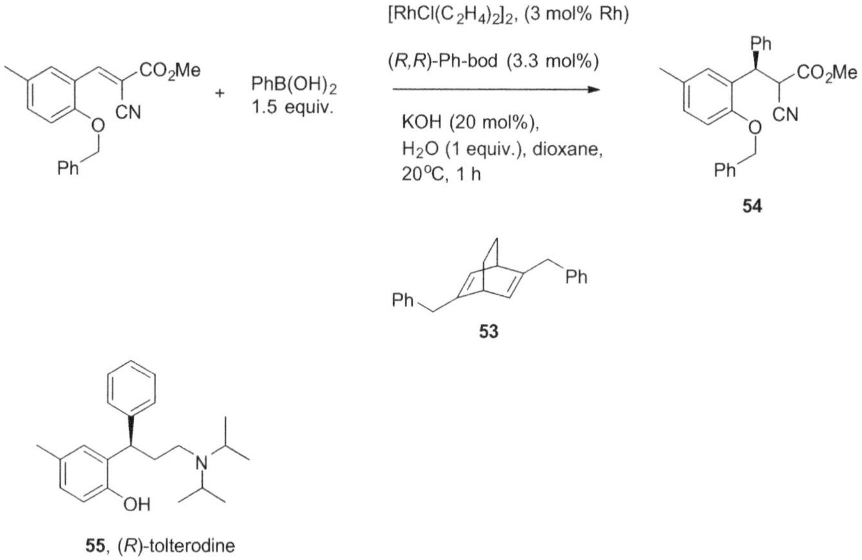

SCHEME 4.18   Asymmetric synthesis of (R)-tolterodine (**55**).

[2 + 1] Cycloadditions   Early examples of asymmetric cyclopropanations using chiral transition metal catalysis appear in the work of Aratani in 1985 [45]. Accordingly, the chiral dimeric copper carbenoid catalysts **56**, prepared by the reaction of optically active α-imino alcohols and cupric acetate, facilitated the [2 + 1] cycloaddition of alkyl diazoacetates with a variety of olefins to produce chiral cyclopropanes, as shown in Scheme 4.19. For example, reaction of isobutylene and ethyl diazoacetate in the presence of **56** gave ethyl (+)-(1S)-2,2-dimethylcyclopropanecarboxylate (**57**) in 92% enantiomeric excess. According to the author, "This reaction was proved to work well in a factory scale." In fact, the parent acid of ester **57** is an intermediate in the synthesis of cilastatin (**58**), a dehydropeptidase inhibitor that suppresses the degradation of certain β-lactam antibiotics in the kidney [46].

The asymmetric cyclopropanation of 1,3-butadienes using chiral copper catalysts was also employed by Itagaki for the preparation of chrysanthemic esters, useful intermediates for the synthesis of pyrethroid insecticides. As shown in Scheme 4.20, reaction of diene **59** with *tert*-butyl diazoacetate in the presence of chiral copper catalysts produced the *tert*-butyl ester of chrysanthemic acid. Initial work [47] suggested that the bis(4R)-oxazoline catalyst **60** (0.2 mol%) gave the best *trans: cis* ratio (88:12) of isomeric esters **61:62** in 92% yield. Of a variety of counterions studied, Ph_3CPF_6 proved to be much easier to handle in air than, for example, $AgPF_6$. It also afforded a higher chemical yield (91%) at lower catalyst loading (0.2 mol% versus 0.5 mol%) than $AgPF_6$ for comparable *trans:cis* ratios. Even though the reaction was reported on a gram scale, the authors concluded that the process was

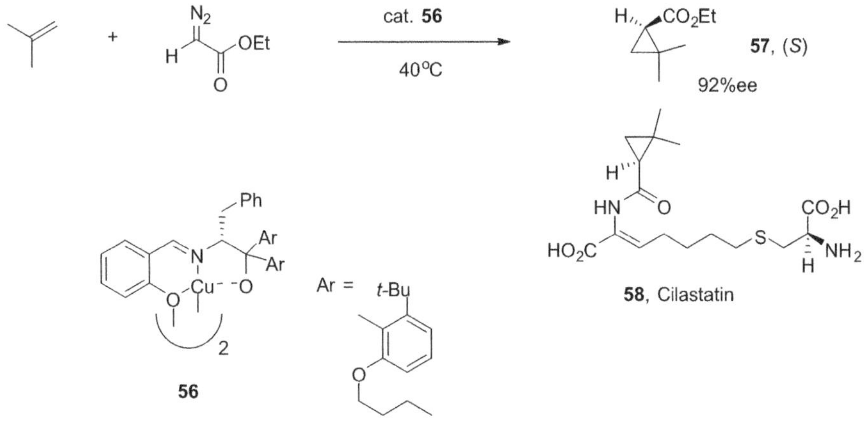

SCHEME 4.19 Asymmetric [2 + 1] cyclopropanation.

"industrially applicable" because it gave high enantio- and *trans*-selectivity, made use of a simple diazoacetate, and had a low catalyst loading. In a subsequent paper [48] Itagaki demonstrated that the same reaction could be scaled up using the catalyst derived from copper salicylaldimine. In the presence of an equimolar amount of the Lewis acid $Al(OEt)_3$ the chrysanthemate esters **61** and **62** were produced in 90% chemical yield and 91%ee (the *trans:cis* ratio was 78:22). In the *absence* of $Al(OEt)_3$ only a 27% yield was obtained. The reaction was thought to proceed through complex **63** (Fig. 4.2). The low (0.1 mol%) catalyst loading of the inexpensive copper catalyst makes this a potentially useful process on an industrial

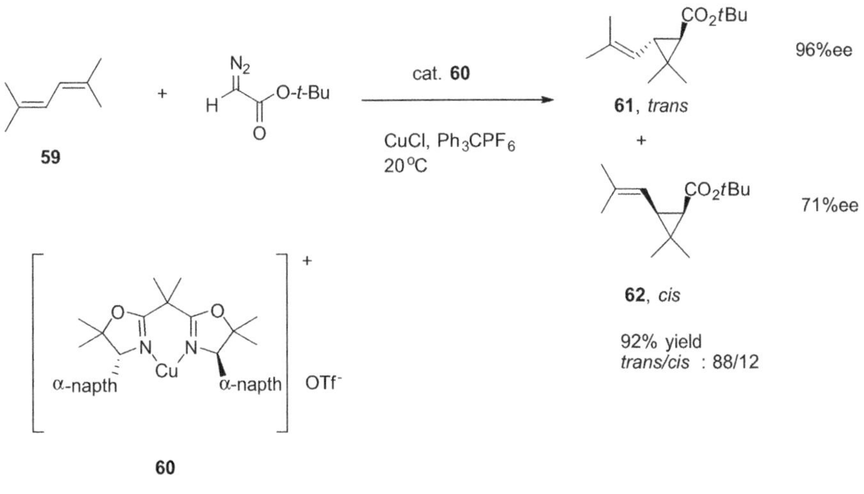

SCHEME 4.20 Asymmetric synthesis of *tert*-butyl chrysanthemate.

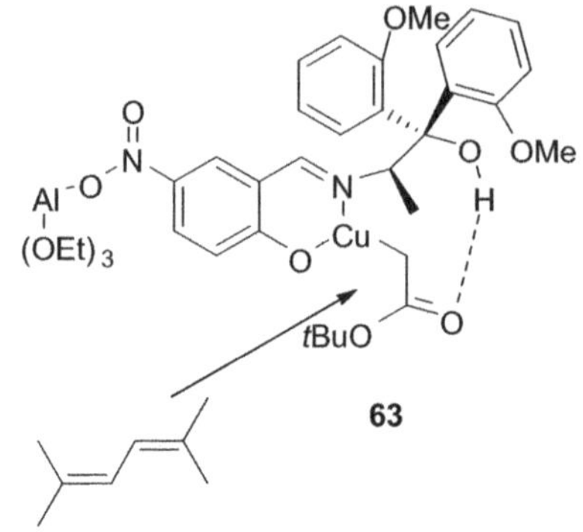

FIGURE 4.2   Copper (I) carbine complex with Lewis acid.

scale. The catalyst ligand is available via a double Grignard addition to D-alanine methyl ester hydrochloride, followed by reaction of the resulting amino alcohol with commercially available 5-nitrosalicylaldehyde.

Chemists at Bristol-Myers Squibb produced a variety of substituted 3-cyclopropylindoles as selective serotonin reuptake inhibitors. For example, the 3-vinylindoles **64** were found to undergo [2 + 1] cyclopropanation with diazo esters in the presence of ip-pybox-Ru(II) (**65**) to afford the *trans*-cyclopropyl carboxylates **66** in 81–88% enantiomeric excess [49]. As shown in Scheme 4.21, these could be further elaborated into conformationally restricted homotryptamine analogs such as BMS-505130 (**67**), a highly potent and selective serotonin reuptake inhibitor

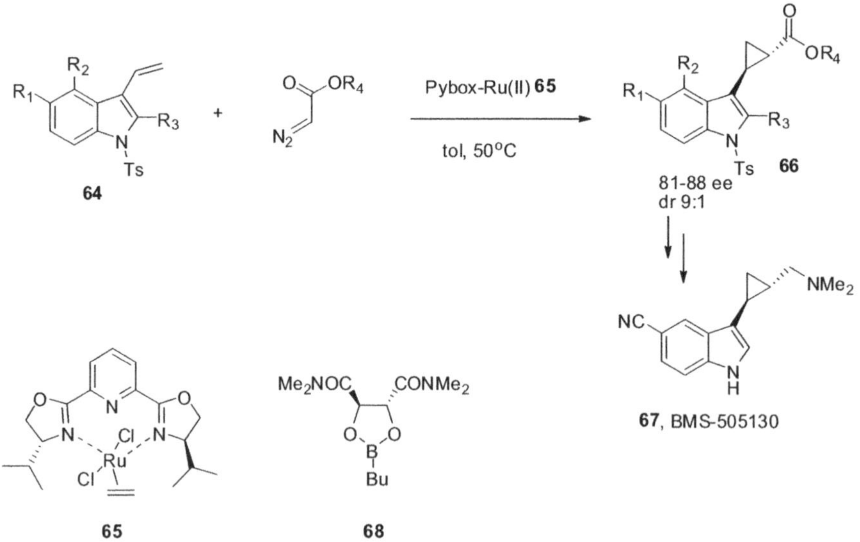

SCHEME 4.21   Synthesis of BMS-505130 (**67**).

(SSRI). Process improvement by another group from BMS [50] using the (R,R)-pybox catalyst (produced from the ligand and $[RuCl_2(p\text{-cymene})]_2$ in THF) led to the synthesis of the chiral, nonracemic acid **66** (for $R_1 = CN$; $R_2 = R_3 = H$; $R_4 = OH$) in quantitative yield from 7.8 kg of **64** ($R_1 = CN$; $R_2 = R_3 = H$). In the paper that followed [51], the same group reported a further process improvement for the synthesis of BMS-505130 (**67**) using Charette's chiral dioxaborolane ligand (**68**) in place of pybox on a laboratory scale. The target was synthesized in an overall yield of 38% in eight steps (five isolations) from the corresponding indole **64**.

The ruthenium ip-pybox catalyst **65** was used in the pilot-scale asymmetric cyclopropanation of styrene **69** (generated *in situ*) as the limiting starting material [52]. As shown in Scheme 4.22, $(1R)$-*trans*-cyclopropyl acid **70** was produced as a 90:10 mixture of *trans:cis* isomers. Selective hydrolysis increased the *trans:cis* ratio to 96:4. Additional separation and enantiomeric enrichment of **70** was achieved by recrystallization of the (+)-dehydroabeitylamine salt of **70** to produce the (R,R)-isomer in $\geq 99.9\%$ enantiomeric excess and chemical yields of 60–65%.

[3 + 2] Cycloaddition   The key step in a synthesis of the hepatitis C virus polymerase inhibitor **71** (Scheme 4.23) was a [3 + 2] cycloaddition of the L-leucine derivative **72** with methyl acrylate. The group from GlaxoSmithKline found that hydroquinine (**73**, 6 mol%) was the most effective of several cinchona alkaloids tried, serving as both a catalyst ligand for the asymmetric reaction and a base [53]. The hydroxyl group of **73** was found to be a critical feature for maximizing the enantioselectivity of the reaction. Silver acetate (3 mol%) proved to be the best metal salt for maximizing enantio- and diastereoselectivity. In the absence of a metal salt the reaction was slow (50% completion after several days). The use of lithium bromide produced no observable stereoselectivity. The 1,3-dipolar cycloaddition

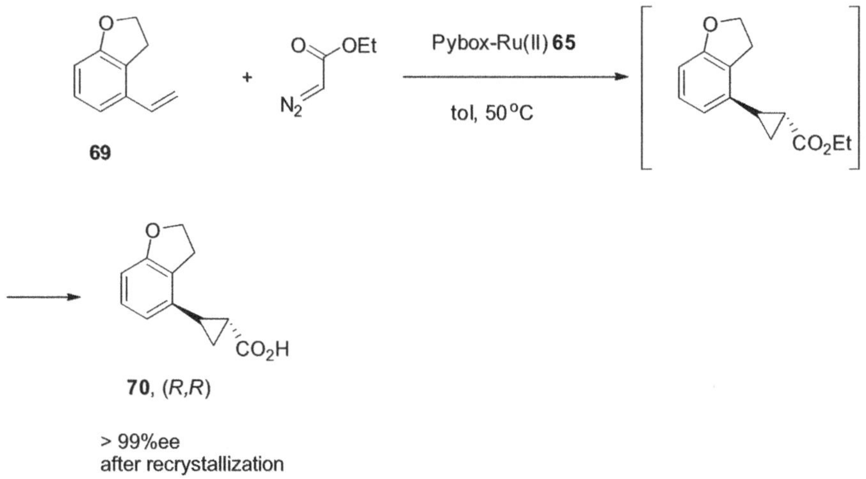

SCHEME 4.22   [2 + 1] Cyclopropanation of a styrene.

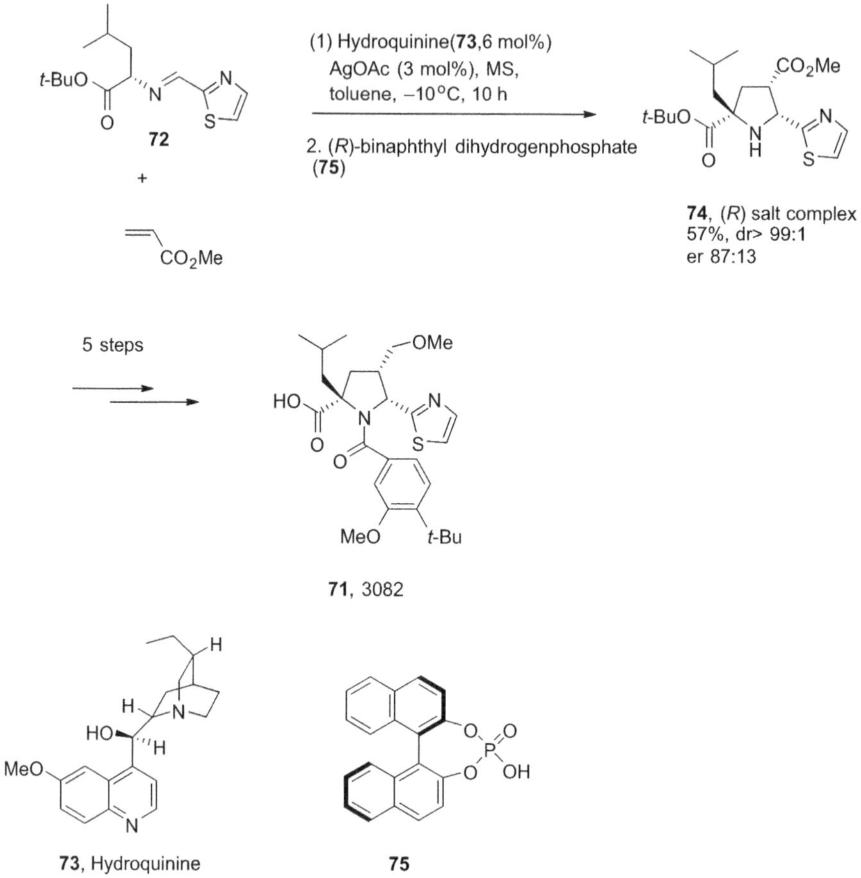

SCHEME 4.23 Synthesis of a hepatitis C virus polymerase inhibitor (**71**).

product **74** was obtained in a diastereomeric ratio of greater than 99:1 (enantiomeric ratio 87:13), isolated as its (R)-binaphthyl hydrogen phosphate salt. The chiral phosphate **75** was fully recovered in the subsequent step. In this manner inhibitor **71** ("3082") could be prepared in seven steps from L-leucine *tert*-butyl ester hydrochloride and 2-thiazolecarboxylic acid on a multikilogram scale in a 300 gallon reactor.

[4 + 2] Cycloadditions   The asymmetric Diels–Alder reaction has figured in the construction of countless single and fused six-membered rings. For example, Shibasaki's group employed a [4 + 2] cyclization using various aryl-substituted (R,R)-pybox ligands (**76**) to produce two contiguous chiral centers in the acylphloroglucinols **77** in enantioselectivities of up to 92% (Scheme 4.24) [54]. The reaction was conducted in the presence of $FeBr_3$ and $AgSbF_6$ in a ratio of 1:2. The authors found that the reaction could be routinely performed on up to a 20 g scale with an

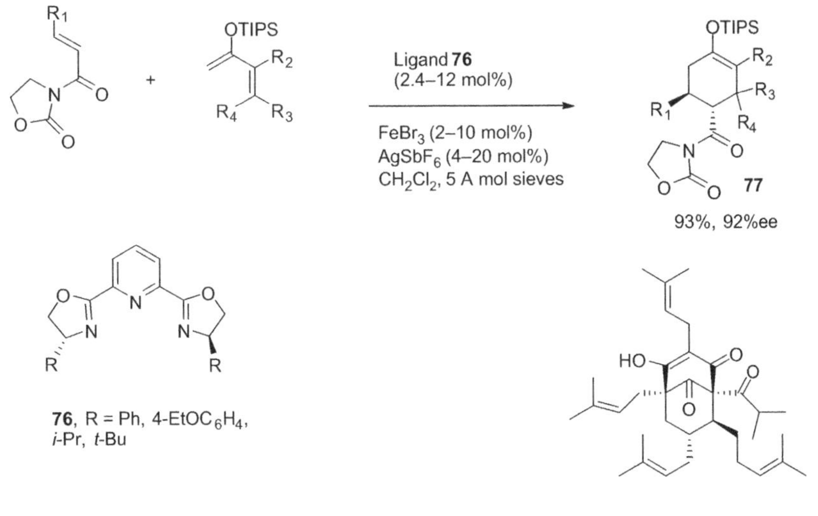

SCHEME 4.24 [4 + 2] Cycloaddition in the synthesis of an *ent*-hyperforin precursor.

average 89% enantiomeric excess, although the enantioselectivity decreased at scales above 10 g, probably due to the slight increase in the reaction temperature on substrate addition [55]. Recently the same group applied the method to "the first catalytic asymmetric total synthesis" of *ent*-hyperforin (**78**), the antipode of a biologically active constituent of the herb St. John's wort [55].

4.2.3.3 Allylic Alkylation

Most of the asymmetric alkylation reactions done in industrial settings rely on the use of well-developed and highly versatile stoichiometric chiral auxiliaries. A large number of examples exist in which Evans' oxazolidinone imides or Myers' pseudoephedrine amides have been used to synthesize APIs on large scales [56]. Chiral phase transfer catalysts are emerging to be very useful alternatives to auxiliary-based alkylation reactions. A number of designer chiral quaternary ammonium salts have proven to be very efficient catalysts in the synthesis of chiral amino acids [57] and they have started to gain the attention of industrial groups [58]. Transition metal-catalyzed allylic alkylations offer a unique and complementary method for forming carbon–carbon bonds compared to the methods mentioned above. The product of allylic alkylation in nature contains an olefin that often serves as a functional handle for subsequent transformations. Numerous chiral transition metal catalysts have been developed, and the following examples illustrate the applications this reaction in API synthesis [59].

Palladium-catalyzed asymmetric allylic alkylation (AAA) served as a starting point in the synthesis of prostaglandin D2 receptor antagonist **79** developed by Merck chemists, as shown in Scheme 4.25 [60]. Using a procedure and catalyst invented by Trost and Bunt [61], racemic cyclopentenyl acetate **80** was converted to the malonate

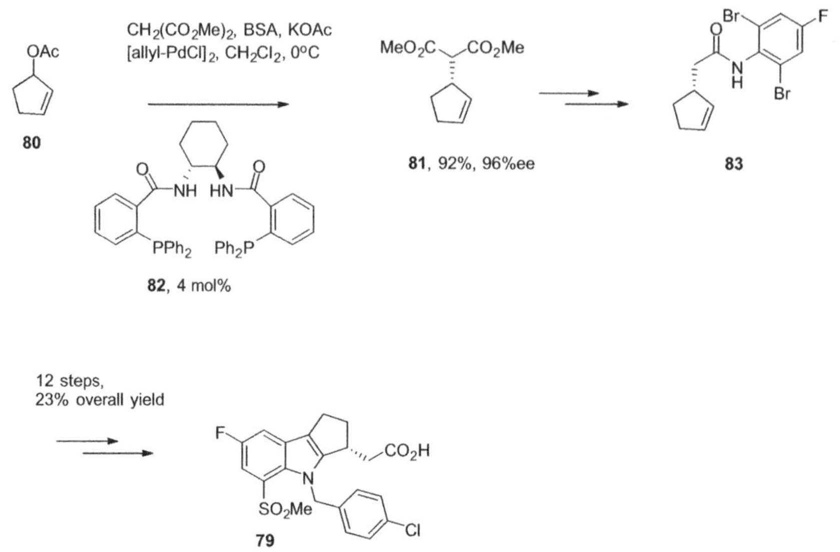

SCHEME 4.25   Asymmetric synthesis of a prostaglandin D2 receptor antagonist.

alkylation product **81** in 92% yield and 96%ee. The palladium catalyst formed an intermediate π-allyl complex in which the bisphosphine ligand **82** controlled the approach of the nucleophile. The reaction was carried out on a 32 g scale using 4 mol% of the Trost ligand and palladium metal at 0°C, and was completed in 7 h. It was found that the replacement of NaH/Hex₄NBr, the more commonly used base/phase transfer catalyst combination for this reaction, with bistrimethylsilylacetamide (BSA) resulted in a more practical and reproducible procedure. The product was carried forward in a hydrolysis step, without purification, to the corresponding bis-acid. During the hydrolysis reaction, the palladium catalyst precipitated as a Pd(II)–Trost ligand complex which was removed by filtration. This observation may present a viable way of recycling the catalyst if the Pd(II) can be reduced back to a Pd(0) species. The synthesis of **79** was accomplished in 12 steps from **83** in 23% overall yield.

In a synthesis of mGluR2 receptor agonist MGS0028 (**84**, Scheme 4.26), a separate group of Merck chemists utilized asymmetric allylic alkylation to set the initial stereogenic center [62]. Alkylation of 2-cyclopentenyl acetate (**80**) with 2-fluoroacetoacetate using the standard (R,R)-Trost ligand gave the alkylation product **85** in >90% yield and 94-96%ee. The enantioselectivity of the reaction was enhanced by reducing the catalyst loading from 5 mol% to 1 mol% as well as using a bulkier n-Hex₄NBr (versus the n-Bu₄NBr) phase-transfer agent. The stereogenic center established by AAA was utilized in the construction of the other stereogenic centers seen in the synthetic intermediate **86**, which was obtained in eight steps from **85**. The use of the AAA route and inexpensive 2-fluoroacetoacetate obviated the need for the Merck chemists to use a much more toxic fluoroacetate from an alternative undisclosed route.

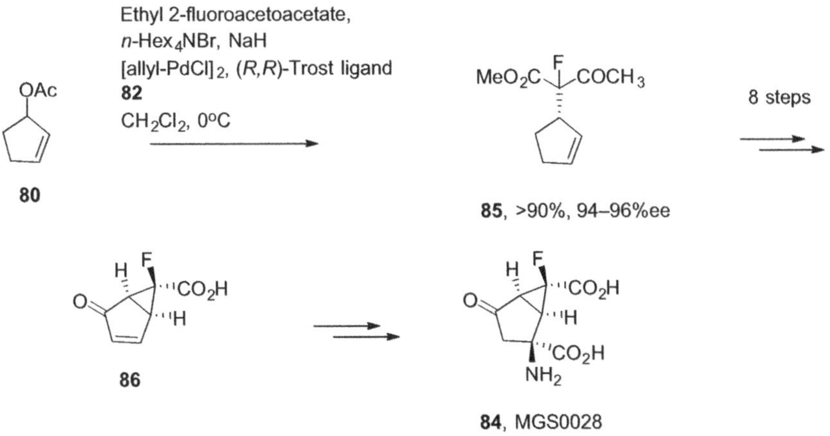

Ethyl 2-fluoroacetoacetate,
n-Hex4NBr, NaH
[allyl-PdCl]2, (R,R)-Trost ligand
82
CH2Cl2, 0°C
OAc
80
MeO2C
F
COCH3
85, >90%, 94–96%ee
8 steps
86
84, MGS0028

SCHEME 4.26 Asymmetric synthesis of MGS0028 (**84**).

It is worthwhile to highlight that the structure-based rational for selectivity in AAA using the Trost ligand was only elucidated in 2009 after nearly 17 years after its discovery in 1992. Trost originally proposed a mechanism that accurately predicted the reaction selectivity outcome based on the hypothesis of a wall-and-flap model of the π-allyl complex in which the phosphorus-linked phenyl rings form the chiral environment and control the direction of nucleophilic attack (Fig. 4.3). In the most recent mechanistic study using a combination of NMR experiments of deuterium labeled ligand–Pd complex and computational studies, Lloyd-Jones and coworkers [63] demonstrated that the catalyst formed a 13-membered chelate with one of the hydrogen atoms of an amide NH group in close proximity to one allyl terminus that helped direct the incoming negatively charged nuclophile via a hydrogen bond (Fig. 4.4). This new mechanism not only explained why the catalyst is a highly activating system versus the original model proposed by Trost, it explained the experimentally known observation of increased enantioselectivity when the

Chiral
scaffold
Wall
Flap
Pd
Flap
Wall
Pro-R
Pro-S

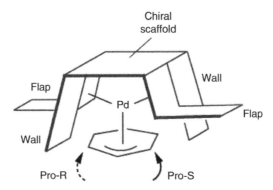

FIGURE 4.3 Trost's wall-and-flap model. Reproduced with permission from Chemical & Engineering News 2009, V87, P31–33. © 2009, American Chemical Society.

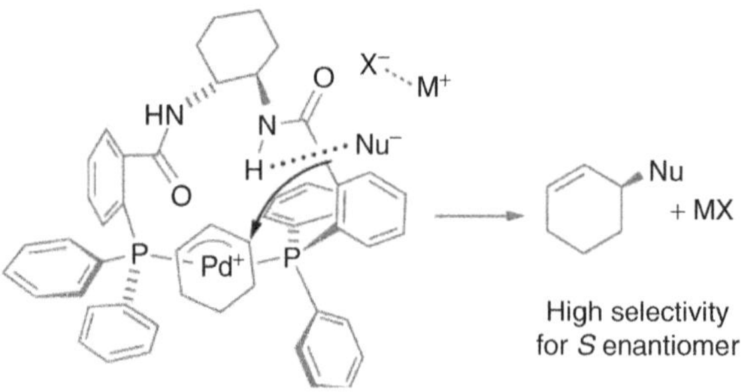

FIGURE 4.4 Thirteen-membered chelate model. Reproduced with permission from Chemical & Engineering News 2009, V87, P31–33. © 2009, American Chemical Society.

counterion is more dissociative (lithium versus cesium) due to the increased nucleophile association with the activating NH bond [63]. The elucidation of the catalyst mechanism can help the design of more-efficient allylic alkylation catalysts and the hydrogen bond directed activation nucleophiles can be incorporated into other catalytic reaction designs.

Transition metals other than palladium have also been used to catalyze AAA [59b]. The molybdenum bis-picolinamide catalyst invented by the Trost group was utilized by chemists at Merck to synthesize a chiral cyclopentanone intermediate **87** for an API (Scheme 4.27) [64]. For an acyclic substrate such as carbonate **88**, molybdenum-catalyzed AAA yielded predominantly a branched enantiomerically enriched alkylation product **89**. Conversely, palladium-catalyzed AAA yielded the linear product **90**. The synthesis of cyclopentanone **87** was achieved by

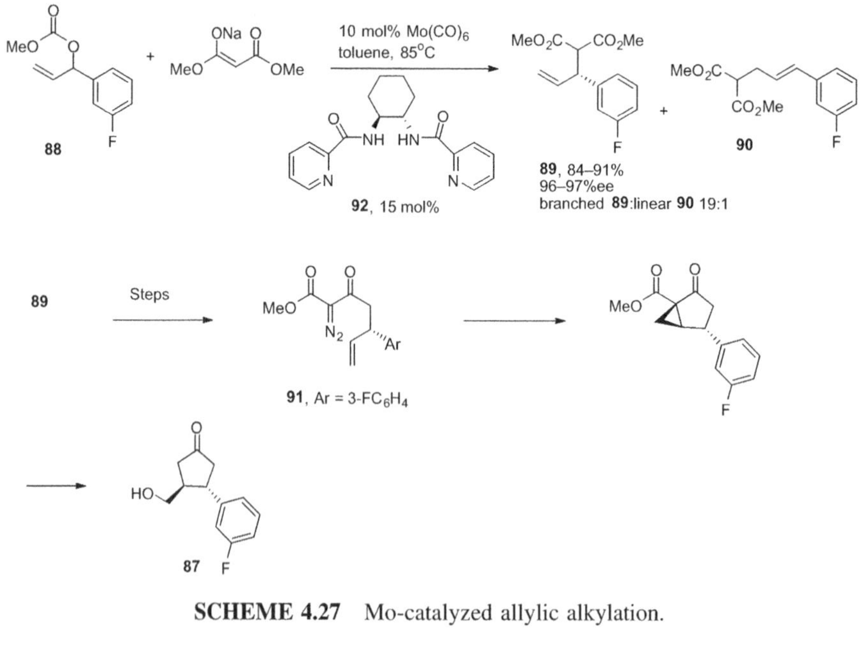

SCHEME 4.27 Mo-catalyzed allylic alkylation.

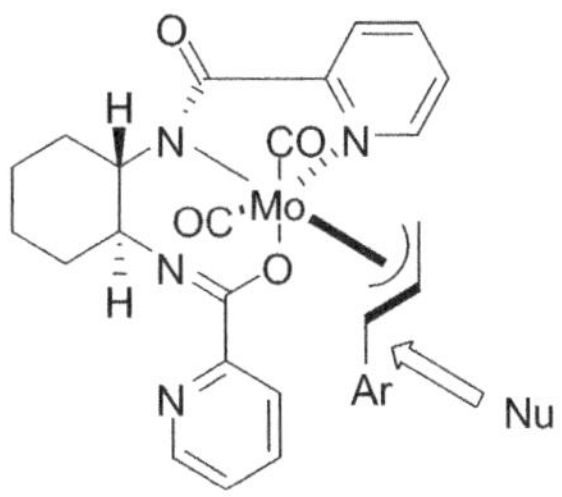

93

FIGURE 4.5 Proposed catalytic complex.

intramolecular cyclopropanation reaction of **91**, which was derived from the branched AAA product **89**. In the original report by Trost, commercially unavailable air-sensitive precatalyst $(EtCN)_3Mo(CO)_3$ was used to form the active catalyst. In the course of the synthetic study of **87**, the Merck chemists found that the active catalyst could be formed from a readily available commercial source of air stable $Mo(CO)_6$ by simply heating in the presence of the ligand **92** in toluene at 85°C for 4 h [65]. With this seemly simple innovation, which greatly enhanced the industrial applicability of the reaction, the reaction was carried out on a multikilogram scale to obtain the desired branched product **89** in >90% yield, 97%ee, and 19:1 branched-to-linear ratio. The yield and enantiomeric excess of the alkylation reaction was reproducible on a 2 kg scale. A study was carried out to determine the nature of the catalytic complex and the structural activity relationship of the chiral ligand. It was found that both of the amide groups on the chiral backbone of the ligand and only one of the picolinamide nitrogen atoms participated in the coordination, with **93** (Fig. 4.5) being a likely structure complex formed during the reaction [66].

4.2.4 Insertion

The formal insertion of carbon-bearing moieties into carbon–hydrogen bonds (Scheme 4.28) has been the subject of several recent reviews [67]. A particularly striking example of C–H insertion is the two-step intermolecular synthesis of the most active enantiomer (**94**) of the ADHD drug *threo*-methylphenidate (Novartis' Ritalin®) [68]. As shown in Scheme 4.29, 4 equivalents of *N*-Boc-piperidine treated with methyl phenyldiazoacetate (**95**) in the presence of ca. 1 mol% of the dirhodium tetraprolinate catalyst $Rh_2(S\text{-biDOSP})_2$ (**96**) produced, after deprotection, the R,R' enantiomer **94** in 52% isolated yield and 86% enantiomeric excess [69]. At the time

$$C\text{—H} \quad + \quad Carbon \quad \xrightarrow{\text{cat.*}} \quad C^*\text{—Carbon}$$

SCHEME 4.28 Asymmetric insertion into carbon–hydrogen bonds.

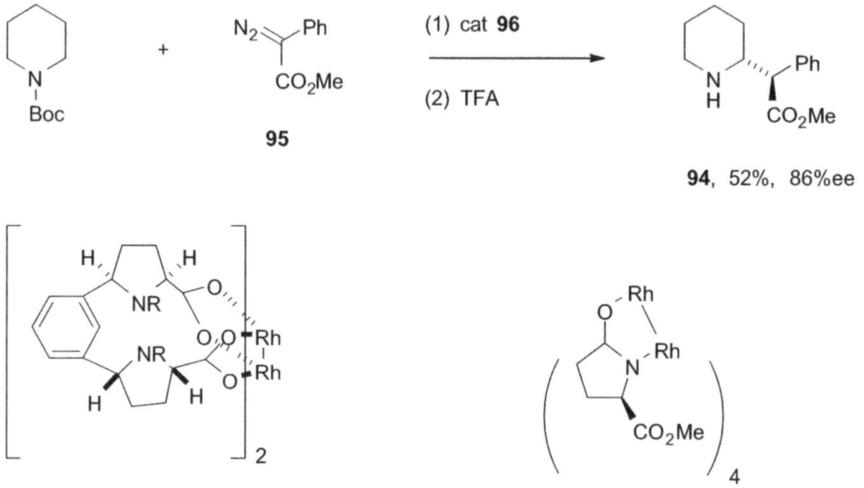

94, 52%, 86%ee

96, R = 4-(C$_{12}$H$_{25}$)C$_6$H$_5$SO$_2$
Rh$_2$(S-biDOSP)$_2$

97, Rh$_2$(5R-MEPY)$_4$

SCHEME 4.29   Synthesis of *threo*-methylphenidate (**94**).

of publication, lead author H. M. L. Davies claimed that "demand for [his] catalysts has gone from grams to kilogram quantities." [69b]

In the paper which immediately followed the publication by Davies, Winkler and coworkers described [70] the same synthetic route to **94** using the commercially available Doyle's catalyst, Rh$_2$(5R-MEPY)$_4$(**97**) [71] at ca. 1 mol% catalyst loading. Accordingly, the (2R,3R)-methylphenidate was produced in 94%de and 69%ee and could be purified by two recrystallizations to 95%de and >95%ee. The yield of the N-Boc-methylphenidate intermediate, after purification, was 64.5% [70b].

4.3   CARBON–HETEROATOM COUPLING REACTIONS

This section examines various methods for the construction of asymmetric molecules by carbon–nitrogen, carbon–oxygen, and carbon–fluorine coupling and insertion reactions mediated by chiral transition metal catalysts.

4.3.1   Addition of Nitrogen Nucleophiles

4.3.1.1   Aziridine Ring Opening   The enantioselective desymmetrization of *meso*-aziridines figured in a synthesis of the anti-influenza drug, oseltamivir phosphate (Tamiflu®; **98**) [72]. As shown in Scheme 4.30, treatment of the dinitrobenzamide **99** with trimethylsilyl azide (1.5 equivalents) in the presence of 2 mol% of yttrium triisopropoxide (to generate the reactive yttrium azide *in situ*) and 4 mol% of the

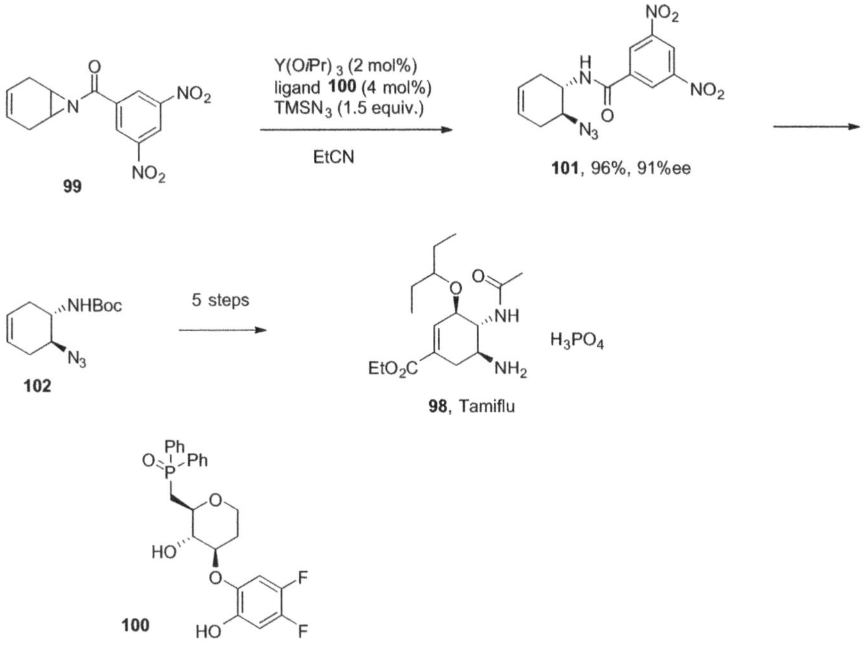

SCHEME 4.30 Synthesis of oseltamivir phosphate (**98**).

chiral ligand **100** provided the azide **101** in 91% enantiomeric excess and 96% yield. The method was applied to the synthesis of Tamiflu (**98**) with the proper absolute stereochemistry.

An improved route to **98** which relied on the same stereospecific aziridine ring opening was described more recently by the same group [73]. Accordingly, intermediate **102**, used in the first Shibasaki synthesis, was converted to Tamiflu as shown in Scheme 4.31. In process improvements to this second synthesis it was found that the aziridine desymmetrization could be performed with only a 1 mol% yttrium catalyst loading (versus 2 mol% previously). Addition of the proton source 2,6-dimethylphenol (1 equivalent) shortened the reaction time from 48 to 12 h. Enantioselectivities were comparable to those obtained from the earlier synthesis

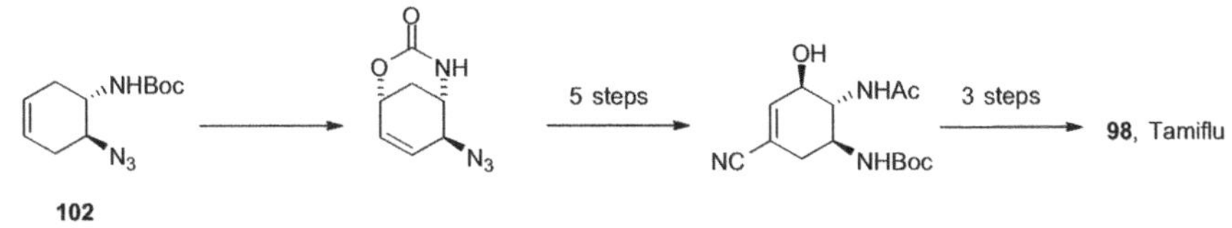

SCHEME 4.31 Improved synthesis of **98**.

(89%ee versus 91%ee), as were yields (94% versus 96%). In addition, it was found that the chiral ligand **100** could be recovered in 81% yield by base extraction. The reaction was run on a 30 g scale.

4.3.1.2 N-Allylation

4.3.1.2 N-Allylation   Trost has employed the palladium-catalyzed asymmetric allylic alkylation in a practical synthesis of the antiepileptic drug, vigabatrin (**103**) and the antimycobacterial agent, ethambutol (**104**), used in the treatment of tuberculosis [74]. As shown in Scheme 4.32, the key step was the asymmetric *N*-allylation of racemic butadiene monoepoxide by phthalimide in the presence of the catalyst produced *in situ* from π-allylpalladium chloride dimer (**105**) and the (*R,R*)-ligand **106**. A dynamic kinetic asymmetric transformation (DYKAT) converted the racemic epoxide into a single enantiomeric product, **107**, which could be converted to **103** in three steps and to **104** in five steps.

As shown in Scheme 4.33, the Trost group later extended the DYKAT methodology to the enantioselective total synthesis of a variety of other natural products [75]. For example, intermediate **107** (Scheme 4.32), produced on greater than 50 g scale, was converted to the chiral oxazolidinone **108** in 71% overall yield for the two-step process. A second addition of butadiene monoepoxide produced **109** in quantitative yield. Intermediate **109** was converted to the glucosidase inhibitor (−)-bulgecinine (**110**) in 18% overall yield for the 12-step process starting from phthalimide.

A third example of *N*-allylation from the Trost group involved the palladium-catalyzed desymmetrization of the cyclopentene bis-benzoate **111**, the initial step in an enantio- and diasterocontrolled synthesis of the potent antiviral agent (−)-neplanocin A (**112**) [76]. Accordingly, **111** was treated with 6-chloropurine in the presence

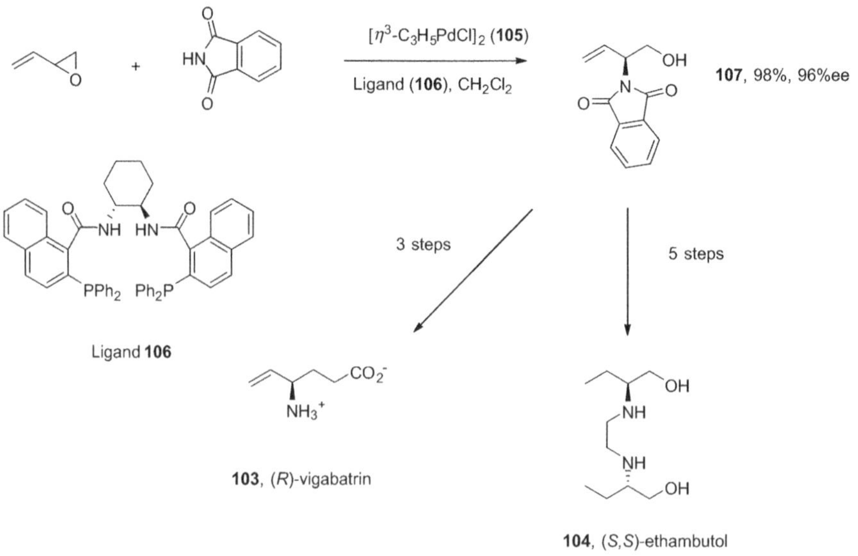

SCHEME 4.32   Asymmetric synthesis of vigabatrin (**103**) and ethambutol (**104**).

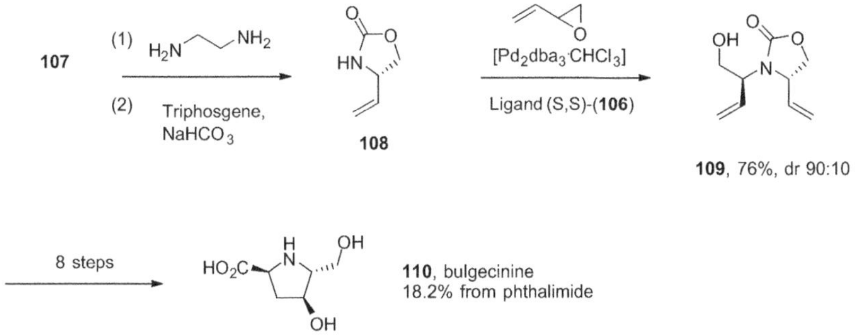

SCHEME 4.33   Synthesis of (−)-bulgecinine (**110**).

of the chiral catalyst produced *in situ* from ligand **113** and Pd₂dba₃ in THF (Scheme 4.34). The monoalkylated product **114** was subsequently converted to the carbanucleoside **112** in 12 steps. Since the approach does not depend on the use of carbohydrate precursors it provides an entry to both enantiomeric carbanucleosides in a series.

The Hartwig group recently reported a method for the *N*-allylation of carbamates **115** with high regioselectivity and enantioselectivity [77]. As shown in Scheme 4.35, the ethylene-bound iridium metacycle **116**, prepared *in situ* from [Ir(COD)Cl]₂ and the corresponding phosphoramidite ligand, served as the chiral catalyst at a 4 mol% loading. Average enantiomeric excesses of 98% were achieved [78]. The reactions were conducted on a 1.0 mmol scale at room temperature in THF.

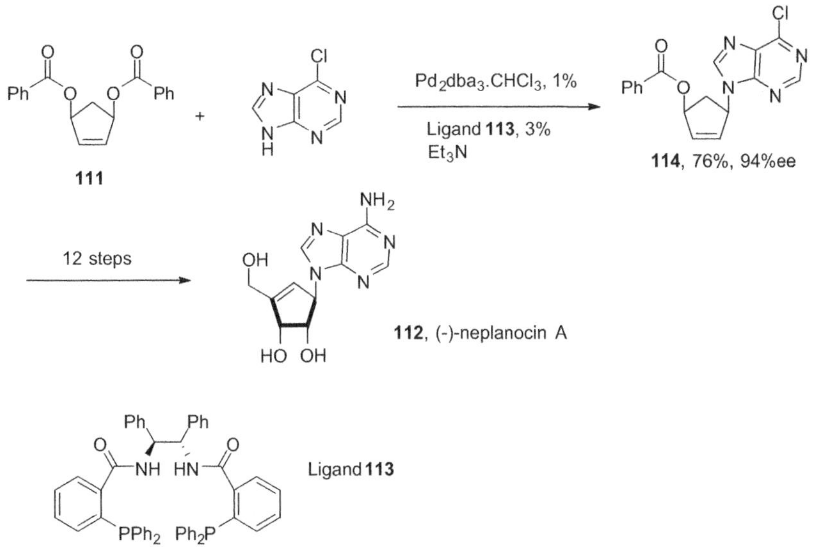

SCHEME 4.34   Synthesis of (−)-neplanocin A (**112**).

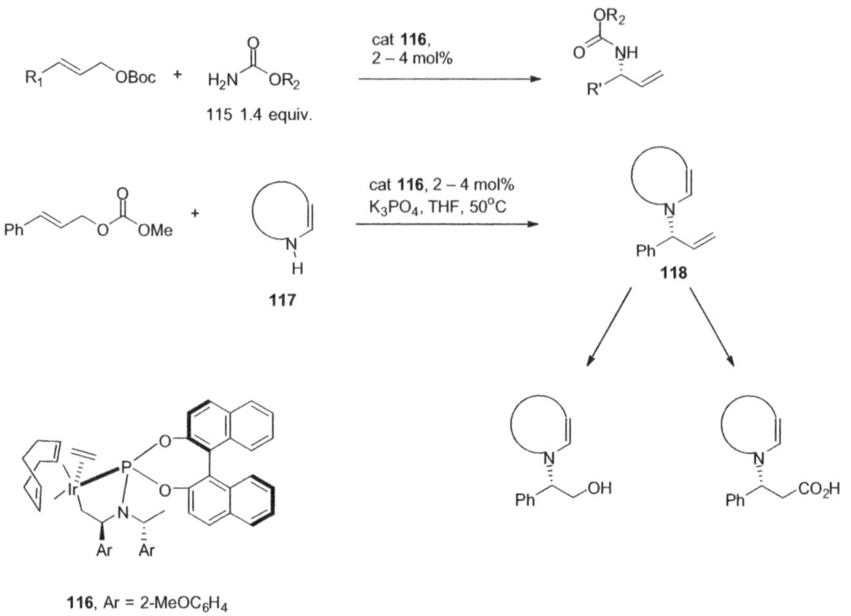

SCHEME 4.35 Asymmetric *N*-allylation of carbamates and azoles.

The reaction was extended to the *N*-allylation of a variety of azoles (**117**) including imidazoles, benzimidazoles, and purines [79]. The same chiral iridium catalyst **116** was employed at 2–4 mol% loading. Again, the reaction proceeded with high regio- and enantioselectivity. The resulting adducts **118** served as useful intermediates for the synthesis of a variety of alcohols, carboxylic acids, and other derivatives. For example, a key initial step in a formal synthesis of the kinase inhibitor **119** was the reaction of carbonate **120** with imidazole **121** to produce the chiral adduct **122** in 91%ee and 64% yield, as shown in Scheme 4.36. The product was predominantly the desired *N*1 alkylation product (*N*1:*N*3::72:28), thus the lower overall yield of **122**. Efforts are underway to optimize the iridium catalysts to extend the method to other useful azole nucleophiles and improve regioselectivity.

4.3.1.3 Conjugate Addition The asymmetric 1,4-addition of nitrogen nucleo-philes to α,β-unsaturated ketones has been employed by several groups to produce the corresponding optically active β-amino ketones. For example, Shibasaki and coworkers prepared amines **123** (Scheme 4.37) by the treatment of enones **124** with *O*-methylhydroxylamine in the presence of chiral heterobimetallic yttrium-BINOL catalysts of type **125** [80].

The products were obtained using as little as 0.5–3 mol% of the catalyst in good yields (80–98%) and enantiomeric excesses (81–96%). In turn, products **123** served as useful precursors for the synthesis of C-acylated aziridines and other versatile chiral building blocks.

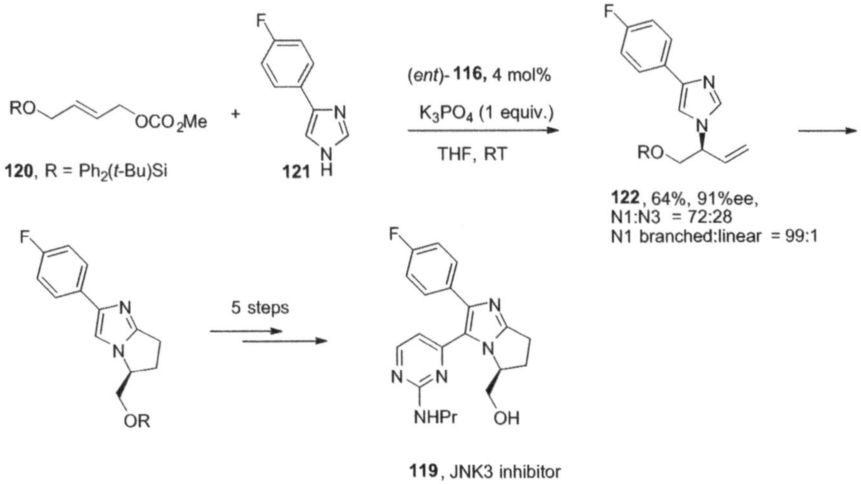

SCHEME 4.36 Synthesis of JNK3 inhibitor **119**.

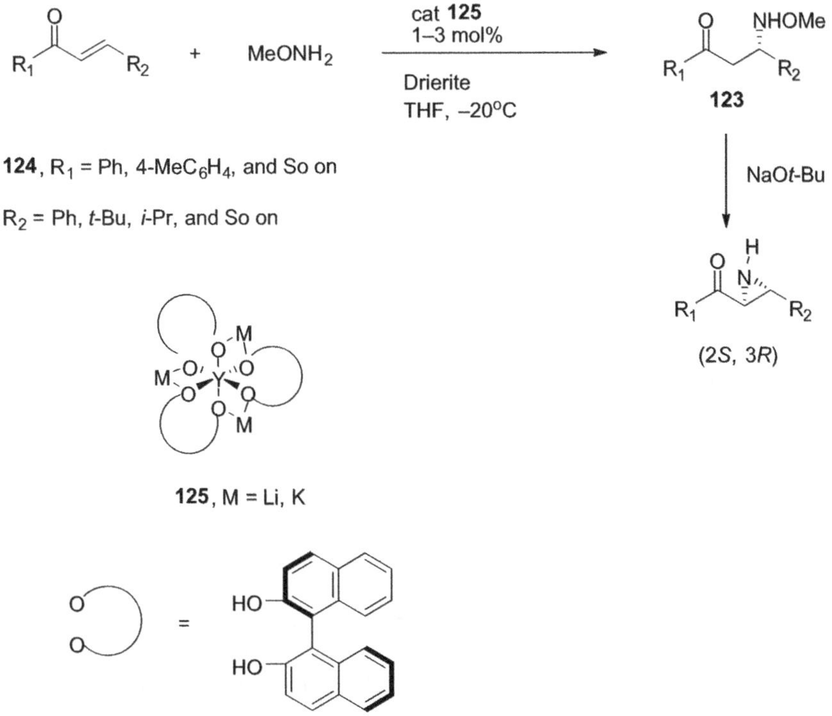

SCHEME 4.37 Conjugate addition of a nitrogen nucleophile.

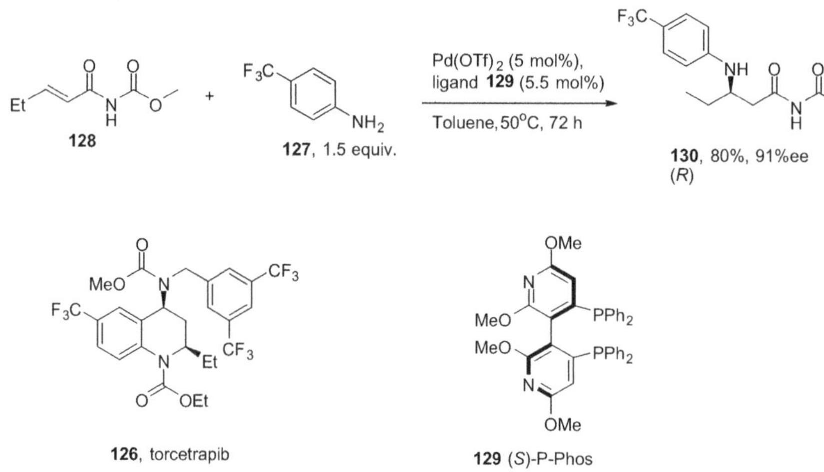

SCHEME 4.38   Synthesis of torcetrapib (**126**).

A concise asymmetric synthesis of torcetrapib (**126**, Scheme 4.38) [81], formerly in development as an inhibitor of cholesterol ester transfer protein (CETP), depended on another stereospecific conjugate addition, in this case between the aniline **127** and the Michael acceptor **128**. The most effective chiral catalyst, based on yield and stereoselectivity, was found to be the complex of (*S*)-P-Phos (**129**) and palladium triflate dihydrate at a 5 mol% loading. The reaction on a 3.8 mmol scale proceeded in 79% yield to produce the *R*-amine **130** in 91% enantiomeric excess.

The authors noted that the synthesis of **126** was achieved in seven steps from a chiral precursors without the need for protecting groups. This route offered advantages over the original Pfizer synthesis, which required several functional group conversions and a classical resolution [82].

4.3.2   Addition of Nitrogen Electrophiles

The electrophilic property of di-*tert*-butyl azodicarboxylate (**131**, DBAD, Scheme 4.39) has been put to use in asymmetric amination reactions using the chiral lanthanum-amide catalyst derived from ligand **132**. For example, the succinimide **133** was treated with DBAD in the presence of 2–4 mol% of catalyst to produce the aminated derivative **134** in >99% yield and 92% enantiomeric excess [83]. The product served as a useful intermediate for the synthesis of the aldose reductase inhibitor (−)-ranirestat (**135**), currently in late-stage clinical studies for the treatment of diabetic neuropathy. The same group reported an improved synthesis of **135** the following year. In that work, the lanthanum triisopropoxide was replaced by lanthanum nitrate hydrate, thereby making the process "reproducible, scalable, and cost-effective." [84] Full details of the process were reported

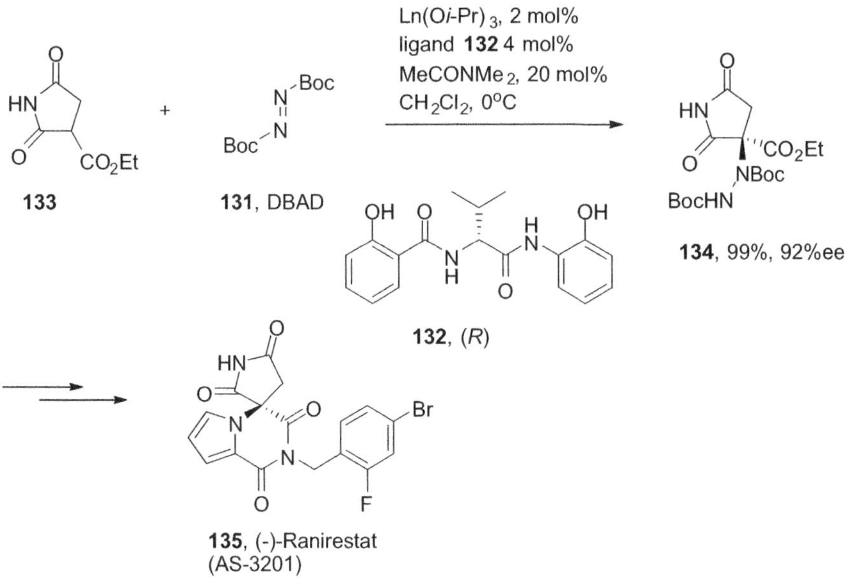

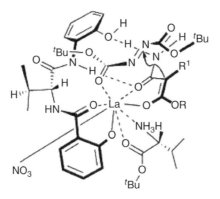

SCHEME 4.39 Synthesis of (−)-ranirestat (**135**).

recently [85]. That paper also described a superior *ternary* catalyst derived from ligand **132**, lanthanum nitrate hydrate, and D-valine *tert*-butyl ester, used in as little as 1 mol% loading. The adduct with DBAD was depicted as **136** (Fig. 4.6). Product **134** was obtained in up to 99 + % yield and over 99% enantiomeric excess.

4.3.3 Addition of Oxygen Nucleophiles

A convergent synthesis of the nonpeptidic protease inhibitor tipranavir (**137**) depended on the preparation of chiral intermediates **138** and **139**, both accomplished

FIGURE 4.6 Proposed ternary lanthanum catalyst. Reproduced with permission from J. Am. Chem. Soc. 2009, V131, P14990–14999. © 2009, American Chemical Society.

using a dynamic kinetic asymmetric transformation [86]. Accordingly, the quaternary stereogenic center of **138** was created by the asymmetric allylic *O*-alkylation of racemic epoxide **140**. As shown in Scheme 4.40, treatment of racemic vinyl epoxide **140** with 4-methoxybenzyl alcohol (as the nucleophile) in the presence of 1 mol% of Pd$_2$(dba)$_3$ and the (*S,S*)-ligand **141** produced adduct **138** in 69% yield and 98% enantiomeric excess. The asymmetric center of **138** eventually became carbon atom 6 of tipranavir.

The second chiral synthon, **139**, was prepared by the *molybdenum*-catalyzed DYKAT reaction of racemic carbamate **142** to create a tertiary stereogenic center,

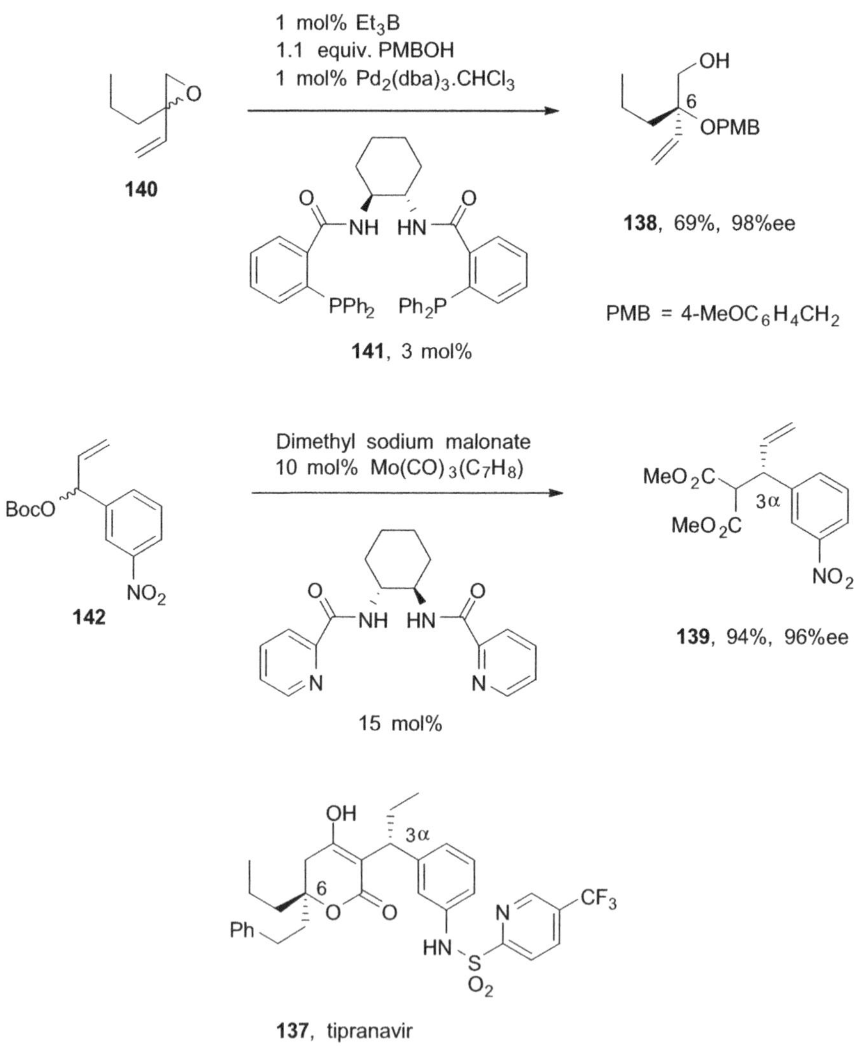

SCHEME 4.40   Synthesis of tipranavir (**137**).

$$\text{C—H} \quad + \quad \text{Hetero} \quad \xrightarrow{\text{cat.*}} \quad \text{C*—Hetero}$$

$$\text{Hetero—H} \quad + \quad \text{C} \quad \xrightarrow{\text{cat.*}} \quad \text{Hetero—C*}$$

SCHEME 4.41   Asymmetric insertion into carbon–hydrogen and hetero-hydrogen bonds.

which ultimately became the 3α carbon atom of the target. Although not an example of the addition of an oxygen nucleophile, this reaction is cited to demonstrate the synthetic utility of the DYKAT strategy in the same synthesis.

The two fragments were further elaborated and combined with a third to produce tipranavir (**137**). This concise synthesis is noteworthy because it does not require the use of chiral auxiliaries or resolutions.

4.3.4   Insertion

The stereospecific replacement of prochiral hydrogen atoms by heteroatoms and, conversely, the insertion of carbon atoms between heteroatom–hydrogen bonds, provide very useful methods for the construction of chiral carbon–heteroatom centers, as shown in Scheme 4.41. Examples of both processes follow.

The ability to discriminate between two hydrogen atoms on a prochiral carbon center was the outcome of a study by the Du Bois group [87]. They discovered a very versatile method for the synthesis of *N*-protected β-amino acids of type **143** (Scheme 4.42) via the enantioselective, intramolecular insertion of nitrogen from sulfamate esters **144** using the chiral rhodium catalyst $Rh_2(S\text{-nap})_4$ (**145**).

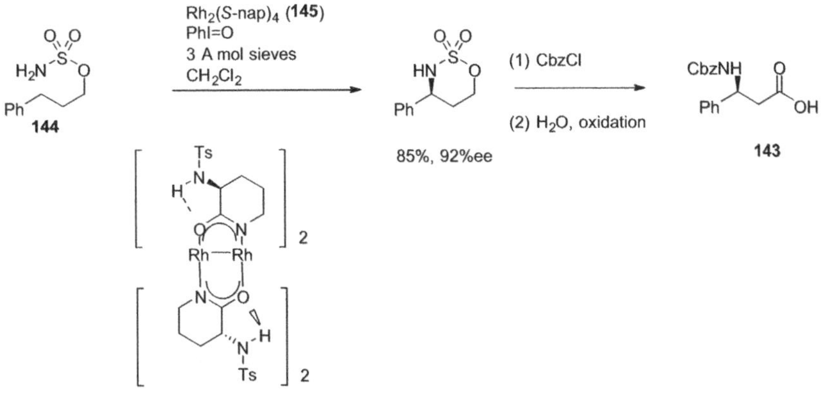

SCHEME 4.42   Asymmetric C–H bond insertion by nitrogen.

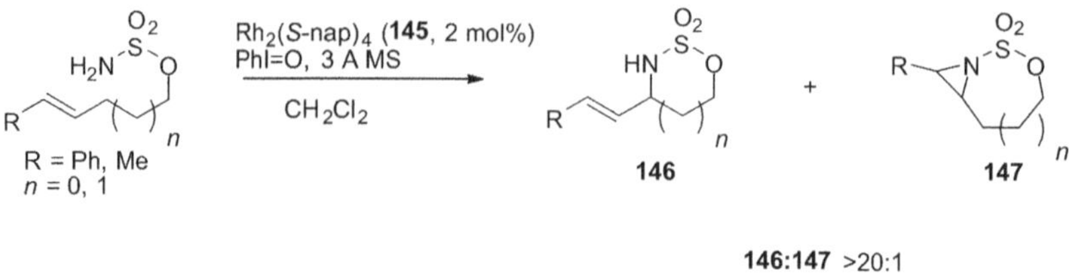

146:147 >20:1

SCHEME 4.43   Asymmetric allylic C–H bond insertion by nitrogen.

The catalyst is prepared from valerolactam and, used at 2 mol% loading, afforded some of the highest levels of asymmetric control in sulfamate ester cyclizations. Enantiomeric excesses of >80% were achieved. As shown in Scheme 4.43, the authors also showed the utility of catalyst **145** in chemoselective *allylic* C–H bond insertion, which favors formation of the analogous allylated cyclic sulfamates **146** over the bicyclic aziridines **147**. Experimental evidence was presented for a concerted nitrene insertion mechanism but did not rule out the possibility of a stepwise pathway.

Asymmetric fluorination has been the subject of several reviews [88]. The "first highly enantioselective fluorination of malonates" was achieved by the stereospecific replacement of an enantiotopic hydrogen atom in **148** by fluorine, as shown in Scheme 4.44. The zinc complex of (R,R)-DBFOX-Ph (**149**) served as the chiral

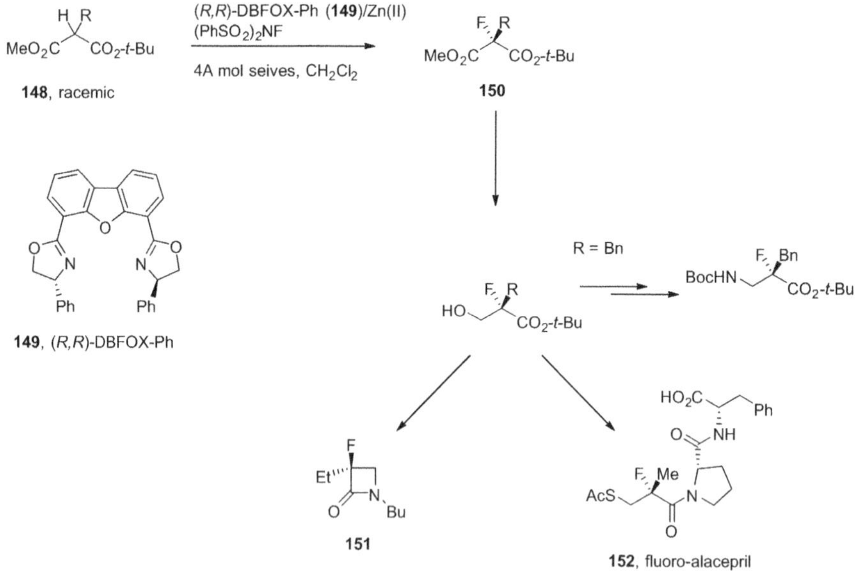

SCHEME 4.44   Asymmetric C–H bond insertion by fluorine.

catalyst [89]. It was found that molecular sieves were essential for stereoselectivity. The resulting fluoromalonates **150** could be produced in up to 99% enantiomeric excess (for R = Me, Bu, or Ph) in isolated yields of over 90%. They were shown to be attractive precursors for the synthesis of a variety of molecules of medicinal interest, including β-lactams (**151**) and the antihypertensive ACE inhibitor fluoro-alacepril (**152**), produced in eight steps from **148** in 18% overall yield.

Examples of the asymmetric insertion of α-diazo esters into heteroatom–hydrogen bonds were published by Fu's group at MIT. In the case of insertion into a hydrogen–nitrogen bond [90], carbamates **153** were treated with α-diazo esters **154** (1.5 equivalents) in the presence of 8.0 mol% of the chiral copper catalyst derived from (−)-BPY (**155**), resulting in the very useful nonracemic, easily deprotected α-amino esters **156** as shown in Scheme 4.45. Enantiomeric excesses of 85–95% were realized and could be enhanced by recrystallization. The stereoselectivity of the reaction decreased as the steric demands of the ester became smaller. Accordingly, replacing *tert*-butyl in ester **155** with methyl saw the enantiomeric excess of the product fall from 94% to 70%. On the other hand, the reaction proceeded with a wide range of aryl esters and Cbz- (in place of Boc-) protected amines. Insertion did not occur in the absence of $AgSbF_6$, and the use of other silver salts resulted in slightly lower stereoselectivities.

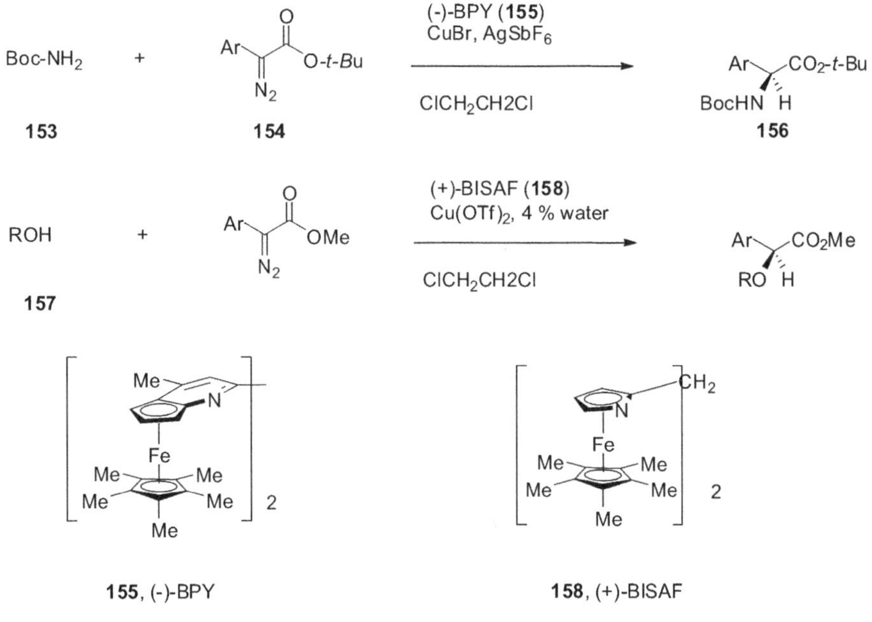

SCHEME 4.45 Asymmetric N–H and O–H bond insertion.

The analogous reaction of α-diazo esters with *alcohols* (**157**) in the presence of the related bis-azaferrocene catalyst (+)-BISAF (**158**) at 3.8 mol% loading resulted in the "first efficient method for catalytic enantioselective insertions into O–H bonds" in up to 98% enantiomeric excess [91].

4.4   EMERGING TECHNOLOGIES

New methodologies continue to emerge from various academic labs at a rapid pace. The selected examples discussed here highlight some of the technologies that significantly expand the current synthetic idiom by either reducing the number of steps to access a certain structural type compared to older methods or introducing new retro-synthetic disconnections. Furthermore, the selected examples also show potential to be applied in an industrial setting owing to their reaction efficiency and robust conditions.

Fu's lab discovered a series of Ni-catalyzed reactions that utilize racemic sp^3 secondary bromides leading to highly enantiomerically enriched C–C coupled products. As shown in Scheme 4.46, racemic α-bromo amide **159** underwent reaction with a Ni-pybox complex and a primary functionalized organozinc bromide **160** to give coupled product **161** in good yields and excellent enantiomeric excesses [92]. The organozinc species was freshly prepared prior to the reaction following established procedures [93]. The reaction was easily performed without the need for a glove box or excess cooling to obtain high selectivities. A number of functional groups tolerated by the reaction include protected alcohols, acetals, phthalimides, and nitriles. Typically enantiomerically enriched products such as **161** arise from chiral auxiliary-aided alkylation using a strong base such as LDA or NaN(TMS)$_2$ to generate a reactive anion. The racemic bromide could be easily accessed, and both enantiomers converged into a single enantiomer of the coupled product without

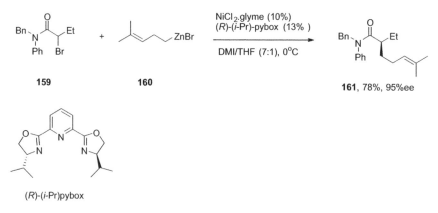

(*R*)-(*i*-Pr)pybox

DMI = 1,3-dimethyl-2-imidazolone

SCHEME 4.46   Ni-catalyzed asymmetric coupling of α-bromo amides and alkylzinc species.

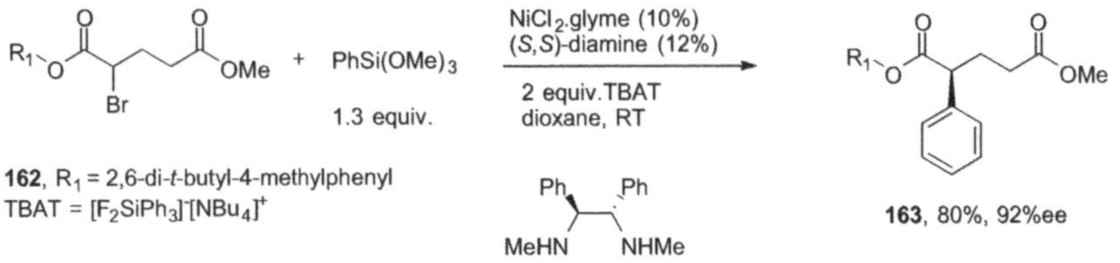

SCHEME 4.47 Ni-catalyzed asymmetric coupling of α-bromo esters and organosilanes.

the need for kinetic resolution. The largest scale of this coupling reaction reported by the authors was 10 mmol.

Racemic α-bromo esters of type **162** have been shown to undergo Ni-catalyzed Hiyama cross-couplings leading to enantiomerically enriched esters **163** (Scheme 4.47) [94]. This reaction is complimentary to the Negishi-type reaction described above but with the added convenience that a number of organosilanes are commercially available. It is noteworthy that the cross-coupling reaction occurred at ambient temperature and the catalyst components were air-stable. The scope of the reaction is not limited to arylation of α-bromo esters; alkenyl silanes also work well as nucleophile coupling partners leading to α-alkenyl esters. The reaction was performed on gram scale with only slight decrease in enantioselectivity.

Racemic propargylic halides can also serve as electrophilic substrates in Ni-catalyzed couplings [95]. The nucleophile employed in this reaction was an arylzinc species generated from readily available arylboronic acids and Et_2Zn. As shown in Scheme 4.48, coupling between TMS-protected alkyne **164** and aryl-Zn **165**, promoted by 3 mol% of nickel complex ligated with pybox **166**, gave the coupled product **167** in 93%ee on a gram scale. Alkyne **167** is a key building block for the synthesis of inhibitors of protozoan dihydrofolate reductase such as **168**. In a

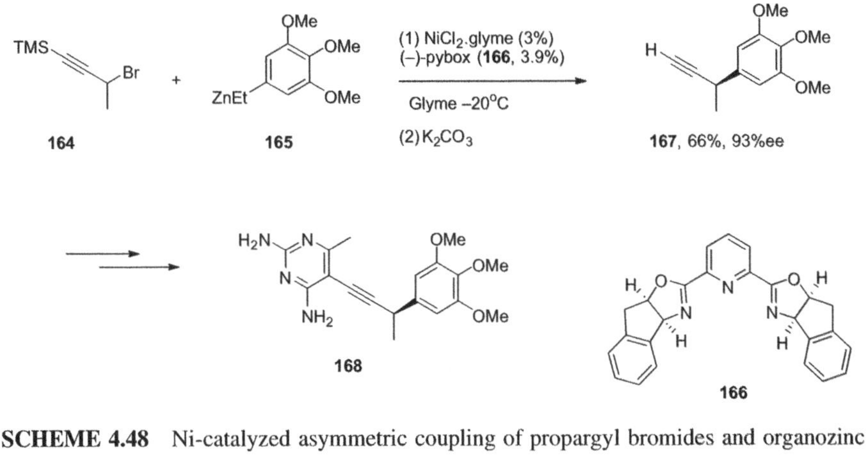

SCHEME 4.48 Ni-catalyzed asymmetric coupling of propargyl bromides and organozinc species.

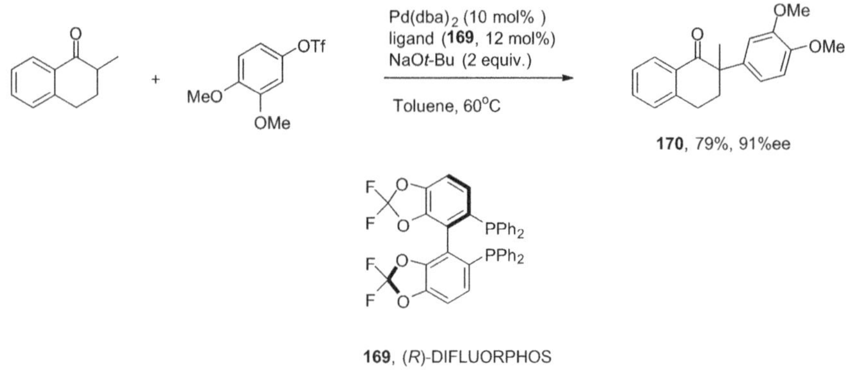

SCHEME 4.49   Enantioselective α-arylation of ketones.

previous synthesis of **168** employing chiral auxiliary-mediated alkylation, the reaction sequence took five steps and proceeded in 22% overall yield. The current coupling method obviates the need for a chiral auxiliary thereby reducing the number of steps to reach the same product.

The mechanism of the reaction has not been fully established. Based on the diverse scope and the practicality of the Ni-catalyzed coupling reaction that Fu has demonstrated, further studies on the mechanism are highly warranted.

The α-arylation of enolates has been extensively studied [96]. Diastereoselective variants are known which rely on chiral auxiliaries to induce the product stereochemistry. However, enantioselective versions are very rare. Hartwig's group has found that by using DIFLUORPHOS™ (**169**) complexes of Pd or Ni, ketones such as α-methyl tetralone underwent cross-coupling with an aryl triflate to give an α-aryl-coupled product **170** in high yield and enantiomeric excess, as shown in Scheme 4.49 [97]. It was found during their ligand screening that the DIFLUOR-PHOS ligand, which possess a smaller dihedral angle than the standard BINAP, gave higher enantioselectivities. The leaving group on the electrophile also played a role in reactivity and selectivity. Coupling reactions with triflate, the most reactive leaving group, gave higher yields than the typical bromides and iodides and the reaction could be carried out at lower temperatures. This reaction was only reported on a milligram scale.

Buchwald's group has published an α-arylation of oxindoles using an axially chiral P-stereogenic ligand [98]. Under the relatively mild reaction conditions shown in Scheme 4.50, oxindole **171** could be converted to the arylated product **172** (which contains a fully substituted quaternary stereogenic center) in good yields and excellent enantiomeric excesses. In addition to aryl bromides, vinyl bromides can also participate in the coupling reaction to give α-vinyl indoles. This reaction has only been reported on a milligram scale. Clearly, the most exotic component of the reaction is the unique P–N ligand (**173**) with a P-stereogenic center. It would be interesting to examine if such a ligand can facilitate the enantioselective α-arylations of other nucleophiles such as esters and amides.

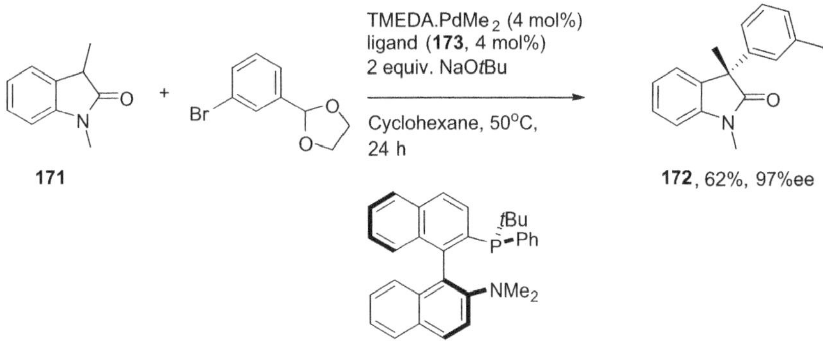

SCHEME 4.50 Enantioselective α-arylations of oxindoles.

4.5 CONCLUSION

The examples in this chapter have served to illustrate the tremendous potential of chiral transition metal catalysis for the asymmetric synthesis of a wide range of small molecules, in particular those of therapeutic interest. Although not necessarily the first option considered, the use of chiral transition metal catalysis is yet another tool in the chemist's synthetic arsenal. In many cases the technology provides the *only* expeditious method for the construction of asymmetric carbon–carbon and carbon–heteroatom bonds.

A number of considerations that we have discussed in the chapter are crucial to the selection of technologies, among them scalability (including technical feasibility and robustness) and the availability of raw materials. The inventors of new technology are increasingly paying attention to these multifaceted challenges to ensure that their technology addresses the unmet needs of the synthetic organic chemist.

We trust that, as more of the technology is adapted to the scale-up and production of chiral intermediates and APIs, it will achieve even greater consideration as a viable synthetic technique.

REFERENCES

[1] For earlier discussions of relevant aspects see, for example: (a) Shibasaki, M.; Yoshikawa, N. *Chem. Rev.* **2002**, *102*, 2187–2209; (b) Hawkins, J. M.; Watson, T. J. N. *Angew. Chem. Int. Ed.* **2004**, *43*, 3224–3228; (c) Federsel, H.-J. *Nat. Rev. Drug Discov.* **2005**, *4*, 685–697; (d) Farina, V.; Reeves, J. T.; Senanayake, C. H.; Song, J. J. *Chem. Rev.* **2006**, *106*, 2734–2793; (e) Shibasaki, M.; Kanai, M.; Matsunaga, S. *Aldrichim. Acta* **2006**, *39*(2), 31–39; (f) Ikunaka, M. *Org. Process Res. Dev.* **2007**, *11*(3), 495–502.

[2] (a) ChiraTecnicsLda., caixa postal 59, Rossio, 7006-802 Evora, Portugal; http://chiratecnics.com; (b) http://blog.hitbarcelona.com/tag/chiral-technology; (c) Pfizer,

Inc., 10-K, filed on 2/26/10, p. 28 therein; accessed from http://www.pfizer.com/ investors/financial_reports/financial_reports.jsp.

[3] http://www.fda.gov/Drugs/GuidanceComplianceRegulatoryInformation/Guidances/ ucm122883.htm.

[4] Hawkins, J. M.; Watson, T. J. N. *Angew. Chem. Int. Ed.* **2004**, *43*, 3225.

[5] Oguni, N.; Omi, T. *Tetrahedron Lett.* **1984**, *25*, 2823.

[6] (a) Pu, L.; Yu, H. B. *Chem. Rev.* **2001**, *101*, 757; (b) Ramon, D. J.; Yus, M. *Chem. Rev.* **2006**, *106*, 2126.

[7] Magnus, N. A.; Anzeveno, P. B.; Coffey, D. S.; Hay, D. A.; Laurila, M. E.; Schkeryantz, J. M.; Shaw, B. W.; Staszak, M. A. *Org. Process Res. Dev.* **2007**, *11*, 560.

[8] Schmidt, F.; Stemmler, R. T.; Rudolph, J.; Bolm, C. *Chem. Soc. Rev.* **2006**, *35*, 454.

[9] Kim, J. G.; Walsh, P. J. *Angew. Chem. Int. Ed.* **2006**, *45*, 4175.

[10] (a) Tokunaga, T.; Hume, W. E.; Umezome, T.; Okazaki, K.; Ueki, Y.; Kumagai, K.; Hourai, S.; Nagamine, J.; Seki, H.; Taiji, M.; Noguchi, H.; Nagata, R. *J. Med. Chem.* **2001**, *44*, 4641; (b) Tokunaga, T.; Hume, W. E.; Nagamine, J.; Kawamura, T.; Taiji, M.; Nagata, R. *Bioorg. Med. Chem. Lett.* **2005**, *15*, 1789.

[11] Di Malta, A.; Garcia, G.; Roux, R.; Schoentjes, B.; Serradeil-le Gal, S.; Tonnerre, B.; Wagnon, J., PCT Int. Appl. No. WO2003008407, 2003.

[12] Shintani, R.; Inoue, M.; Hayashi, T. *Angew. Chem. Int. Ed.* **2006**, *45*, 3353.

[13] Tomita, D.; Yamatsugu, K.; Kanai, M.; Shibasaki, M. *J. Am. Chem. Soc.* **2009**, *131*, 6946.

[14] Gröger, H. *Chem. Rev.* **2003**, *103*, 2795–2827.

[15] Hamashima, Y.; Kanai, M.; Shibasaki, M. *J. Am. Chem. Soc.* **2000**, *122*, 7412.

[16] Kanai, M.; Kato, N.; Ichikawa, E.; Shibasaki, M. *Synlett* **2005**, 1491.

[17] Yabu, K.; Masumoto, S.; Yamasaki, S.; Hamashima, Y.; Kanai, M.; Du, W.; Curran, D. P.; Shibasaki, M. *J. Am. Chem. Soc.* **2001**, *123*, 9908.

[18] Thompson, I. M.; Lauvetz, R. *Urology* **1976**, *8*, 452.

[19] Masumoto, S.; Suzuki, M.; Kanai, M.; Shibasaki, M. *Tetrahedron* **2004**, *60*, 10497.

[20] (a) Trost, B. M.; Yeh, V. S. C.; Ito, H.; Bremeyer, N. *Org. Lett.* **2002**, *4*, 2621; (b) Shibasaki, M.; Sasai, H.; Arai, T. *Angew. Chem. Int. Ed.* **1997**, *36*, 1236.

[21] Roche, D.; Prasad, K.; Repic, O.; Blacklock, T. J. *Tetrahedron Lett.* **2001**, *42*, 1459–1462.

[22] Shibasaki, M.; Yoshikawa, N. S. *Chem. Rev.* **2002**, *102*, 2187.

[23] Ellman, J. A.; Owens, T. D; Tang, T. P. *Acc. Chem. Res.* **2002**, *35*, 984.

[24] Charette, A. B.; Boezio, A.; Cote, E. *Pure Appl. Chem.* **2005**, *77*, 1259.

[25] Hayashi, T.; Kawai, N. *Angew. Chem. Int. Ed.* **2004**, *43*, 6125.

[26] Ueda, M.; Saito, A.; Miyaura, N. *Synlett*, **2000**, 1637.

[27] (a) Weix, D. J.; Shi, Y.; Ellman, J. A. *J. Am. Chem. Soc.* **2005**, *127*, 1092; (b) Trincado, M.; Ellman, J. A. *Angew. Chem. Int. Ed.* **2008**, *47*, 5623; (c) Storgaard, M.; Ellman, J. A. *Org. Synth.* **2009**, *86*, 360.

[28] (a) Rishel, M. J.; Amarasinghe, K. K. D.; Dinn, S. R.; Johnson, B. F. *J. Org. Chem.* **2009**, *74*, 4001; (b) Sasamoto, N.; Dubs, C.; Hamashima, Y.; Sodeoka, M. *J. Am. Chem. Soc.* **2006**, *128*, 14010.

[29] Comins, D. L.; Huang, S.; McArdle, C. L.; Ingalls, C. L. *Org. Lett.* **2001**, *3*, 467.

[30] Takamaru, M.; Funabashi, K.; Kanai, M.; Shibasaki, M. *J. Am. Chem. Soc.* **2001**, *123*, 6801.

[31] (a) Funabashi, K.; Ratni, H.; Kanai, M.; Shibasaki, M. *J. Am. Chem. Soc.* **2001**, *123*, 10784; (b) Ichikawa, E.; Suzuki, M.; Yabu, K.; Albert, M.; Kanai, M.; Shibasaki, M. *J. Am. Chem. Soc.* **2004**, *126*, 11808.

[32] Masumoto, S.; Usuda, H.; Suzuki, M.; Kanai, M.; Shibasaki, M. *J. Am. Chem. Soc.* **2003**, *125*, 5634.

[33] Kato, N.; Suzuki, M.; Kanai, M.; Shibasaki, M. *Tetrahedron Lett.* **2004**, *45*, 3147.

[34] Kato, N.; Suzuki, M.; Kanai, M.; Shibasaki, M. *Tetrahedron Lett.* **2004**, *45*, 3153.

[35] (a) For recent reviews on asymmetric Michael reactions see: Christoffers, J.; Baro, A. *Angew. Chem. Int. Ed.* **2003**, *42*, 1688; and Christoffers, J.; Koripelly, G.; Rosiak, A.; Rössle, M. *Synthesis* **2007**, 1279; (b) Barnes, D. M.; Ji, J.; Fickes, M. G.; Fitzgerald, M. A.; King, S. A.; Morton, H. E.; Plagge, F. A.; Preskill, M.; Wagaw, S. H.; Wittenberger, S. J.; Zhang, J. *J. Am. Chem. Soc.* **2002**, *124*, 13097.

[36] Evans, D. A.; Mito, S.; Seidel, D. *J. Am. Chem. Soc.* **2007**, *129*, 11583.

[37] (a) Xu, Y.; Ohori, K.; Ohshima, T.; Shibasaki, M. *Tetrahedron* **2002**, *58*, 2585; (b) Ohshima, T.; Xu, Y.; Takita, R.; Shibasaki, M. *Tetrahedron* **2004**, *60*, 9569.

[38] Hayashi, T.; Yamasaki, K. *Chem. Rev.* **2003**, *103*, 2829.

[39] (a) Lukin, K.; Zhang, Q.; Leanna, M. R. *J. Org. Chem.* **2009**, *74*, 929; (b) Wallace, G. A.; Gordon, T. D.; Hayes, M. E.; Konopaki, D. B.; Fix-Stenzel, S. R.; Zhang, X.; Grongsaard, P.; Cusack, K. P.; Schaffter, L. M.; Henry, R. F.; Stoffel, R. H. *J. Org. Chem.* **2009**, *74*, 4886.

[40] Review: Glorius, F. *Angew. Chem. Int. Ed.* **2004**, *43*, 3364.

[41] (a) Hayashi, T.; Ueyama, K.; Tokunaga, N.; Yoshida, K. *J. Am. Chem. Soc.* **2003**, *125*, 11508; (b) Chen, F.-X.; Kina, A.; Hayashi, T. *Org. Lett.* **2006**, *8*, 341; and references cited therein.

[42] (a) Fischer, C.; Defieber, C.; Suzuki, T.; Carreira, E. M. *J. Am. Chem. Soc.* **2004**, *126*, 1628; (b) Paquin, J.-F.; Defieber, C.; Stephenson, C. R. J.; Carreira, E. M. *J. Am. Chem. Soc.* **2005**, *127*, 10850; and references cited therein.

[43] Wang, Z.-Q.; Feng, C.-G.; Xu, M.-H.; Lin, G.-Q. *J. Am. Chem. Soc.* **2007**, *129*, 5336.

[44] Sörgel, S.; Tokunaga, N.; Sasaki, K.; Okamoto, K.; Hayashi, H. *Org. Lett.* **2008**, *10*, 589.

[45] Aratani, T. *Pure Appl. Chem.* **1985**, *57*(12), 1839–1844.

[46] Keynan, S.; Hooper, N. M.; Felici, A.; Amicosante, G.; Turner, A. J. *Antimicrob. Agents Chemother.* **1995**, *39*(7), 1629–1631; for another synthesis of cilastatin using a chiral *iron* catalyst, see Wang, Q.; Yang, F.; Du, H.; Hossain, M. M.; Bennett, D.; Grubisha, D. S. *Tetrahedron Asymm.* **1998**, *9*(22), 3917–3977.

[47] Itagaki, M.; Masumoto, K.; Suenobu, K.; Yamamoto, Y. *Org. Process Res. Dev.* **2006**, *10*(2), 245–250.

[48] Itagaki, M.; Suenobu, K. *Org. Process Res. Dev.* **2007**, *11*(3), 509–518.

[49] Marcin, L. R.; Denhart, D. J.; Mattson, R. J. *Org. Lett.* **2005**, *7*(13), 2651–2654.

[50] Anthes, R.; Bello, O.; Benoit, S.; Chen, C.-K.; Corbett, E.; Corbett, R. M.; DelMonte, A. J.; Gingras, S.; Livingston, R.; Sausker, J.; Soumeillant, M. *Org. Process Res. Dev.* **2008**, *12*(2), 168–177.

[51] Anthes, R.; Benoit, S.; Chen, C.-K.; Corbett, E. A.; Corbett, R. M.; DelMonte, A. J.; Gingras, S.; Livingston, R. C.; Pendri, Y.; Sausker, J.; Soumeillant, M. *Org. Process Res. Dev.* **2008**, *12*(2), 178–182.

[52] Simpson, J. H.; Godfrey, J.; Fox, R.; Kotnis, A.; Kacsur, D.; Hamm, J.; Totelben, M.; Rosso, V.; Mueller, R.; Delaney, E.; Deshpande, R. P. *Tetrahedron Asymm.* **2003**, *14*(22), 3569–3574.

[53] Agbodjan, A. A.; Cooley, B. E.; Copley, R. C. B.; Corfield, J. A.; Flanagan, R. C.; Glover, B. N.; Guidetti, R.; Haigh, D.; Howes, P. D.; Jackson, M. M.; Matsuoka, R. T.; Medhurst, K. J.; Millar, A.; Sharp, M. J.; Slater, M. J.; Toczko, J. F.; Xie, S. *J. Org. Chem.* **2008**, *73*, 3094–3102.

[54] Usuda, H.; Kuramochi, A.; Kanai, M.; Shibasaki, M. *Org. Lett.* **2004**, *6*(23), 4387–4390.

[55] Shimizu, Y.; Shi, S.-L.; Usuda, H.; Kanai, M.; Shibasaki, M. *Angew. Chem. Int. Ed.* **2010**, *48*, 1103–1106.

[56] Farina, V.; Reeves, J. T.; Senanayake, C. S.; Song, J. J. *Chem. Rev.* **2006**, *106*, 2734.

[57] Hashimoto, T.; Maruoka, K. *Chem. Rev.* **2007**, *107*, 5656.

[58] Maeda, K.; Miller, R. A.; Szumigala, R. H.; Shafiee, A.; Karady, S.; Armstrong, J. D. III *Tetrahedron Lett.* **2005**, *46*, 15451.

[59] For reviews see (a) Lu, Z.; Ma, S. *Angew. Chem. Int. Ed.* **2008**, *47*, 258; (b) Trost, B. M.; Crawley, M. L. *Chem. Rev.* **2003**, *103*, 2921.

[60] Campos, K. R.; Journet, M.; Lee, S.; Grabowski, E. J. J.; Tillyer, R. D. *J. Org. Chem.* **2005**, *70*, 268.

[61] Trost, B. M.; Bunt, R. C. *J. Am. Chem. Soc.* **1994**, *116*, 4089.

[62] Zhang, F.; Song, Z. J.; Tschaen, D.; Volante, R. P. *Org. Lett.* **2004**, *6*, 3775.

[63] Butts, C. P.; Filali, E.; Lloyd-Jones, G. C.; Norrby, P.-O.; Sale, D. A.; Schramm, Y. *J. Am. Chem. Soc.* **2009**, *131*, 9945.

[64] Palacki, M.; Um, J. M.; Yasuda, N.; Conlon, D. A.; Tsay, F.-R.; Hartner, F. W.; Hsiao, Y.; Marcune, B.; Karady, S.; Hughes, D. L.; Dormer, P. G.; Reider, P. J. *J. Org. Chem.* **2002**, *67*, 5508.

[65] (a) Palucki, M.; Um, J. M.; Conlon, D. A.; Yasuda, N.; Hughes, D. L.; Mao, B.; Wang, J.; Reider, P. J. *Adv. Synth. Catal.* **2001**, *343*, 46; (b) Trost, B. M.; Hildbrand, S.; Dogra, K. *J. Am. Chem. Soc.* **1999**, *121*, 10416.

[66] Trost, B. M.; Dogra, K.; Hachiya, I.; Emura, T.; Hughes, D. L.; Krsha, S.; Reamer, R. A.; Paluki, M.; Yasuda, N.; Reider, P. J. *Angew. Chem. Int. Ed.* **2002**, *41*, 1929.

[67] (a) Doyle, M. P.; Duffy; R. Ratnikov, M.; Zhou, L. *Chem. Rev.* **2010**, *110*(2), 704–724; (b) Davies, H. M. L.; Manning, J. R. *Nature* **2008**, *451*, 417–424.

[68] Davies, H. M. L.; Hansen, T.; Hopper, D. W.; Panaro, S. A. *J. Am. Chem. Soc.* **1999**, *121*, 6509–6510.

[69] (a) Davies, H. M. L.; Panaro, S. A. *Tetrahedron Lett.* **1999**, *40*, 5287–5290; (b) http://www.buffalo.edu/news/9097.

[70] Axten, J. M.; Ivy, R.; Krim, L.; Winkler, J. D. *J. Am. Chem. Soc.* **1999**, *121*, 6511–6512.

[71] Doyle, M. P.; Winchester, W. R.; Hoorn, J.; Lynch, V.; Simonsen, S.; Ghosh, R. *J. Am. Chem. Soc.* **1993**, *115*, 9968–9978.

[72] Fukuta, Y.; Mita, T.; Fukuda, N.; Kanai, M.; Shibasaki, M. *J. Am. Chem. Soc.* **2006**, *128*, 6312–6313.

[73] Mita, T.; Fukuda, N.; Roca, F. X.; Kanai, M.; Shibasaki, M. *Org. Lett.* **2007**, *9*(2), 259–262.

[74] Trost, B. M.; Bunt, R. C.; Lemoine, R. C.; Calkins, T. L. *J. Am. Chem. Soc.* **2000**, *122*, 5968–5976.

[75] Trost, B. M.; Horne, D. B.; Woltering, M. J. *Chem. Eur. J.* **2006**, 6607–6620.

[76] Trost, B. M.; Madsen, R.; Guile, S. D. *Tetrahedron Lett.* **1997**, *38*(10), 1707–1710.

[77] Weix, D. J.; Marković, D.; Ueda, M.; Hartwig, J. F. *Org. Lett.* **2009**, *11*(13), 2944–2947.

[78] Leitner, A.; Shu, C.; Hartwig, J. F. *Org. Lett.* **2005**, *7*, 1093–1096.

[79] Stanley, L. M.; Hartwig, J. F. *J. Am. Chem. Soc.* **2009**, *131*, 8971–8983.

[80] (a) Yamagiwa, N.; Matsunaga, S.; Shibasaki, M. *J. Am. Chem. Soc.* **2003**, *125*, 16178–16179; (b) for recent reviews of the transformations of *rare earth—alkali metal—BINOL* ("REMB") catalysts see: Shibasaki, M.; Kanai, M.; Matsunaga, S.; Kumagai, N. *Accts. Chem. Res.* **2009**, *42*(8), 1117–1127; (c) Shibasaki, M.; Kanai, M.; Matsunaga, S. *Aldrichim. Acta* **2006**, *39*, 31–39.

[81] Guinó, M.; Phua, P. H.; Caille, J.-C.; Hii, K. K. *J. Org. Chem.* **2007**, *72*, 6290–6293.

[82] Damon, D. B.; Dugger, R. W.U.S. Patent 6,313,142, November 6, 2001.

[83] Mashiko, T.; Hara, K.; Tanaka, D.; Fujiwara, Y.; Kumagai, N.; Shibasaki, M. *J. Am. Chem. Soc.* **2007**, *129*, 11342–11343.

[84] Mashiko, T.; Kumagai, N.; Shibasaki, M. *Org. Lett.* **2008**, *10*(13), 2725–2728.

[85] Mashiko, T.; Kumagai, N.; Shibasaki, M. *J. Am. Chem. Soc.* **2009**, *131*, 14990–14999.

[86] Trost, B. M.; Andersen, N. G. *J. Am. Chem. Soc.* **2002**, *124*, 14320–14321.

[87] Zalatan, D. N.; Du Bois, J. *J. Am. Chem. Soc.* **2008**, *130*, 9220–9221.

[88] Ma, J.-A.; Cahard, D. *Chem. Rev.* **2008**, *108*(9), PR1–PR43, and references cited therein.

[89] Reddy, D. S.; Shibata, N.; Nagai, J.; Nakamura, S.; Toru, T; Kanemasa, S. *Angew. Chem. Int. Ed.* **2008**, *47*, 164–168.

[90] Lee, E. C.; Fu, G. C. *J. Am. Chem. Soc.* **2007**, *129*, 12066–12067.

[91] Maier, T. C.; Fu, G. C. *J. Am. Chem. Soc.* **2006**, *128*, 4594–4595.

[92] Fischer, C.; Fu, G. C. *J. Am. Chem. Soc.* **2005**, *127*, 4594.

[93] Huo, S. *Org. Lett.* **2003**, *5*, 423.

[94] Dai, X.; Strotman, N. A.; Fu, G. C. *J. Am. Chem. Soc.* **2008**, *130*, 3302.

[95] Smith, S. W.; Fu, G. C. *J. Am. Chem. Soc.* **2008**, *130*, 12645.

[96] (a) Fox, J. M.; Huang, X.; Chieffi, A.; Buchwald, S. L. *J. Am. Chem. Soc.* **2000**, *122*, 1360; (b) Kawatsura, M.; Hartwig, J. F. *J. Am. Chem. Soc.* **1999**, *121*, 1473.

[97] Liao, X.; Weng, Z.; Hartwig, J. F. *J. Am. Chem. Soc.* **2008**, *130*, 195.

[98] Taylor, A. M.; Altman, R. A.; Buchwald, S. L. *J. Am. Chem. Soc.* **2009**, *131*, 9900.

5

METATHESIS REACTIONS

Oliver R. Thiel

5.1 INTRODUCTION

The term olefin metathesis defines the mutual exchange of alkylidene fragments between two alkenes. The intramolecular version of the reaction leads to cyclic products and is therefore described as ring-closing metathesis (RCM). The inter-molecular reaction between two olefins is described as cross metathesis (CM) (Scheme 5.1). Since the discovery of the reaction in 1955, metathesis has found industrial application in the Shell Higher Olefin Process (SHOP) and in the ring-opening metathesis polymerization of strained olefins [1]. These early applications relied on heterogeneous early transition metal precatalysts that were activated with

Applications of Transition Metal Catalysis in Drug Discovery and Development: An Industrial Perspective,
First Edition. Edited by Matthew L. Crawley and Barry M. Trost.
© 2012 John Wiley & Sons, Inc. Published 2012 by John Wiley & Sons, Inc.

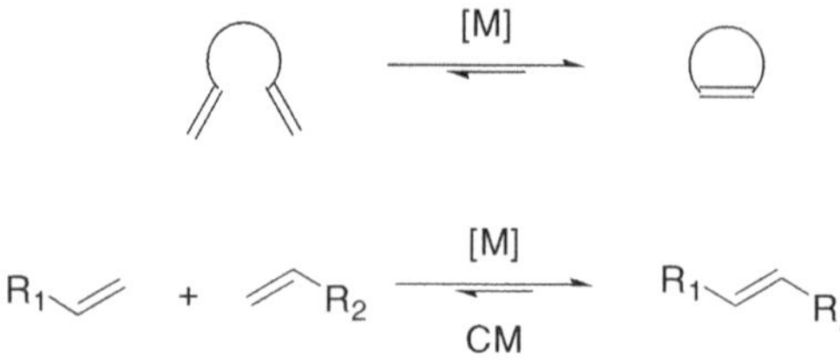

SCHEME 5.1   Metathesis reactions.

alkylating agents. Since the discovery of the first highly active and well-defined homogeneous catalysts for olefin metathesis the reaction has had a significant impact in organic synthesis. The broad versatility of the catalysts and the functional group tolerability enabled numerous applications in organic synthesis, especially natural product synthesis [2]. The impact of this reaction on the field was recognized with the Nobel Prize in Chemistry 2005 for Robert H. Grubbs, Richard R. Schrock, and Yves Chauvin [3].

The introduction of readily available homogeneous catalysts enabled the widespread application of the reaction. Early examples included RCM for the synthesis of five- and six-membered rings by Fu and Grubbs [4] and for the synthesis of macrocycles by Fürstner and Langemann [5]. The first application for the synthesis of a natural product was reported by Hoveyda with the synthesis of SCH38516 [6]. Not surprisingly the new tools also found ample application in the synthesis of new biologically active scaffolds within medicinal chemistry groups of pharmaceutical companies. Similar to the impact of palladium-catalyzed coupling reactions that led to an easier access to compounds with direct aryl–aryl bonds, olefin metathesis has transformed drug discovery by offering easier access to aliphatic rings, especially macrocycles. With the advancement of some of these molecules into development the reaction also was introduced into process chemistry laboratories, where it has shown to be practical on large scale.

The field has been extensively reviewed in the past; therefore this review only summarizes the basic principles of the most common catalyst systems [7]. The main focus is then on developments and applications from laboratories within the pharmaceutical industry. It is of note that olefin metathesis has been applied both for the synthesis of medium and large rings by ring-closing metathesis as well as for the formation of olefins by cross metathesis.

5.2   CATALYST SYSTEMS

The metathesis reaction can be catalyzed by various transition metals, but in organic synthesis the use of molybdenum and ruthenium complexes is prevalent (Scheme 5.2). The broad utilization of metathesis in organic synthesis started with the discovery of well-defined molybdenum complexes by Schrock and co-workers in 1990 [8]. Catalyst **1** displays high activity and is suitable for sterically demanding substrates. Limited tolerability of polar functional groups as well as its

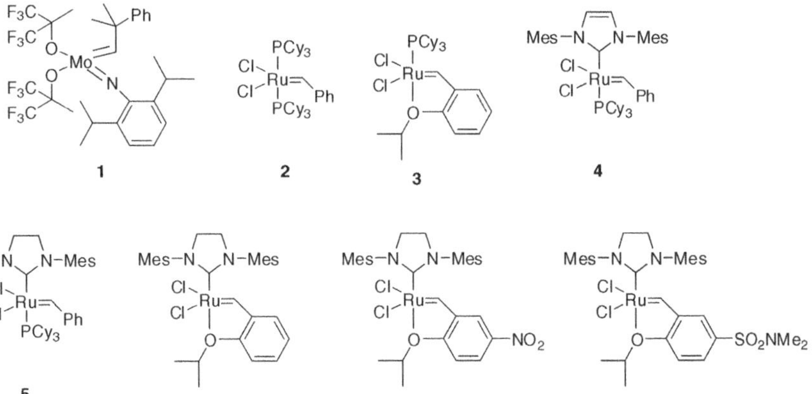

SCHEME 5.2 Most common catalyst systems for metathesis reactions.

sensitivity toward air and moisture precluded more frequent utilization especially in industrial laboratories. This was strikingly different from the ruthenium-based systems introduced by Grubbs and coworkers in 1992 [9]. Catalyst **2** is commonly referred to as the "first generation" Grubbs catalyst. The catalyst system is very robust, but it does show limitations with electron-poor and electron-rich double bonds as well as with sterically hindered systems. A variation of this catalyst was introduced by Hoveyda, where one of the phosphine ligands is replaced by a chelating ether ligand leading to catalyst **3** [10].

The introduction of *N*-heterocyclic carbene ligands elevated the performance of ruthenium-based systems. Most commonly used examples are the "second generation" Nolan catalyst **4** [11] and Grubbs catalyst **5** [12]. These catalyst systems allow formation of sterically encumbered double bonds and are also suitable for reaction with olefins that carry electron-withdrawing substituents [13]. Similar to the first generation catalysts the replacement of the electron-rich phosphine ligand with a more labile functionality was shown to be beneficial for certain applications. This is exemplified by the Grubbs–Hoveyda catalyst **6** [14]. The activity of this catalyst system can be fine-tuned by the introduction of electron-withdrawing groups on the phenyl-ring of the carbene substituent. This concept led to the discovery of Grela catalyst **7** [15], and Zhan catalyst **8** [16]. Comparative studies on the performance of various catalyst systems have been published [17]. Most of these catalyst systems are commercially available or relatively easily accessible, thereby allowing application both on small and large scale.

Most common applications within the pharmaceutical industry require catalyst loadings of 5–10 mol% of ruthenium complexes. It has, however, been demonstrated that reactions can be carried out with <1 mol% catalyst if reaction conditions and substrate are thoroughly optimized. The removal of toxic ruthenium by-products from the reaction mixtures can be challenging. A deactivation of the catalyst is often necessary in order to eliminate potential for isomerization or polymerization reactions in the workup of metathesis reactions. The removal of heavy metal contamination from the reaction products is required in order to avoid confounding results in the biological testing of the final pharmaceutically active substances. For application in the synthesis of clinical trial materials stringent specifications with regard to heavy metal content exist, often requiring levels of less than 10 ppm ruthenium in the final product. While those levels can be obtained through repeated chromatography on lab scale, more cost-efficient methods are required for large-scale production. The issue of catalyst deactivation and control of ruthenium levels has been approached by different means including workup with lead tetraacetate [18a], dimethylsulfoxide [18b], triphenylphosphine oxide [18b], tris(hydroxymethyl) phosphine [18c], activated carbon [18d], mesoporous silicates [18e], and di(ethylene glycol) vinyl ether [18f]. The optimal system for removal is dependent on the substrate and catalyst system and this aspect of the reaction needs to be individually optimized, especially for large-scale applications.

5.3 APPLICATIONS IN MEDICINAL CHEMISTRY

The development of well-defined homogeneous catalysts systems in combination of reaction scope studies in academic laboratories triggered widespread application of metathesis reactions for the synthesis of pharmaceuticals. This section will focus on various applications of the reaction within Medicinal Chemistry.

5.3.1 Synthesis of Standard Ring Sizes (5–7) by RCM

Hayes and coworkers at Abbott Laboratories reported an efficient synthesis of a conformationally constrained S1P$_1$ agonist in which a five-membered ring was formed by RCM [19]. The diene could be cyclized in good yield using catalyst 6 (Scheme 5.3). After further functionalization by coupling at the arylbromide, diastereoselective hydrogenation and deprotection, the target compound was obtained. The sequence could be performed on gram scale and the RCM product lends itself to analog synthesis. This example showcases formation of a sterically hindered double bond that would be challenging to obtain via alternative pathways.

Wilson and coworkers from Johnson & Johnson reported on the use of microwave heating in the synthesis of five-membered rings with external carboxymethyl substituents (Scheme 5.4) [20,21]. Using catalyst 5 conversions of 97% could be obtained within 5 min at 150°C, whereas traditional heating resulted in 75% conversion within 5 h at 50°C. The reaction could be extended to other related

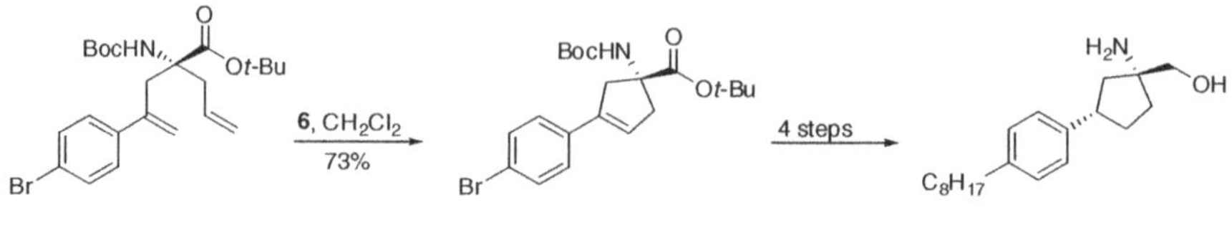

SCHEME 5.3 Synthesis of an S1P$_1$ agonist.

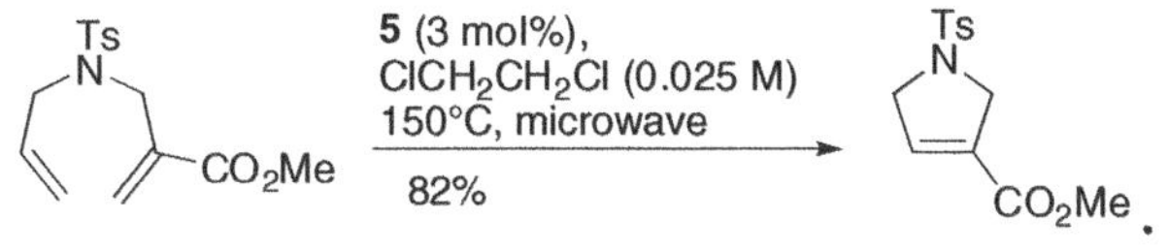

SCHEME 5.4 RCM under microwave conditions.

five-membered rings. In some cases oxidation and aromatization to pyrroles was observed.

Meng and Parker from Merck described the synthesis of cycloalkenones as estrogen receptor ligands through a sequence of RCM and oxidative rearrangement (Scheme 5.5) [22]. The reaction was unsuccessful in the presence of catalyst 2, but the use of catalyst 5 allowed for the formation of the simple product (R = Me) in good yield (45%) from a 1:1 mixture of diastereomers. The reaction was also investigated with a triene (R = allyl). Initially, formation of the less sterically hindered spirocyclic product was observed, but exposure to the catalyst under an atmosphere of ethylene resulted in the formation of the desired thermodynamic product. In this case RCM is used to install a double bond from a readily accessible substrate. This double bond then undergoes oxidative allylic transposition to a more sterically hindered double bond, which would be very challenging to access directly by RCM.

Nilsson and coworkers from AstraZeneca utilized RCM to construct cyclopentene- and cyclohexenedicarboxylic acid derivatives as thrombin inhibitors (Scheme 5.6) [23]. The use of catalyst 5 was required for a successful outcome of the RCM reaction on the electron-poor olefin, use of catalyst 2 failed to yield cyclized material. Quantitative yields were obtained in the formation of the five-membered ring.

Sturino and Yong from Merck Frosst reported on the RCM of vinyl ethers with catalyst 2 [24]. High catalyst loadings were required and the substrate scope with this catalyst is rather limited. The substitution pattern has a significant impact on the success of the reaction (Scheme 5.7). In general vinyl ethers are more readily cyclized

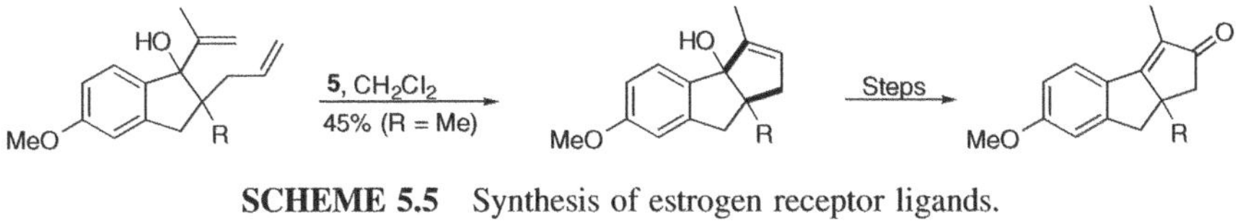

SCHEME 5.5 Synthesis of estrogen receptor ligands.

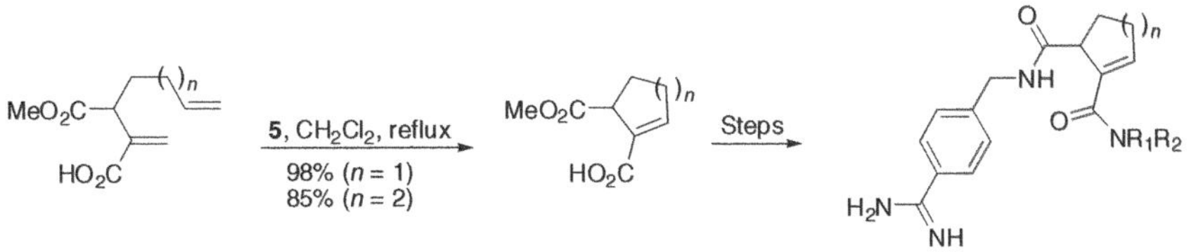

SCHEME 5.6 Synthesis of thrombin inhibitors via RCM.

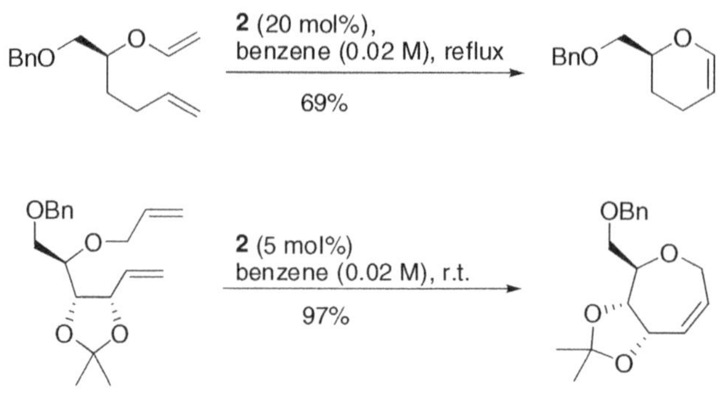

SCHEME 5.7   Synthesis of cyclic enol ethers and oxepanes.

with molybdenum-based catalyst **1** [25]. In contrast to vinyl ethers, allyl ethers are ideal substrates for RCM leading to oxepanes in high yields [26].

In an unrelated application Sturino reported on the synthesis of quinolinones and benzothiazine 1,1-dioxides (Scheme 5.8) [27]. These heterocycles are useful fragments for biologically active compounds. A broad substrate scope was demonstrated in reactions with catalyst **5**, highlighting the usefulness of RCM for the synthesis of aromatic heterocycles.

Renaud and Ouellet from Merck Frosst reported on the synthesis of five-, six-, and seven-membered carbocyclic alkenylboronates in high yields via RCM with catalyst **2** (Scheme 5.9) [28]. The products are useful for further functionalization through C–C coupling reactions. This application represented one of the first examples of RCM employing a functionalized olefin, where the substituent is neither carbon nor oxygen. In an extension of this methodology enyne metathesis was used for the

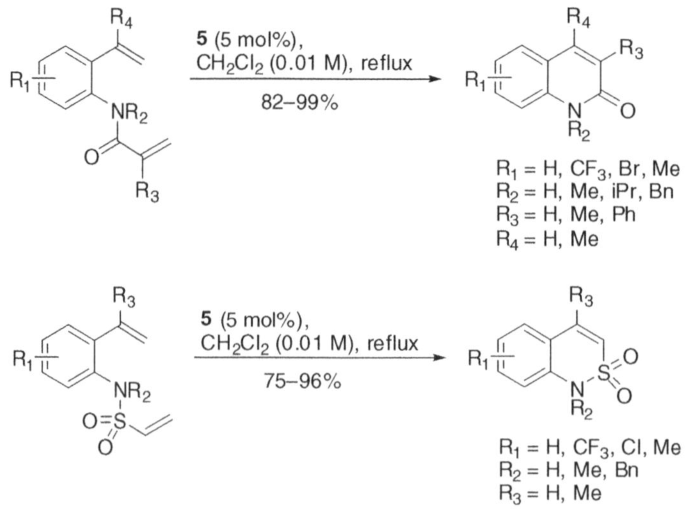

SCHEME 5.8   Synthesis of quinolinones and benzothiazine 1,1-dioxides.

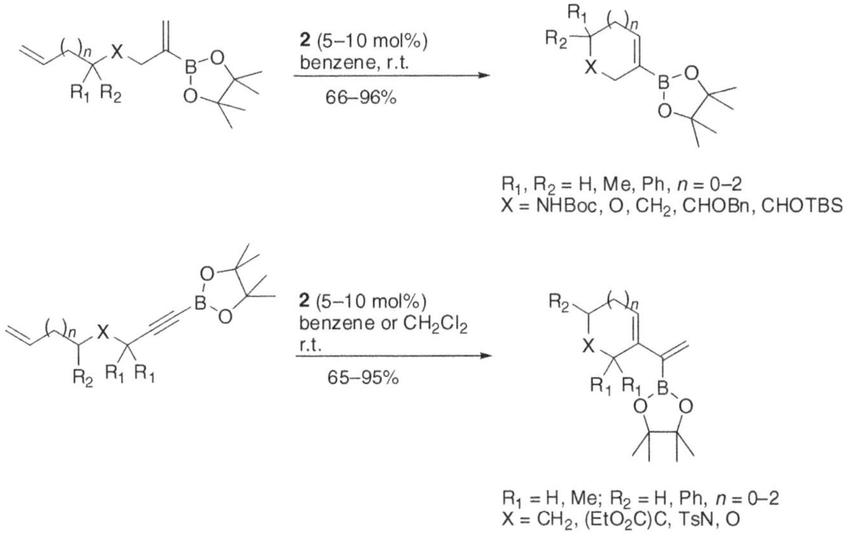
2 (5–10 mol%)
benzene, r.t.
66–96%
R1, R2 = H, Me, Ph, n = 0–2
X = NHBoc, O, CH2, CHOBn, CHOTBS
2 (5–10 mol%)
benzene or CH2Cl2
r.t.
65–95%
R1 = H, Me; R2 = H, Ph, n = 0–2
X = CH2, (EtO2C)C, TsN, O

SCHEME 5.9 Synthesis of alkenylboronates via ring-closing metathesis.

construction of 1,3-dialkenylboronates [26]. The diene products are suitable substrates for Diels–Alder cycloadditions.

Piscopio and coworkers from Amgen described the synthesis of functionalized carbo- and heterocycles employing an RCM approach [30]. The substrates were accessed by an ester enolate Claisen rearrangement. Subsequent RCM with catalyst **1** or **2** delivered the products in good yields. A broad variety of synthetically useful groups were tolerated as substituents ($SiMe_3$, $SnBu_3$) in the substrates (Scheme 5.10).

Walters et al. from Pfizer reported the synthesis of an exploratory library of spirocyclic piperidines with RCM using catalyst **2** (Scheme 5.11) [31]. Epoxidation of the double bond followed by further functionalization allowed synthesis of a library of compounds. While this RCM–reaction is not particularly challenging, it highlights the usefulness of the method by allowing installation of a remote double bond, which can undergo diverse hetero-functionalization.

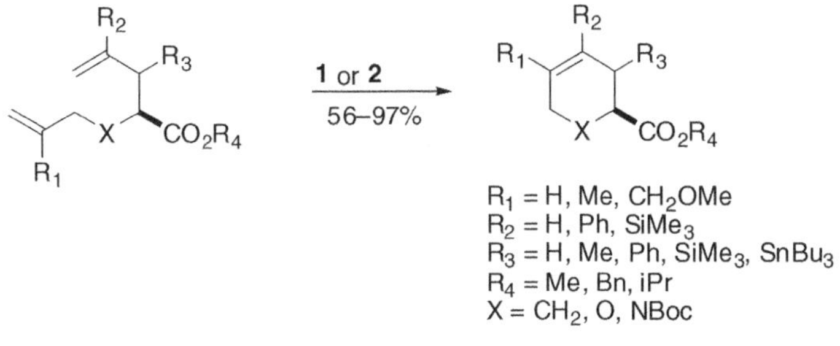
1 or 2
56–97%
R1 = H, Me, CH2OMe
R2 = H, Ph, SiMe3
R3 = H, Me, Ph, SiMe3, SnBu3
R4 = Me, Bn, iPr
X = CH2, O, NBoc

SCHEME 5.10 RCM for synthesis of six-membered rings.

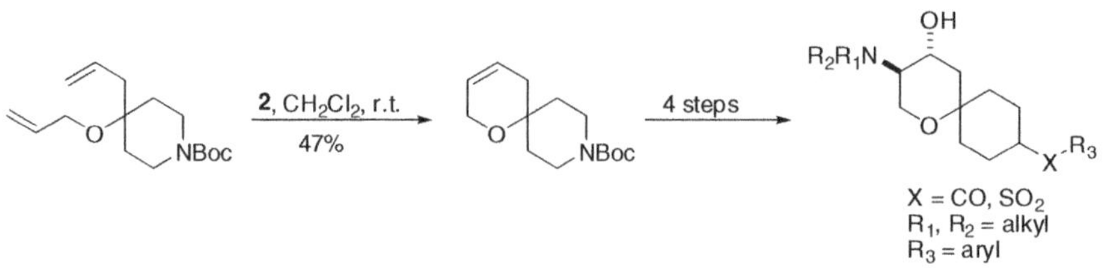

SCHEME 5.11 Synthesis of an exploratory library.

Liras et al. from Pfizer reported on the synthesis of opioid receptor ligands using an RCM approach (Scheme 5.12) [32]. The cyclization proceeded in high yield utilizing catalyst **2**. Alternatively, the unprotected amine substrate could be cyclized only after transformation into a hydrochloride salt. Elaboration of the double bond allowed access to the target regioisomeric ketones. Similarly to the previous example this approach installs a remote double bond from a relatively simple substrate.

Le Bourdonnec et al. from Adolor reported on the synthesis of μ-opioid receptor antagonists using RCM [33]. Closure of a sterically hindered six-membered ring was achieved in high efficiency using catalyst **5** (Scheme 5.13). Diastereoselective reduction of the double bond and amide reduction led to the target compounds. Regio- and stereoisomers were also prepared using the same methodology.

Hu et al. from Procter & Gamble Pharmaceuticals utilized a metathesis approach for the synthesis of chiral piperidine derivatives (Scheme 5.14) [34]. The six-membered heterocycles were obtained uneventfully by action of catalyst **2**. Diastereoselective hydrogenation of the double bond afforded chiral piperidine derivatives that were converted to various quinolone-based antibacterial agents. Alternative

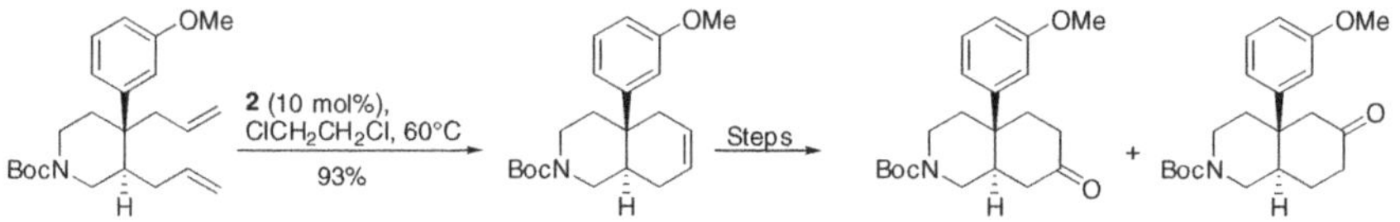

SCHEME 5.12 Synthesis of opioid receptor ligands via RCM.

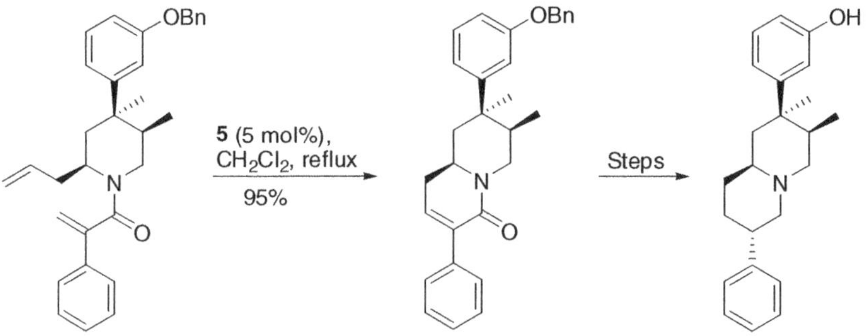

SCHEME 5.13 Synthesis of μ-opioid receptor antagonists.

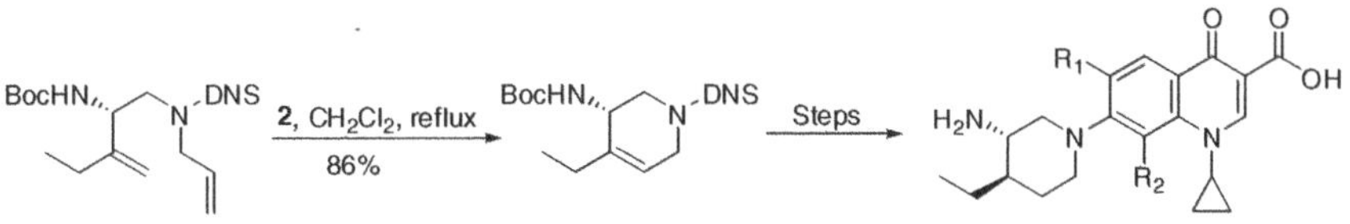

SCHEME 5.14   Synthesis of chiral piperidines via RCM.

protecting groups on the ring nitrogen where investigated with respect to cyclization efficiency. For sterically demanding substrates with a phenyl-substituent on the olefin a trifluoroacetyl-group let to superior results compared to a Boc-group [35].

Cases and coworkers from Eli Lilly utilized RCM for the synthesis of fused polycyclic amines [36]. Closure of the strained six-membered ring was achieved with catalyst **1** (Scheme 5.15). The free amine functionality in the substrate precluded successful application of ruthenium-based catalysts. The RCM product was transformed into the final target compound by a radical cyclization.

Gross from Wyeth described the synthesis of repinotan (BAY × 3702), a potent 5-hydroxytryptamine (5-HT$_{1A}$) antagonist (Scheme 5.16) [37]. The RCM reaction was leveraged for the construction of a chromene. Cyclization in presence of catalyst **2** afforded the desired compound in good yield. Based on ease in the substrate synthesis a nonterminal olefin was used as cyclization precursor, and propylene was generated as by-product in the RCM reaction. Subsequent introduction of the primary amine and alkylation afforded repinotan.

Wagner and coworkers from Novartis utilized RCM for the synthesis of unsaturated bicyclic lactams (Scheme 5.17) [38]. The reaction was performed with catalyst **2** (10–15 mol%) in refluxing dichloromethane, and good yields were obtained for the closure of six- and seven-membered rings, while lower yields were observed for the formation of eight-membered rings. The trend in yields highlights the increasing difficulty in synthesizing 8- to 10-membered rings via RCM.

Mazur and coworkers from Procter & Gamble reported on the synthesis of an oxepin-4-one [39]. The synthesis relied on closure of the seven-membered ring via

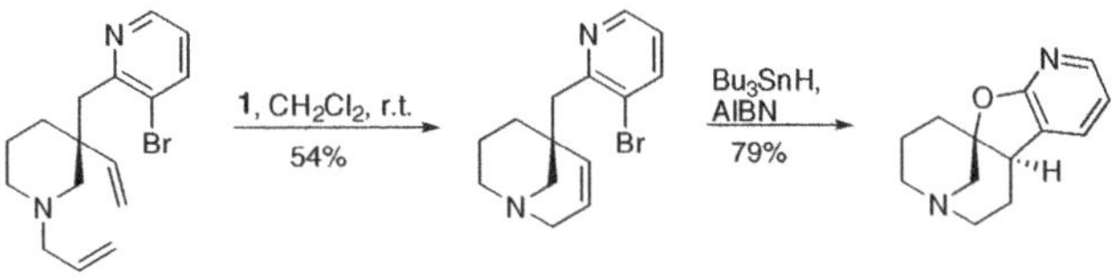

SCHEME 5.15   Synthesis of polycyclic amines via RCM.

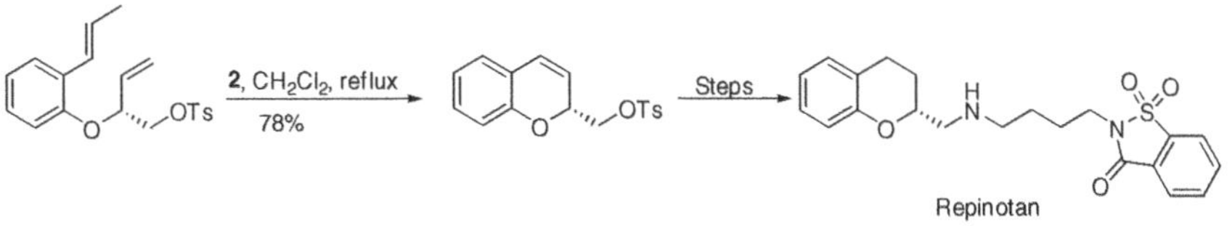

SCHEME 5.16   Synthesis of repinotan via RCM.

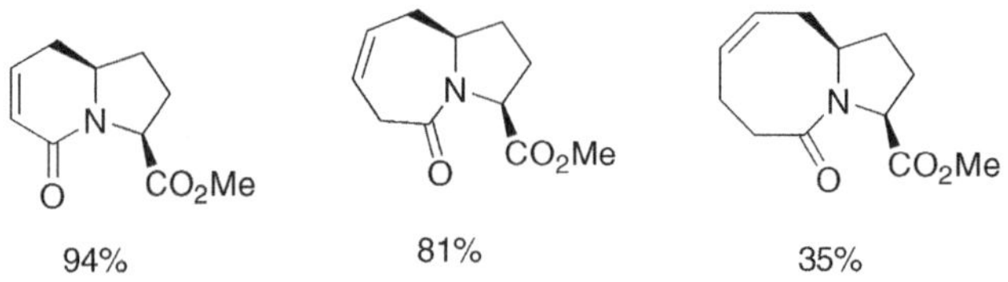

N
O
CO2Me
94%
N
O
CO2Me
81%
N
O
CO2Me
35%

SCHEME 5.17 Unsaturated bicyclic lactams obtained by RCM.

RCM (Scheme 5.18). Due to the electron-withdrawing ketone substituent on the olefin catalyst **5** had to be employed in the reaction. Use of this catalyst resulted in product formation in excellent yield (95%).

Hoyt et al. from Merck reported on the synthesis of 1-benzazepin-2-ones (Scheme 5.19) [40]. The RCM for the synthesis of these compounds can be employed in the formation of the C4–C5 bond or the C3–C4 bond. The latter bond formation was more efficient when reduced to practice. The cyclization was most efficiently conducted with Zhan catalyst **8**. The method tolerated a variety of aromatic substituents.

Marquis and Yamashita from GlaxoSmithKline utilized RCM in the synthesis of the core fragment of cathepsin K inhibitors [41]. Cyclization with catalyst **2** led to a simple azepine (R = H) in good yield (Scheme 5.20). Functionalization of the double bond by epoxidation, epoxide opening with sodium azide, and reduction of the azide, followed by peptide bond formation allowed for extensive SAR studies. The introduction of an additional ring substituent (R = Me) allowed for a further improvement in potency and pharmacokinetic parameters. Compared to acyclic inhibitors the central seven-membered ring is thought to lock the inhibitor in a bioactive conformation (increase in potency) as well as disfavoring potential conformations that are more accessible to metabolizing enzymes (improved oral

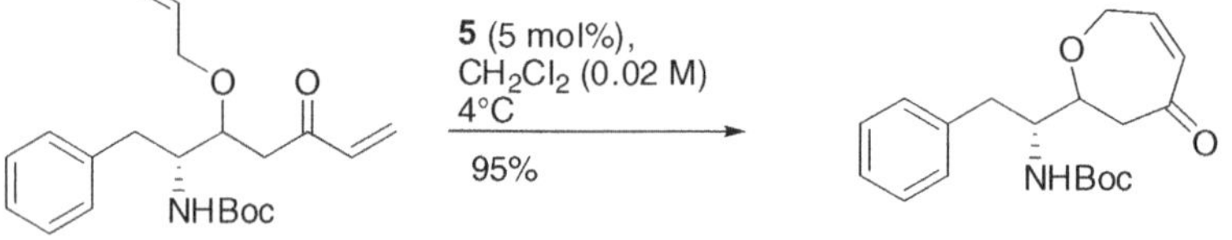

O
O
NHBoc
5 (5 mol%),
CH2Cl2 (0.02 M)
4°C
95%
O
O
NHBoc

SCHEME 5.18 Synthesis of oxepin-4-one.

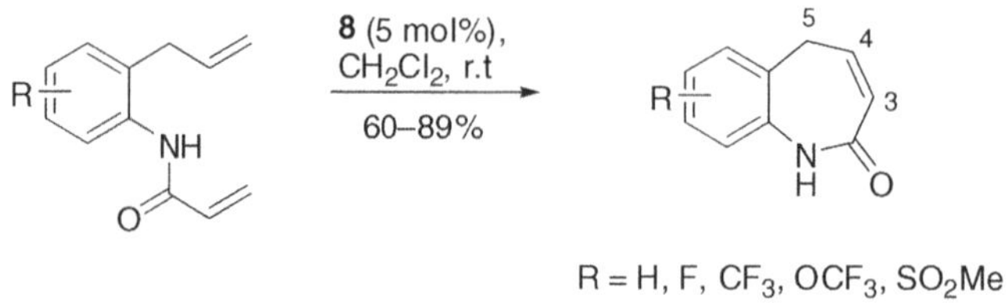

R
NH
O
8 (5 mol%),
CH2Cl2, r.t
60–89%
5
4
3
R
N
H
O
R = H, F, CF3, OCF3, SO2Me

SCHEME 5.19 Synthesis of 1-benzazepin-2-ones.

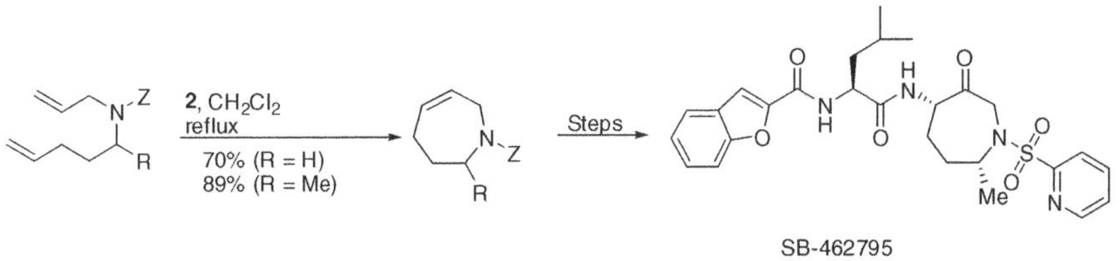

SCHEME 5.20   Synthesis of an azepanone-based cathepsin K inhibitor.

bioavailability). Ultimately SB-462795 was identified as a highly potent and orally bioavailable cathepsin K inhibitor.

The original synthesis had relied on a racemic approach to the seven-membered ring. After identification of the desired enantiomeric series Marquis reported on an improved RCM for the synthesis of a des-methyl analogue (Scheme 5.21) [42]. RCM with catalyst **2** provided the desired product in 75% yield. The synthesis of SB-462795 was also further optimized using alternative RCM approaches (*vide infra*).

Paone et al. from Merck described the application of RCM for the synthesis of calcitonin gene-related peptide (CGRP) receptor antagonists (Scheme 5.22) [43]. RCM of vinylbromide in the presence of relatively high loadings of catalyst **5** afforded the desired compound in low yield (15%), showcasing the limited scope for the use of halogenated olefins in RCM reactions. This approach, however, had the advantage to allow formation of a wide variety of analogues through subsequent

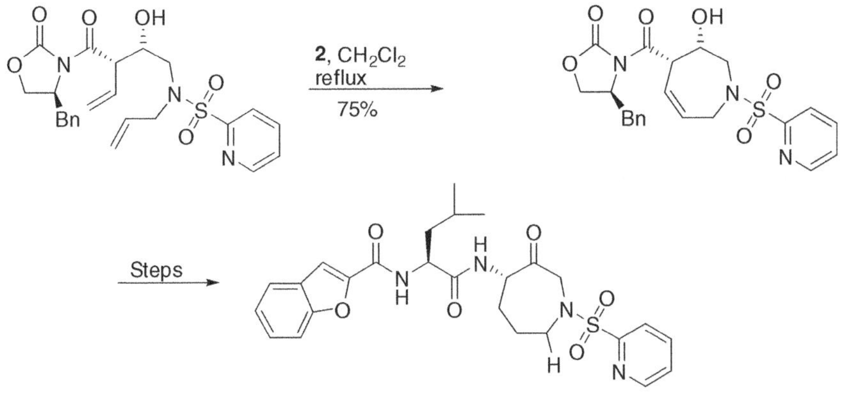

SCHEME 5.21   Second-generation synthesis of a cathepsin K inhibitor.

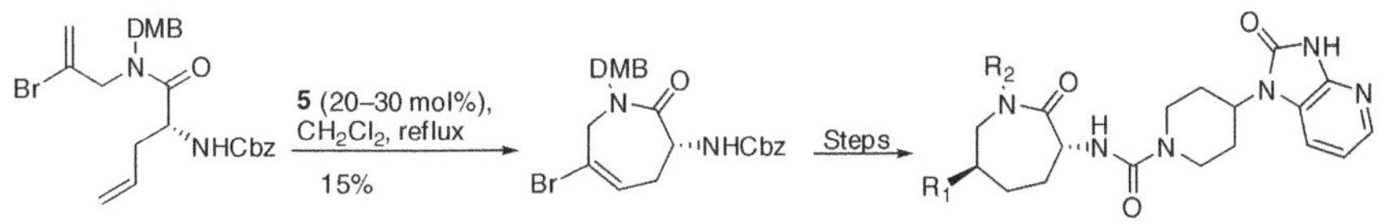

SCHEME 5.22   Synthesis of CGRP receptor via RCM.

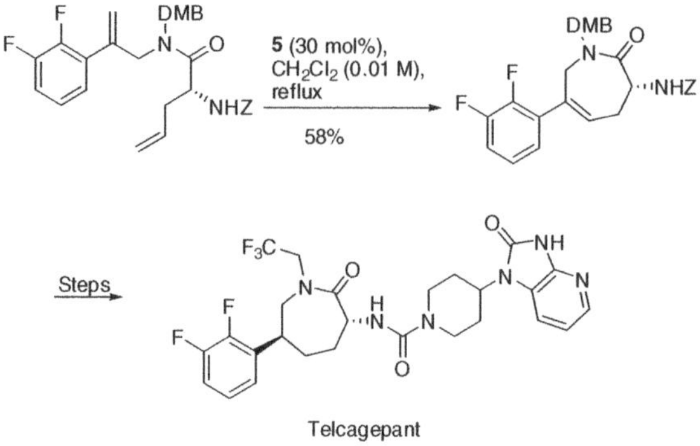

SCHEME 5.23 Synthesis of telcagepant via RCM.

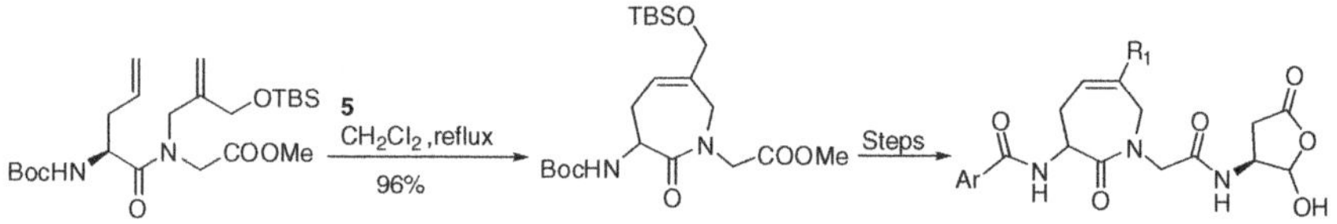

SCHEME 5.24 Synthesis of caprolactams as ICE inhibitors via RCM.

functionalization of the alkenylbromide by Suzuki-coupling reactions. Further manipulation led to potent CGRP receptor antagonists.

An analogue from this compound series was selected for further development and improvements to the original synthesis were desired. Toward this end Burgey from Merck utilized an alternative RCM for the synthesis of the left-hand fragment of the CGRP receptor antagonist telcagepant (MK-0974) (Scheme 5.23) [44]. In contrast to the prior approach the aryl-moiety was introduced via Suzuki-reaction prior to the RCM. The cyclization to the seven-membered ring was challenging since a sterically hindered trisubstituted double bond needed to be formed. Use of catalysts **2** and **3** was unsuccessful, whereas the cyclization could be achieved in reasonable yield by application of a high catalyst loading of **5** (30 mol%). Side-products from dimerization and cross metathesis with styrene (from catalyst **5**) were observed. Through multiple manipulations involving a diastereoselective hydrogenation of the double bond the seven-membered building block could be transformed into telcagepant.

Wos and coworkers from Procter & Gamble adapted an RCM approach to the synthesis of unsaturated caprolactams as interleukin-1β converting enzyme (ICE) inhibitors (Scheme 5.24) [45]. The trisubstituted double bond was formed with catalyst **5**. Simplified analogues with a disubstituted double bond were also prepared.

5.3.2 Synthesis of Medium-Sized Rings (8–9) by RCM

O'Neil et al. from Procter & Gamble reported on the synthesis of 8,6-fused bicyclic peptidomimetic compounds as ICE inhibitors (Scheme 5.25) [46]. Closure of the eight-membered ring was achieved with catalyst **2**. Alternative formation of this

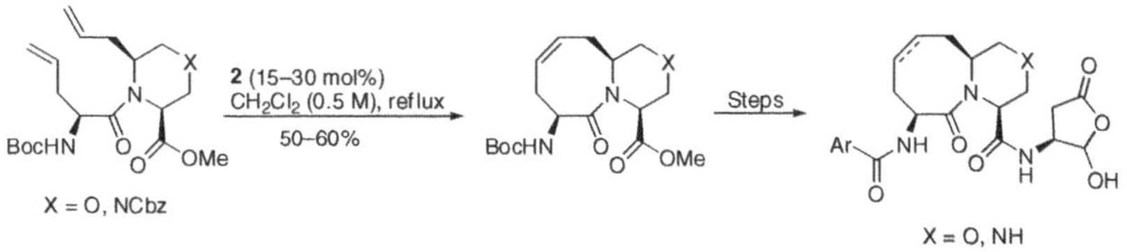

SCHEME 5.25   Synthesis of bicyclic ICE inhibitors via RCM.

medium-sized ring via lactamization would have been challenging. The final inhibitors were obtained by functionalization of the N- and C-terminus of the peptide. A similar approach was used for the synthesis of the 8,5-fused ICE inhibitors [47].

Creighton and Reitz from R. W. Johnson Pharmaceuticals described the synthesis of an eight-membered cyclic pseudo dipeptide by RCM (Scheme 5.26) [48]. The unprotected amide does not undergo cyclization due to preference for the *trans*-amide bond rotamer. Transient alkylation allows for a conformational change to the *cis*-amide rotamer, which undergoes intramolecular metathesis in good yield.

Wilson et al. from Johnson & Johnson reported the use of RCM for the synthesis of indole carbazole alkaloids, which have potential as protein kinase inhibitors (Scheme 5.27) [49]. Cyclization of a simple substrate ($R_1 = Me$, $X = O$, R_2, $R_3 = H$) proceeded with catalyst **2**, delivering the eight- and nine-membered product in high yields (85–96%). The formation of these otherwise challenging to obtain

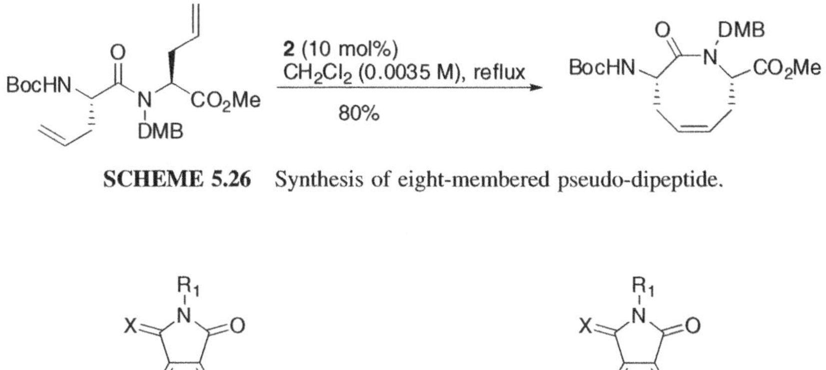

SCHEME 5.26   Synthesis of eight-membered pseudo-dipeptide.

$R_1 = Me$, DMB
R_2, $R_3 = H$, Me, CO_2Me
$X = O$, H_2
m, $n = 1,2$

SCHEME 5.27   Synthesis of indole carbazole alkaloids.

medium-sized rings was facilitated by the templating of the double bonds that is imposed by the aromatic backbone. Cyclization of a substrate with an electron-withdrawing substituent on the double bond ($R_2 = H$, $R_3 = CO_2Me$) required use of catalyst **5** under microwave conditions in order to obtain satisfactory results [20]. Synthesis of a sterically encumbered product (R_2, $R_3 = Me$) was best accomplished with catalyst **5** in refluxing toluene. Further elaboration of the double bond by oxidation allowed access to a variety of analogues which are potent Janus kinase 3 (JAK3) inhibitors [50].

5.3.3 Synthesis of Large-Sized Rings by RCM

While the RCM can excel in the formation of standard- and medium-sized rings by allowing access to targets that would be otherwise difficult to obtain, it becomes particularly useful for the synthesis of large rings. Formation of ring-extended analogues becomes very facile by synthesis of substrates with terminal olefin bond-bearing alkylchains of different length. In the absence of specific conformational effects the yields of the ring-closed products are fairly independent of ring-size.

Rojo et al. from Lilly reported the synthesis of macrocyclic peptidomimetic BACE-inhibitors (Scheme 5.28) [51,52]. Based on computational docking-studies it was concluded that a 10-membered ring would be ideal. The RCM proceeded uneventfully with catalyst **2**. Hydrogenation of the double bond and elaboration of the side-chain led to BACE inhibitors.

Ring-closing metathesis was extensively utilized by Boeringher Ingelheim for the synthesis of hepatitis C virus NS3 protease inhibitors. In an initial application Tsantrizos reported the use of catalyst **2** for the formation of the 15-membered ring (Scheme 5.29) [53]. A good yield (98% of crude macrocyclic products) was achieved, and the Z-isomer was formed exclusively. Alternative locations of the double bond within the 15-membered macrocycle were also explored, in which case mixtures of E- and Z-isomers were obtained. This study as well as the studies shown below was facilitated by the ease of access to the substrates with varied chain-lengths.

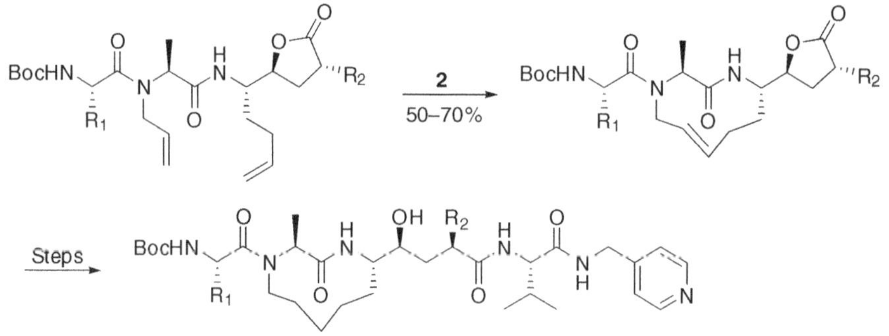

SCHEME 5.28 Synthesis of peptidomimetic BACE-inhibitors.

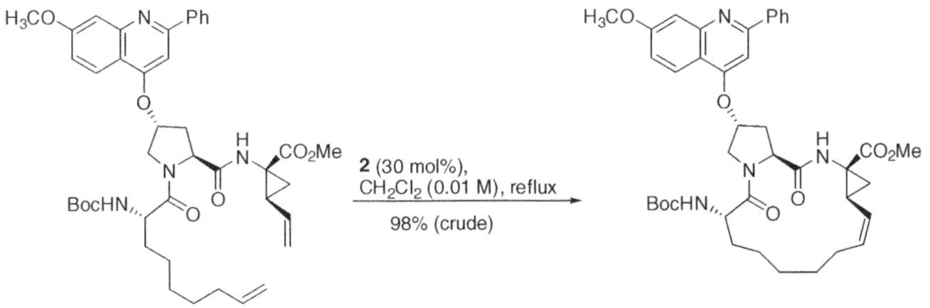

SCHEME 5.29   Synthesis of a 15-membered HCV NS3 inhibitor.

Alternative ring-sizes and substitution patterns were also explored [54]. Substrates without the bulky NHBoc-substituent reacted sluggishly in the RCM. With these compounds the amide exists as a 1:1 mixture of *cis/trans* rotamers thereby making cyclization more difficult. In contrast the substrate with an NHBoc-substituent showed a *cis/trans*-ratio of 1:9, and the cyclization proceeded in higher yield. Interesting observations were made when smaller ring-sizes were examined (13- and 14-membered). A complete racemization of the β-carbon of the vinyl moiety by a metal-catalyzed process was observed. Only the stereochemically inverted substrate led to the formation of the cyclic product as an *E*-isomer (Scheme 5.30).

Based on the initially discovered inhibitors, optimization of the side-chains through SAR-studies led to the discovery of BILN 2061. This HCV NS3 protease inhibitor was advanced into clinical trials [55,56]. An initial optimization of the synthesis resulted in a route that was capable of delivering multigram quantities for preclinical evaluation (Scheme 5.31) [57]. The catalyst choice was based on an optimization study [54]. The desired cyclized product was obtained in 83% yield with catalyst **2** in dichloromethane, but the product with epimerized β-carbon of the vinyl moiety was also observed (17%). Alternative solvent systems did not suppress epimerization. Application of catalyst **6** resulted in a slower reaction with incomplete conversion, but the epimerization was completely suppressed. Under the optimized conditions the product could be obtained in 85% yield.

A related analogue was prepared by Randolph from Abbott. The target molecule was derived from a one-atom modification resulting in an aza-peptide analogue [58]. Similar conditions as reported for BILN 2061 were utilized for the closure of the 15-membered ring. Interestingly the reaction is significantly more sluggish (62% yield) and the compound is obtained as a mixture of *E*- and *Z*-isomers

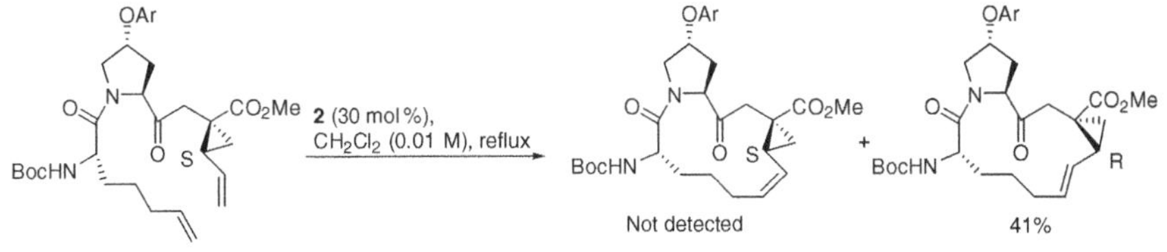

SCHEME 5.30   Attempted synthesis of a 13-membered HCV NS3 inhibitor.

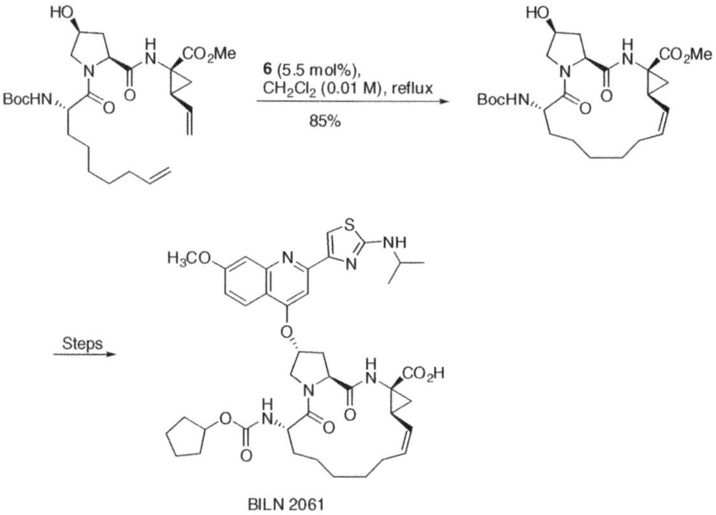

SCHEME 5.31 Synthesis of BILN 2061.

(Scheme 5.32). The more sluggish reactivity can be explained by a conformational preference that placed the two olefin moieties of the substrate at a significant distance apart from one another.

RCM was also employed in the discovery of HCV NS3 protease inhibitors by Velazquez et al. from Schering-Plough [59]. The 16-membered carbocyclic macrocycle $(X = CH_2)$ was obtained in excellent yield, and similar results were obtained for the 14-, 15-, and 17-membered macrocycles (Scheme 5.33). Similarly the 16-membered ether analogue $(X = O)$ could be obtained in slightly diminished yield (63%), a result of competing O-deallylation. The products were obtained as a mixture of olefin isomers, which was inconsequential since the fully saturated macrocycle was targeted. After assembly of the core structure the amino and carboxy-function could be appropriately derivatized in order to obtain nanomolar HCV inhibitors. Structure–activity relationships were established for the compound series, and the optimized compounds are shown in Scheme 5.34 [60,61]. Binding efficiency was

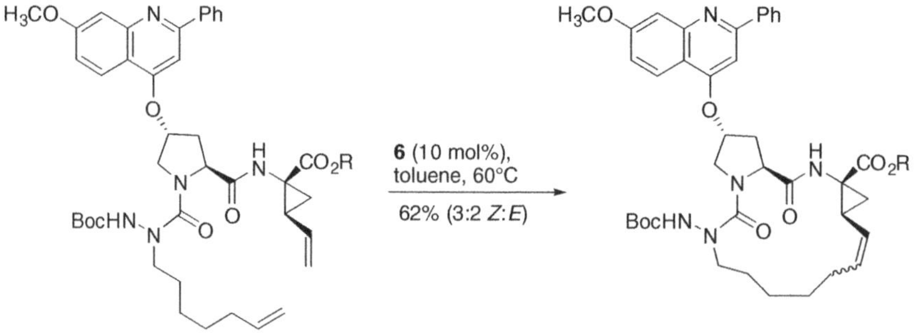

SCHEME 5.32 Synthesis of a 15-membered aza-peptide HCV NS3 inhibitor.

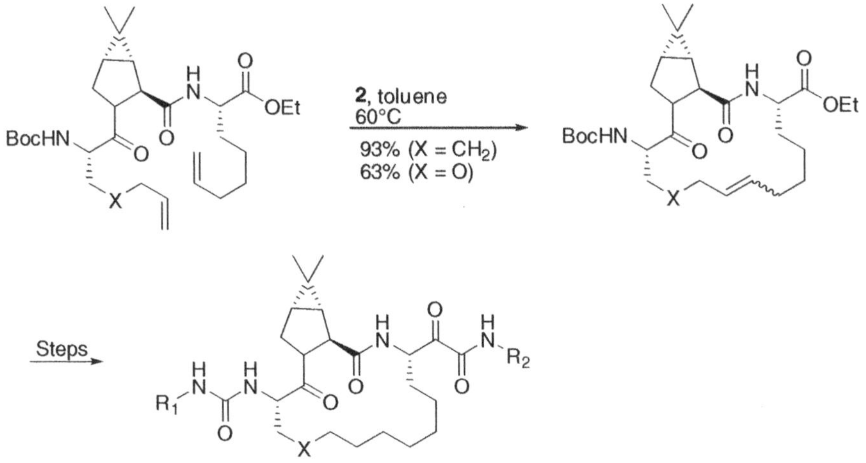

SCHEME 5.33 Synthesis of 16-membered HCV NS3 inhibitor.

increased with the saturated macrocyclic RCM product as compared to the acyclic RCM precursor. This can be explained by a preorganization of the binding conformation of the inhibitor, thereby leading to an increase in potency.

A successful RCM reaction also positively impacted the HCV inhibitor program at Merck as reported by Liverton et al. [62] Installation of a P2–P4 linker was achieved in good yields (84–93%) by application of catalyst **8** (Scheme 5.35). The compounds could be transformed into the final target compounds by hydrogenation and installation of the P1-fragment. Macrocyclization had a positive impact on enzyme activity and the five- and six-carbon linkers were found to be ideal. As in the previous examples, the power of RCM lies in the easy access to analogues with varying ring size.

Liverton et al. from Merck also described the synthesis of the HCV NS3 inhibitor clinical candidate MK-7009 and related analogues [63]. Cyclization was achieved again with Zhan catalyst **8** (Scheme 5.36). As in the previous example, a styrene double bond participated in the reaction, and the thermodynamically favored E-isomer was formed.

McCauley et al. from Merck expanded the macrocyclization approach to inhibitors in which two macrocycles are formed [64]. Different synthetic approaches with

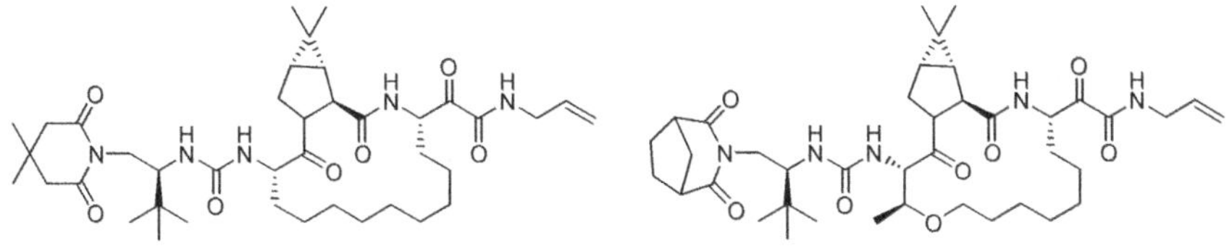

SCHEME 5.34 Optimized HCV NS3 inhibitors.

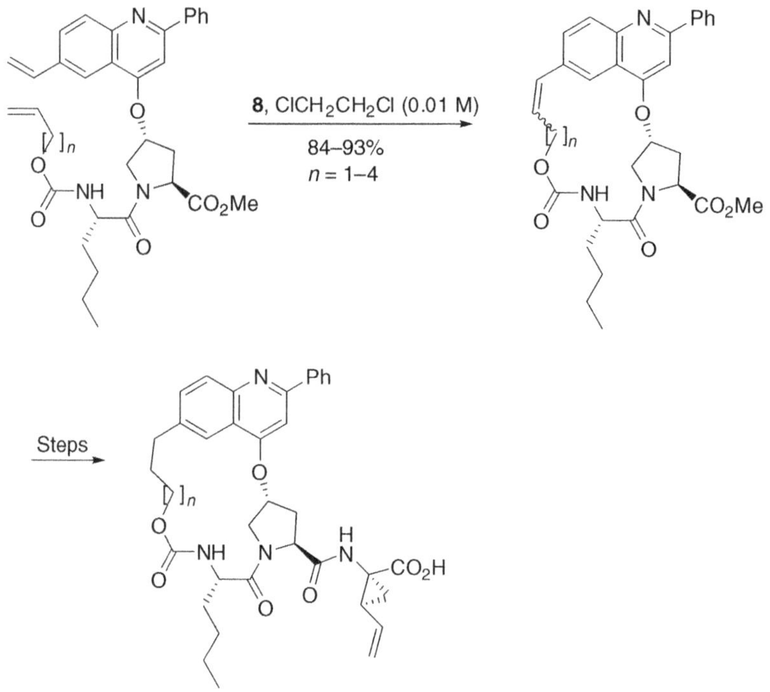

SCHEME 5.35   Synthesis of a P2–P4 macrocyclic HCV NS3 inhibitor.

sequential construction of the two macrocycles were considered, but the most straightforward approach involved two RCM reactions on a tetraene, to form both olefins in the desired geometry. Based on the tether lengths it was conceivable that not only the desired 18- and 15-membered macrocycles would form, but that also the product with 16- and 17-membered rings or with one 14-membered ring would form. In practice the desired compound was formed in 66% yield when tetraene was subjected to RCM conditions with catalyst **8** (Scheme 5.37). Mechanistic studies showed that the desired product is the thermodynamic product under RCM conditions, but it was also shown that the substrates exhibit a high kinetic preference for cyclization of P1–P3 chains and P2–P4 chains over other possible combinations. This example highlights the selectivity of the metathesis reaction, as a complex bicycle is formed in high efficiency from a relatively simple open-chain precursor.

Stachel et al. from Merck reported the synthesis of macrocyclic inhibitors of β-secretase (BACE-1) [65]. A 14-membered macrocycle could be obtained via RCM by employing catalyst **5** under standard conditions (Scheme 5.38). The cyclization substrate is synthetically relatively simple, as it is comprised of two allyl-substituted aryl-moieties. The *E*-isomer was formed exclusively in this reaction. Further elaboration by hydrogenation and side-chain modification allowed access to BACE-inhibitors, which showed improved potency and functional activity as compared to acyclic analogues.

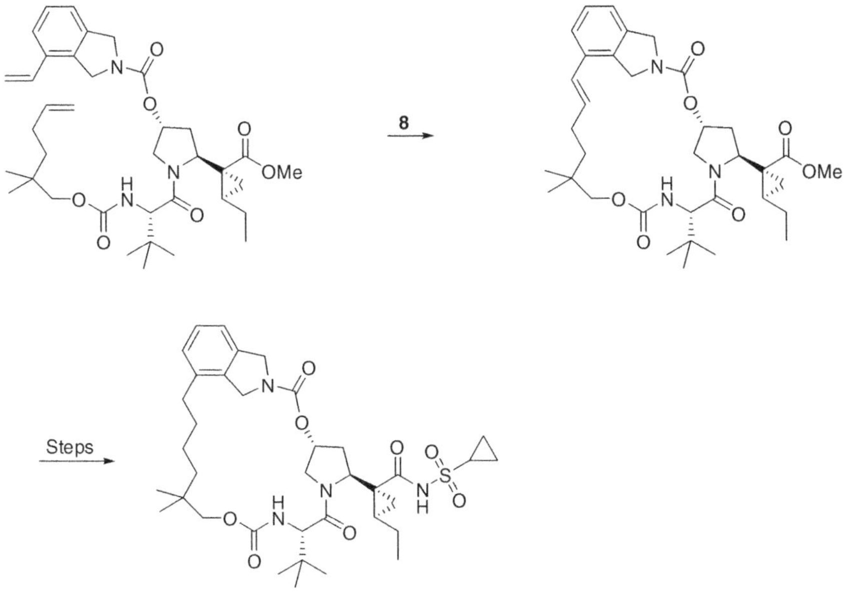

MK-7009

SCHEME 5.36 Synthesis of HCV NS3 inhibitor MK-7009.

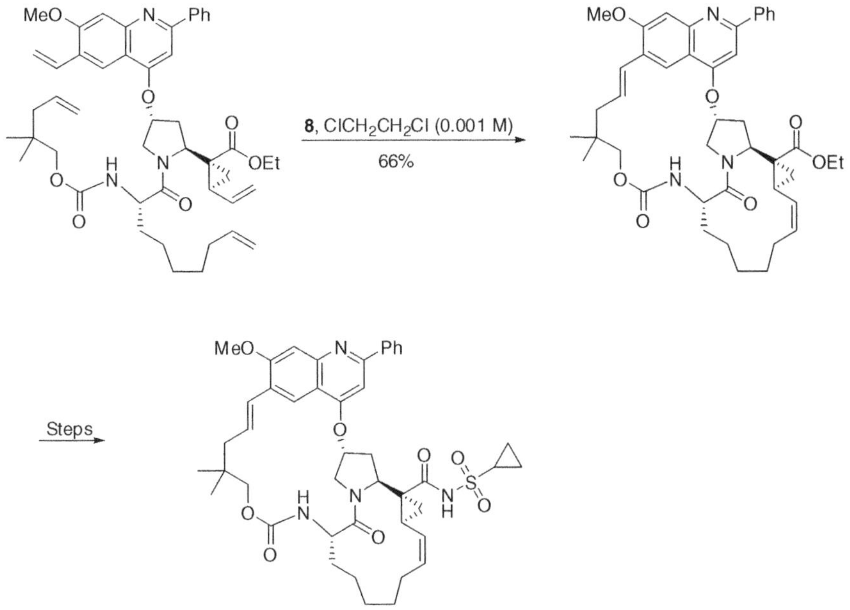

SCHEME 5.37 Synthesis of a bicyclic HCV NS3 inhibitor.

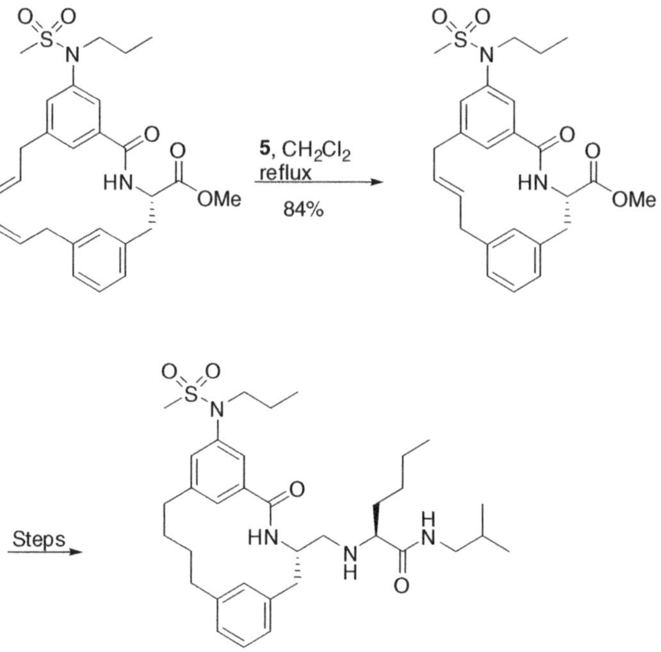

5, CH$_2$Cl$_2$
reflux

84%

Steps

SCHEME 5.38 Synthesis of BACE inhibitor.

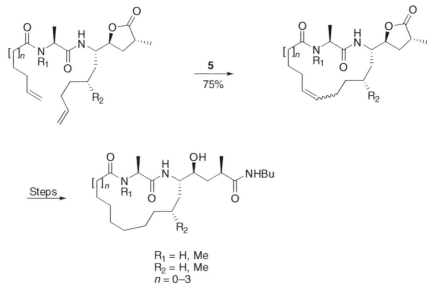

5

75%

Steps

R$_1$ = H, Me
R$_2$ = H, Me
n = 0–3

SCHEME 5.39 Alternative synthesis of peptidomimetic BACE inhibitors.

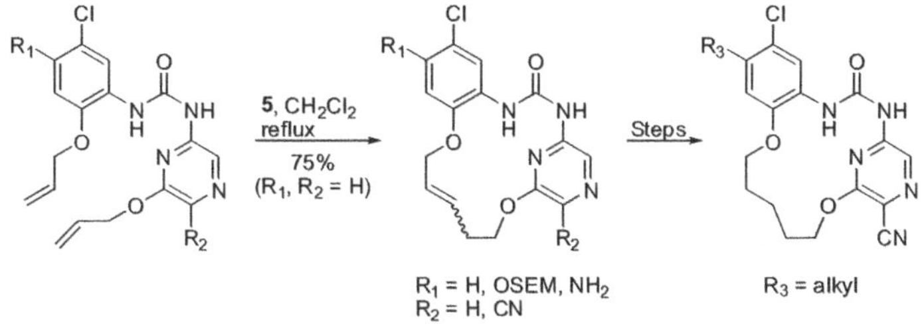

SCHEME 5.40   Synthesis of macrocyclic checkpoint kinase 1 inhibitors.

Alternative macrocyclic peptidomimetic BACE inhibitors were also described by Machauer et al. from Novartis [66]. Two side chains were linked, varying the ring sizes from 14 to 17 (Scheme 5.39) using catalyst **5**. Variation of ring-size was facile by application of different long chain carboxylic acid with terminal double bond. The final products were obtained after hydrogenation of the olefin, and side chain elaboration.

Tao et al. from Abbott reported the synthesis of macrocyclic checkpoint kinase 1 inhibitors as possible potentiators for DNA-damaging agents in the treatment of cancer [67]. Cyclization was uneventfully using catalyst **5** providing the desired 15-membered macrocycle in reasonable yield (Scheme 5.40). In close analogy the 14- and 16-membered macrocycles were prepared, by facile modification of the substrate. Incorporation of an oxygen or nitrogen heteroatom on one of the phenyl rings allowed exploration of the SAR.

5.3.4   RCM in the Synthesis of Natural Products and Analogues

Dong et al. from Kosan Biosciences reported the use of metathesis in the synthesis of epothilone analogues [68]. The epothilones were one of the earliest examples of the significant impact of RCM on the synthesis of macrocyclic natural products, as numerous reported syntheses relied on the reaction for the closure of the 16-membered ring. Synthesis of epothilones A and C relied on formation of a disubstituted Z-olefin via RCM utilizing catalyst **2** [69], while synthesis of epothilones B and D was accomplished via RCM formation of a trisubstituted double bond with molybdenum catalyst **1** [70]. In the analogue synthesis (Scheme 5.41) a 1:1 mixture of olefin isomers was obtained by utilizing catalyst **5**.

Wagner et al. from Novartis reported the synthesis of truncated analogues of the immunosuppressant sanglifehrin (Scheme 5.42) [71]. The ring closure reaction to form the 22-membered macrolide is especially noteworthy since it was the first application of RCM for the synthesis of a macrocyclic diene. The reaction occurs exclusively at the terminal olefin of the diene when catalyst **2** is employed. The *E,E*-isomer, which is also found in the natural product, is formed preferentially and only low amounts of the *E,Z*-isomer were detected in the reaction mixture. A different reactivity pattern was observed when the more active catalyst **4** was employed [72]. RCM occurred at the internal double bond with loss of butadiene to afford the contracted 20-membered macrocycle. Cyclization of a simpler substrate, in which the

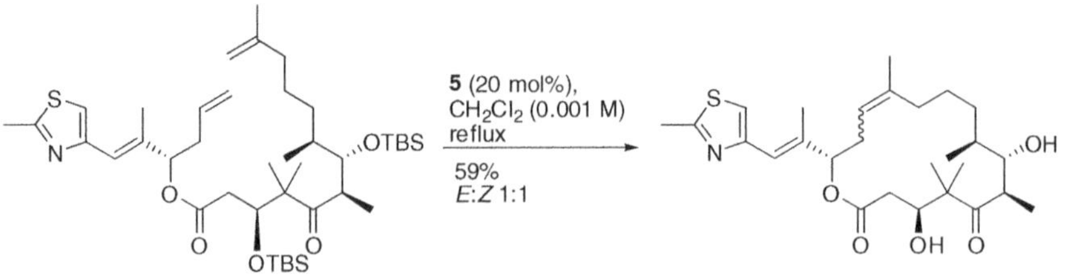

SCHEME 5.41 Synthesis of an epothilone analogue by RCM.

diene moiety was replaced by an isolated double bond, led to the target macrocycle
without issue. Attempts to form a triene in the product by cyclization of a substrate
with two diene moieties were unsuccessful. In an extension of this work additional
analogues were prepared and their biological properties were tested [73].

Pladienolide B belongs to a family of natural products containing 12-membered
macrolides that are potent antitumor agents. Kotake and coworkers from Eisai
described the first total synthesis of these compounds employing RCM as key step
for the closure of the macrocycle (Scheme 5.43) [74]. After optimization of
catalyst, solvent, and additives, a 46% yield was obtained for cyclization employing catalyst **6**. The relatively low yield was explained by the severe steric
hindrance close to the side of ring-closure. Significant olefin isomerization of
the allylic ether double bond was also observed under the reaction conditions. The

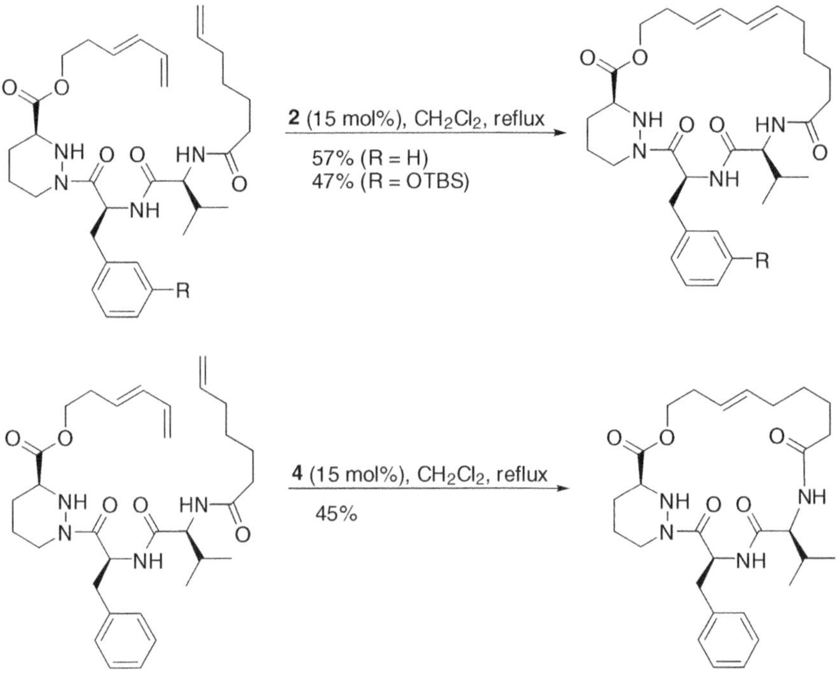

SCHEME 5.42 Synthesis of sanglifehrin analogues.

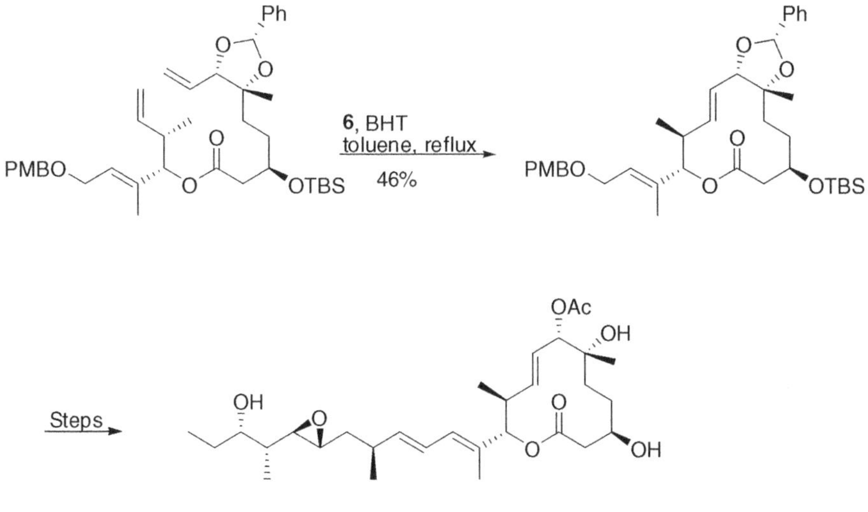

Pladienolide B

SCHEME 5.43   Synthesis of pladienolide B.

obtained product could be further elaborated to pladienolide B. A synthetic intermediate could also be transformed into pladienolide D employing a cross-metathesis approach (*vide infra*).

Labrecque et al. from Biochem Pharma reported the synthesis of salicylihalamides A and B using an RCM approach [75]. These compounds are of interest due to their

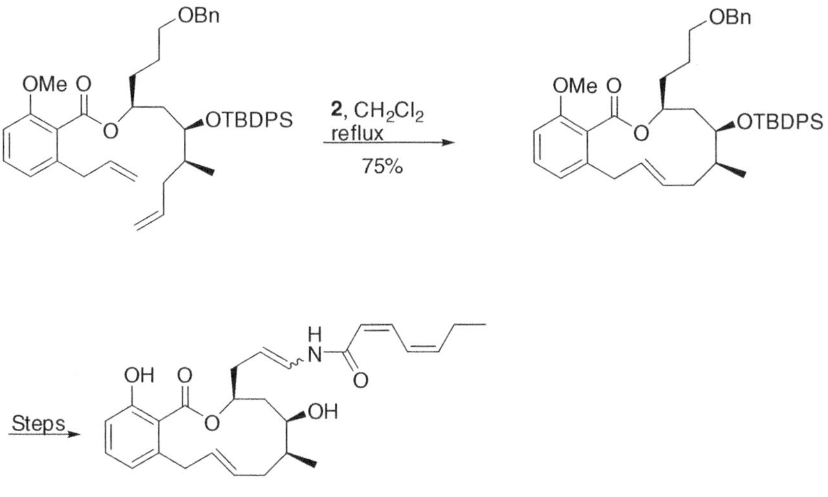

Salicylihalamide A (*E*)
Salicylihalamide B (*Z*)

SCHEME 5.44   Synthesis of salicylihalamides.

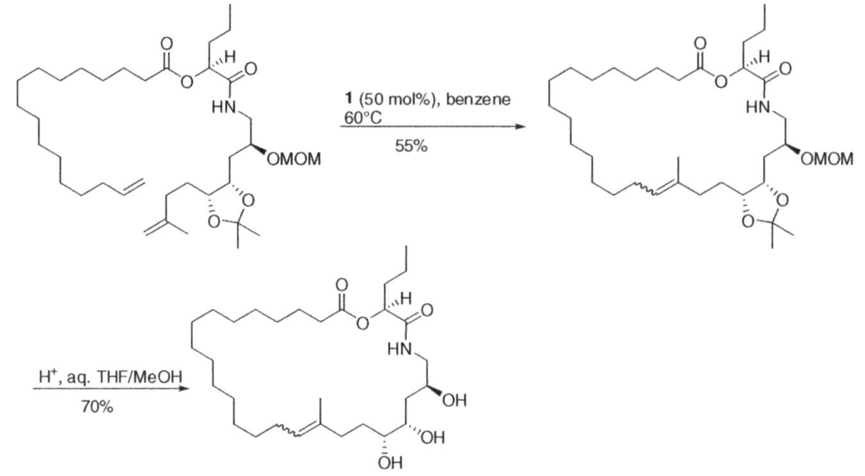

SCHEME 5.45   Synthesis of a myxovirescin analogue by RCM.

unique cytotoxicity. Catalyst **2** was employed in the closure of the 12-membered ring, the product was obtained in good yield and selectivity favoring the desired *E*-isomer (9:1) (Scheme 5.44). The intermediate could be transformed into the natural products by functionalization of the side chain. Related RCM approaches to the salicyliha- lamides have been reported by several academic laboratories [76].

Dutton and coworkers from Pfizer reported on the use of RCM for the synthesis of myxovirescin analogues [77]. The target compounds are of biological interest due to their antibacterial properties. RCM closure yielding a trisubstituted olefin was achieved using a high catalyst loading of catalyst **1** (Scheme 5.45). Lower catalyst loadings led to reduced yields, indicating the sensitivity of the molybdenum catalyst toward polar functional groups. The 28-membered macrocycle was obtained in as inseparable mixture of *E/Z* isomers. Separation was possible after removal of the protecting groups. A simplified analogue with a disubstituted double bond could be prepared by action of catalyst **2**.

5.3.5   Formation of Olefins by Cross Metathesis

Hoveyda and Vezina from Tranzyme Pharma described a selective cross metathesis with allyl cyanide as coupling partner (Scheme 5.46) [78]. Optimization of catalyst (5 mol% **5**), reaction concentration (0.5 M) and work-up protocol (quench with vinyl

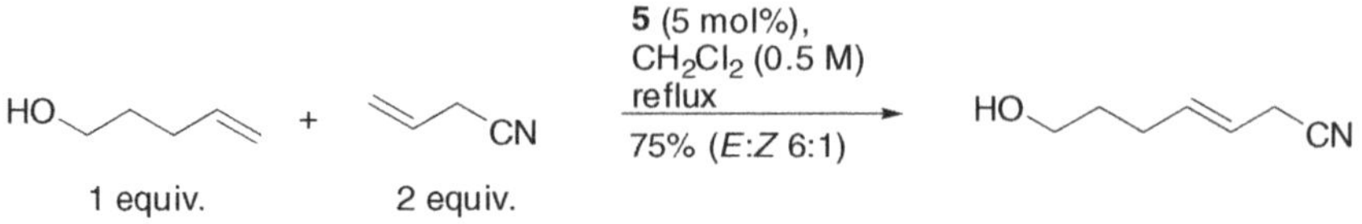

SCHEME 5.46   Cross metathesis with allyl cyanide.

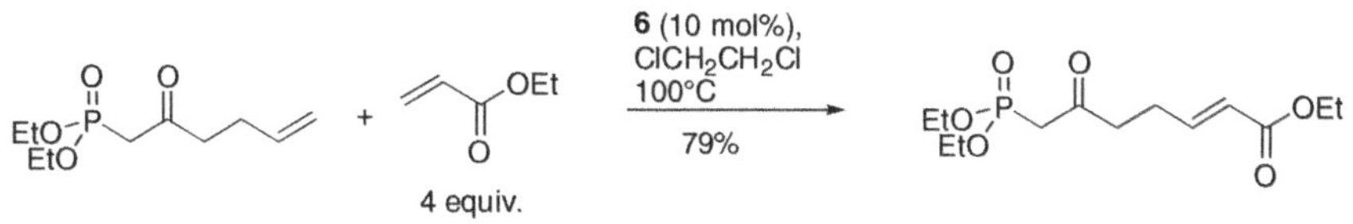

SCHEME 5.47   Cross metathesis under microwave conditions.

ether and DMSO) was required in order to obtain good yields of the desired coupling product. Lower concentrations resulted in lower yields.

In an extension of their methodology for RCM [20], Wilson from Johnson & Johnson reported the use of microwave heating to facilitate cross-metathesis reactions (Scheme 5.47) [79]. While yields were similar to thermal heating, shorter reaction times (15 min versus 6 h) could be realized. In general higher yields were obtained with catalyst **6** compared to catalyst **5**.

Kotake from Eisai reported the use of cross metathesis for the synthesis of pladienolide D [74]. Approaches to install the desired double bond via a Julia–Kocienski olefination had failed. An intermediate that had been obtained via RCM could be transformed into a requisite diene. This diene underwent a selective cross metathesis in which pladienolide D was obtained in good yield and stereoselectivity employing catalyst **5** (Scheme 5.48).

Lazarova et al. from Enanta Pharmaceuticals reported on the use of cross metathesis for the derivatization of cyclosporine A [80]. This natural product is a potent immunosuppressant, in which the unsaturated side-chain is essential for the biological activity. Using catalyst **6** and conducting the cross metathesis with different maleate diesters three derivatives of cyclosporine A could be prepared (Scheme 5.49). The target compounds showed similar biological activity compared

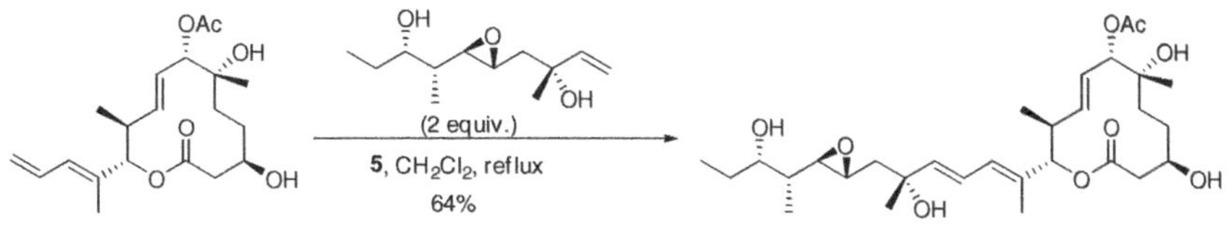

SCHEME 5.48   Synthesis of pladienolide D.

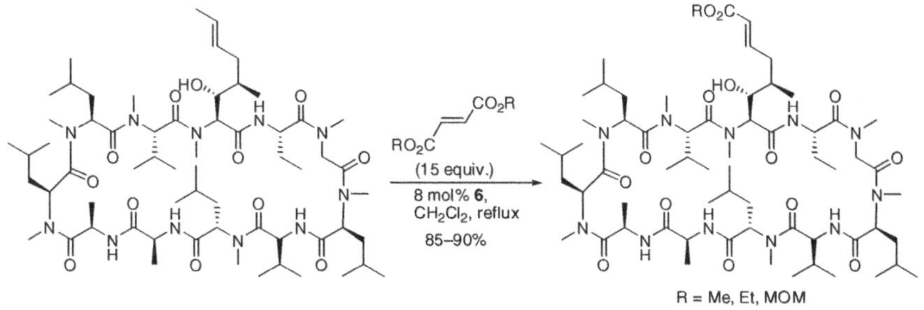

SCHEME 5.49   Derivatization of cyclosporine A by cross metathesis.

to the natural product. This example highlights an interesting concept in which the double bond of a known natural product offers a handle for diverse derivatization via cross metathesis.

Palmer et al. from Nycomed reported on the synthesis of potassium-competitive acid blockers [81]. Various aromatic olefins could be introduced via cross metathesis with catalyst **5** (Scheme 5.50). Protection of the phenol was necessary in order to achieve good conversions. The products were obtained as the thermodynamically favored *E*-isomers. Treatment of the products under acidic conditions led to cleavage of the protecting group and cyclization. Methyl or fluoro-substituted styrenes were successfully employed in the reaction. The reaction could also be expanded to 1,2-disubstituted alkenes thereby allowing access to spirocyclic compounds [82]. For this application, use of catalyst **6** was preferred. While reactions with catalyst **5** achieved higher conversions and yields, purification was complicated by formation of cross-metathesis products derived from the styrene moiety of the catalyst.

Nussbaumer et al. from Novartis utilized a cross-metathesis derivatization of shingolipids for the introduction of fluorescent labels (Scheme 5.51) [83]. The

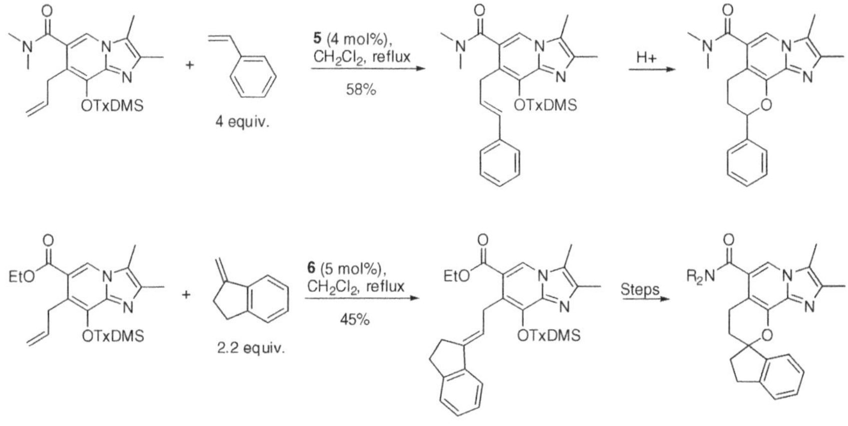

SCHEME 5.50 Synthesis of potassium-competitive acid blockers.

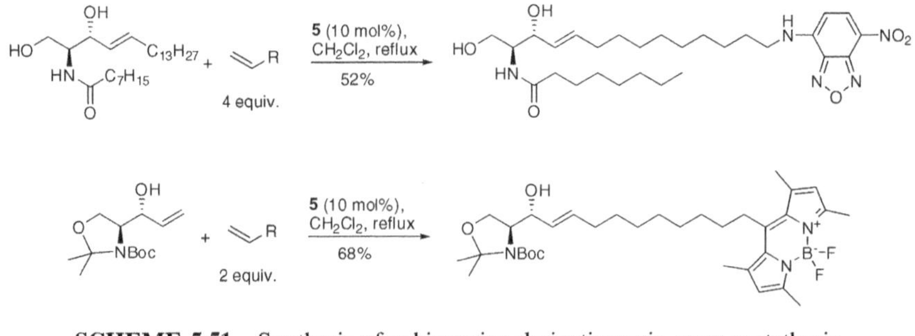

SCHEME 5.51 Synthesis of sphingosine derivatives via cross metathesis.

reactions proceeded in good yield when 4 equivalents of the fluorescent-labeled olefin were used. Notably the *E*-isomer was formed exclusively. The reaction could also be utilized for the synthesis of benzophenone and biotin-labeled sphingolipids, although the products were obtained in lower yields with these labels. The reaction was subsequently applied to the synthesis of boron dipyrromethene-labeled sphingosine derivatives [84]. In this case the CM was performed on a synthetically derived terminal olefin. The same substrate was utilized for the introduction of long-chain amines via cross metathesis. These amines could then be immobilized on an affinity matrix thereby enabling access to tools for affinity chromatography [85].

5.3.6 Metathesis Applications in Solid-Phase Synthesis

While most applications of metathesis reactions in the pharmaceutical industry are performed as solution-phase reactions, some examples of solid-phase synthesis in the context of combinatorial chemistry have been reported.

Piscopio et al. from Amgen described one of the first examples of use of RCM in solid-phase synthesis as traceless linking and cyclative cleavage method [86,87]. The utility of the method was demonstrated in the synthesis of a Freidinger lactam. The combination of solid-phase peptide synthesis with the mild cleavage allowed for the construction of diverse heterocycles (Scheme 5.52) [88].

Heerding et al. from SmithKline Beecham reported on the use of enyne metathesis on solid phase for the generation of a compound library [89]. After exploration of chemistry in solution the enyne metathesis could be performed on substrates that were immobilized on a Merrifield resin (Scheme 5.53). The resulting dienes could be subjected to Diels–Alder reactions to access hexahydro isoindolines. A library of 4200 compounds was prepared by application of this methodology.

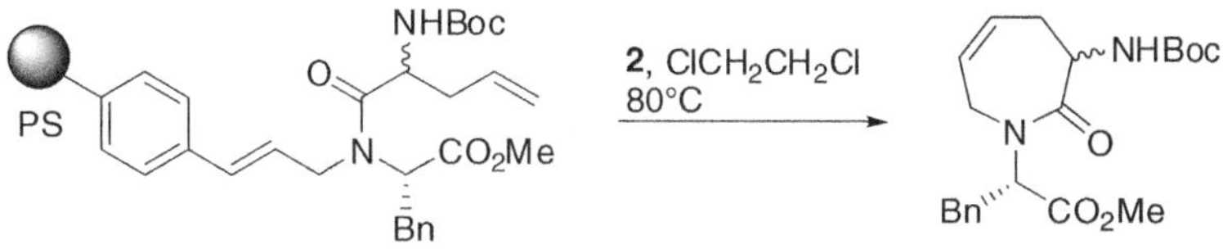

SCHEME 5.52 Solid-phase RCM-based synthesis of a Freidinger lactams.

SCHEME 5.53 Solid-phase enyne-metathesis.

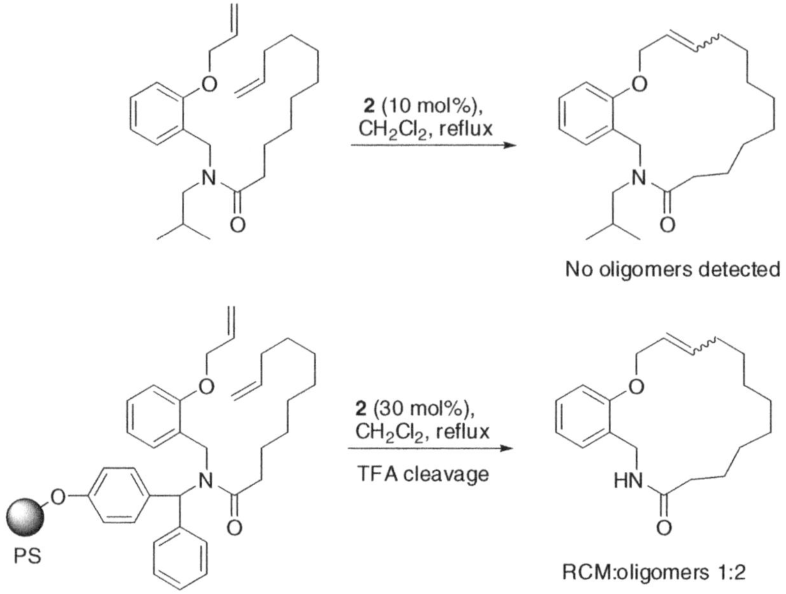

SCHEME 5.54 Mechanistic studies related to solid-phase RCM.

Tang and Wareing from Novartis studied the ring-closing metathesis of resin-bound olefins (Scheme 5.54) [90]. Cyclization of an amide in solution phase afforded the 17-membered ring without evidence of formation of dimers or oligomers. In contrast when the same reaction was carried out on a substrate in which the amide nitrogen was attached to a polystyrene-resin the product after cleavage from the resin was obtained as a mixture with oligomers. This finding suggests that the substrates were not fully site-separated on the resin and that intermolecular cross-metathesis reactions can occur during solid-phase RCM reactions. Conformational differences in the amide could be a potential alternative explanation for this finding.

Marsault et al. from Tranzyme Pharma reported the synthesis of macrocyclic antagonists for the human motilin receptor [91]. Linear tripeptides on a PS-chlorotrityl resin were subjected to catalyst **5** or **6** to obtain the macrocyclic product after cyclative cleavage (Scheme 5.55). The synthesis and purification was performed as a combinatorial library with tethers that contained aliphatic chains as well as phenyl rings.

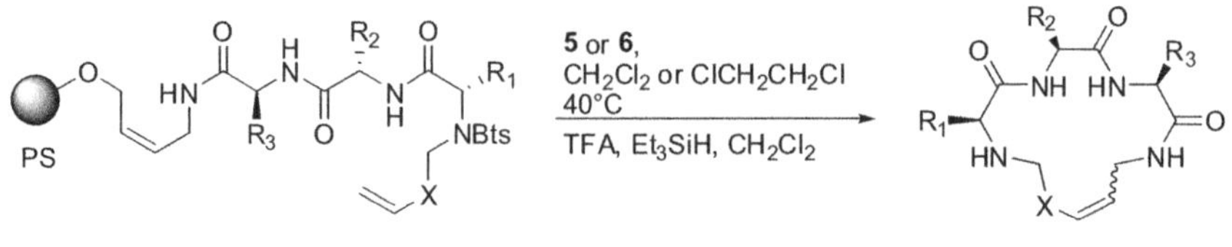

SCHEME 5.55 Solid-phase RCM-based synthesis of antagonists of the human motilin receptor.

5.4   APPLICATIONS IN PROCESS CHEMISTRY

The application of metathesis in the discovery of pharmaceutically active compounds within the medicinal chemistry laboratories has afforded ready access to new scaffolds. Several of the compounds were evaluated preclinically and after passing safety and animal efficacy hurdles where advanced into clinical development. This usually requires access to the target compounds in kilogram quantities. While it is potentially suitable to construct the compounds of interest via different synthetic routes, the ring-closing metathesis has established itself as a valuable tool for the synthesis of medium- and large-size rings on large scale. The utility of the reaction is further evidenced by the fact that it can become an integral part of route selection for process chemists. This leads to applications of metathesis reactions in Process Chemistry, for which the discovery synthesis did not involve metathesis. Thus far three applications of the metathesis reaction from Process Chemistry laboratories have been described in the literature, but it is reasonable to assume that the number of applications is already larger and that further publications will occur in the future.

Wallace et al. from Merck reported the use of a double ring-closing metathesis reaction in the synthesis of a spirocyclic neurokinin 1 (NK-1) receptor antagonist [92]. The approach relies on the diastereoselective ring closure of a five- and six-membered ring with catalyst **2** (Scheme 5.56). Good diastereoselectivities were observed with aliphatic substituents, whereas the diastereoselectivity is lower with the phenyl-substituent. Control experiments and molecular modeling studies suggest that the reaction selectivity is based on kinetic control and not on a thermodynamic equilibration of the products [93]. The reaction is thought to proceed via initial formation of the five-membered ring, followed by ring-closure of the six-membered ring. This was confirmed by isolating products at shorter reaction times. Initially the formation of two monocyclic five-membered products (1:1 ratio) and one monocyclic six-membered ring was observed (Scheme 5.57). Subjection of the isolated intermediates to RCM conditions revealed that the desired product is formed from both five-membered ring intermediates thereby implicating a ring-opening ring-closing process through the six-membered intermediate. The reaction scope was expanded to cyclic ethers in the double RCM of spirocyclic compounds (Scheme 5.58, eq. 1). Notably the opposite diastereomer as compared to the nitrogen-heterocycles was observed as the major product [94]. Ultimately the concept was extended to a

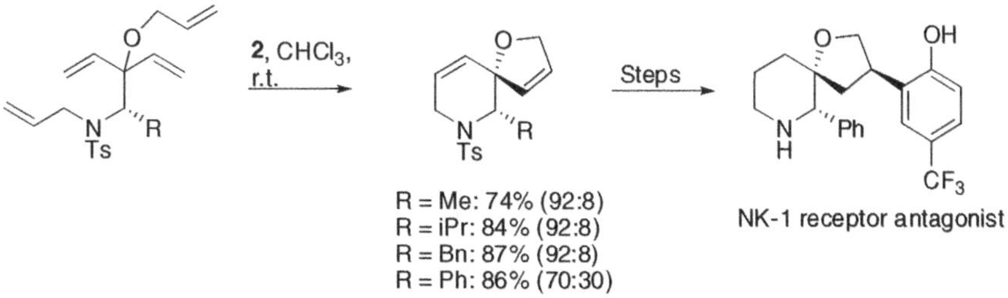

R = Me: 74% (92:8)
R = iPr: 84% (92:8)
R = Bn: 87% (92:8)
R = Ph: 86% (70:30)

NK-1 receptor antagonist

SCHEME 5.56   Synthesis of a NK-1 receptor antagonist.

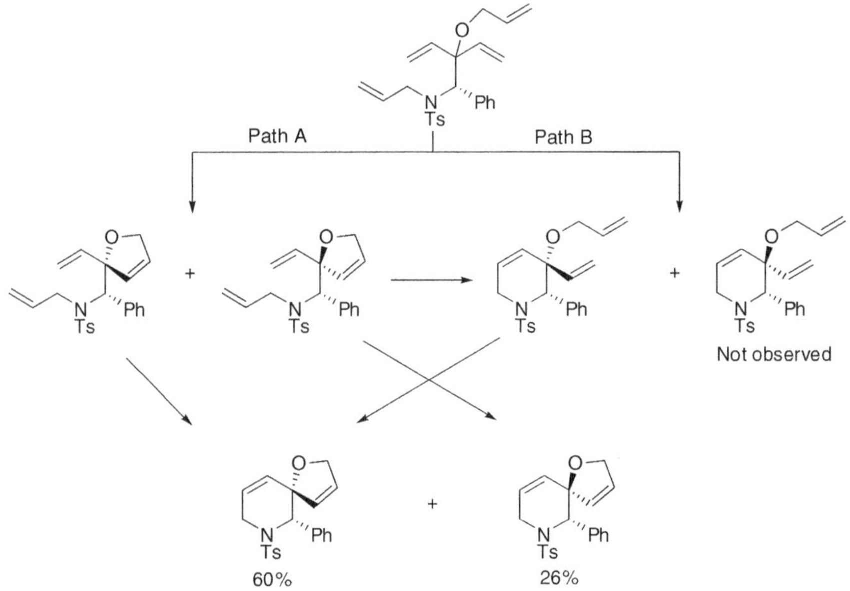
Path A
Path B
Not observed
60%
26%

SCHEME 5.57 Mechanism of the double RCM reaction.

quadruple RCM in which a bis-spirocycle was obtained in 65% yield (Scheme 5.58, eq. 2) [95].

The cathepsin K inhibitor SB-462795 was advanced into clinical studies by GlaxoSmithKline as a potential treatment of osteoporosis and osteoarthritis. The original synthesis had relied on RCM for the construction of the seven-membered ring (Scheme 5.20). Wang et al. described four different RCM-based routes that were

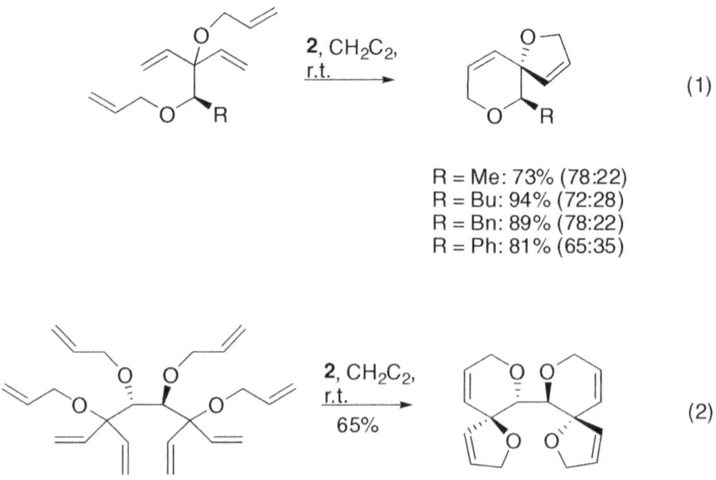
2, CH2C2,
r.t.
(1)
R = Me: 73% (78:22)
R = Bu: 94% (72:28)
R = Bn: 89% (78:22)
R = Ph: 81% (65:35)
2, CH2C2,
r.t.
65%
(2)

SCHEME 5.58 Synthesis of spirocyclic and bisspirocyclic ethers.

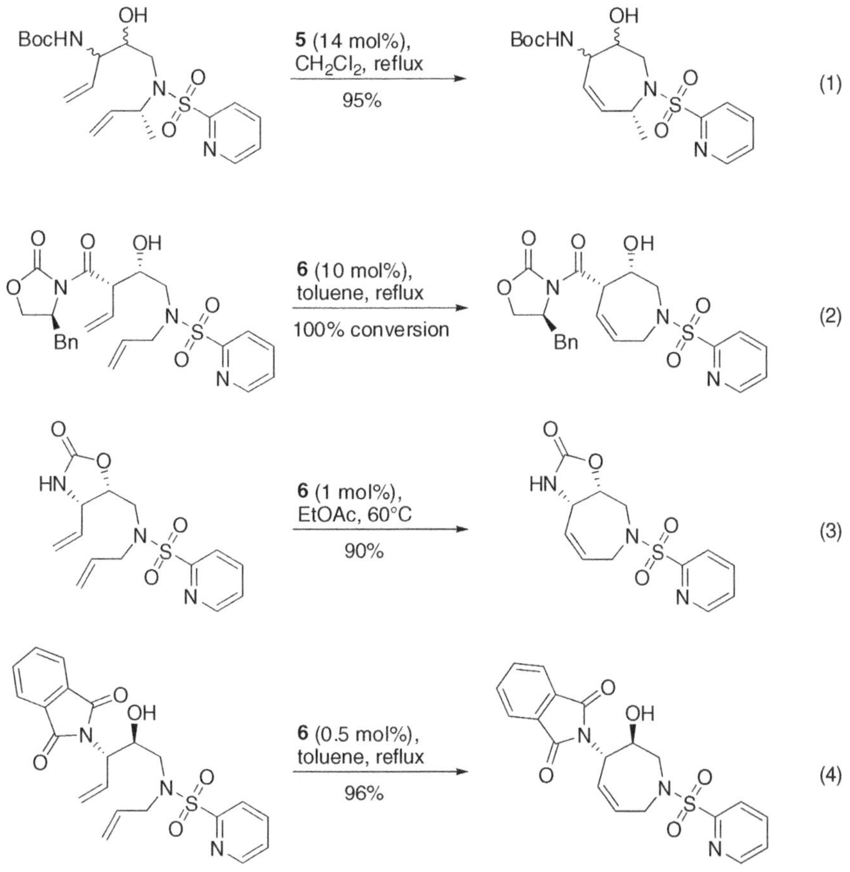

SCHEME 5.59 RCM approaches in the synthesis of cathepsin K inhibitor SB-462795.

explored for large-scale manufacturing of this compound (Scheme 5.59) [96]. In the first approach, a diastereomeric mixture was subjected to RCM with catalyst **5**, and the desired product was obtained in high yield (Scheme 5.59, eq. 1). The stereochemical mixtures at C3 and C4 are inconsequential for the synthesis of the target compound since the C4 stereocenter is readily epimerizable after oxidation of the alcohol to the ketone. This approach, however, was not pursued further, since the presence of diastereomeric mixtures hindered the isolation of synthetic intermediates by crystallization. An improved approach relied on construction of the vicinal stereocenters through a diastereoselective aldol reaction. This approach had been successfully applied to the synthesis of the des-methyl analogue (Scheme 5.21). The reaction proceeded to completion with 10 mol% of catalyst **6**, but lowering the catalyst loading to 5 mol% reduced the conversion significantly (60–79%) (Scheme 5.59, eq. 2). Alternative catalyst systems led to inferior results. The removal of the ruthenium by-products was also challenging, as some treatments resulted in decomposition of the product. This approach was therefore abandoned and a modified substrate was pursued. The aldol-product could be subjected to a cleavage of the auxiliary and a Curtius-rearrangement affording a carbamate. Cyclization of the hydrochloride salt of this carbamate with catalyst **6** (1 mol%) proceeded cleanly, and the ruthenium by-products could be removed by washes with a basic solution of cysteine (Scheme 5.59, eq. 3). Cyclization of the corresponding freebase resulted in significant amounts of olefin isomerization. While this approach was successful it required too many steps for the construction of the substrate and therefore was not pursued further. The optimal substrate for the RCM could be constructed by setting the stereochemistry through a Sharpless epoxidation. The cyclization could be performed with low catalyst loading (0.5 mol% **6**) at relatively high concentration (0.2 M) and the ruthenium residues could be removed by treatment with tetrakis (hydroxymethyl)phosphonium chloride (Scheme 5.59, eq. 4). The reaction is some-what sensitive to impurities that are related to the synthesis of the substrate [97], but after a clear understanding of the impact of impurity profile on catalyst performance the reaction could be performed on 80 kg scale.

Ring-closing metathesis has been applied in the discovery synthesis of HCV protease inhibitor BILN 2061 from Bohringer-Ingelheim [56]. The reaction was also applied for large-scale manufacture of this compound. Different substrates were investigated for the RCM reaction, the overall strategy sought to avoid the metathesis in the final steps of the API assembly due to expected challenges with removal of catalyst residues from the target compound (Scheme 5.60) [98]. Initial cyclization attempts were made with a substrate that carried a PNB-protecting group on the hydroxyl-substituent (Scheme 5.60, eq. 1). Cyclizations with catalyst **2** produced the desired product none of the undesired *E*-double bond isomer was formed. Variable amounts of epimerization at the β-carbon of the vinyl moiety on the adjacent cyclopropyl group was, however, observed. This isomerization was especially pronounced in the presence of small amounts of basic impurities such as phosphines and amines [99]. The presence of tricyclohexylphosphine had a detrimental effect on the reaction and cleaner reactions were observed with catalyst **3** (> 90% yields). Purification of the substrate entailed passing it through a pad of charcoal and silica in order to reduce catalyst loadings to <5 mol%. Despite high loadings relatively long reaction times (20 h) were required in refluxing dichloromethane or in toluene at 60°C and the reaction had to be performed under high dilution as an increase in concentration led to the formation of oligomers. The reversibility of the metathesis reaction also leads to complications during the removal of solvent at the end of the reaction. Specifically product loss due to oligomerization was observed upon concentration of the reaction mixtures. 2-Mercaptonicotinic acid was identified as effective treatment for the inactivation and removal of **3**. The catalyst loadings could be reduced by application of the second-generation catalysts **5** and **6**, but increased formation of dimers through thermodynamic equilibration was observed with these catalyst systems. These initial studies had identified a suitable catalyst system for the RCM but the overall number of steps and protecting group manipulations was deemed inappropriate. An alternative route to the substrate with free hydroxyl-group was devised, but this substrate led to significant dimer formation (Scheme 5.60, eq. 2). Due to challenges in the removal of the dimers without chromatography

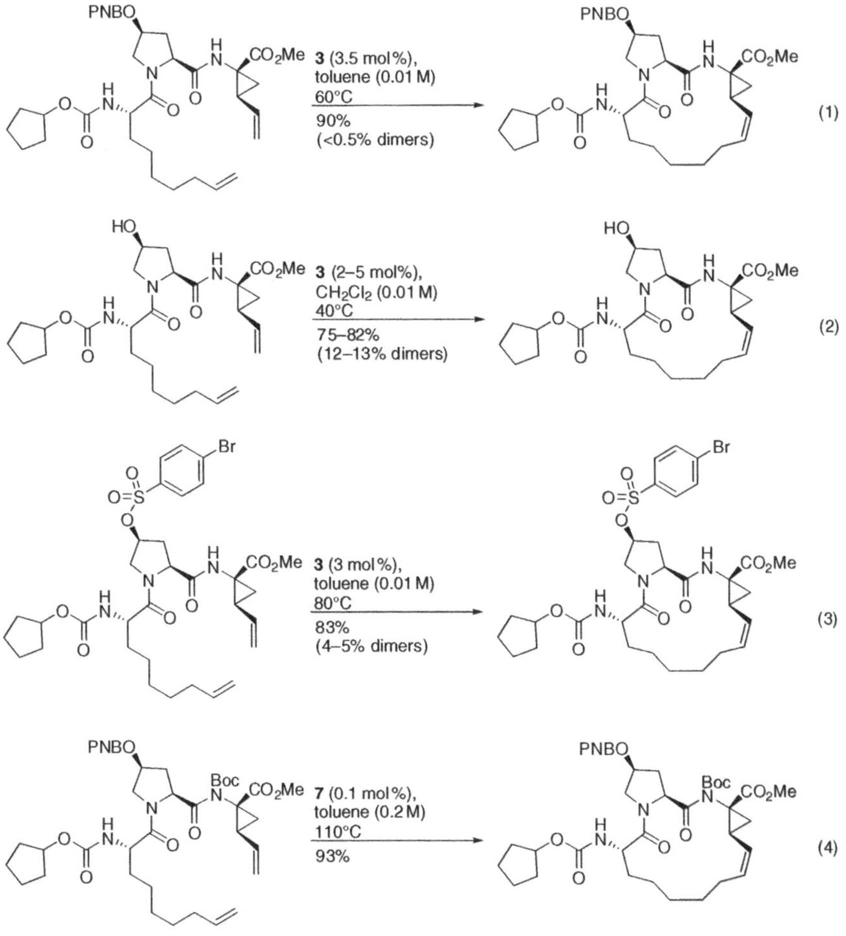

SCHEME 5.60   Key RCM reactions in the synthesis of BILN 2061.

further studies on this substrate were abandoned. The initial large-scale batches were therefore obtained by cyclization of the corresponding brosylate in the presence of catalyst **3** (Scheme 5.60, eq. 3) [100]. The reaction was performed in toluene to avoid use of halogenated solvents and to potentially allow for an increase in reaction temperature. The catalyst was charged in three portions and the reaction mixture was purged with nitrogen throughout the reaction in order to remove dissolved ethylene from the solvent. A relatively high-dilution (0.01 M) was required for a successful reaction outcome. The reaction was sensitive to small impurities in the solvent, as isomerization of the cyclopropane moiety was observed through traces of morpholine in technical grade toluene. The product was isolated by precipitation and high-levels of ruthenium (500–1000 ppm) were entrained. These levels were reduced to < 10 ppm in subsequent intermediates by application of silica and charcoal filtrations [101]. The reaction was successfully performed on 20 kg scale, producing more than 400 kg of the RCM product. Further advances in catalyst development and a careful substrate optimization resolved some major shortcomings of the initial large-scale production related to catalyst loading and reaction concentration [102]. Intro-duction of a Boc-substituent onto the PNB substrate and utilization of catalyst **7** allowed the reaction to be performed at desirable plant concentrations (0.2 M) with low catalyst loadings (0.1 mol%), while still achieving excellent yields of the desired product (93%) (Scheme 5.60, eq. 4). No special precautions for the removal of ruthenium were required with this modified process. Both kinetic and thermody-namic reasons can be put forward for the explanation of the Boc-effect. Introduction of the nitrogen substituent resulted in a change of the initial initiation side. While the unprotected substrate showed initial reaction at the vinyl-cyclopropane moiety the protected substrate showed initiation at the nonenoic acid moiety. This switch in initiation site resulted in a complete suppression of the epimerization reactions at the cyclopropyl ring. Through molecular modeling it was shown that the protecting group also leads to a conformational change in the substrate that results in a lower energy difference between open-chain and cyclized compound, thereby facilitating the RCM. Overall, careful selection of catalyst, reaction conditions and substrate allowed for improving the efficiency and robustness of the RCM reaction in the synthesis of BILN 2061 significantly. The final process meets the main criteria that would be necessary for the implementation on a commercial scale.

5.5   SUMMARY AND OUTLOOK

As shown in this chapter, metathesis has had an impact on drug discovery and development in the pharmaceutical industry in multiple ways. The number of publications in the area has increased steadily since the early reports a decade ago. While early applications were mostly related to RCM for the synthesis of small rings and natural product analogues the reaction most recently has had a profound impact on the synthesis of macrocycles. These compounds can have advantageous pharmacodynamic and pharmacokinetic properties as compared to their noncyclic congeners. This is evident by advancement of at least two of these compounds into clinical testing (BILN 2061 and MK-7009).

It is reasonable to expect that metathesis will continue to have a significant impact in the discovery of new drugs. As space for intellectual property becomes tighter, the reaction offers the opportunity to access novel scaffolds in an efficient manner. Cross metathesis could expand the impact of the reaction by allowing straightforward synthesis of analogues.

Based on the advancement of some of the target molecules from preclinical into clinical studies, there was a need for accessing them on a larger scale. Significant contributions from process chemistry groups in the industry allowed for successful scale-up of RCM reactions. This is evidenced by at least two compounds for which clinical supplies were made using a metathesis-based route. Further applications are expected, as it has been demonstrated that some of the main challenges preventing a scale-up of the reaction can be overcome. Low catalyst loadings (< 1 mol%) already can be applied with many of the current catalyst systems, but it has been demonstrated that newer ruthenium-based catalyst can operate at even lower loadings (0.1 mol%) [103]. Reaction concentration will remain dependent on the specific substrate, but fine tuning of the substrate toward conformations that favor cyclization can allow the reaction to be performed at relatively high concentrations (0.2 M). Efficient systems for the inactivation of the catalyst and removal of the ruthenium residues have also been established. Therefore, metathesis reactions and especially RCM are expected to remain efficient and practical reactions for the large-scale synthesis of active pharmaceutical ingredients up to commercial scale. Large-scale applications of cross metathesis may remain more limited, since this reaction type usually requires significant excess of one coupling partner. Depending on the structural complexity of the alkene this could become an efficiency issue.

REFERENCES

[1] Ivin, K. J.; Mol, J. C. *Olefin Metathesis and Metathesis Polymerisation*; Academic Press: San Diego, **1997**.

[2] For a detailed overview: Grubbs, R. H.Ed.; *Handbook of Metathesis* Wiley-VCH: Weinheim, **2003**.

[3] (a) Chauvin, Y. *Angew. Chem. Int. Ed.* **2006**, *45*, 3740; (b) Schorck, R. R. *Angew. Chem. Int. Ed.* **2006**, *45*, 3748; (c) Grubbs, R. H. *Angew. Chem. Int. Ed.* **2006**, *45*, 3760.

[4] (a) Fu, G. C.; Grubbs, R. H. *J. Am. Chem. Soc.* **1992**, *114*, 5426; (b) Fu, G. C.; Grubbs, R. H. *J. Am. Chem. Soc.* **1992**, *114*, 7324; (c) Fu, G. C.; Grubbs, R. H. *J. Am. Chem. Soc.* **1993**, *115*, 3800; (d) Fu, G. C.; Grubbs, R. H. *J. Am. Chem. Soc.* **1993**, *115*, 9856.

[5] Fürstner, A.; Langemann, K. *J. Org. Chem.* **1996**, *61*, 8746.

[6] Houri, A. F.; Xu, Z.; Cogan, D. A.; Hoveyda, A. H. *J. Am. Chem. Soc.* **1995**, *117*, 2943.

[7] (a) For leading reviews: Samojloowicz, C.; Bieniek, M.; Grela, K. *Chem. Rev.* **2009**, *109*, 3708; (b) van Otterlo, W. A. L.; de Koning, C. B. *Chem. Rev.* **2009**, *109*, 3743; (c) Monfette, S.; Fogg, D. E. *Chem. Rev.* **2009**, *109*, 3783; (d) Hoveyda, A. H.; Zhugralin, A. R. *Nature* **2007**, *450*, 243; (e) Gradillas, A.; Perez-Castells, J. *Angew. Chem. Int. Ed.* **2006**, *45*, 6086; (f) Nicolaou, K. C.; Bulger, P. G.; Sarlah, D. *Angew. Chem. Int. Ed.* **2005**, *44*, 4490; (g) Van de Weghe, P.; Eustache, J.; Cossy, J. *Curr. Top. Med. Chem.* **2005**, *5*, 1461; (h) Grubbs, R. H. *Tetrahedron* **2004**, *60*, 7117; (i) Diver, S. T.; Giessert, A. J. *Chem. Rev.* **2004**, *104*, 1317; (j) Deiters, A.; Martin, S. F. *Chem. Rev.* **2004**, *104*, 2199; (k) Piscopio, A. D.; Robinson, J. E. *Curr. Opin. Chem. Biol.* **2004**, *8*, 245; (l) Hoveyda, A. H.; Schrock, R. R. *Angew. Chem. Int. Ed.* **2003**, *42*, 4592; (m) Gannon, S. J.; Blechert, S. *Angew. Chem. Int. Ed.* **2003**, *42*, 1900; (n) Fürstner, A. *Angew. Chem. Int. Ed.* **2000**, *39*, 3012; (o) Yet, L. *Chem. Rev.* **2000**, *100*, 2963.

[8] Schrock, R. R.; Murdzek, J. S.; Bagan, G. C.; Robbins, J.; DiMare, M.; O'Regan, M. *J. Am. Chem. Soc.* **1990**, *112*, 3875.

[9] (a) Nguyen, S. T.; Johnson, L. K.; Grubbs, R. H.; Ziller, J. W. *J. Am. Chem. Soc.* **1992**, *114*, 3974; (b) Schwab, P.; France, M. B.; Ziller, J. W.; Grubbs, R. H. *Angew. Chem. Int.*

Ed. **1995**, *34*, 2039; (c) Schwab, P.; Grubbs, R. H.; Ziller, J. W. *J. Am. Chem. Soc.* **1996**, *118*, 100.

[10] Kingsbury, J. S.; Harrity, J. P. A.; Bonitatebus, P. J.; Hoveyda, A. H. *J. Am. Chem. Soc.* **1999**, *121*, 791.

[11] Scholl, M.; Ding, S.; Lee, S. W.; Grubbs, R. H. *Org. Lett.* **1999**, *1*, 953.

[12] Huang, J.; Stevens, E. D.; Nolan, S. P.; Peterson, J. L. *J. Am. Chem. Soc.* **1999**, *121*, 2674.

[13] (a) Chatterjee, A. K.; Morgan, J. P.; Scholl, M.; Grubbs, R. H. *J. Am. Chem. Soc.* **2000**, *122*, 3783; (b) Fürstner, A.; Thiel, O. R.; Ackermann, L.; Schanz, H.-J.; Nolan, S. P. *J. Org. Chem.* **2000**, *65*, 2204.

[14] Garber, S. B.; Kingsbury, J. S.; Gray, B. L.; Hoveyda, A. H. *J. Am. Chem. Soc.* **2000**, *122*, 8168.

[15] Michrowska, A.; Bujok, R.; Harutyunyan, S.; Sashuk, V.; Dolgonos, G.; Grela, K. *J. Am. Chem. Soc.* **2004**, *126*, 9318.

[16] Zhan, Z.-Y. J. U.S. Patent 0,043,180 (**2007**).

[17] (a) Clavier, H.; Nolan, S. P. *Chem. Eur. J.* **2007**, *13*, 8029; (b) Bieniek, M.; Michrowska, A.; Usanov, D. L.; Grela, K. *Chem. Eur. J.* **2008**, *14*, 806.

[18] (a) Paquette, L. A.; Schloss, J. D.; Efremov, L.; Fabris, F.; Gallou, F.; Mendez-Andino, J.; Yang, J. *Org. Lett.* **2000**, *2*, 1259; (b) Ahn, Y. M.; Yang, K.; Georg, G. I. *Org. Lett.* **2001**, *3*, 1411; (c) Maynard, H. D.; Grubbs, R. H. *Tetrahedron Lett.* **1999**, *40*, 4137; (d) Cho, J. H.; Kim, B. M. *Org. Lett.* **2003**, *5*, 531; (e) McEleney, K.; Allen, D. P.; Holliday, A. E.; Crudden, C. M. *Org. Lett.* **2006**, *8*, 2663; (f) Liu, W.; Nichols, P. J.; Smith, N. *Tetrahedron Lett.* **2009**, *50*, 6103.

[19] Fix-Stenzel, S. R.; Hayes, M. E.; Zhang, X.; Wallace, G. A.; Grongsaard, P.; Schaffter, L. M.; Hannick, S. M. Franczyk, T. S.; Stoffel, R. H.; Cusack, K. P. *Tetrahedron Lett.* **2009**, *50*, 4081.

[20] Yang, C.; Murray, W. V.; Wilson, L. J. *Tetrahedron Lett.* **2003**, *44*, 1783.

[21] (a) For earlier examples of microwave heating of metathesis reactions: Mayo, K. G.; Nearhoof, E. H.; Kiddle, J. *Org. Lett.* **2002**, *4*, 1567; (b) Varray, S.; Gauzy, C.; Lamaty, F. *J. Org. Chem.* **2000**, *65*, 6787.

[22] Meng, D.; Parker, Jr., D. L. *Tetrahedron Lett.* **2002**, *43*, 9035.

[23] Thorstensson, F.; Kvarnström, I.; Musil, D.; Nilsson, I.; Samuelsson, B. *J. Med. Chem.* **2003**, *46*, 1165.

[24] Sturino, C. F.; Yong, J. C. Y. *Tetrahedron Lett.* **1998**, *39*, 9623.

[25] Fujimura, O.; Fu, G. C.; Grubbs, R. H. *J. Org. Chem.* **1994**, *59*, 4029.

[26] Yong, J. C. Y.; Lacombe, P.; Sturino, C. F. *Tetrahedron Lett.* **1999**, *40*, 8751.

[27] Minville, J.; Poulin, J.; Dufresne, C.; Sturino, C. F. *Tetrahedron Lett.* **2008**, *49*, 3677.

[28] Renaud, J.; Ouellet, S. G. *J. Am. Chem. Soc.* **1998**, *120*, 7995.

[29] Renaud, J.; Graf, C.-D.; Oberer, L. *Angew. Chem. Int. Ed.* **2000**, *39*, 3101.

[30] Miller, J. F.; Termin, A.; Koch, K.; Piscopio, A. D. *J. Org. Chem.* **1998**, *63*, 3158.

[31] Walters, M. A.; La, F.; Deshmukh, P.; Omecinsky, D. O. *J. Comb. Chem.* **2002**, *4*, 125.

[32] Liras, S.; Allen, M. P.; Blake, J. F. *Org. Lett.* **2001**, *3*, 3483.

[33] Le Bourdonnec, B.; Goodman, A. J.; Michaut, M.; Ye, H.-F.; Graczyk, T. M.; Belanger, S.; Herbertz, T.; Yap, G. P. A.; DeHaven, R. N.; Dolle, R. E. *J. Med. Chem.* **2006**, *49*, 7278.

[34] Hu, X. E.; Kim, N. K.; Gray, J. L.; Almstead, J. K.; Seibel, W. L.; Ledoussal, B. *J. Med. Chem.* **2003**, *46*, 3655.

[35] Hu, X. E.; Kim, N. K.; Ledoussal, B. *Org. Lett.* **2002**, *4*, 4499.

[36] Baker, S. R.; Cases, M.; Keenan, M.; Lewis, R. A.; Tan, P. *Tetrahedron Lett.* **2003**, *44*, 2995.

[37] Gross, J. L. *Tetrahedron Lett.* **2003**, *44*, 8563.

[38] Grossmith, C. E.; Senia, F.; Wagner, J. *Synlett* **1999**, 1660.

[39] Kulesza, A.; Ebetine, F. H.; Mazur, A. W. *Tetrahedron Lett.* **2003**, *44*, 5511.

[40] Hoyt, S. B.; London, C.; Park, M. *Tetrahedron Lett.* **2009**. *50*, 1911.

[41] (a) Marquis, R. W.; Ru, Y.; LoCastro, S. M.; Zeng, J.; Yamashita, D. S.; Oh, H.-J.; Erhard, K. F.; Davis, L. D.; Tomaszek, T. A.; Tew, D.; Salyers, K.; Proksch, J.; Ward, K.; Smith, B.; Levy, M.; Cummings, M. D.; Haltiwanger, R. C.; Trescher, G.; Wang, B.; Hemling, M. E.; Quinn, C. J.; Cheng, H.-Y.; Lin, F.; Smith, W. W.; Janson, C. A.; Zhao, B.; McQueney, M. S.; D'Alessio, K.; Lee, C.-P.; Marzulli, A.; Dodds, R. A.; Blake, S.; Hwang, S.-M.; James, I. E.; Gress, C. J.; Bradley, B. R.; Lark, M. W.; Gowen, M.; Veber, D. F. *J. Med. Chem.* **2001**, *44*, 1380; (b) Yamashita, D. S.; Marquis, R. W.; Xie, R.; Nidamarthy, S. D.; Oh, H.-J.; Jeong, J. U.; Erhard, K. F.; Ward, K.; Roethke, T. J.; Smith, B.; Cheng, H.-Y.; Ceng, X.; Lin, F.; Offen, P. H.; Wang, B.; Nevins, N.; Head, M. S.; Haltiwanger, R. C.; NarducciSarjeant, A. A.; Liable-Sands, L. M.; Zhao, B.; Smith, W. W.; Janson, C. A.; Gao, E.; Tomaszek, T. A.; McQueney, M. S.; James, I. E.; Gress, C. J.; Zembryki, D. L.; Lark, M. W.; Veber, D. F. *J. Med. Chem.* **2006**, *49*, 1597.

[42] Trout, R. E. L.; Marquis, R. W. *Tetrahedron Lett.* **2005**, *46*, 2799.

[43] Paone, D. V.; Shaw, A. W.; Nguyen, D. N.; Burgey, C. S.; Deng, J. Z.; Kane, S. A.; Koblan, K. S.; Salvatore, C. A.; Mosser, S. D.; Johnston, V. K.; Wong, B. K.; Miller-Stein, C. M.; Hershey, J. C.; Graham, S. L.; Vacca, J. P.; Williams, T. M. *J. Med. Chem.* **2007**, *50*, 5564.

[44] Burgey, C. S.; Paone, D. V.; Shaw, A. W.; Deng, J. Z.; Nguyen, D. N.; Potteiger, C. M.; Graham, S. L.; Vacca, J. P.; Williams, T. M. *Org. Lett.* **2008**, *10*, 3235.

[45] Wang, Y.; O'Neil, S. V.; Wos, J. A.; Oppong, K. A.; Laufersweiler, M. C.; Soper, D. L.; Ellis, C. D.; Baize, M. W.; Fancher, A. N.; Lu, W.; Suchanek, M. K.; Wang, R. L.; Schwecke, W. P.; Cruze, C. A.; Buchalova, M.; Belkin, M.; De, B.; Demuth, Jr., T. P. *Bioorg. Med. Chem.* **2007**, *15*, 1311.

[46] O'Neil, S. V.; Wang, Y.; Laufersweiler, M. C.; Oppong, K. A.; Soper, D. L.; Wos, J. A.; Ellis, C. D.; Baize, M. W.; Bosch, G. K.; Fancher, A. N.; Lu, W.; Suchanek, M. K.; Wang, R.L.; De, B.; Demuth, Jr., T. P. *Bioorg. Med. Chem. Lett.* **2005**, *15*, 5434.

[47] Soper, D. L.; Sheville, J. X.; O'Neil, S. V.; Wang, Y.; Laufersweiler, M. C.; Oppong, K. A.; Wos, J. A.; Ellis, C. D.; Baize, M. W.; Chen, J. J.; Fancher, A. N.; Lu, W.; Suchanek, M. K.; Wang, R. L.; Schwecke, W. P.; Cruze, C. A.; Buchalova, M.; Belkin, M.; Wireko, F.; Ritter, A.; De, B.; Wang, D.; Demuth, Jr., T. P. *Bioorg. Med. Chem.* **2006**, *14*, 7880.

[48] Creighton, C. J.; Reitz, A. B. *Org. Lett.* **2001**, *3*, 893.

[49] Wilson, L. J.; Yang, C.; Murray, W. V. *Tetrahedron Lett.* **2007**, *48*, 7399.

[50] Wilson, L. J.; Malaviya, R.; Yang, C.; Argentieri, R.; Wang, B.; Chen, X.; Murray, W. V.; Cavender, D. *Bioorg. Med. Chem. Lett.* **2009**, *19*, 3333.

[51] Rojo, I.; Martin, J. A.; Broughton, H.; Timm, D.; Erickson, J.; Yang, H.-C.; McCarthy, J. R. *Bioorg. Med. Chem. Lett.* **2006**, *16*, 191.

[52] Peptidomimetic macrocyclic BACE inhibitors were also prepared by RCM in collaboration of Hanessian et al. with researchers at Novartis: Hanessian, S.; Yang, G.; Rondeau, J.-M.; Neumann, U.; Betschart, C.; Tintelnot-Blomley, M. *J. Med. Chem.* **2006**, *49*, 4544.

[53] Goudreau, N.; Brochu, C.; Cameron, D. R.; Duceppe, J.-S.; Faucher, A.-M.; Ferland, J. M.; Grand-Maitre, C.; Poirier, M.; Simoneau, B.; Tsantrizos, Y. S. *J. Org. Chem.* **2004**, *69*, 6185.

[54] Poirier, M.; Aubry, N.; Boucher, C.; Ferland, J. M.; LaPlante, S.; Tsantrizos, Y. S. *J. Org. Chem.* **2005**, *70*, 10765.

[55] Llinas-Brunet, M.; Bailey, M. D.; Bolger, G.; Brochu, C.; Faucher, A.-M.; Ferland, J. M.; Garneau, M.; Ghiro, E.; Gorys, V.; Grand-Maitre, C.; Halmos, T.; Lapeyre-Paquette, N.; Liard, F.; Poirier, M.; Rheaume, M.; Tsantrizos, Y. S.; Lamarre, D. *J. Med. Chem.* **2004**, *47*, 1605.

[56] Tsantrizos, Y. S.; Ferland, J.-M.; McClory, A.; Poirier, M.; Farina, V.; Yee, N. K.; Wang, X; Haddad, N.; Wei, X.; Xu, J.; Zhang, L. *J. Organomet. Chem.* **2006**, *691*, 5163.

[57] Faucher, A.-M.; Bailey, M. D.; Beaulieu, P. L.; Brochu, C.; Duceppe, J.-S.; Ferland, J. M.; Ghiro, E.; Gorys, V.; Halmos, T.; Kawai, S. H.; Poirier, M.; Simoneau, B.; Tsantrizos, Y. S.; Llinas-Brunet, M. *Org. Lett.* **2004**, *6*, 2901.

[58] Randolph, J. T.; Zhang, X.; Huang, P. P.; Klein, L. L.; Kurtz, K. A.; Konstantinidis, A. K.; He, W.; Kaiti, W. M.; Kempf, D. J. *Bioorg. Med. Chem. Lett.* **2008**, *18*, 2745.

[59] Velazquez, F.; Venkatraman, S.; Wu, W.; Blackman, M.; Prongay, A.; Girijavallabhan, V.; Neng-Yang, S.; Njoroge, F. G. *Org. Lett.* **2007**, *9*, 3061.

[60] Venkatraman, S.; Velazquez, F.; Wu, W.; Blackman, M.; Chen, K. X.; Bogen, S.; Nair, L.; Tong, X.; Chase, R.; Hart, A.; Agrawal, S.; Pichardo, J.; Prongay, A.; Cheng, K.-C.; Girijavallabhan, V.; Piwinski, J.; Neng-Yang, S.; Njoroge, F. G. *J. Med. Chem.* **2009**, *52*, 336.

[61] Velazquez, F.; Venkatraman, S.; Blackman, M.; Pinto, P.; Bogen, S.; Sannigrahi, M.; Chen, K.; Pichardo, J.; Hart, A.; Tong, X.; Girijavallabhan, V.; Njoroge, F. G. *J. Med. Chem.* **2009**, *52*, 700.

[62] Liverton, N. J.; Holloway, M. K.; McCauley, J. A.; Rudd, M. T.; Butcher, J. W.; Carroll, S. S.; DiMuzio, J.; Fandozzi, C.; Gilbert, K. F.; Mao, S.-S.; McIntyre, C. J.; Nguyen, K. T.; Romano, J. J.; Stahlhut, M.; Wan, B.-L.; Olsen, D. B.; Vacca, J. P. *J. Am. Chem. Soc.* **2008**, *130*, 4607.

[63] (a) Liverton, N. J.; Carroll, S. S.; DiMuzio, J.; Fandozzi, C.; Graham, D. J.; Hazuda, D.; Holloways, K.; Ludmerer, S. W.; McCauley, J. A.; McIntyre, C. J.; Olsen, D. B.; Rudd, M. T.; Stahlhut, M.; Vacca, J. P. *Antimicrob. Agents Chemother.* **2010**, *54*, 305; (b) Harper, S.; Ferrara, M.; Crescenzi, B.; Pompei, M.; Pumbi, M. C.; DiMuzio, J.; Donghi, M.; Fiore, F.; Koch, U.; Liverton, N. J.; Pesci, S.; Petrocchi, A.; Rowley, M.; Summa, V.; Gardelli, C. *J. Med. Chem.* **2009**, *52*, 4820.

[64] McCauley, J. A.; Rudd, M. T.; Nguyen, K. T.; McIntyre, C. J.; Romano, J. J.; Bush, K. J.; Varga, S. L.; Ross, C. W.; Carroll, S. S.; DiMuzio, J.; Stahlhut, M.; Olsen, D. B.; Lyle, T. A.; Vacca, J. P.; Liverton, N. J. *Angew. Chem. Int. Ed.* **2008**, *47*, 9104.

[65] Stachel, S. J.; Coburn, C. A.; Sankaranarayanan, S.; Price, E. A.; Pietrak, B. L.; Huang, Q.; Lineberger, J.; Espeseth, A. S.; Jin, L.; Ellis, J.; Holloway, M. K.; Munshi, S.; Allison, T.; Hazuda, D.; Simon, A. J.; Graham, S. L.; Vacca, J. P. *J. Med. Chem.* **2006**, *49*, 6147.

[66] Machauer, R.; Veenstra, S.; Rondeau, J.-M.; Tintelnot-Blomley, M.; Betschart, C.; Neumann, U.; Paganetti, P. *Bioorg. Med. Chem. Lett.* **2009**, *19*, 1361.

[67] Tao, Z.; Wang, L.; Stewart, K. D.; Chen, Z.; Gu, W.; Bui, M.; Merta, P.; Zhang, H.; Kovar, P.; Johnson, E.; Park, C.; Judge, R.; Rosenberg, S.; Sowin, T.; Lin, N. *J. Med. Chem.* **2007**, *50*, 1514.

[68] Dong, S. D.; Sundermann, K.; Smith, K. M. J.; Petryka, J.; Liu, F.; Myles, D. C. *Tetrahedron Lett.* **2004**, *45*, 1945.

[69] (a) Yang, Z.; He, Y.; Vourloumis, D.; Vallberg, H.; Nicolaou, K. C. *Angew. Chem. Int. Ed.* **1997**, *36*, 166; (b) Balog, A.; Meng, D.; Kamenecka, T.; Bertinato, P.; Su, D.-S.; Sorensen, E. J.; Danishefsky, S. J. *Angew. Chem. Int. Ed.* **1996**, *35*, 2801; (c) Meng, D.; Su, D.-S.; Balog, A.; Bertinato, P.; Sorensen, E. J.; Danishefsky, S. J.; Zheng, Y.-H.; Chou, T.-C.; He, L.; Horowitz, S. B. *J. Am. Chem. Soc.* **1997**, *119*, 2733; (d) Schinzer, D.; Limberg, A.; Bauer, A.; Böhm, O. M.; Cordes, M. *Angew. Chem. Int. Ed.* **1997**, *36*, 523.

[70] Meng, D.; Bertinato, P.; Balog, A.; Su, D.-S.; Kamenecka, T.; Sorensen, E. J. *J. Am. Chem. Soc.* **1997**, *119*, 10073.

[71] Wagner, J.; Cabrejas, L. M. M.; Grossmith, C. E.; Papageorgiou, C.; Senia, F.; Wagner, D.; France, J.; Nolan, S. P. *Angew. Chem. Int. Ed.* **1999**, *38*, 2443.

[72] Cabrejas, L. M. M.; Rohrbach, S.; Wagner, D.; Kallen, J.; Zenke, G.; Wagner, J. *J. Org. Chem.* **2000**, *65*, 9255.

[73] Sedrani, R.; Kallen, J.; Cabrejas, L. M. M.; Papageorgiou, C. D.; Senia, F.; Rohrbach, S.; Wagner, D.; Thai, B.; JutziEme, A.-M.; France, J.; Oberer, L.; Rihs, G.; Zenke, G.; Wagner, J. *J. Am. Chem. Soc.* **2003**, *125*, 3849.

[74] Kanada, R. M.; Itoh, D.; Nagai, M.; Niijima, J.; Asai, N. Mizui, Y. Abe, S.; Kotake, Y. *Angew. Chem. Int. Ed.* **2007**, *46*, 4350.

[75] Labrecque, D.; Charron, S.; Rej, R.; Blais, C.; Lamothe, S. *Tetrahedron Lett.* **2001**, *42*, 2645.

[76] (a) Fürstner, A.; Thiel, O. R.; Blanda, G. *Org. Lett.* **2000**, *2*, 3731; (b) Fürstner, A.; Dierkes, T.; Thiel, O. R.; Blanda, G. *Chem. Eur. J.* **2001**, *7*, 5286; (c) Snider, B. B.; Song, F. *Org. Lett.* **2001**, *3*, 1817; (d) Wu, Y.; Liao, X.; Wang, R.; Xie, X.-S.; De Brabander, J. K. *J. Am. Chem. Soc.* **2002**, *124*, 3245; (e) Wu, Y.; Esser, L.; De Brabander, J. K. *Angew. Chem. Int. Ed.* **2000**, *39*, 4308; (f) Georg, G. I.; Ahn, Y. M.; Blackman, B.; Farokhi, F.; Flaherty, P. T.; Mossman, C. J.; Roy, S.; Yang, K. *J. Chem. Soc. Chem. Commun.* **2001**, 255; (g) Smith, A. B., III; Zheng, J. *Tetrahedron* **2002**, *58*, 6455; (h) Smith, A. B., III; Zheng, J. *Synlett* **2001** 1019; (i) Herb, C.; Bayer, A.; Maier, M. E. *Chem. Eur. J.* **2004**, *10*, 5649; (j) Holloway, G. A.; Huegel, H. M.; Rizzacasa, M. A. *J. Org. Chem.* **2003**, *68*, 2200.

[77] Content, S.; Dutton, C. J.; Roberts, L. *Bioorg. Med. Chem. Lett.* **2003**, *13*, 321.

[78] Hoveyda, H. R.; Vezina, M. *Org. Lett.* **2005**, *7*, 2113.

[79] Bargiggia, F. C.; Murray, W. V. *J. Org. Chem.* **2005**, *70*, 9636.

[80] Lazarova, T.; Chen, J. S.; Hamann, B.; Kang, J. M.; Homuth-Trombino, D.; Han, F.; Hoffmann, E.; McClure, C.; Eckstein, J.; Or, Y. S. *J. Med. Chem.* **2003**, *46*, 674.

[81] Palmer, A. M.; Grobbel, B.; Jecke, C.; Brehm, C.; Zimmermann, P. J.; Buhr, W.; Feth, M. P.; Simon, W.-A.; Kromer, W. *J. Med. Chem.* **2007**, *50*, 6240.

[82] Palmer, A. M.; Christmann, S.; Münch, G.; Brehm, C.; Zimmermann, P. J.; Buhr, W.; Senn-Bilfinger, J.; Feth, M. P.; Simon, W.-A. *Bioorg. Med. Chem.* **2009**, *17*, 368.

[83] Nussbaumer, P.; Ettmayer, P.; Peters, C.; Rosenbeiger, D.; Högenauer, K. *Chem. Commun.* **2005**, 5086.

[84] Peters, C.; Billich, A.; Ghobrial, M.; Högenauer, K.; Ullrich, T.; Nussbaumer, P. *J. Org. Chem.* **2007**, *72*, 1842.

[85] Ullrich, T.; Ghobrial, M.; Peters, C.; Billich, A.; Guerini, D.; Nussbaumer, P. *Chem-MedChem* **2008**, *3*, 356.

[86] Piscopio, A. D.; Miller, J. F.; Koch, K. *Tetrahedron Lett.* **1997**, *38*, 7143.

[87] (a) For related examples from academic laboratories: van Maarseveen, J. H.; den Hartog, J. A.; Engelen, V.; Finner, E.; Visser, G.; Kruse, C. G. *Tetrahedron Lett.* **1996**, *37*, 8249; (b) Nicolaou, K. C.; Winssinger, N.; Pastor, J.; Ninkovic, S.; Sarabia, F, He, Y.; Vourloumis, D.; Yang, Z.; Li, T.; Giannakakou, P.; Hamel, E. *Nature* **1997**, *387*, 268.

[88] (a) Piscopio, A. D.; Miller, J. F.; Koch, K. *Tetrahedron Lett.* **1998**, *39*, 2667; (b) Piscopio, A. D.; Miller, J. F.; Koch, K. *Tetrahedron* **1999**, *55*, 8189.

[89] Heerding, D. A.; Takata, D. T.; Kwon, C.; Huffman, W. F.; Samanen, J. *Tetrahedron Lett.* **1998**, *39*, 6815.

[90] Tang, Q.; Wareing, J. R. *Tetrahedron Lett.* **2001**, *42*, 1399.

[91] Marsault, E.; Hoveyda, H. R.; Peterson, M. L.; Saint-Louis, C.; Landry, A.; Vezina, M.; Ouellet, L.; Wang, Z.; Ramaseshan, M.; Beaubien, S.; Benakli, K.; Beauchemin, S.; Deziel, R.; Peeters, T.; Fraser, G. L. *J. Med. Chem.* **2006**, *49*, 7190.

[92] (a) Wallace, D. J.; Cowden, C. J.; Kennedy, D. J.; Ashwood, M. S.; Cottrell, I. F.; Dolling, U.-H. *Tetrahedron Lett.* **2000**, *41*, 2027; (b) Wallace, D. J.; Goodmann, J. M.; Kennedy, D. J.; Davies, A. J.; Cowden, C. J.; Ashwood, M. S.; Cottrell, I. F.; Dolling, U.-H.; Reider, P. *J. Org. Lett.* **2001**, *3*, 671.

[93] Wallace, D. J. *Tetrahedron Lett.* **2005**, *46*, 591.

[94] Wallace, D. J.; Bulger, P. G.; Kennedy, D. J.; Ashwood, M. S.; Cottrell, I. F.; Dolling, U.-H. *Synlett 2000 41*, **2027**.

[95] Wallace, D. J. *Tetrahedron Lett.* **2003**, *44*, 2145.

[96] Wang, H.; Matsuhashi, H.; Doan, B. D.; Goodmann, S. N.; Ouyang, X.; Clark, W. M. *Tetrahedron* **2009**, *65*, 6291.

[97] Wang, H.; Goodmann, S. N.; Dai, Q.; Stockdale, G. W.; Clark, W. M. *Org. Process Res. Dev.* **2008**, *12*, 226.

[98] Yee, N. K.; Farina, V.; Houpis, I. N.; Haddad, N.; Frutus, R. P.; Gallou, F.; Wang, X.; Wei, X.; Simpson, R. D.; Feng, X.; Fuchs, V.; Xu, Y.; Tan, J.; Zhang, L.; Xu, J.; Smith-Keenan, L. L.; Vitous, J.; Ridges, M. D.; Spinelli, E. M.; Johnson, M. *J. Org. Chem.* **2006**, *71*, 7133.

[99] (a) Zeng, X.; Wei, X.; Farina, V.; Napolitano, E.; Xu, Y.; Zhang, L.; Haddad, N.; Yee, N. K.; Grinberg, N.; Shen, S.; Senanayake, C. H. *J. Org. Chem.* **2006**, *71*, 8864; (b) Farina, V.; Zeng, X.; Wei, X.; Xu, Y.; Zhang, L.; Haddad, N.; Yee, N. K.; Senanayake, C. H. *Catalysis Today* **2009**, *140*, 74.

[100] Nicola, T.; Brenner, M.; Donsbach, K.; Kreye, P. *Org. Process Res. Dev.* **2005**, *9*, 513.

[101] Supercritical fluid extraction with carbon dioxide was described as an alternative means for reduction of ruthenium levels in the product: Gallou, F.; Saim, S.; Koenig, K.; Bochniak, D. Horhota, S. T.; Yee, N. K.; Senanayake, C. H. *Org. Process Res. Dev.* **2006**, *10*, 937.

[102] (a) Shu, C.; Zeng, X.; Hao, M.-H.; Wei, X.; Yee, N. K.; Busacca, C. A.; Han, Z.; Farina, V.; Senanayake, C. H. *Org. Lett.* **2008**, *10*, 1303; (b) Farina, V.; Shu, C.; Zeng, X.; Wei, X.; Han, Z.; Yee, N. K.; Senanayake, C. H. *Org. Process Res. Dev.* **2009**, *13*, 250.

[103] Kuhn, K. M.; Champagne, T. M.; Hong, S. H.; Wei, W.-H.; Nickel, A.; Lee, C.W.; Virgil, S. C.; Grubbs, R. H.; Pederson, R. L. *Org. Lett.* **2010**, *12*, 984.

6

TRANSITION METAL-CATALYZED SYNTHESIS OF FIVE- AND SIX-MEMBERED HETEROCYCLES

CHEOL K. CHUNG AND MATTHEW L. CRAWLEY

6.1   INTRODUCTION AND BACKGROUND

Efficient and scalable synthesis of compounds containing five- and six-membered heterocycles is a critical process in both drug discovery and development. While by far the most common approach to this challenge is functionalization of commercially available heterocyclic building blocks, an important and growing strategy involves building these ring systems through transition metal-catalyzed processes [1,2]. The diversity of rings that can be assembled through this approach as well as the scope of the reactions employed is quite broad. As a result, some of the examples discussed

Applications of Transition Metal Catalysis in Drug Discovery and Development: An Industrial Perspective, First Edition. Edited by Matthew L. Crawley and Barry M. Trost.
© 2012 John Wiley & Sons, Inc. Published 2012 by John Wiley & Sons, Inc.

herein will have overlap with transition metal-catalyzed reactions discussed elsewhere in this book. However, the primary focus here is to highlight the processes specifically as applied to heterocyclic synthesis, selected either because of the proven scalability and efficiency of the transformation or the synthetic promise as applied to a drug or drug-like molecule total synthesis.

Practically to best capture the current state of the art, the drug targets reviewed are divided into two categories: those synthesized utilizing industrially proven processes, including many examples with the Heck reaction and Sonagashira reactions, and those with promise that have not yet been reported on kilogram scale. It is important to note that the scope and depth of these methods is undoubtedly greater than the examples selected; however, only those examples reported to date in peer-reviewed journals were eligible for inclusion in this chapter. Furthermore, the processes highlighted are not intended to represent a comprehensive review of this topic. Rather, the objective is to underscore the power of these processes and the direction in which the field is evolving.

6.2 HECK CYCLIZATION

In pharmaceutical manufacturing, the Heck reaction is increasingly recognized as a method of choice to functionalize aromatic and heteroaromatic compounds with alkenes. As evident in earlier academic research examples, the intramolecular Heck reaction is capable of constructing highly congested carbon–carbon bonds, including tetrasubstituted double bonds and quaternary carbon bonds [3,4]. One such example is illustrated in GSK's synthesis of a glycine antagonist for the development of an oral treatment for nicotine craving, whereby the intramolecular Heck reaction was used as the key step to build the tetrahydroquinoline core of the target compound (Scheme 6.1) [5].

The mildness of palladium catalysis was evident as no racemization was observed during the reaction. The reaction was highly selective for E olefin, but control of *endo/exo* olefin formation required extensive optimization. The authors found that the use of at least 2.5 equivalents of triphenylphosphine with respect to palladium in aromatic hydrocarbon solvent was required to minimize the formation of undesired *endo* olefin. These conditions are thought to stabilize the alkyl-palladium intermediate allowing it to undergo a rotation so that the subsequent *beta*-hydride elimination could occur in *exo* fashion (Scheme 6.2). It is also notable that excellent *exo*

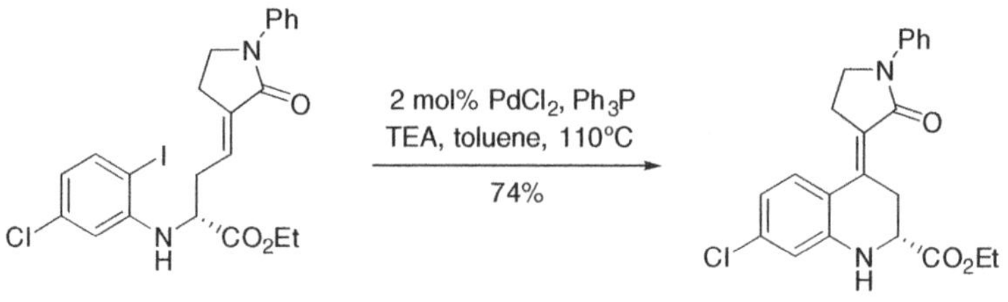

SCHEME 6.1 Tetrahydroquinoline synthesis via Heck cyclization.

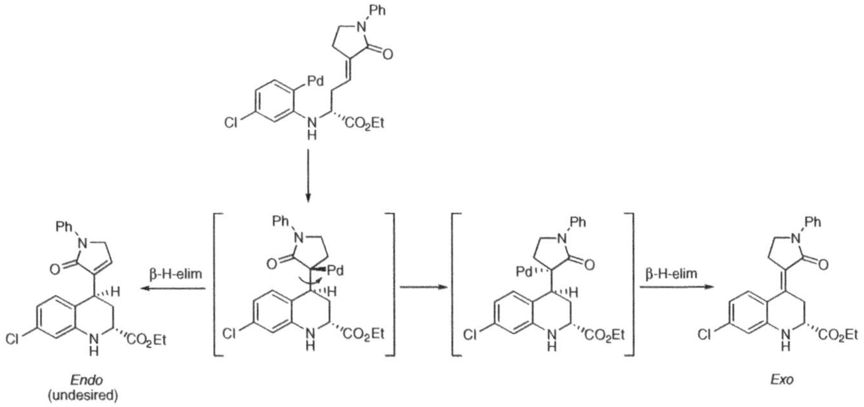

SCHEME 6.2 On *endo/exo* olefin selectivity.

selectivity could be achieved when the reaction was run under "ligandless" conditions [6]. This cyclization was demonstrated on a massive 87 kg scale using 2 mol% of palladium chloride.

Indole analogues are one of the most important pharmacophores, showing wide spectrum of biological activity [7]. While classical indole synthesis such as Fischer, Bartoli, and Reissert methods are still widely used in pharmaceutical research, transition metal-catalyzed indole synthesis is rapidly gaining ground providing rapid access to structurally diverse indole derivatives. The intramolecular Heck reaction is one such example. For instance, chemists at Idenix Pharmaceuticals prepared the indole-2-carboxylate intermediate by using a palladium-catalyzed annulation between 2-iodoaniline and pyruvic acid in their preparation of an HIV NNRTI drug candidate (Scheme 6.3) [8]. This reaction proceeded in an excellent 95% yield.

Originally developed by Chen et al. at Merck, the annulation occurs via *in situ* enamine formation followed by a Heck coupling reaction [9]. Since this method tolerates a variety of functional groups including acid-sensitive groups such as acetal, it could be a good alternative to Fischer indole synthesis. It was noted that the Idenix's approach resulted in a significant overall yield increase of up to 10-fold compared to the classical indole syntheses.

The versatility as well as scalability of Heck coupling strategy is illustrated in the large scale synthesis of DG-041, an antagonist of the EP$_3$ receptor developed by scientists at deCODE Chemistry (Scheme 6.4) [10]. In this synthesis, the indole core of the target compound was constructed via a tandem, one-pot intra- and intermolecular Heck reaction. The initial Heck cyclization could be achieved using 0.2 mol% of palladium acetate, and subsequent intermolecular Heck coupling was performed in the

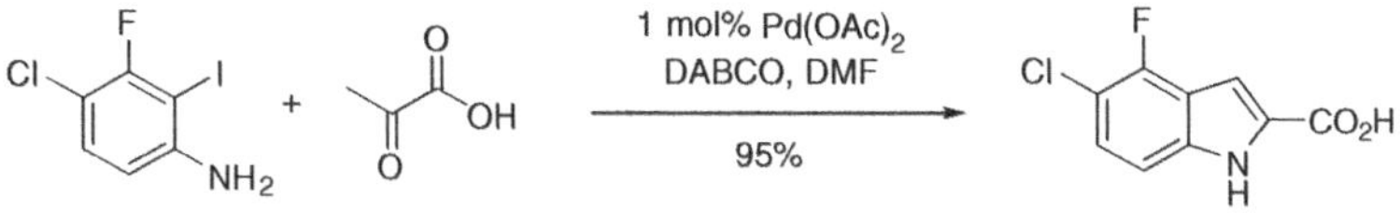

SCHEME 6.3 Indole synthesis via Pd-catalyzed annulation.

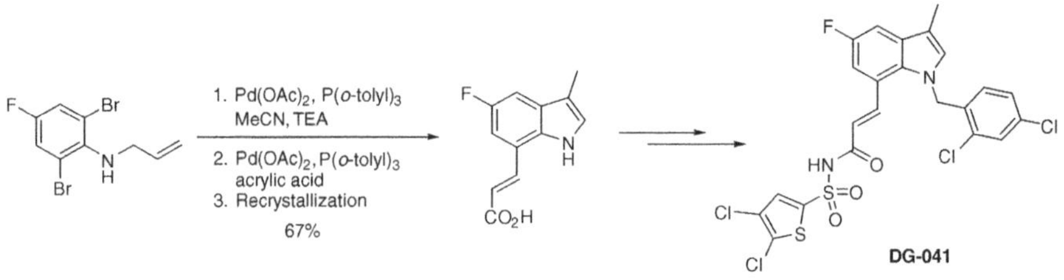

SCHEME 6.4   Indole synthesis via Heck cyclization.

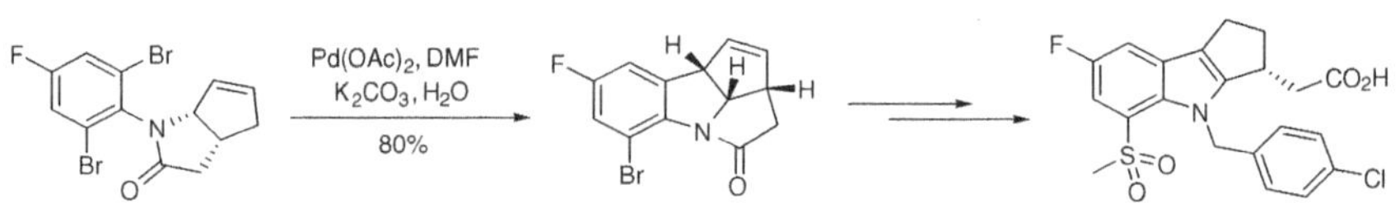

SCHEME 6.5   Fused indole synthesis via Heck cyclization.

same pot simply by charging additional palladium acetate (1 mol%) and olefin coupling partner affording the indole core of DG-041 in 67% isolated yield on 5 kg scale.

In yet another example, to gain access to a fused indole derivative, researchers at Merck applied Heck cyclization during their process development for a prostaglandin D_2 receptor antagonist (Scheme 6.5) [11]. The cyclization to the strained tetracyclic indole core was best achieved under ligandless conditions in aqueous DMF, which allowed over 60% increase in productivity in comparison to the reactions run in the presence of phosphine ligands.

Heck cyclization has been used as one of the key steps in the synthesis of camptothecin and its various derivatives (Scheme 6.6) [12,13]. In the process development of a DNA topoisomerase I inhibitor, scientists at GSK applied Heck's conditions to convert the crotyl ether starting material to the pyranopyridine product. The cyclization, however, afforded the product as a mixture of olefin isomers albeit the desired *endo* isomer was the major product (8/1 initially, then 11/1 after distillation). In a separate step, the *exo* isomer could be converted to the desired olefin isomer when treated with Wilkinson's catalyst. Intriguingly, when the Heck cyclization was conducted in the presence of the Wilkinson's catalyst, the *endo/exo* ratio could be improved significantly up to 29:1. In addition, this palladium/rhodium bimetallic catalytic system allowed the reaction to complete much faster (16 ∼ 20 h versus 4 h) [14]. Based on the absence of 5-*exo-trig* pathway, it was proposed that the

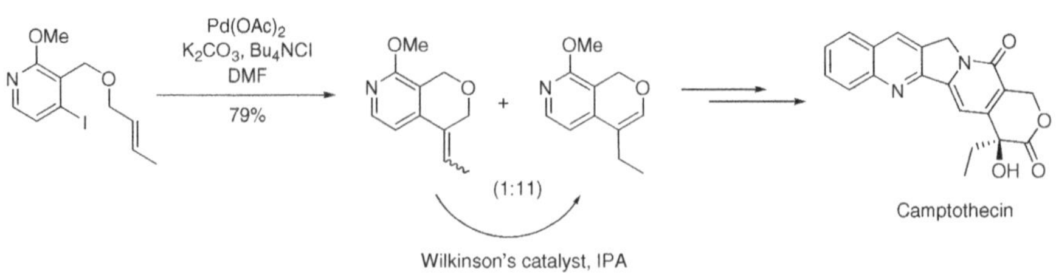

SCHEME 6.6   Pyranopyridine core synthesis via Heck cyclization.

olefin isomerization occurs after the cyclization is completed. The mechanism for the rate enhancement is not clear. The authors suggested that the Wilkinson's catalyst may form a η^4-metal complex with the crotyl ether moiety preorganizing the substrate for cyclization.

6.3 CYCLIZATION INDUCED BY CROSS-COUPLING

Tandem reaction whereby an initial cross-coupling adduct subsequently undergoes intramolecular cyclization under the reaction conditions has become an increasingly popular approach in the large-scale synthesis of drugs containing heterocycles. An eloquent and representative example of the power of cyclization-induced cross-coupling is evident in Pfizer's process synthesis of PHA-529311, a pyrroloquinoline that is highly active against herpes virus DNA polymerases [15]. The critical step in this sequence is a classic Sonogashira coupling followed by a sequential cyclization to afford the pyrrole (Scheme 6.7). In this route a substituted iodo-quinolone is coupled in an asynchronous concerted Sonogashira coupling with 3-butynol followed by further palladium-catalyzed ring closure.

The ratio of palladium catalyst to copper iodide needed optimization both to maximize yield and minimize product metal contamination. The optimum ratio established was 10:1 copper to palladium with 5 mol% copper employed. This enabled the synthesis of PHA-529311 in seven steps with 15.5% overall yield, which in the pilot plant was run on 20 kg scale and allowed pyrrole adduct purity with less than 50 ppm of either copper or palladium.

Similar utility of the Sonogashira cross-coupling/cyclization pathway was found in another Pfizer synthesis of a functionalized benzofuran core [16]. Utilizing copper iodide and triphenylphosphine palladium dichloride as co-catalysts, efficient conversion (68% yield after crystallization) of a bis-phenol and propyne to product was

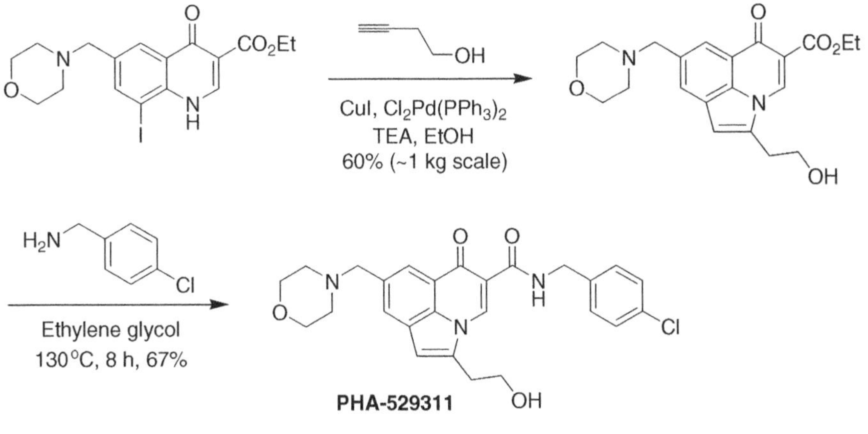

SCHEME 6.7 Synthesis of PHA-529311.

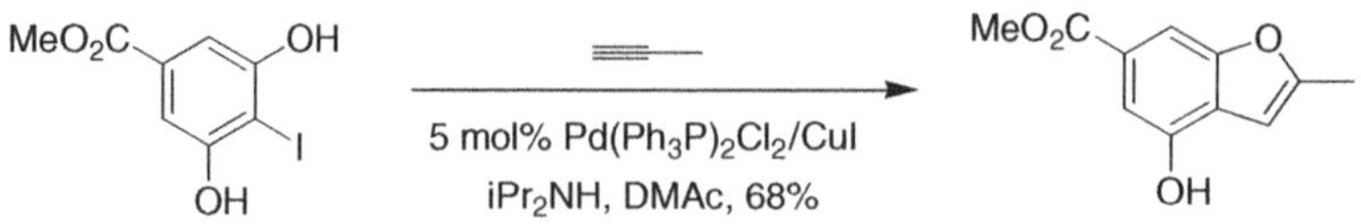

SCHEME 6.8 Synthesis of a functionalized benzofuran drug intermediate.

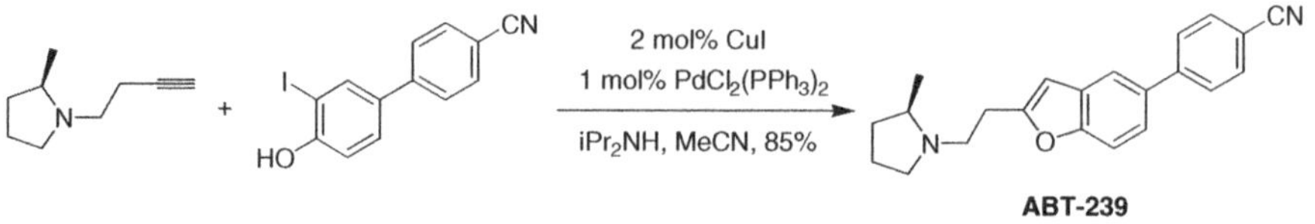

ABT-239

SCHEME 6.9 An efficient Sonogashira–Stevens reaction to access ABT-239.

achieved on 1.67 kg scale (Scheme 6.8). The key building block was isolated with less than 49 ppm of palladium with sufficient purity to use in subsequent steps. Of interest, while the final process gave good yield and allowed for crystallization of high purity, an eight by seven matrix screening of solvents and bases was required to find acceptable conditions. Other common solvents and bases resulted in significantly higher yields of by-products, lower conversions, or no yield of desired product at all.

While on a smaller scale (~20 g starting material) than the previous two examples, Abbott also successfully employed a Sonogashira cross-coupling/cyclization protocol (Stevens modification) in their synthesis of a potent and selective H$_3$ antagonist, ABT-239 [17]. The reaction proceed cleanly with a chromatography-free purification and isolation protocol and then was crystallized as the tartrate salt with >99% purity and <10 ppm of both palladium and copper metal residue (Scheme 6.9). The high yield is significant as synthesis by this method of similar benzofuran analogue reactions had been reported to give under 20% yield.

Complimentary to the Sonogashira coupling-cyclizations is an efficient Stille-coupling cyclization protocol that was recently reported. In this small-scale discovery example, reaction of a protected iodo-aniline with 1,1-disubstituted allene in the presence of palladium and copper catalysts produced desired adduct in 54% yield (Scheme 6.10) [18]. The intermediate was subsequently transformed in three steps to target indomethacin, a biologically active indole nucleus with anti-inflammatory activity. In order for the authors to get the cyclization to occur in tandem with the Stille coupling (or at least *in situ* under reaction conditions), careful selection of

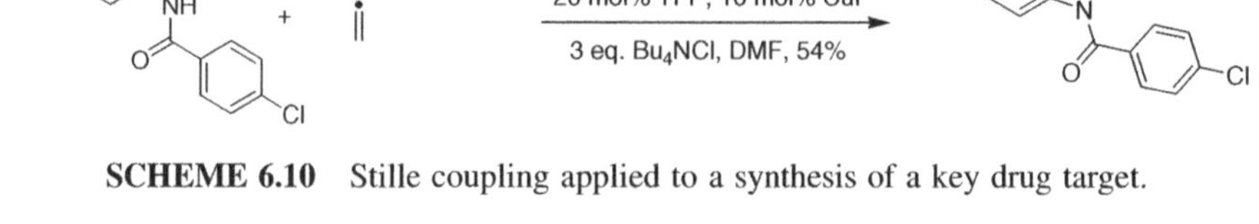

SCHEME 6.10 Stille coupling applied to a synthesis of a key drug target.

additive was critical. In fact, on model systems use of standard lithium salts (LiCl, LiBr, LiI) or other tetra-alkyl ammonium salts produced high yields of uncyclized product, where tetrabutylammonium chloride afforded only cyclized product.

6.4 MIGRATORY INSERTION-INDUCED CYCLIZATION

Migratory insertion is one of the fundamental processes in organometallic chemistry where typically an anionic ligand is combined with a neutral ligand such as carbonyl group and unsaturated carbon–carbon bond to form a new anionic ligand. While this insertion process often leads directly to cyclic structures, it can also lead to the structures that are amenable to cyclization via other mechanisms. A quinolone synthesis developed by Boehringer Ingelheim chemists is an excellent example for the latter case (Scheme 6.11) [19]. In this annulation process, an ynone intermediate was generated via a palladium-catalyzed carbonylative Sonogashira reaction, which set the stage for the regioselective cyclization by conjugate addition/elimination process to form the desired 6-*endo* cyclic product. The resultant quinolone compound containing an aminothiazole group is a substructure of BILN 2061, an exploratory drug for HCV infection.

One of the most attractive methods for the synthesis of 2,3-disubstituted indoles is Larock's indole synthesis [20]. In this palladium-catalyzed reaction internal alkynes typically undergo annulation with *o*-iodoaniline in regioselective fashion wherein the larger group ends up on the 2-position of the indole product. This selectivity is thought to be determined in the migratory insertion step in which the aryl group insert alkyne distal to the larger substituent to avoid unfavorable steric interactions. Due to

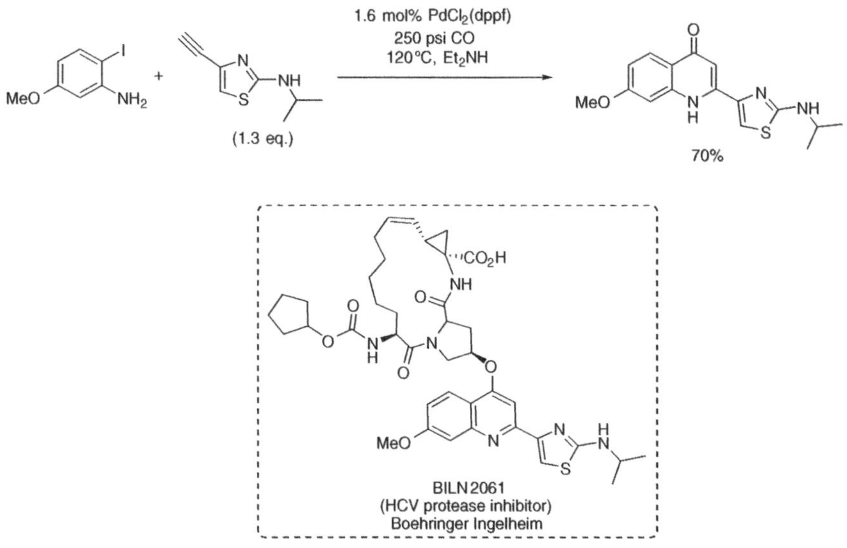

SCHEME 6.11   Quinolone synthesis via carbonylative Sonogashira reaction.

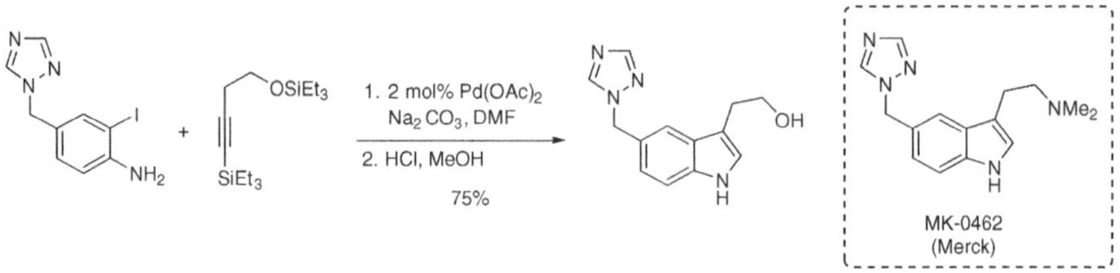

SCHEME 6.12   Indole synthesis via Larock annulation.

the mildness of the reaction conditions and its useful regioselectivity, Larock's annulation method was adopted in many pharmaceutical preparations. In a representative example, Merck's scientists utilized the method to synthesize MK-0462, a 5-HT$_{1D}$ receptor antagonist for the treatment of migraine (Scheme 6.12) [21]. Under ligandless palladium-catalyzed conditions, the two reactants undergo smooth cyclization to afford the bis-triethylsilylated product, which upon acidic deprotection gave the tryptophol product in 75% yield. It is noteworthy that Fischer-indole method was not practical for the synthesis of MK-0462 owing to the instability of triazole group.

A palladium-catalyzed cascade reaction adopted by Lilly scientists is another example where cyclization is achieved via a migratory insertion (Scheme 6.13) [22,23]. In the campaign to support nuclear hormone receptor modulator program, a series of dibenzoxapines containing an exocyclic tetrasubstituted olefin was prepared efficiently and stereoselectively through tandem cyclization-Suzuki coupling reactions. Remarkably, the cyclization could be achieved with 0.1 mol% of palladium acetate, which gave cleaner conversion than with higher catalyst loading. Also, ligandless conditions proved to be most effective at suppressing the impurity that arises from the direct Suzuki coupling without cyclization. This cyclization was demonstrated on a 300 g scale affording the dibenzoxapine product in 83% yield after isolation.

Treprostinil, a stable form of prostacyclin was developed for the treatment of advanced pulmonary hypertension by United Therapeutics. In 2004, an improved preparation was published wherein the Pauson–Khand reaction was used as the key step (Scheme 6.14) [24]. With stoichiometric $Co_2(CO)_8$ or under Livinghouse's catalytic method [25], the enyne substrate smoothly cyclized with the insertion of carbon monoxide to provide the tricyclic enone core in high yield and with excellent stereoselectivity. This cobalt-mediated reaction was conducted on multikilogram

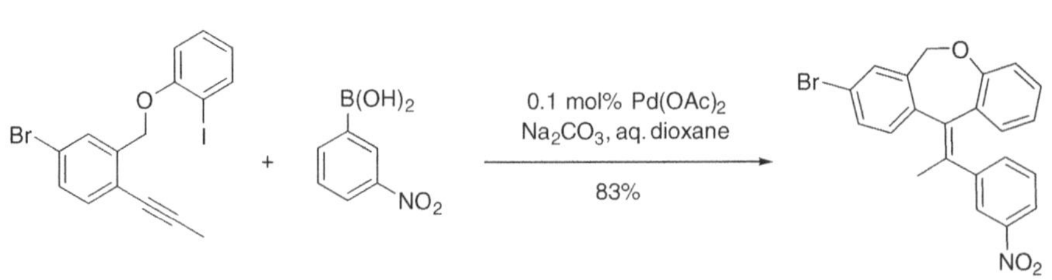

SCHEME 6.13   Dibenzoxapine synthesis via Pd-catalyzed Tanden reaction.

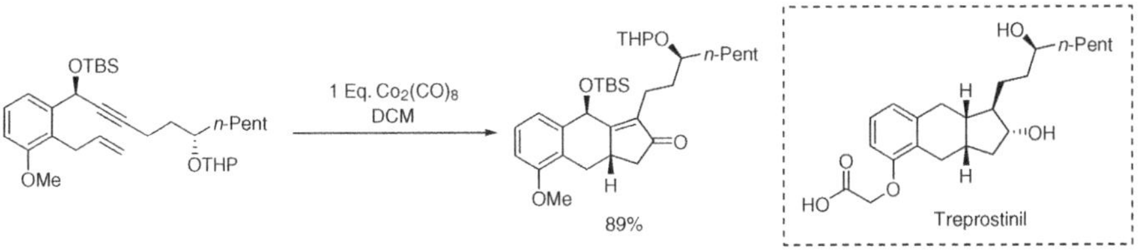

SCHEME 6.14 Pauson–Khand approach to Treprostinil.

scale (>4 kg), ultimately providing the active pharmaceutical ingredient (API) in 15 steps and 6% overall yield.

6.5 CYCLIZATION VIA CARBON–HETEROATOM COUPLING

Formation of heterocyclic drug intermediates through direct coupling/cyclization protocols is a strategy that has definitely been used in large-scale synthesis of pharmaceutical intermediates but has been sparsely reported in literature. The earlier discovery efforts that set the stage for larger scale applications have more descriptive reports and for the purpose of this chapter offer some insight as to the potential of this approach to access key cores in drug synthesis. As such, a handful of representative discovery examples are reviewed herein, with one process scale route.

Typically oxazepine cores are prepared in a five-step process starting with two commercially available building blocks. In this discovery example, the authors took advantage of recently developed palladium-catalysis conditions to efficiently form tricycle core in an excellent 93% yield (Scheme 6.15) [26]. While this reaction was not run on process scale to create this core, which is an integral part of several drugs with angiogenic, antiarrhythmic, and CNS activities, it was run reproducibly on 125 mmol (20 g) scale. In this case the best results required the use of the highly sterically encumbered tri-*tert*-butylphosphine ligand under strongly basic (sodium *tert*-butoxide) conditions.

The promazine series of compounds are a class of low-potency antipsychotic drugs that have been on the market for quite some time [27]. Efficient assembly of the tricyclic parent promazine X remains a topic of research as more analogues of the original series are tested and brought through clinical trials. Retrosynthetically one could envision a one pot, three component coupling of two aryl groups and an amine to give final product; this was nicely demonstrated in the synthesis of

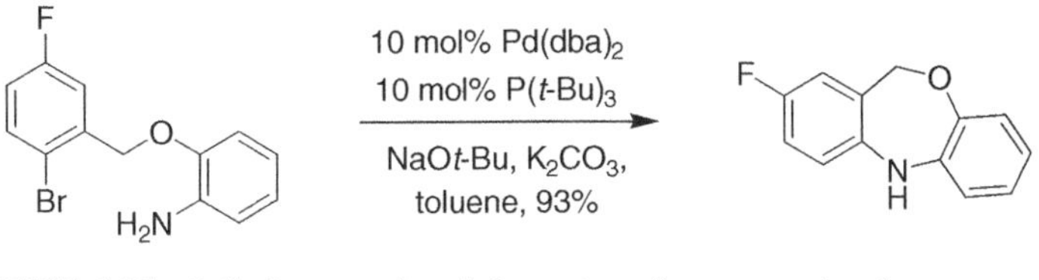

SCHEME 6.15 Palladium-catalyzed formation of an oxazepine drug target core.

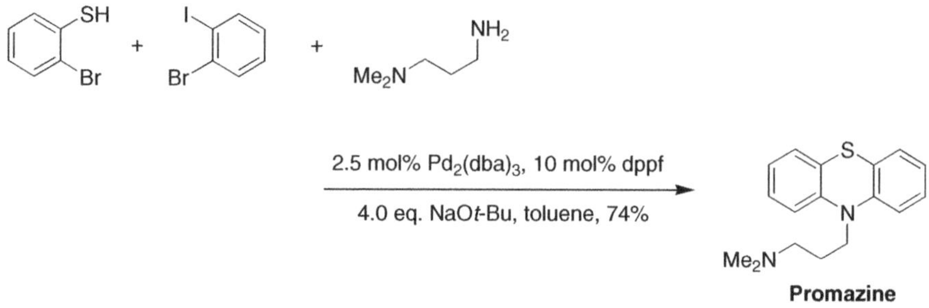

SCHEME 6.16 Palladium-catalyzed synthesis of the bioactive promazine core.

promazine (Scheme 6.16). Generation of a palladium–dppf complex *in situ* allowed for catalytic cascade cyclization, presumably first through formation of the C–S bond and then reaction of both bromides with the amine. This hypothesis was supported by isolation of intermediate by-products, which contained the C–S bond formed with other monoamine and dehalogenated derivatives.

The promazine core was also accessed through a sequentially controlled cascade reaction; this time instead it is catalyzed by a copper–proline metal–ligand system [28]. With the amine tail already attached to the iodoaromatic fragment, reaction in 2-methoxyethanol with potassium carbonate base afforded trifluoropromazine in 83% yield. In addition to this psychotropic drug, the chloropromazine and acepromazine derivatives were also assembled utilizing this approach in 78 and 75% yield, respectively (Scheme 6.17).

Scientists at GlaxoSmithKline chemical development employed a similar copper catalyzed approach in their large-scale synthesis of a 4*H*-imidazo[5,1-*c*][1.4]benzoxazine-3-carboxamide derivative for the treatment of depression and anxiety [29]. While the report is relatively devoid of details for the reaction shown (Scheme 6.18) as it was simply a scale-up of the medicinal chemistry route, the authors reported the process was run successfully on multikilogram scale. In this case, the authors use a bidentate amine ligand instead of proline in the conversion of bromoaromatic to the cyclic target. That target was further converted in several steps to the tricyclic target utilized in Phase I clinical trials.

Another small scale but representative example involved the three-component coupling reaction toward the imidazo-heterocycles [30]. In this method, 5-chloro-2-aminopyridine, aromatic aldehyde, and the alkyne are reacted in a one-pot

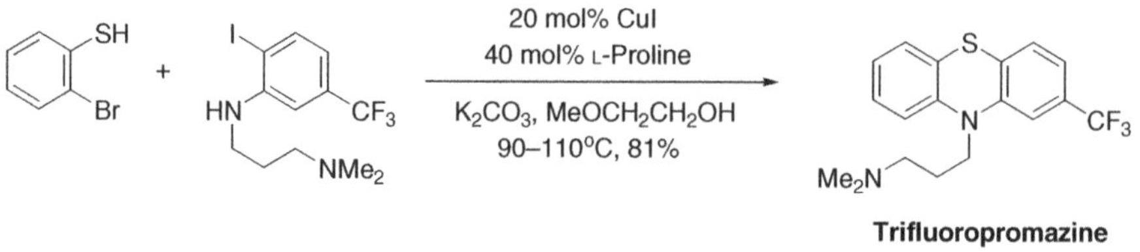

SCHEME 6.17 Synthesis of the antipsychotic drug trifluoropromazine.

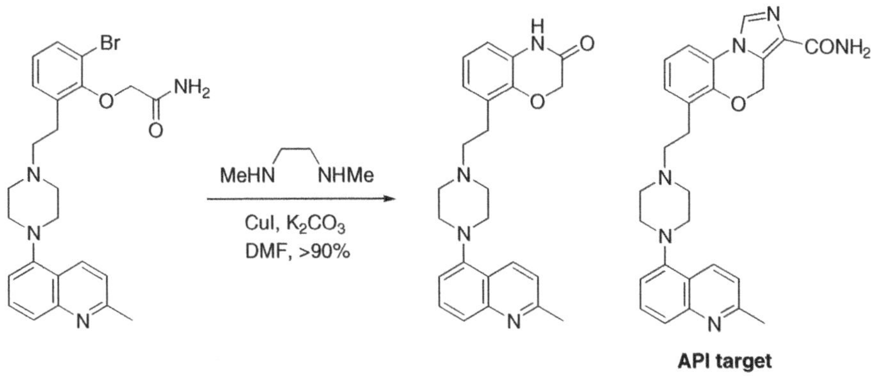

SCHEME 6.18   Process route toward an antidepressant target.

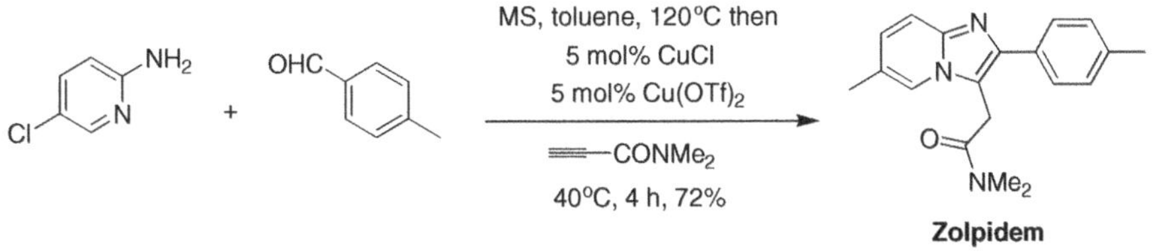

SCHEME 6.19   Synthesis of the anti-insomnia target zolpidem.

asynchronous coupling process whereby the initial imine is generated *in situ* and then reacted under copper-catalyzed conditions to give the imidazole product. It is of note that copper chloride or copper triflate alone resulted in significantly inferior yields of product, and a combination of the two was required. While this protocol was performed on discovery chemistry scale, it further builds in scope on the previous examples presented (Scheme 6.19).

6.6   CYCLOISOMERIZATION

Cycloisomerization is an atom economical transformation to access ring structures since all the atoms in the starting material are present in the product. In particular, the intramolecular trapping of metal vinylidenes by a tethered nucleophile is a powerful method for the synthesis of five- and six-membered heterocycles [31,32]. These vinylidenes are typically generated by a transition metal-catalyzed isomerization of terminal alkynes. Many transition metals including Ru, Rh, Cr, Mo, and W are known to catalyze such transformation, and are successfully applied to the synthesis of oxygen, nitrogen, and sulfur-containing heterocycles [33–36]. Researchers at Merck recently applied the Ru- and Rh-catalyzed cycloisomerization to prepare a key

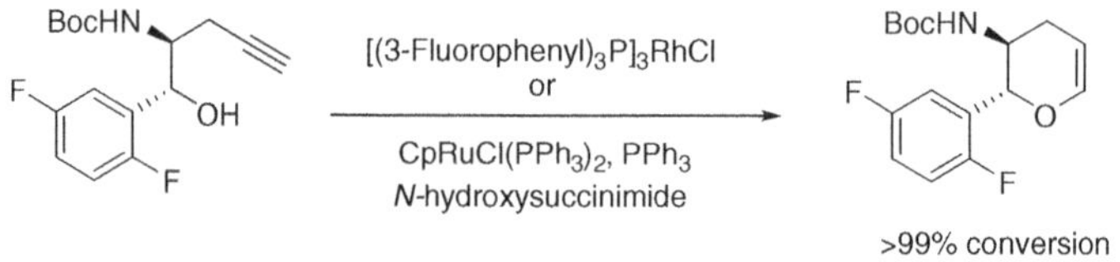

SCHEME 6.20   Dihydropyran synthesis via metal-catalyzed cycloisomerization.

intermediate for the synthesis of dipeptidyl peptidase IV inhibitors for the treatment of Type 2 diabetes (Scheme 6.20) [37].

It is notable that the reaction provided only six-membered ring oxacycle although the alkynol potentially can cyclize on the amino group to form a pyrroline instead. In a similar circumstance, the tungsten carbonyl complex has favored a five-membered azacycle formation [34]. In a related study, Zacuto et al. revealed that the efficiency of Ru-catalyzed cyclization is significantly affected by the substitution pattern on the substrate [38]. As can be seen in Scheme 6.21, the threo substrate displayed drastically lower reactivity compared to the erythro even at higher catalyst loading. A steric model involving a chair-like transition state has been proposed to explain the reactivity difference. However, it appears that the Rh-catalysis does not present the similar behavior [39].

During the process development for the synthesis of a microtubule inhibitor STA-5312, Sun and coworkers at Synta Pharmaceuticals applied a transition metal-catalyzed cycloisomerization reaction to overcome the problems encountered during the large-scale preparation of the clinical candidate following the original discovery route. The Cu-catalyzed cycloisomerization of 2-alkynylpyridine allowed them a rapid and reliable access to the key indolizine intermediate in high yield up to 50 g scale eliminating the need for column purification (Scheme 6.22) [40].

Originally developed by Gevorgyan, the Cu-catalysis is believed to proceed via an allene intermediate **1**, which undergoes cyclization induced by the coordination of the pi-philic copper species. The proposed mechanism shown in Scheme 6.23 is supported by deuterium labeling and electrophile trapping experiments [41].

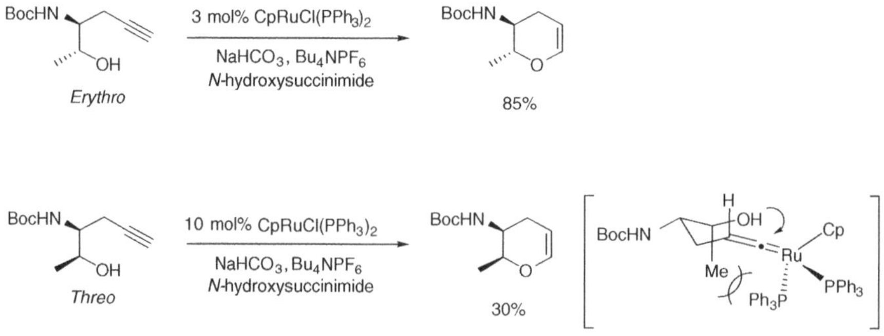

SCHEME 6.21   Influence of substituent on reactivity.

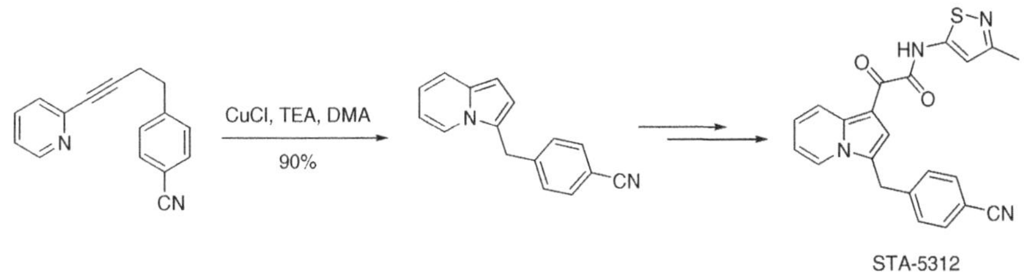

SCHEME 6.22 Indolizine synthesis via Cu-catalyzed cycloisomerization.

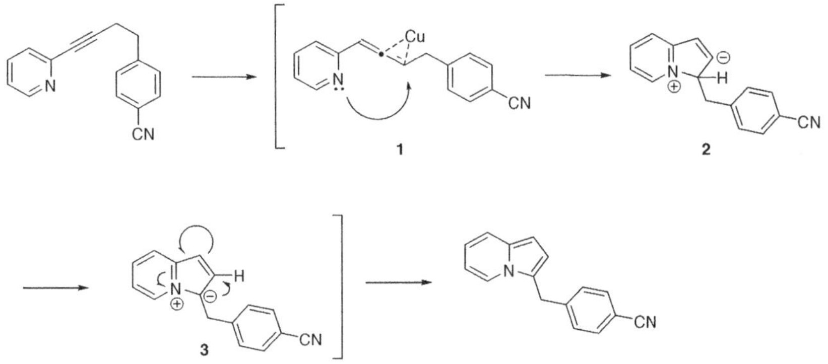

SCHEME 6.23 Proposed mechanism for the copper-catalyzed process.

A closely related copper-catalyzed cyclization strategy was used to develop an alternative synthesis of Zolpidem, an insomnia treatment [42]. Under Cu(I)/Cu(II) binary catalytic conditions, the imidazopyridyl core could be assembled in high yield from three simple starting materials via a condensation followed by 5-*exo-dig* cyclization (Scheme 6.24). This one-pot operation allows the synthesis of Zolpidem in one step, a significant improvement from previous synthesis that require 4–7 steps [43].

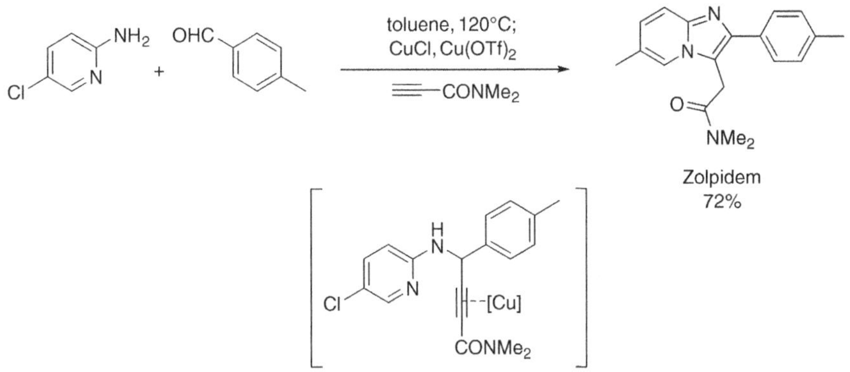

SCHEME 6.24 Zolpidem synthesis via Cu-catalyzed annulation.

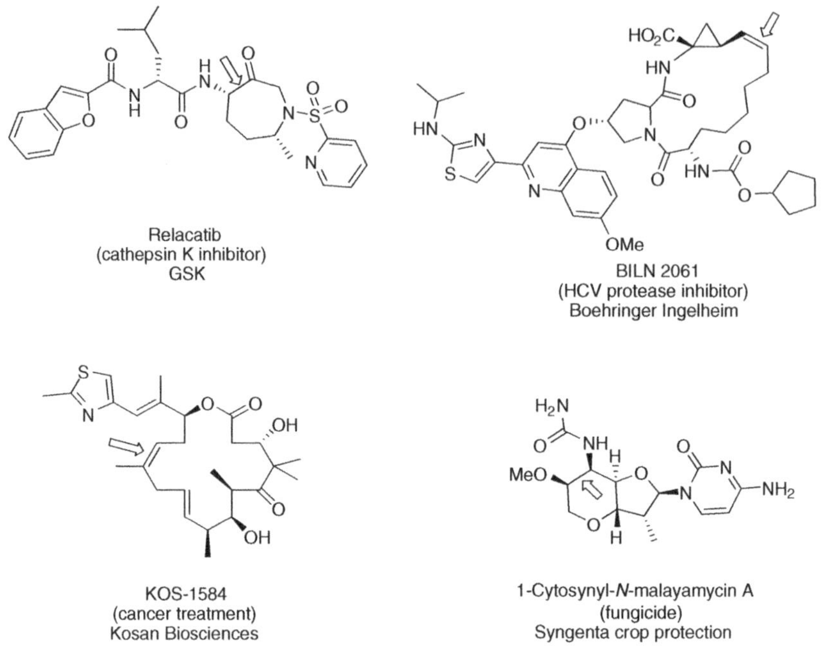

FIGURE 6.1   Targets of RCM in drug synthesis.

Relacatib
(cathepsin K inhibitor)
GSK

BILN 2061
(HCV protease inhibitor)
Boehringer Ingelheim

KOS-1584
(cancer treatment)
Kosan Biosciences

1-Cytosynyl-N-malayamycin A
(fungicide)
Syngenta crop protection

6.7   RING-CLOSING METATHESIS

Ring-closing metathesis (RCM) has emerged as an indispensible tool for the preparation of small-to-medium sized heterocycles, and has been widely used in academic and discovery space [44]. However, its application to the production of pharmaceuticals is still in its infancy, and there are only several published examples, some representative ones of which are shown in Figure 6.1. Nonetheless, RCM approach is expected to find more application in pharmaceutical industry considering its many advantages including its environmental and atom economical benefits. As RCM is truly its own unique reaction, a separate later chapter contains a detailed discussion on metathesis (this can be found in Chapter 4).

6.8   OTHER CYCLIZATIONS OF INTEREST

Recently, metal-catalyzed C–H bond functionalization has gained a significant attention in synthetic community. This approach is considered one of the most desirable methods in organic synthesis because it allows the synthesis of complex molecules starting from relatively simple and minimally activated precursors. One of the first examples of industrial applications of transition metal-catalyzed C–H functionalization is illustrated in the Pfizer's synthesis of an oxazolidinone antibac- terial compound (Scheme 6.25) [45]. The palladium-catalyzed oxindole formation

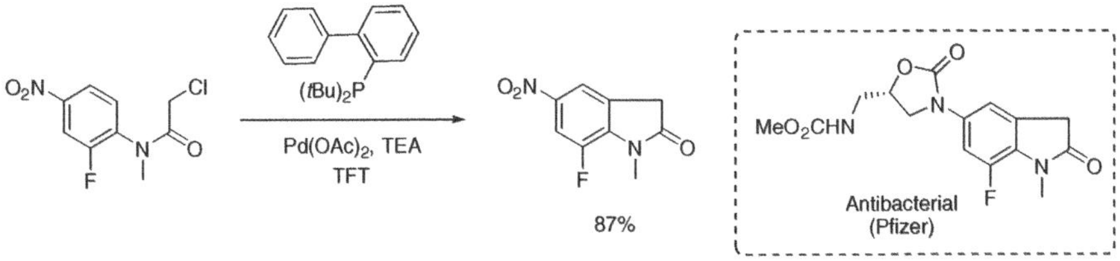

SCHEME 6.25   Oxindole synthesis via Pd-catalyzed C–H functionalization.

originally developed by Hennessey and Buchwald allows smooth cyclization without the need for harsh acidic conditions often required for Friedel–Crafts procedure, but has relatively narrow substrate scope requiring N-alkyl or N-aryl chloroacetanilides [46]. For example, primary amide or N-acyl substrates fail to form oxindole products. Also, bromoacetanilides instead of chloro substrate did not participate in the reaction. Nonetheless, Buchwald's original conditions served well in Pfizer's example affording the oxindole intermediate in 87% yield on multikilogram scale.

The success of C–H functionalization approach hinges on the selective activation of C–H bond at the desired position, which is often achieved by strategically placing metal-directing groups [47]. Since many biologically active compounds possess heteroatoms, which can also act as metal-directing groups, C–H functionalization approach is expected to find ample opportunity in pharmaceutical industry to streamline the synthesis of drug molecules. For instance, Bergman and Ellman's approach to c-jun N-terminal kinase 3 (JNK3) inhibitor took advantage of imidazole nitrogen's ability to bring the metal species to close proximity to the desired reaction center thereby lowering the entropic barrier of the cyclization process (Scheme 6.26) [48]. Even with some drawbacks such as the requirement of high catalyst loading (10 mol% in Rh) and high temperature (180 °C), the C–H activation approach accomplished the synthesis of JNK3 inhibitor in 11 linear steps and 13% overall

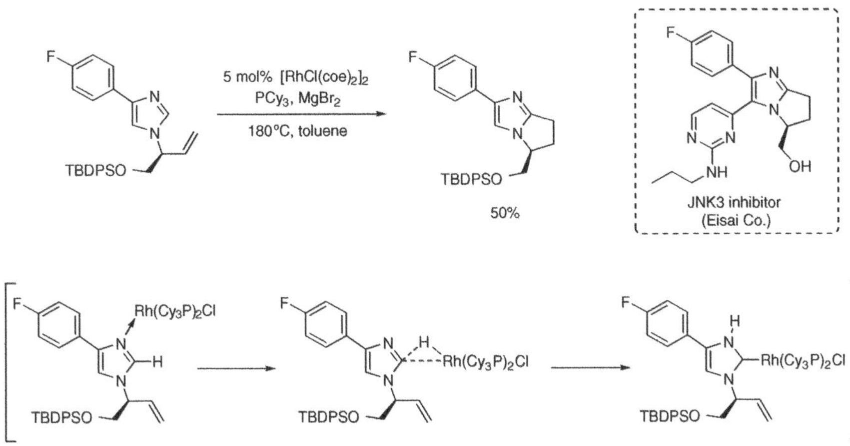

SCHEME 6.26   Rh-catalyzed C–H functionalization.

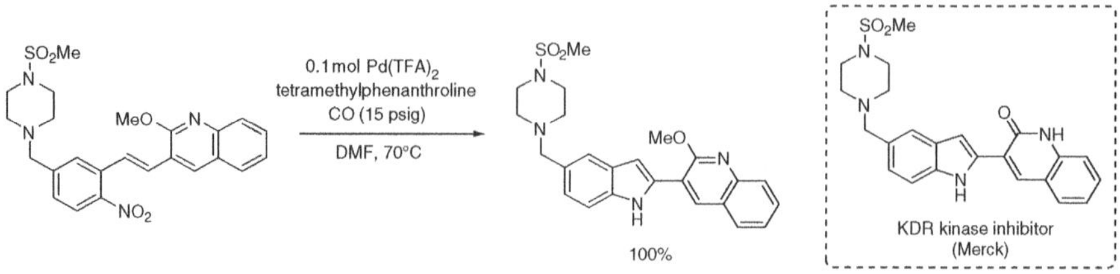

SCHEME 6.27   Indole synthesis via Pd-catalyzed reductive cyclization.

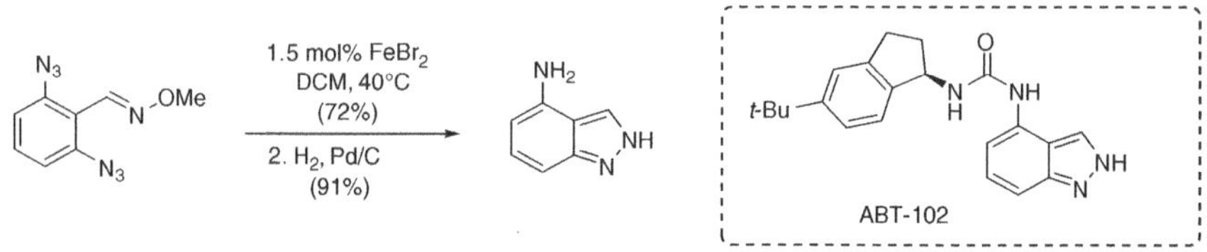

SCHEME 6.28   Indazole synthesis via Fe-catalyzed cyclization.

yield, a significant improvement from Eisai's original synthesis, which required 14 linear steps and about 6% overall yield [49].

A metal-catalyzed reductive cyclization aproach was developed in Merck's effort to design an efficient synthesis of an inhibitor for the kinase insert domain receptor (KDR) (Scheme 6.27). A number of approaches to the indole core of the KDR kinase inhibitor were evaluated, including Fischer-indole chemistry, palladium-catalyzed annulation, and reductive cyclizations, among which the palladium-catalyzed reductive cyclization with CO as the terminal reductant proved most efficient providing the desired indole product in quantitative yield [50,51].

Along with indoles, the indazole ring system is a popular structural feature found in many pharmacophores [52]. Indazole rings are typically constructed under harsh thermal conditions with varying yields [53]. Recently, Driver and coworkers developed an iron-catalyzed reaction that allowed a mild access to 4-aminoindazole, a substructure of Abbott Laboratories' pain management drug, ABT-102 [54]. It is notable that only one azido group participated in the cyclization leaving the second azide intact. Under thermal conditions, on the other hand, the bisazido starting material turned to a tarry material. Based on the absence of aniline by-product, the authors proposed that iron bromide acts as a Lewis acid activating the azide group toward N–N bond formation via a nucleophilic attack (Scheme 6.28).

6.9   CONCLUSION

Molecules containing heterocycles constitute especially important targets in the pharmaceutical industry. While there are a limited number of examples of direct transition metal-catalyzed reactions to access these molecules on a process scale, the methods highlighted illustrate the power of direct heterocyclic synthesis on scale for drug intermediates and targets. These tandem and direct processes nicely complement the array of other cross-coupling processes and enhance the toolbox of synthetic chemists. The recent nature of a majority of the reported kilogram scale reactions combined with the representative discovery examples show the future promise of this type of synthetic strategy in the scalable synthesis of bioactive molecules.

REFERENCES

[1] Nakamura, I.; Yamamoto, Y. *Chem. Rev.* **2004**, *104*, 2127.

[2] Schlummer, B.; Scholz, U. *Adv. Synth. Catal.* **2004**, *346*, 1599.

[3] Overman, L. E. *Pure Appl. Chem.* **1994**, *66*, 1423.

[4] Dounay, A. B.; Overman, L. E. *Chem. Rev.* **2003**, *103*, 2945.

[5] Banks, A.; Breen, G. F.; Caine, D.; Carey, J. S.; Drake, C.; Forth, M. A.; Gladwin, A.; Guelfi, S.; Hayes, J. F.; Maragni, P.; Morgan, D. O.; Oxley, P.; Perboni, A.; Popkin, M. E.; Rawlinson, F.; Roux, G. *Org. Proc. Res. Dev.* **2009**, *13*, 1130.

[6] Di Fabio, R.; Alvaro, G.; Bertani, B.; Donati, D.; Giacobbe, S.; Marchioro, C.; Palma, C.; Lynn, S. M. *J. Org. Chem.* **2002**, *67*, 7319.

[7] Sundberg, R. J. *Indoles*; Academic Press: London, 1996.

[8] Mayes, B. A.; Chaudhuri, N. C.; Hencken, C. P.; Jeannot, F.; Latham, G. M.; Mathieu, S.; McGarry, F. P.; Stewart, A. J.; Wang, J. Y.; Moussa, A. *Org. Proc. Res. Dev.* **2010**, *14*, 1248.

[9] Chen, C. Y.; Lieberman, D. R.; Larsen, R. D.; Verhoeven, T. R.; Reider, P. J. *J. Org. Chem.* **1997**, *62*, 2676.

[10] Zegar, S.; Tokar, C.; Enache, L. A.; Rajagopol, V.; Zeller, W.; O'Connell, M.; Singh, J.; Muellner, F. W.; Zembower, D. E. *Org. Proc. Res. Dev.* **2007**, *11*, 747.

[11] Campos, K. R.; Journet, M.; Lee, S.; Grabowski, E. J. J.; Tillyer, R. D. *J. Org. Chem.* **2005**, *70*, 268.

[12] Fang, F. G.; Bankston, D. D.; Huie, E. M.; Johnson, M. R.; Kang, M. C.; LeHoullier, C. S.; Lewis, G. C.; Lovelace, T. C.; Lowery, M. W.; McDougald, D. L.; Meerholz, C. A.; Partridge, J. J.; Sharp, M. J.; Xie, S. P. *Tetrahedron* **1997**, *53*, 10953.

[13] Fang, F. G.; Xie, S. P.; Lowery, M. W. *J. Org. Chem.* **1994**, *59*, 6142.

[14] Bankston, D.; Fang, F.; Huie, E.; Xie, S. P. *J. Org. Chem.* **1999**, *64*, 3461.

[15] Dorow, R. L.; Herrinton, P. M.; Hohler, R. A.; Maloney, M. T.; Mauragis, M. A.; McGhee. W. E.; Moeslein, J. A.; Strohbach, J. W.; Veley, M. F. *Org. Proc. Res. Dev.* **2006**, *10*, 493.

[16] Berliner, M. A.; Cordi, E. M.; Dunetz, J. R.; Price, K. E. *Org. Proc. Res. Dev.* **2010**. *14*, 180.

[17] Ki, Y.-Y.; Pu, Y.-M.; Grieme, T.; Sharma, P.; Bhatia, A. V.; Cowart, M. *Tetrahedron* **2006**, *62*, 4584.

[18] Mukai, C.; Takahashi, Y. *Org. Lett.* **2005**, *7*, 5793.

[19] Haddad, N.; Tan, J.; Farina, V. *J. Org. Chem.* **2006**, *71*, 5031.

[20] Larock, R. C.; Yum, E. K. *J. Am. Chem. Soc.* **1991**, *113*, 6689.

[21] Chen, C. Y.; Lieberman, D. R.; Larsen, R. D.; Reamer, R. A.; Verhoeven, T. R.; Reider, P. J.; Cottrell, I. F.; Houghton, P. G. *Tetrahedron Lett.* **1994**, *35*, 6981.

[22] Richey, R. N.; Yu, H. *Org. Proc. Res. Dev.* **2009**, *13*, 315.

[23] Yu, H.; Richey, R. N.; Mendiola, J.; Adeva, M.; Somoza, C.; May, S. A.; Carson, M. W.; Coghlan, M. J. *Tetrahedron Lett.* **2008**, *49*, 1915.

[24] Moriarty, R. M.; Rani, N.; Enache, L. A.; Rao, M. S.; Batra, H.; Guo, L.; Penmasta, R. A.; Staszewski, J. P.; Tuladhar, S. M.; Prakash, O.; Crich, D.; Hirtopeanu, A.; Gilardi, R. *J. Org. Chem.* **2004**, *69*, 1890.

[25] Pagenkopf, B. L.; Livinghouse, T. *J. Am. Chem. Soc.* **1996**, *118*, 2285.

[26] Margolis, B. J.; Swidorski, J. J.; Rogers, B. N. *J. Org. Chem.* **2003**, *68*, 644.

[27] Bogeso, K. P.; Bang-Anderson, B. *Textbook of Drug Design and Discovery*, 3rd. edn., Taylor and Francis Publications: London, UK, **2002**, Chapter 11.

[28] Ma, D.; Geng, Q.; Zhang, H.; Jiang, Y. *Angew. Chem. Int. Ed.* **2010**, *49*, 1291.

[29] Giubellina, N.; Stabile, P.; Laval, G.; Perboni, A. D.; Cimarosti, Z.; Westerduin, P.; Cooke, J. W. B. *Org. Proc. Res. Dev.* **2010**, *14*, 859.

[30] Chernayak, N.; Gevorgyan, V. *Angew. Chem. Int. Ed.* **2010**, *49*, 2743.

[31] McDonald, F. E. *Chem.-Eur. J.* **1999**, *5*, 3103.

[32] Trost, B. M.; Frederiksen, M. U.; Rudd, M. T. *Angew. Chem. Int. Ed.* **2005**, *44*, 6630.

[33] McDonald, F. E.; Burova, S. A.; Huffman, L. G. *Synthesis-Stuttgart* **2000**, 970.

[34] McDonald, F. E.; Chatterjee, A. K. *Tetrahedron Lett.* **1997**, *38*, 7687.

[35] McDonald, F. E.; Zhu, H. Y. H. *J. Am. Chem. Soc.* **1998**, *120*, 4246.

[36] Trost, B. M.; Rhee, Y. H. *J. Am. Chem. Soc.* **1999**, *121*, 11680.

[37] Xu, F.; Kim, M. M.; Kohmura, Y.; Sladicka, T.; Rosen, J. D.; Zacuto, M., J. U.S. Patent 2009/0187028 A1 (2009).

[38] Zacuto, M. J.; Tomita, D.; Pirzada, Z.; Xu, F. *Org. Lett.* **2010**, *12*, 684.

[39] Trost, B. M.; Rhee, Y. H. *J. Am. Chem. Soc.* **2003**, *125*, 7482.

[40] Li, H.; Xia, Z. Q.; Chen, S. J.; Koya, K.; Ono, M.; Sun, L. *J. Org. Proc. Res. Dev.* **2007**, *11*, 246.

[41] Kel'in, A. V.; Sromek, A. W.; Gevorgyan, V. *J. Am. Chem. Soc.* **2001**, *123*, 2074.

[42] Chernyak, N.; Gevorgyan, V. *Angew. Chem. Int. Ed.* **2010**, *49*, 2743.

[43] Sumalatha, Y.; Reddy, T. R.; Reddy, P. P.; Satyanarayana, B. *Arkivoc* **2009**, 315.

[44] Deiters, A.; Martin, S. F. *Chem. Rev.* **2004**, *104*, 2199.

[45] Choy, A.; Colbry, N.; Huber, C.; Pamment, M.; Van Duine, J. *Org. Proc. Res. Dev.* **2008**, *12*, 884.

[46] Hennessy, E. J.; Buchwald, S. L. *J. Am. Chem. Soc.* **2003**, *125*, 12084.

[47] Lyons, T. W.; Sanford, M. S. *Chem. Rev.* **2010**, *110*, 1147.

[48] Rech, J. C.; Yato, M.; Duckett, D.; Ember, B.; LoGrasso, P. V.; Bergman, R. G.; Ellman, J. A. *J. Am. Chem. Soc.* **2007**, *129*, 490.

[49] Graczyk, P. P.; Khan, A.; Bhatia, G. S.; Palmer, V.; Medland, D.; Numata, H.; Oinuma, H.; Catchick, J.; Dunne, A.; Ellis, M.; Smales, C.; Whitfield, J.; Neame, S. J.; Shah, B.; Wilton, D.; Morgan, L.; Patel, T.; Chung, R.; Desmond, H.; Staddon, J. M.; Sato, N.; Inoue, A. *Bioorg. Med. Chem. Lett.* **2005**, *15*, 4666.

[50] Kuethe, J. T.; Wong, A.; Qu, C. X.; Smitrovich, J.; Davies, I. W.; Hughes, D. L. *J. Org. Chem.* **2005**, *70*, 2555.

[51] Davies, I. W.; Smitrovich, J. H.; Sidler, R.; Qu, C. X.; Gresham, V.; Bazaral, C. *Tetrahedron* **2005**, *61*, 6425.

[52] Cerecetto, H.; Gerpe, A.; Gonzalez, M.; Aran, V. J.; de Ocariz, C. O. *Mini-Rev. Med. Chem.* **2005**, *5*, 869.

[53] Schmidt, A.; Beutler, A.; Snovvdovych, B. *Eur. J. Org. Chem.* **2008**, 4073.

[54] Stokes, B. J.; Vogel, C. V.; Urnezis, L. K.; Pan, M.; Driver, T. G. *Org. Lett.* **2010**, *12*, 2884.

7

OXIDATIVE CATALYSIS

LAMONT TERRELL

7.1   INTRODUCTION

Oxidation reactions are a powerful and often used synthetic transformation in organic synthesis. Despite their versatile synthetic utility, in the pharmaceutical industry on a preparative scale oxidative reactions encompass only 3–5% of chemical reactions performed [1]. The lack of use in the pharmaceutical industry on a large scale is likely attributed to several factors. The first drawback concerns synthetic efficiency (i.e., step count). With the development of a manufacturing process, there is a strong drive to decrease the number of chemical transformations in order to reduce the cost of goods. The movement up of an oxidation state is often viewed as a "wasteful" synthetic step and more than often the use of protecting group chemistry is also needed. Consequently, routes are often developed that introduce functionality in the correct oxidation state without the use of protecting groups. Probably, the greatest

Applications of Transition Metal Catalysis in Drug Discovery and Development: An Industrial Perspective, First Edition. Edited by Matthew L. Crawley and Barry M. Trost.
© 2012 John Wiley & Sons, Inc. Published 2012 by John Wiley & Sons, Inc.

reservation stands from a safety issue since many of the oxidation reactions involve highly reactive intermediates or are environmentally unfriendly. The apparent safety concern can be seen in comparing the synthetic toolbox between a discovery chemist and a process chemist. For example, discovery chemists have at their disposal oxidation reagents that are known to be incompatible on a process scale due to cost and/ or safety concerns. One such reagent is the oxidant Dess–Martin periodinane [2]. Although the reagent is a powerful and selective oxidant of alcohols to aldehydes and ketones, on a large scale, the reagent preparation is potentially explosive and costly. If the use of unfavorable oxidation protocols is required to synthesize the desired target molecule, a significant amount of safety data and supporting calorimetric studies are needed to understand the profile of the reaction prior to using the oxidation on scale.

Despite the many challenges of oxidation reactions, they are routinely run in pilot plants and production facilities to make the required active pharmaceutical ingredient (API). In recent years, transition metal-mediated oxidative catalysis has grown steadily as a subclass of oxidation reactions in the pharmaceutical industry. The use of metal catalysis allows one to potentially control the high oxidation potential, reduce unwanted by-products and waste, and introduce chirality stereoselectively. This chapter will illustrate and highlight only examples of transition metal-catalyzed chemical transformations ran in pilot plants or developed for manufacturing. Many of the synthetic examples were found in the journal *Organic Process Research & Development*. The *Mini-Medicinal Chemistry Review* journal is an excellent source to readily identify new drugs marketed yearly and the synthetic route likely used. For a general overview of large-scale oxidations in the pharmaceutical industry, one should read the review by Brown Ripin and colleagues [3].

7.2 CARBON–OXYGEN BOND OXIDATION

Oxidation of alcohols to aldehydes, ketones, or carboxylic acids is one of the most often utilized reactions in organic chemistry and thus a plethora of reagents have been developed for these transformations, such as Moffatt–Swern oxidation, Jones oxidation, Dess–Martin oxidation, PCC, and PDC. Besides the aforementioned stoichiometric oxidant reagents, catalytic protocols that utilize a "safe" primary oxidant have also been developed, for example, TEMPO. Although there are many examples in the pharmaceutical industry on the kilogram scale that utilize both the Moffatt–Swern and TEMPO protocols [3], the use of transition metal-catalyzed oxidation reactions is fewer but is gaining popularity in recent years with the demand for "green" chemistry, selectivity, and waste reduction. One such metal with increasing use in oxidation chemistry is ruthenium with the advent of the tetra-propylammonium perruthenate (TPAP) protocol [4] and *in situ* methods to generate RuO_4 catalytically.

TPAP is a nonvolatile, air-stable solid and typically used with N-methylmorpholine N-oxide (NMO) as the cooxidant. The oxidation operates under neutral conditions at room temperature with only 5 mol% of TPAP that normally gives 250 turnovers. The oxidation is mild enough such that it shows excellent compatibility with highly sensitive acid- and base-labile functionality. As illustrated in Figure 7.1, TPAP shuttles between Ru(VII) and Ru(IV) with NMO serving as the reoxidant. With TPAP being an overall three-electron oxidant, 3 mol of an alcohol reacts with 2 mol of the oxidant. Mechanistic studies have also shown that primary alcohols react faster than secondary alcohols, which suggest that steric factors affect the rate of reaction.

In the process development of rosuvastatin calcium (Crestor®), a prescribed agent for lowering high cholesterol and to slow the buildup of plaque in the arteries, researchers at Shionogi Research Laboratories [5] utilized the TPAP oxidation to prepare a key aldehyde intermediate as illustrated in Scheme 7.1. Crestor® is part of a class of compounds that inhibit 3-hydroxy-3-methylglutaryl CoA reductase (HMG-CoA reductase), a major rate-limiting enzyme in cholesterol biosynthesis. The synthesis of compound **1** began with benzaldehyde **2** and, in six steps, core ester **3** was prepared. The ester was reduced with DIBAL giving the primary alcohol and subsequent oxidation with TPAP gave the required aldehyde intermediate **4** in 71%

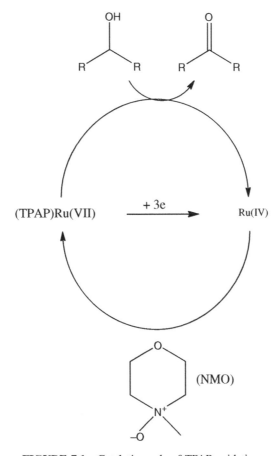

FIGURE 7.1   Catalytic cycle of TPAP oxidation.

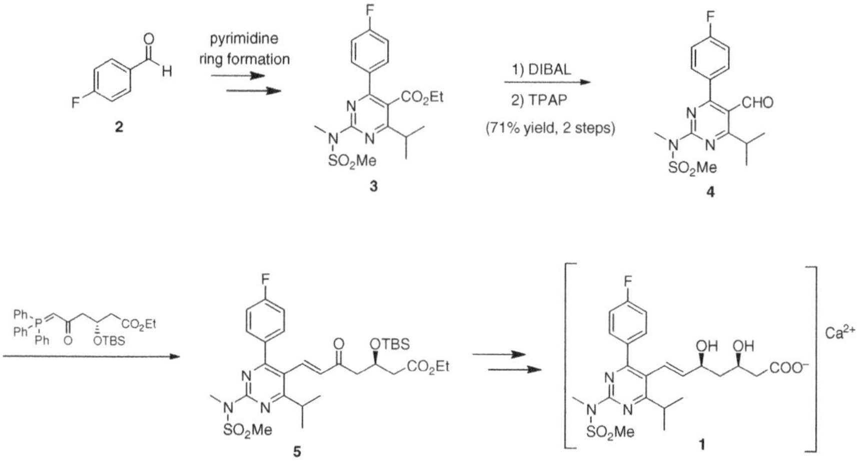

SCHEME 7.1   TPAP oxidation of a key intermediate in Crestor®.

yield over two steps. With aldehyde **4** in hand, the side chain was installed with a Wittig olefination procedure giving optically pure intermediate **5**. Four additional steps of protecting group manipulation and a ketone reduction gave compound **1**. The process chemists noted that care must be taken when using TPAP on large scale since it is necessary to moderate the reaction by cooling and by slow, portionwise addition of TPAP.

Although ruthenium tetroxide is less selective than TPAP in the oxidation of substrates, *in situ* methods to generate RuO$_4$ have also found utility in the pharmaceutical industry on large scale. In particular for secondary alcohols that do not include any incompatible functional group such as alkenes, RuO$_4$ is an excellent oxidant choice with the advent of the sodium bromate protocol [6]. The use of NaBrO$_3$ as the cooxidant offers several advantages over other cooxidants (i.e., cost and substoichiometric quantities are used). For this catalytic reaction, the initial ruthenium source is either RuO$_2$ or RuCl$_3$ and the catalytic cycle is illustrated in Figure 7.2. The Ru(III) or Ru(IV) starting material (**A**) is oxidized in the aqueous phase by NaBrO$_3$ to RuO$_4$ (**B**) that then partitions between the two phases, favoring strongly the organic phase. The water-insoluble secondary alcohol in the organic phase is oxidized by RuO$_4$, which is in return reduced to an insoluble Ru(IV) (**C**) species, probably a partially hydrated form of RuO$_2$. The Ru(IV) (**C**) species partitions back to aqueous phase where it is reoxidized to RuO$_4$ (**B**) by NaBrO$_3$ and completes the catalytic cycle. The above biphasic nature of the reaction protocol was an early drawback due to slow rates of oxidations. However, with the use of acetonitrile, an anticomplexing solvent, this limitation was overcome.

The following two examples will highlight the utility of the RuO$_4$ oxidation for large-scale preparation of an API. In the synthesis of the potent 5-lipoxygenase inhibitor **6** (Fig. 7.3), researchers at Merck needed a practical and enantioselective synthesis of the key 1,6-anhydro-2,4-dideoxyhexopyranose (**9**). Despite the widespread usage of dideoxyhexopyranoses as chiral synthons in the synthesis of

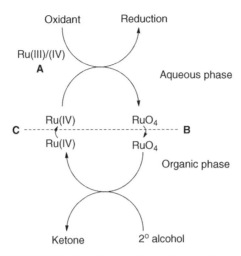

FIGURE 7.2   Catalytic cycle for RuO_4 oxidation.

biologically active compounds, few efficient methods existed for their large-scale preparation. The first synthesis of ketone **9** was reported by Cerny and coworkers as illustrated in Scheme 7.2 [7]. Tosylation of levoglucosan **7** at C-2 and C-4 gave intermediate **10**. Oxidation of the alcohol at C-3 gave ketone **11** and reduction of the tosylate groups with Raney nickel afforded ketone **9** in low overall yield ($\sim$24%). Prior to the significant improvements by the Merck researchers, several modifications to the Cerny procedure had been reported but all were insufficient for multikilogram scale. Most notable of the improvements made by the Merck researchers was the use of $NaBrO_3$ with $RuCl_3$ for the alcohol oxidation at C-3 and the reductive cleavage of the α,α'-substituted dideoxygenation of a ketone by a metal or metal salt other than Raney Ni. The optimized tosylation conditions of levoglucosan **7** provided ditosylate

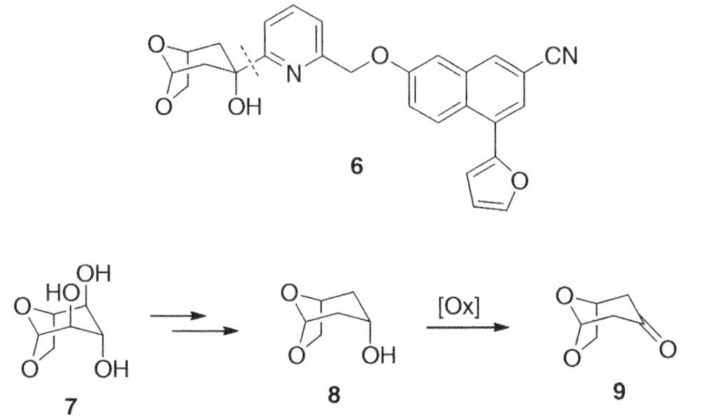

FIGURE 7.3   Retrosynthetic analysis of the 5-lipoxygenase inhibitor **6**.

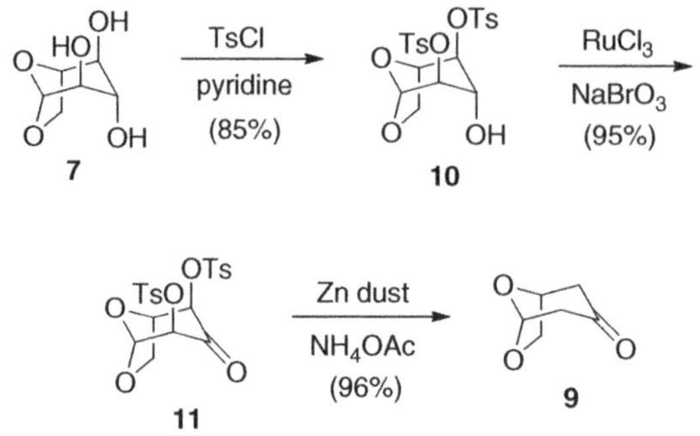

SCHEME 7.2 Ruthenium oxidation to generate ketone **9**.

10 in excellent yield (85%) with the major by-product being the tritosylate (about 14%) that was easily removed by crystallization in the final step. The next step in the synthesis was the oxidation of **10** to ketone **11**. The initial oxidation with RuCl$_3$/ NaOCl in acetonitrile and acetic acid gave the ketone in only a 15% yield, which was attributed to incomplete conversion. With a large excess use of NaOCl or NaIO$_4$, complete conversion was achievable but at the expense of epimerization at C-2 and C-4. Optimization of the reaction resulted in the substitution of NaBrO$_3$ as the oxidant. With this change, oxidation of **10** to ketone **11** was accomplished in a solvent system of CH$_3$CN/AcOH/H$_2$O using NaBrO$_3$ (65 mol%) in the presence of 1 mol% RuCl$_3$ in 95% yield. In comparison to the original chromic anhydride oxidation, the RuCl$_3$/NaBrO$_3$ oxidation provided ketone **11** in at least 40% higher yield and without the need for further purification. The last step is the reduction of the ditosylates and was achieved in 93% yield employing Zn dust and solid NH$_4$OAc as the proton source. The practicality of the above three-step chromatography-free synthesis of 1,6-anhydro-2,4-dideoxy-β-D-glycero-hexopyranos-3-ulose (**9**) was demonstrated on a 23 g product scale in 75% overall yield.

Another example highlighting the RuCl$_3$/NaBrO$_3$ oxidation is illustrated in the kilogram synthesis of the pyrazinone acetic acid core **20** of an orally efficacious thrombin inhibitor. The retrosynthetic analysis of the thrombin inhibitor **12** is shown in Figure 7.4. The initial route (R$_1$/R$_2$ = Cl, R$_3$ = Bn) was deemed impractical for larger scale syntheses. This was due to the use of the highly toxic trimethylsilyl cyanide in the initial step, a Strecker reaction, and dechlorination of the aromatic core that was sensitive to the reductive conditions of the hydrogenolysis. An attractive alternative based on the cyclization of an unsymmetrically substituted oxalic diamide to give the pyrazinone core was reported in the literature [8]. With this approach, the scalable synthesis of **20b** (Fig. 7.4, R$_1$ = H, R$_2$ = OH, R$_3$ = OEt) required the acetal of 1-amino-2-propanone. Although a few published methods are known for this intermediate [8a,9], the syntheses were not amenable for scale-up and thus an alternative strategy was explored using the known monoamide **17** [10].

The synthesis of the pyrazinone core (Scheme 7.3) began with the acylation of ethyl glycinate free base with oxalyl chloride to give the desired monoamide **17** in excellent yield (93% assay). The incorporation of oxalyl chloride was a modification of the

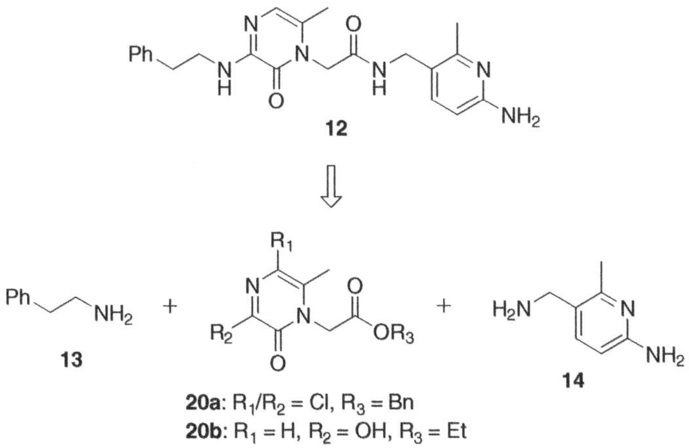

FIGURE 7.4 Retrosynthetic analysis of thrombin inhibitor **12**.

original synthesis that employed diethyl oxalate but resulted in a cleaner reaction profile and thus avoided either a distillation at high vacuum and high temperature or column chromatography to remove the excess diethyl oxalate and diamide impurity, respectively. Treatment of crude monoamide **17** with 1-aminopropanol provided alcohol **18** in an 86% isolated yield (80% over two steps) after crystallization and filtration of the reaction mixture. For the oxidation of **18** to **19**, several conditions were explored with only the Jones, NBS/aq. DME [11], and RuCl$_3$ procedures giving the desired ketone in appreciable yield and sufficient purity. The Jones oxidation proceeded readily to give ketone **19** but the necessity for large-scale synthesis of the ketone prohibited the use of stoichiometric heavy metal oxidants. Oxidation with NBS/aq. DME was clean but sluggish and the removal of succinimide appeared to be problematic due to the high solubility of the ketone in water. Therefore, efforts were focused on the ruthenium-based oxidation. Optimization of the temperature, oxidant (NaOCl, NaIO$_4$, NaBrO$_3$), stoichiometry of oxidant, and the solvent led to a synthetic protocol of a 1:1 mixture of acetonitrile/water with a slow addition of 0.5 equiv of aqueous sodium bromate keeping the temperature below 40°C.[1] In addition, in efforts to lower the residual ruthenium metal,[2] a series of variables such as change of pH (oxidation of RuO$_4$ to the more water-soluble RuO$_4{}^{2-}$), addition of complexing ligands (NH$_3$, Bu$_3$P), absorbents (silica gel, neutral, acidic, and basic alumina), addition of reductants (Na$_2$S$_2$O$_3$ and isopropanol to give less organic soluble low oxidation state ruthenium), and azeotropic drying of organic extracts (to precipitate low-valent ruthenium) were investigated. This study resulted in the use of NaHCO$_3$ (pH 8) during workup, azeotropic drying of the acetonitrile solution of the ketone to < 3% water, and treatment with neutral alumina to remove most of the ruthenium. The optimized conditions resulted in the isolation of

---

[1] Although an internal temperature greater than 30°C appeared necessary for complete conversion, temperatures above 40°C during oxidant addition or the use of more than 0.5 equiv of NaBrO$_3$ gave lower yields, presumably due to product decomposition.

[2] For bulk drug production, no more than 10 ppm of any heavy metal may be present in the final product.

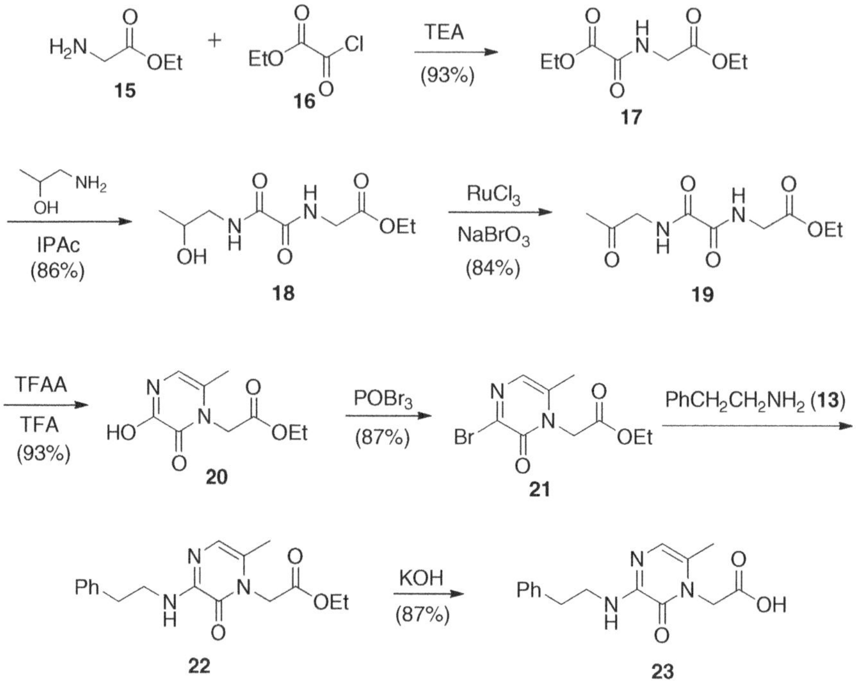

SCHEME 7.3 Synthesis of the advanced left-hand acid **23**.

ketone **19** in 84% isolated yield (after crystallization) containing less than 30 ppm residual Ru, which was further reduced to <1 ppm with a subsequent crystallization. Cyclization to the pyrazinone **20** was easily achieved with treatment of ketone **19** with TFA/TFAA and heating for 24 h. Concentration of the reaction mixture and crystallization gave pyrazinone **20** in excellent yield (93%). Activation of the pyrazinone core for displacement with phenethylamine was found to work better as imidate bromide **21** (formed by treatment of **20** with POBr₃ in 87% yield). Subsequent displacement with phenethylamine **13** resulted in the formation of ester **22** and hydrolysis of the ester with aqueous KOH in MeOH provided the target pyrazinoic acid **23** cleanly in 87% yield. The robustness of the RuCl₃ oxidation and the scalability of the above highlighted synthetic route were demonstrated on a 4.3 kg product scale.

Tungsten has seen increasing use in oxidation chemistry over time because of its potential of being "green." Early oxidation protocols with tungsten and hydrogen peroxide were limited to strongly acidic conditions and required a commercially unavailable phase transfer catalyst [12]. Recently, researchers at Shionogi have developed an oxidation procedure to overcome the above drawbacks and shown its applicability for large-scale manufacturing [13]. In the optimization of the reaction procedure, the authors showed that the acidity of the aqueous hydrogen peroxide solution and catalytic Na₂WO₄·2H₂O was 5.4. The acidic nature of the mixture limited the tungsten reagent solubility in organic solvents and greatly affected the oxidation of 2-octanol to 2-octanone in their model system. In optimizing the reaction conditions, it was found that strong donor solvents were preferred and *N,N*-dimethylacetamide (DMA) was the solvent of choice.

Secondly, in their investigation, it was found that the addition of additives and the pH of the reaction mixture had a dramatic impact on the reaction. Under basic conditions with NaOH, no oxidation occurred due to the unproductive decomposition of H_2O_2. Adjustment of the pH to 6.5 with NaOH resulted in a moderate reaction rate. On the other hand, the addition of a phosphate salt accelerated the oxidation and adjustment to pH 6.5 by the addition of $Na_2HPO_4 \cdot 12H_2O$ was found optimal.

The mechanism for this "neutral" tungsten peroxide oxidation is postulated to be totally different from conventional tungsten oxidations since the catalyst isolated and determined by single-crystal X-ray analysis was disodium tetraperoxotungstate dihydrate $Na_2[W(O_2)_4] \cdot 2H_2O$. Catalysts for conventional oxidations are diperoxo species and have an exchangeable water as a ligand. The exchange of a water molecule with a molecule of the alcohol is a key step in the oxidation. The $Na_2[W(O_2)_4]$ complex does not have a water molecule as a ligand and thus the water/alcohol exchange is inoperable. It is known that aqueous solutions of $Na_2[W(O_2)_4]$ are heat sensitive and decompose even at 25°C. However, in the course of the reaction, the thermal deactivation was not observed. Further investigation by DSC showed that the exothermic onset temperature of $Na_2[W(O_2)_4]$ in DMA was over 105°C. It is hypothesized that DMA solvent coordinates to tungsten and this coordination stabilizes the catalyst. Accordingly, the following reaction mechanism is proposed as illustrated in Figure 7.5, Na_2WO_4 (**24**) and H_2O_2 give $Na_2[W(O_2)_4]$ (**25**).

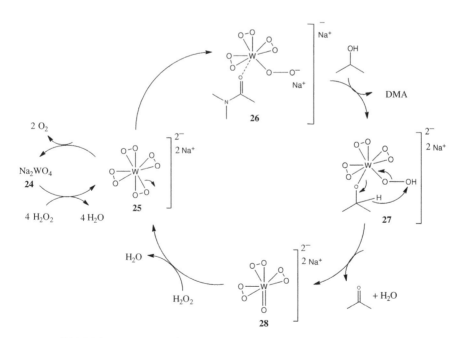

FIGURE 7.5 Proposed mechanism for tetraperoxotungstate oxidation.

Coordination of DMA stabilizes catalyst **26** and slows down the thermal cleavage of the W–O bond. Without the coordination of DMA, complex **25** readily decomposes to 2 mol of oxygen and 1 mol of Na_2WO_4. Next, the ligand exchange of DMA/alcohol occurs to give complex **27**. The alkoxide ligand in **27** is dehydrogenated by hydroperoxo ligand to give the desired ketone and complex **28**. The addition of hydrogen peroxide and loss of water to complex **28** regenerates catalyst **25** and completes the catalytic cycle.

The practicability of this novel Na_2WO_4–H_2O_2 oxidation on pilot plant scale is nicely illustrated in the synthesis of 12α-hydroxy-3-oxooleano-28,13-lactone (**31**), a key intermediate of endothelin A receptor antagonist S-0139 (**32**) [14]. The synthesis began with oleanolic acid **29** (Scheme 7.4). In the discovery route, oxidation of the secondary alcohol to ketone **30** was carried out using Jones oxidation. However, due to the toxicity of chromium(VI) and the large quantity of metal waste, Jones oxidation was not suitable for bulk chemical development. Alternative oxidations were screened, such as the synthetically preferred TEMPO and Moffatt–Swern oxidations, but the desired ketone was not obtained cleanly. Oxidations using hypochlorites such as NaOCl (TEMPO) gave the chlorinated by-product at 12-en and the Moffatt–Swern oxidation gave the methyl thioether at the C-3 position as a by-product. This critical and serious oxidation issue was overcome after optimization of the catalytic Na_2WO_4–H_2O_2 system using the design of experiment (DoE) approach. Factors considered included the reaction temperature, the concentration, the amount of H_2O_2 and $Na_2WO_4 \cdot 2H_2O$, and the pH of the phosphate buffer. The optimized conditions (90°C, DMA, 5 mol% Na_2WO_4, 1.2 equiv H_2O_2, pH 6.8) allowed for a 98% conversion to the ketone **30**. Ketone **30** was not isolated but instead ozonized to give lactone **31** in 77% yield over the two steps. With all the necessary

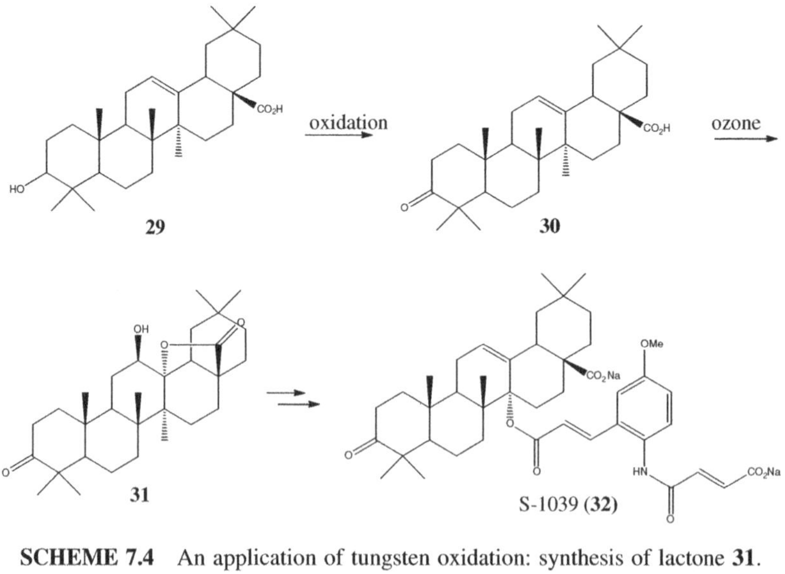

SCHEME 7.4 An application of tungsten oxidation: synthesis of lactone **31**.

safety evaluation conducted on the oxidation and subsequent ozonolysis reaction, the scale-up process was achieved on a 22 kg scale.

A challenging conversion in organic synthesis is the direct conversion of a primary alcohol to the corresponding carboxylic acid in the presence of other functional groups. Two requirements are that the oxidizing reagent show chemoselectivity, that is, complete conversion and functional group compatibility. In the laboratory, this transformation is frequently carried out using Jones oxidation [15]. Although the Jones oxidation has been utilized since the very beginning of organic chemistry, one major drawback is that the metallic by-products are toxic and their presence complicates reaction workup. With the toxicity associated with chromium oxidations, oxidations using the less toxic transition metal ruthenium ($RuCl_3/H_5IO_6$, RuO_4) [16] have gained popularity as a reagent of choice. Alternatively, non-transition metal oxidations such as TEMPO/NaOCl or a two-step protocol of oxidation to the aldehyde and subsequent oxidation to the carboxylic acid can be used to achieve the desired transformation. However, all of these methods have some limitations and disadvantages that limit their use on a process scale. With the apparent void in the chemistry toolbox for this transformation, process chemists at Merck developed a chromium trioxide-catalyzed oxidation utilizing periodic acid (H_5IO_6) as the stoichiometric oxidant [17]. In the optimization of the reaction procedure, it was found that the catalyst loading of CrO_3 could be reduced to 1.1 mol% in the presence of 2.5 equiv of H_5IO_6 and the presence or absence of water had a profound effect on the reaction rate. It was found that under anhydrous conditions, the reaction was complete in less than 15 min at room temperature, and the presence of small amounts of water (0.75 vol%) in acetonitrile attenuated the oxidation strength of the system and provided cleaner reactions. This oxidation protocol is also applicable for the oxidation of secondary alcohols to the corresponding ketone and mild enough for the oxidation of chiral alcohols without racemization at the adjacent chiral center.

Researchers at Merck utilized this novel CrO_3 oxidation in the synthesis of their cathepsin K (Cat K) inhibitors for the treatment of osteoporosis, a disease characterized by excessive bone loss causing skeletal fragility and an increased risk of fracture. The medicinal route for the synthesis of odanacatib (39), a highly selective inhibitor and currently in phase III, is illustrated in Scheme 7.5 [18]. The synthesis of the required primary alcohol began with commercially available aspartic acid derivative 33. Reduction of the acid, tosic anhydride-mediated cyclic carbamate formation, methyl Grignard addition to the benzyl ester, and fluorination of the resulting alcohol gives fluoride 34. Hydrolysis of the carbamate proceeded in quantitative yield and silylation of the amino alcohol facilitated isolation and provided stereochemical control in the subsequent imine addition step. Condensation of the primary amine with trifluoroacetaldehyde hemiacetal provided imine 35. Treatment of imine 35 with bromophenyllithium gave the trifluoroethylamine in a 10:1 ratio of (S,S)- to (R,S)-diastereomers. Desilylation with TBAF gave the required primary alcohol 36 in nine linear steps. Catalytic CrO_3 oxidation of 36 in the presence of periodic acid under anhydrous conditions gave carboxylic acid 37 in 60% yield with >99% de after recrystallization as the dicyclohexylamine salt. HATU-mediated amide formation with 1-amino-1-cyanocyclopropane and subsequent

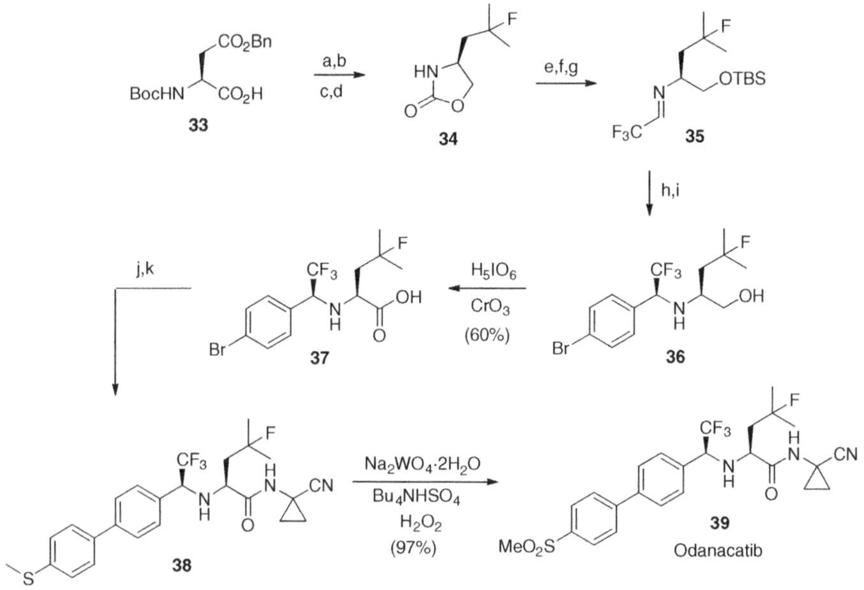

Reagents and conditions: (a) ClCOO*i*Bu, NMM, NaBH$_4$, DME, 85%; (b) Ts$_2$O, pyr, DCE, 83%;
(c) MeMgBr, PhCH$_3$/THF, 85%; (d) DAST, DCM, 60%; (e) Ba(OH)$_2$, EtOH/H$_2$O, 100%; (f) TBSCl, Et$_3$N;
(g) CF$_3$C(OH)OEt, PhH, 88% (two steps); (h) BrPhLi, THF; (i) TBAF, THF, 75% (two steps);
(j) 1-amino-1-cyanocyclopropane, DIEA, HATU, DMF, 80%; (k) MeSPhB(OH)$_2$, PdCl$_2$ dppf, Na$_2$CO$_3$, 70%

SCHEME 7.5   Synthesis of the cathepsin K inhibitor odanacatib **39**.

Suzuki coupling provided sulfide **38**. The last synthetic step in the synthesis of odanacatib is the oxidation of sulfide **38** to sulfone **39** using tungsten in near-quantitative yield and further highlights the utility of transition metals in the pharmaceutical industry (oxidation of sulfur).

The medicinal chemistry route elegantly illustrates the power of transition metal-mediated catalysis in the synthesis of biologically active agents. The synthesis comprises 13 steps with an overall yield of 8%. However, the above route was found not to be suitable for synthesis on a multikilogram scale and thus reminds us of some drawbacks (i.e., chromatography, changing oxidation state, length of synthesis) associated with scaling up a synthetic sequence. As mentioned earlier, the change of oxidation states is often seen as an unwarranted step and adds to the cost of goods for the API. In the process route, the Merck researchers eliminated the need for the above two oxidation steps and synthesized odanacatib in six steps in 61% overall yield [19].

7.3   ALLYLIC OXIDATION (CARBON–HYDROGEN BOND)

As often is the case, the development of many organic reactions is in direct response for the need of a particular chemical transformation. In steroidal chemistry, one long existing problem was the allylic C–H oxidation at C-7 of a Δ-5-steroid to the

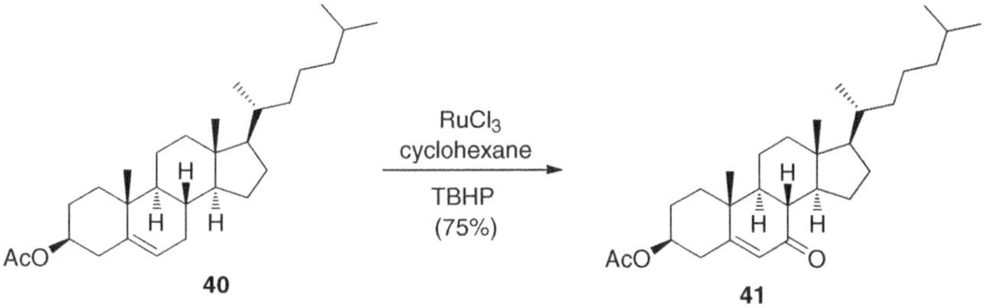

SCHEME 7.6   Ruthenium-catalyzed oxidation of steroidal alkene **40** to enone **41**.

corresponding unsaturated 7-ketone [20]. Several methods for this type of chemical transformation are reported in the literature [21] with chromium(VI)-mediated oxidations [22] being the most prevalent. As mentioned earlier, on process scale these oxidations are often avoided due to the toxicity associated with chromium. With the advent of the ruthenium-catalyzed allylic oxidation procedure, this age-old problem was solved. In the process development of Merck's type I 5α-reductase inhibitor, MK-386 [23], researchers developed a practical procedure [24] as illustrated in Scheme 7.6 for the oxidation of alkene **40** to enone **41** in 75% yield (kilogram scale) based on Murahashi's ruthenium-catalyzed oxidation of alkanes with *tert*-butyl hydroperoxide [25]. In the development of the alkene protocol using 0.7 mol% $RuCl_3$, the choice of solvent was found to have a dramatic impact on the yield of the reaction as shown in Table 7.1. A decrease in solvent polarity corresponded to an increase in product yield (entry 1 versus entry 4). However, due to solubility constraints of some steroidal systems in these nonpolar solvents, dichloroethane or chlorobenzene was the preferred solvent and provided the desired product with no loss of yield. The operational simplicity of this ruthenium-catalyzed oxidation has been proven through the production of kilogram quantities of the desired enone **41** and similar analogs.

A group of researchers at Magainin Pharmaceuticals have also demonstrated the practicality of the above Ru–TBHP oxidation in their process synthesis of squalamine (**45**), a novel polyaminosteroidal sulfate with both antibacterial and tumor-inhibiting properties. As illustrated in Scheme 7.7, the synthesis of **45** began with commercially available 20-hydroxy-4-pregnen-3-one (**42**) and in a few steps the key intermediate

TABLE 7.1   Effect of Solvent on the Allylic Oxidation of 40

Entry	Solvent	Yield (%)
1	Cyclohexane	75
2	Heptane	75
3	Toluene	63
4	*t*-BuOH	24
5	MEK	48
6	Ethyl acetate	68
7	Dichloroethane	75

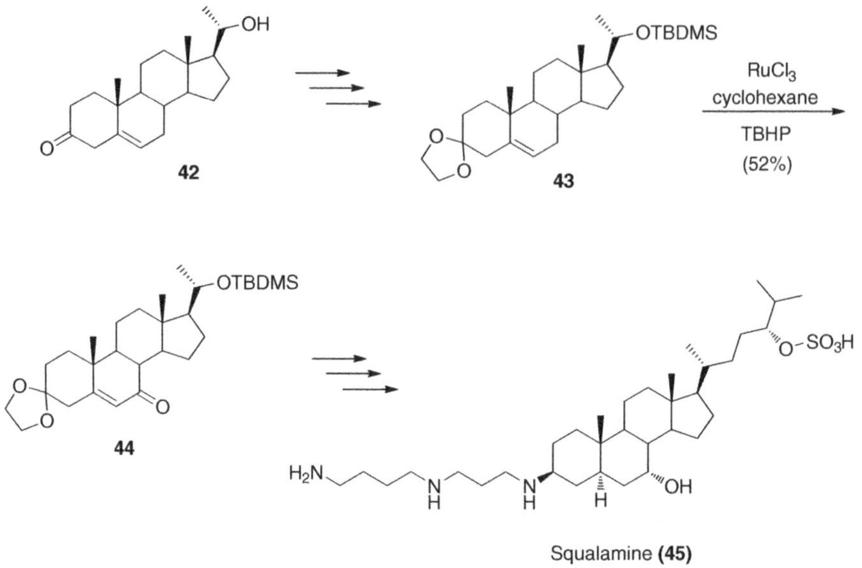

Squalamine (45)

SCHEME 7.7   Ruthenium-catalyzed oxidation of steroidal alkene **43** to enone **44**.

Δ-5-ketal **43** was prepared. Optimization [26] of the original RuCl$_3$–TBHP oxidation procedure by Miller et al. resulted in the isolation of enone **44** (52% yield) as a white solid on a 17 kg pilot plant run. The moderate yield in the pilot plant compared to the laboratory scale (85%) is probably connected with the inefficient mixing of the heterogeneous reaction mixture. Importantly, the mildness of the reaction conditions is demonstrated since the structure of the Δ-5-steroid **43** contains a fragile ketal and *tert*-butyldimethylsilyl ether moiety and both remarkably survive the oxidation.

7.4   CARBON–CARBON DOUBLE BOND OXIDATION

The oxidation of carbon–carbon double bonds is a widely used chemical transformation by organic chemists since the resulting functionalities (e.g., epoxides, vicinal diols, carbonyl compounds, 1,2-amino alcohols) are highly versatile synthetic intermediates and/or themselves found in the API. Furthermore, the asymmetric modifications of the epoxidation, dihydroxylation, and aminohydroxylation reactions by Sharpless and others pushed these oxidation reactions to the forefront of organic chemistry. The asymmetric variants of the aforementioned reactions enable the organic chemist to synthesize target compounds with both high yields and, more importantly, high asymmetric induction.

7.4.1   Epoxidation

Although ideally an asymmetric epoxidation is preferred, nonchiral variants have often been utilized in the process development of API. For instance, researchers at

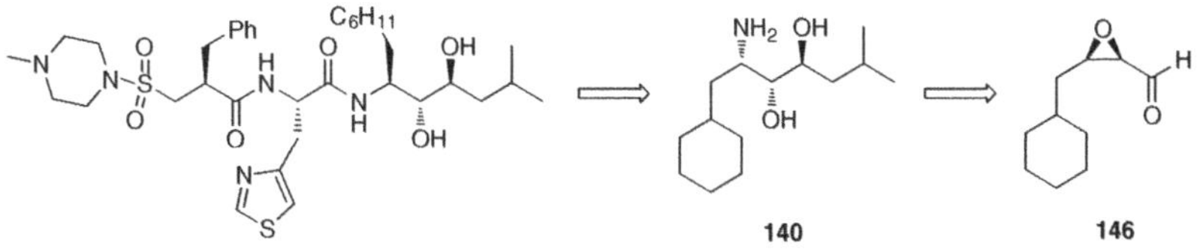

FIGURE 7.6 Retrosynthetic analysis of renin inhibitor **139**.

Abbott Laboratories required an efficient and stereoselective synthesis of the dihydroxyethylene dipeptide isostere subunit **140** of the orally active renin inhibitor **139** (Fig. 7.6) [27]. Several routes were published [28] for the preparation of subunit **140** but none were suitable for their needs due to cost and availability of starting goods. In their route development, the retrosynthetic analysis of fragment **140** led to the construction of α,β-epoxy aldehyde **146** as a key intermediate, which is derived from the epoxidation of an allylic alcohol.

The synthesis of fragment **140** (Scheme 7.8) began with the TEMPO-mediated oxidation of alcohol **141** to corresponding aldehyde that was smoothly converted to propargylic alcohol **143** using a modification of the Corey–Fuchs reaction [29]. Reduction of the alkyne with Pd/BaSO$_4$ in the presence of hydrogen provided Z-olefin **144** in excellent yield (>97%) with <5% unwanted E-olefin. At this stage, both racemic and enantioselective epoxidations were examined. Treatment of **144** with magnesium monoperoxyphthalate (MMPP) [30] provided racemic epoxide **145** in 88% yield. While, on the other hand, the Sharpless asymmetric epoxidation (SAE) gave the (2S,3R)-**145** isomer in only a modest 59% yield (unoptimized) with low enantioselectivity. Oxidation of racemic alcohol **145** with modified Parikh–Doering

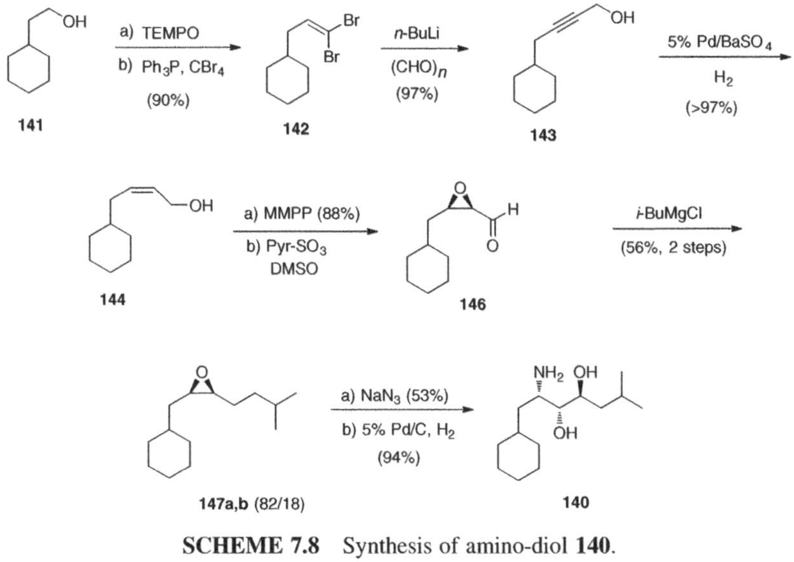

SCHEME 7.8 Synthesis of amino-diol **140**.

conditions [31] provided the sensitive α,β-epoxy aldehyde **146** that was used immediately upon workup. Treatment of **146** with isobutylmagnesium chloride led to a 82/18 mixture of diastereomeric epoxy alcohols **147a,b** in a moderate 55% yield over two steps. The stereochemistry was assigned based on a Felkin–Ahn controlled addition of the isobutyl appendage and subsequently confirmed by conversion of **147a** to **140**. As literature precedent suggested [32], the regioselective addition of sodium azide to the epoxy alcohols occurred uneventful and yielded two diastereomeric azido diols **148a,b** in 65% yield. The desired (2*S*,3*R*,4*S*) epimer **148a** was obtained diastereomerically pure (>95%) by crystallization from hot heptane. The synthesis of the key intermediate was completed by catalytic reduction of the azide to amino-diol **140** in excellent yield. For the nonracemic route, the disappointing enantiomeric excess of only 70% for fragment **140** via the Sharpless asymmetric epoxidation was not completely unexpected since (Z)-allylic alcohols are not ideal substrates. Fortunately, formation of the tartrate salt of either racemic or enantiomerically enriched **140** followed by crystallization provides enantiomeric excess >99%. The overall route is reasonably short (10 steps, including resolution) and efficient (13% overall yield via Sharpless asymmetric epoxidation and 10% via resolution) and has been executed on multigram scale. Moreover, it makes no use of either chiral pool synthons or protecting groups and employs commercially available reagents and materials.

Another example to highlight the utility of an achiral epoxidation on process scale is in the preparation of the 5HT$_{1a}$ receptor agonist **151** by researchers at Lilly (Fig. 7.7). The synthetic route is based on the first reported synthesis of lysergic acid **150** by Kornfeld, Woodward, and colleagues from the Lilly Research Laboratories in 1954 [33]. A critical intermediate in the Kornfeld–Woodward synthesis was a compound that became affectionately known as Kornfeld's ketone **152**. Reduction of the ketone with NaBH$_4$ and subsequent dehydration provides crystalline olefin **153** in excellent yield (Scheme 7.9). The epoxidation of this olefin with peracids was known since the early lysergic acid synthesis days of Kornfeld, but the stereochemical outcome remained unknown for the four decades that followed the first disclosure. Upon reinvestigation of the epoxidation, it was determined that the epoxidation of olefin **153** was highly diastereoselective, affording primarily the *anti*-epoxides **155** with >96% de [34]. The rationale for this observation was based upon a torsional model combined with transition state modeling. The importance of this

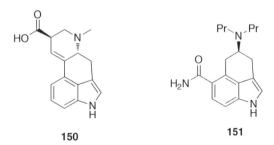

150   **151**

FIGURE 7.7   Lysergic acid **150** and 5HT$_{1a}$ receptor agonist **151**.

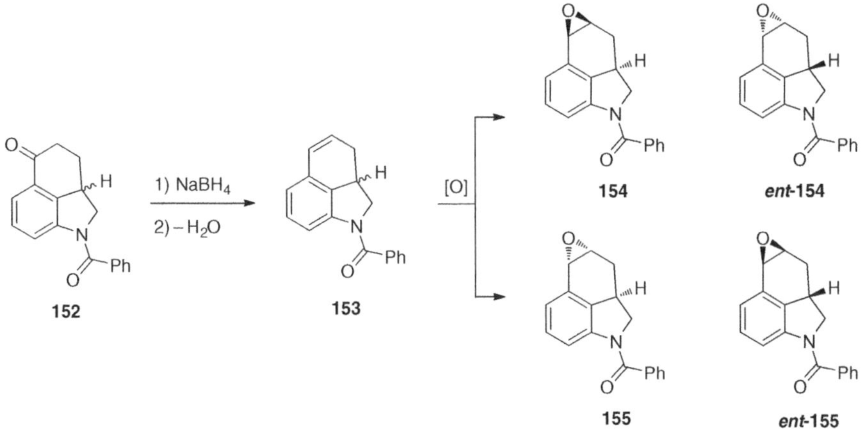

SCHEME 7.9 Epoxidation of olefin **153**.

insight into the chemical process suggested that the epoxidation stereochemistry should be reagent independent and allowed the flexibility to choose the optimal oxidant with confidence that the reaction could be reliably scaled-up.

Initially, *m*-CPBA was employed in the epoxidation studies with great success but due to the availability of relatively pure *m*-CPBA an alternative oxidant was needed. As illustrated in Scheme 7.10, MMPP was chosen and was initially met with the difficulties associated with this reagent, namely limited solubility in organic solvents. With the development of a 50% aqueous *n*-BuOH biphasic reaction medium, olefin **153** was readily oxidized to generate epoxide **155**. Contrary to the earlier work of Kornfeld when using trisubstituted epoxides, the epoxide ring opening of **155** with (*S*)-phenethylamine in *n*-BuOH at 110°C occurred at the more electrophilic benzylic position to afford a diastereomeric mixture of amino alcohols **156** and **157**, and upon cooling of the reaction the desired isomer **156** crystallizes out in a modest 31% yield. To circumvent the disappointment of the regioselectivity in the epoxide ring opening, transposition of the nitrogen around the ring through the intermediacy of an aziridine was considered. As such, subjection of amino alcohol **156** to methanesulfonyl chloride and TEA provided aziridine **158** and avoided the necessary chromatography to remove impurities if the alternative Mitsunobu reaction was used. The use of mesyl chloride for this transformation is remarkable since *N*-mesylation was the expected primary course of reaction; however, it was not possible to isolate any sulfonamide derivative, nor mesylate from this reaction. After optimization, a tandem benzylic hydrogenolysis of the aziridine bond and the auxiliary bond was achieved by exposure to hydrogen in the presence of 10% palladium on carbon to generate amine **159** in 75% yield. Subjection of indoline **159** to iodine and periodic acid provided regioselectively the aryl iodide in 85% yield. Regioselective aromatic electrophilic *para*-substitution on an indoline moiety is well precedented [35] and the mildly directing and activating effects of the *p*-benzamide and *o*-alkyl moieties were essential for regiocontrol. Subsequent alkylation with excess iodopropane and

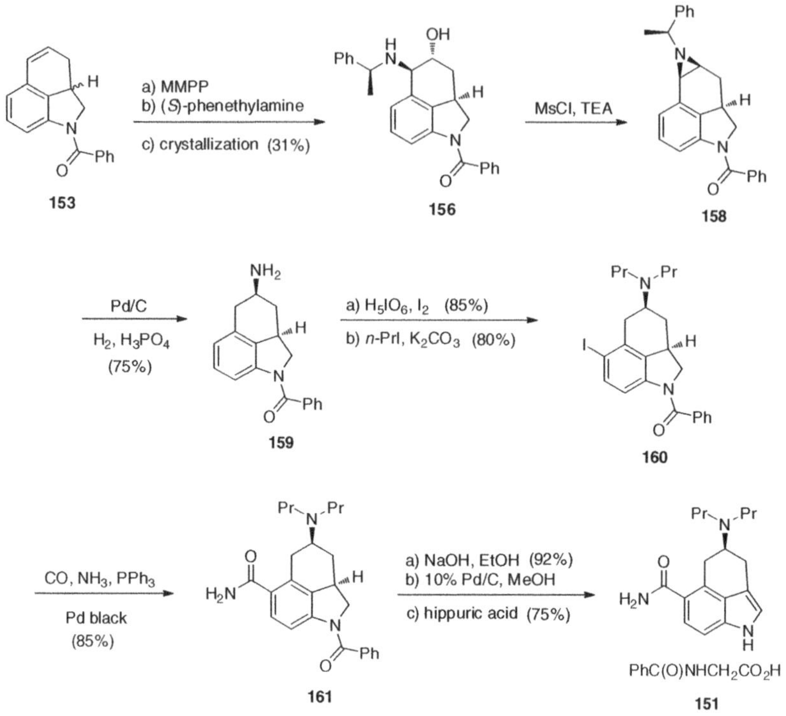

SCHEME 7.10   Synthesis of 5HT$_{1a}$ agonist **151**.

K$_2$CO$_3$ in acetonitrile at 80°C afforded tertiary amine **160** in 80% yield.[3] The primary amide of **161** was introduced by a novel carboxamidation reaction using catalytic palladium black and PPh$_3$ in the presence of carbon monoxide/ammonia in 80% yield. Exposure of **161** to KOH in EtOH resulted in cleavage of the benzamide in 92% yield and subsequent oxidation of the indoline core to the indole was achieved with palladium on carbon in MeOH. The 5HT$_{1a}$ agonist **151** was isolated as the hippurate salt that offered a convenient method for purification by recrystallization in 75% yield. In summary, the synthesis of agonist **151** was accomplished in nine steps in 9% overall yield from commercially available olefin **153** and was the most viable long-term route of choice.

Although achiral epoxidations are often utilized in the preparations of key intermediates in process synthetic routes as illustrated above, one major drawback is the maximum 50% yield for that synthetic step. As a consequence, several asymmetric variants have been developed and used in process routes. The development of Jacobsen–Katsuki epoxidation allowed for the incorporation of unfunctionalized olefins as substrates in asymmetric epoxidations. The epoxidation protocol was a significant achievement in organic synthesis since the main limitation (i.e., a

---

[3] Quaternization of the amine group, rather than incomplete alkylation, was the preferred option since the quaternary ammonium salt could be easily removed by aqueous extraction.

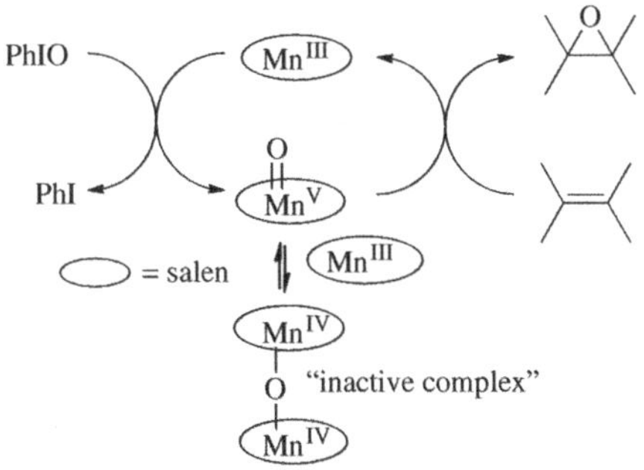

FIGURE 7.8 Catalytic cycle for Jacobsen–Katsuki epoxidation.

required pendant alcohol) of the Sharpless asymmetric epoxidation was overcome. The proposed mechanism based on the work of Kochi et al. for the Mn(salen)-catalyzed epoxidation is illustrated in Figure 7.8 [36]. The Mn(III) intermediate is oxidized to a Mn(V) oxide by the stoichiometric oxidant. Next, the oxygen transfer to the olefin gives rise to the epoxide and regenerates the Mn(III) species. Mechanistic studies have also shown a competing reversible dimer adduct of the active Mn(V) oxide intermediate and a Mn(III) species. The dimer adduct is an inactive epoxidation intermediate and with the addition of N-oxides the equilibrium is shifted in favor of the active Mn(V)=O complex and results in a rate acceleration due to the increased concentration of Mn(V)=O [37]. The higher yields and increased catalytic turnovers could also be attributed to the fact that the N-oxides sequester the Lewis acidic Mn(III) complexes from irreversible reactions with either the substrate or epoxide product. The oxygen transfer to the olefin is postulated to occur via a radical process and the direction of attack is explained by the model illustrated in Figure 7.9. Generally, a side-on approach parallel to the salen ligand by the double bond is accepted. Directional attack by pathways b, c, and d are highly disfavored by the bulky t-butyl groups and thus pathway a is favored. The larger substituent of the double bond is directed away from the axial hydrogen of the chiral diamine bridge.

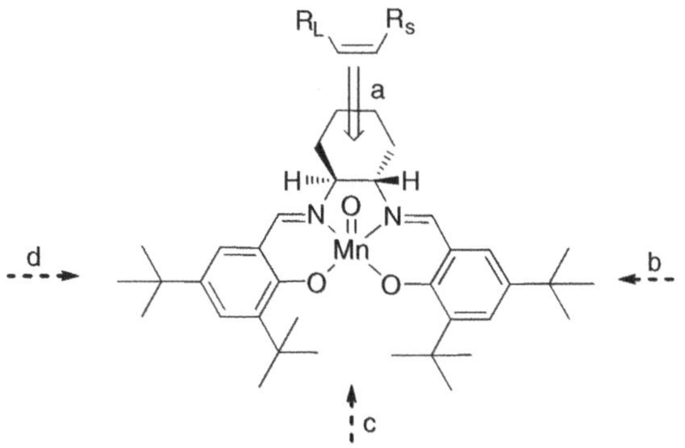

FIGURE 7.9 Model for stereocontrol in Jacobsen–Katsuki epoxidation.

Both isomers of the diamine and/or catalyst are commercially available and thus allow for total asymmetric control. A noted limitation of the model is that it is practical only to rigid *cis*-alkenes, that is, notably cycloalkenes. The optimization and fine-tuning of the reaction protocol has led to the application of the Jacobsen–Katsuki epoxidation for the synthesis of several drug candidates on pilot plant scale.

Researchers at SmithKline Beecham illustrated the practicality of the Mn(salen)-catalyzed epoxidation in the synthesis of their selective potassium channel activator BRL 55834 (**50**) [38]. Their three-step synthesis began with known iodochromene **47** (Scheme 7.11). Installation of the pentafluoroethyl appendage gave epoxide precursor **48**. Optimization of the Jacobsen's epoxidation conditions (0.2 mol% (*S,S*)-salen (**46**) catalyst, NaOCl, 0.1 equiv isoquinoline *N*-oxide (IQNO)) provided epoxide **49** in 84% yield with 93% ee after recrystallization from hexanes. More importantly, their studies also showed that IQNO is an effective donor ligand for the rate acceleration of the asymmetric epoxidation. To complete the synthesis, epoxide ring opening with 2-piperidone afforded BRL 55834 (**50**) in 81% isolated yield.

Separately, the following example illustrates the comparison of two different oxidative catalysis reactions in the quest for a large-scale synthesis of a key intermediate. In the evaluation of a process route for the synthesis of the melatonin agonist **51** [39], researchers at Bristol-Myers Squibb (BMS) identified olefin **52** as a key intermediate *en route* to the *trans*-cyclopropanecarboxylic ester **53a** (Scheme 7.12) [40]. Although direct methods for the asymmetric cyclopropanation of alkenes are known and were under investigation, large quantities of drug substance were needed expeditiously and thus alternative methods to obtain the cyclopropane

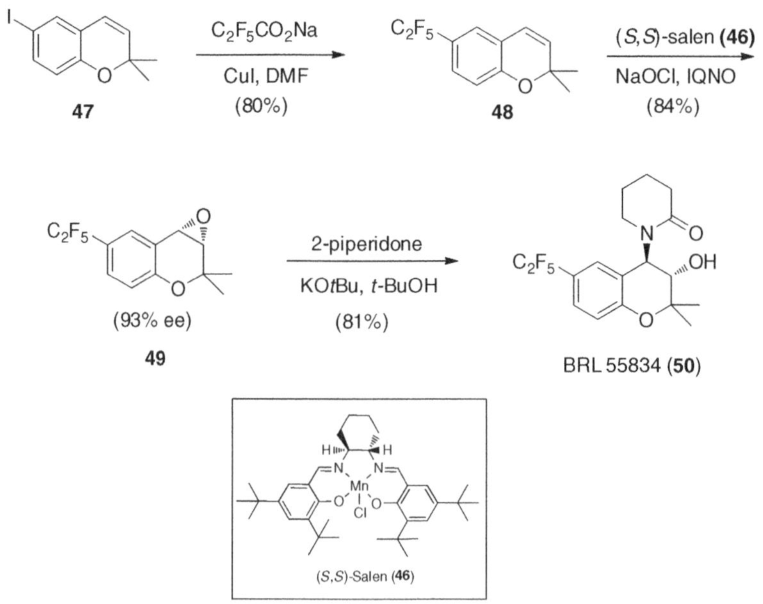

SCHEME 7.11   Synthesis of BRL 55834.

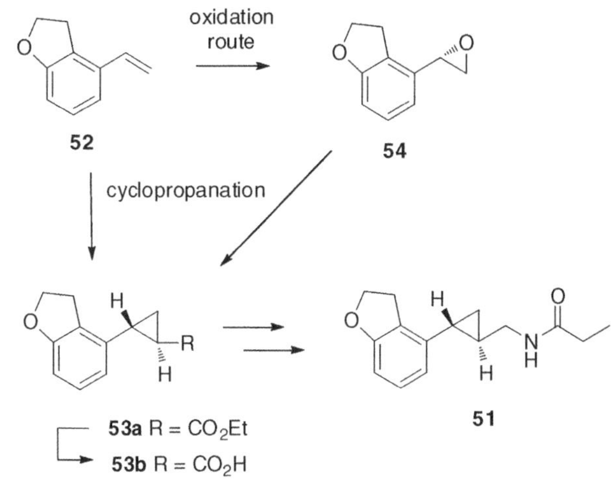

SCHEME 7.12 Synthetic approaches to cyclopropane **53**.

intermediate via an epoxide were also investigated. In their investigation, researchers at BMS compared the Jacobsen asymmetric epoxidation and the two-step process involving asymmetric dihydroxylation (AD) followed by cyclization for the large-scale preparation of epoxide **54**.

In the asymmetric epoxidation (Scheme 7.13) route from olefin **52**, epoxide **54** was isolated in an excellent yield (89%) after considerable optimization of the reaction conditions. However, the enantiomeric excess was consistently only 70% under all reaction conditions and thus necessitated a resolution step later in the reaction sequence to obtain high enantiomeric purity of **51**. In addition to the moderate enantioselectivity, formation of the benzofuran (via overoxidation) by-product **8** was problematic since it reacted similarly to **54** in subsequent reactions and was only removed with multiple crystallizations at the penultimate step. Subjection of epoxide **54** to the anion of triethyl phosphonoacetate (TEPA) resulted in the formation of *trans*-cyclopropanecarboxylic ester **53a** in 85% yield. Hydrolysis of the ester with NaOH and isolation of the acid as the dehydroabietylamine (DAA) salt **53b** occurred with >99% ee. This route provided the initial supply of drug candidate (800 g) in 22% overall yield from olefin **52**, with the low-yielding steps being the ester hydrolysis of **53a** followed by the acid purification via the DAA salt.

In the asymmetric dihydroxylation protocol of styrene **52** (Scheme 7.14), chiral diol **56** was isolated after crystallization in 85% yield with 99% ee. Diol **56** was converted to epoxide **54** via a modified Kolb–Sharpless method with little or no loss of enantiomeric purity in excellent yields (>90%). This dihydroxylation/cyclization sequence was scaled to 16 kg and provided the drug candidate **51** in 43% overall yield. This example clearly illustrates that the shorter reaction route may not always be the route of choice since many variables (i.e., scalability, safety, process ruggedness, and ease of product isolation) must be considered in developing a scale-up or manufacturing route. In this case, the difference in enantioselectivity and

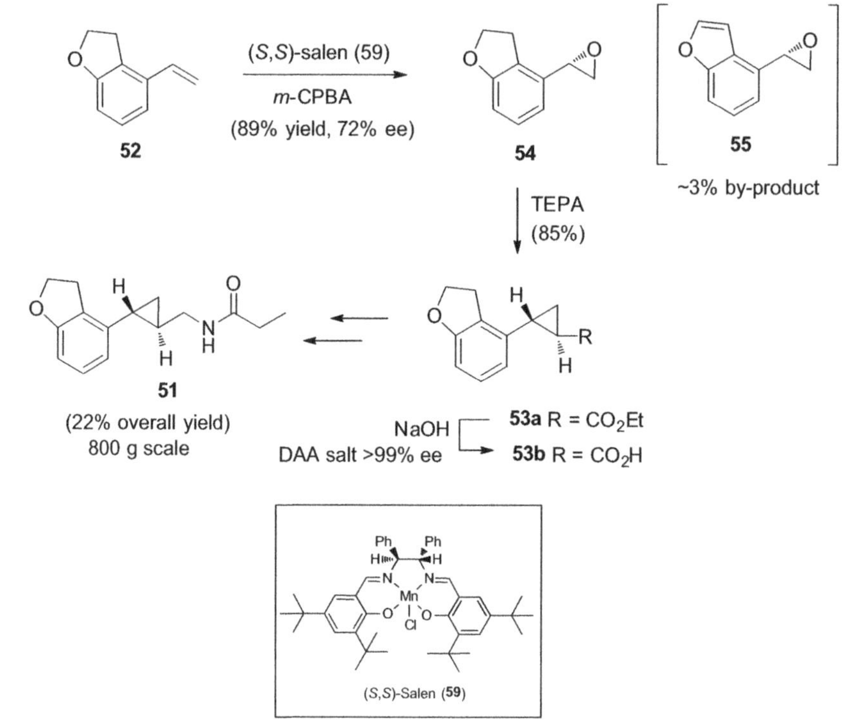

SCHEME 7.13   Synthesis of melatonin agonist **51** via Jacobsen–Katsuki epoxidation.

yield between the two routes was the critical factor in deciding which route was ultimately chosen.

Another efficient example in utilizing the Jacobsen's catalyst in large-scale preparation is in the synthesis of the marketed HIV protease inhibitor Crixivan. As illustrated in Figure 7.10, the retrosynthetic analysis of **60** provides three

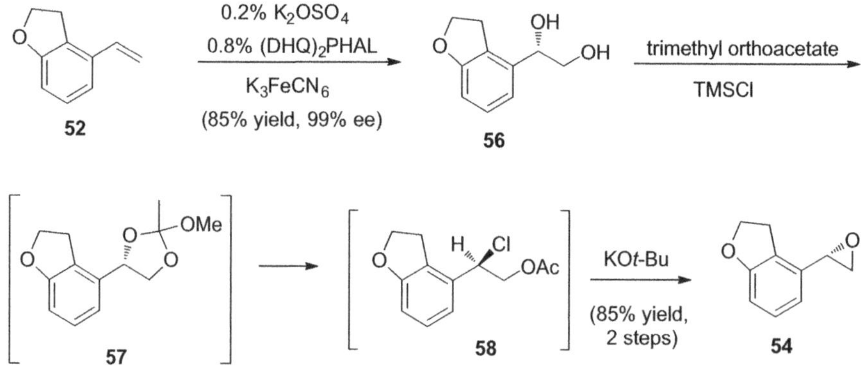

SCHEME 7.14   Synthesis of epoxide **54** via diol **56**.

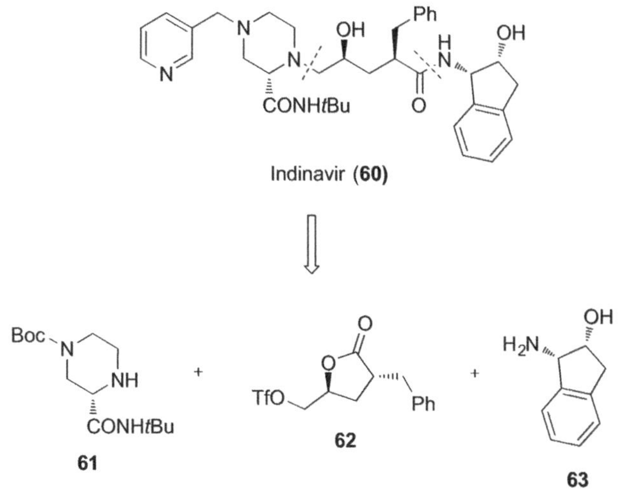

Indinavir (60)

FIGURE 7.10 Retrosynthetic analysis of indinavir 60.

fragments, piperazine fragment **61**, the Phe–Gly hydroxyethylene fragment **62**, and chiral *cis*-aminoindanol **63**. In route to the synthesis of fragment **63**, chiral indene oxide (**68**) was identified as a key intermediate and would provide the opportunity for the application of the Jacobsen's epoxidation reaction.

The synthesis of piperazine fragment **61** (Scheme 7.15) began with the condensation of *N*-Boc-ethylenediamine with dichloroacetaldehyde to give the corresponding imine. Addition of *tert*-butyl isocyanide and formic acid resulted in the formation of the Ugi adduct and TEA addition resulted in the loss of HCl to give intermediate **64** in quantitative yield over three steps. The cyclization to give the piperazine adduct was challenging and realized only with alkoxide bases with potassium *tert*-butoxide being preferred. Catalytic hydrogenation of the enamine with Rh–BINAP and subsequent formyl deprotection with hydrazine gave the desired piperazine fragment **61** with 98% ee. With fragment **61** in hand, the synthesis continued with the preparation of fragment **62** in four steps from furanone **65** (Scheme 7.16). TBS protection of the alcohol, followed by selective benzylation of the lactone, gave intermediate **66**. Deprotection of the alcohol with HF and triflate formation gave fragment **62** in an overall 66% yield.

In the preparation of the *cis*-aminoindanol or similar derivatives, several routes were investigated and the route chosen for the large-scale synthesis incorporated the

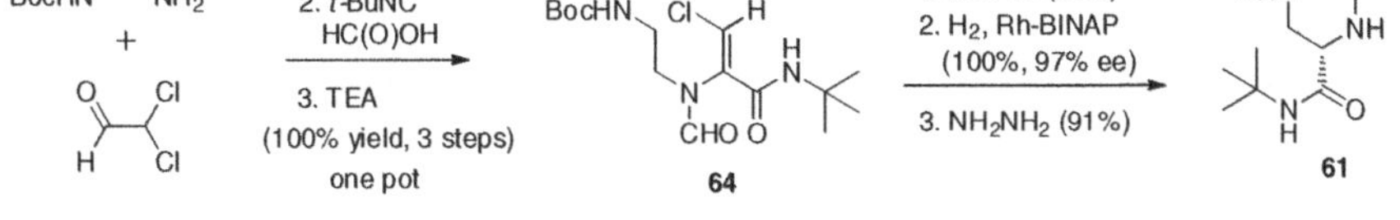

SCHEME 7.15 Synthesis of fragment **61**.

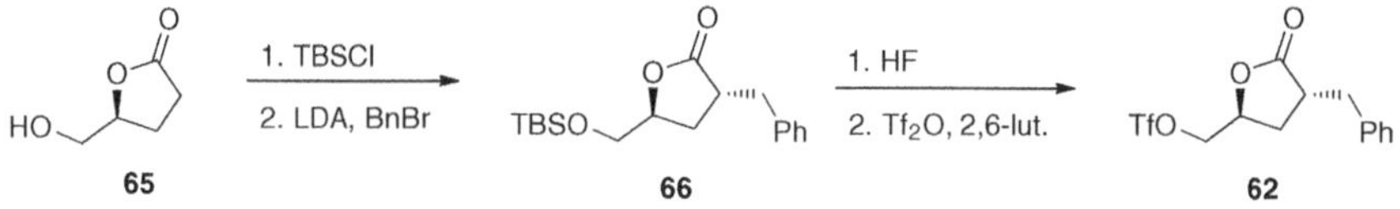

SCHEME 7.16 Synthesis of fragment **62**.

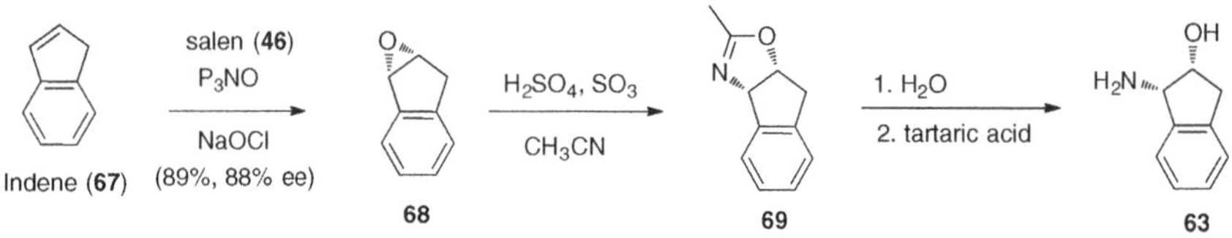

SCHEME 7.17 Preparation of amino alcohol **63**.

Jacobsen epoxidation. The synthesis began (Scheme 7.17) with the asymmetric epoxidation of indene with Jacobsen's (*S,S*)-salen catalyst (**46**) giving the desired indene oxide in 90% isolated yield with 88% ee. During their epoxidation investigation, the Merck researchers noted that lipophilic *N*-oxides led to a two- to threefold increase in the reaction rate with no loss in enantioselectivity. It was also noted that the electronics at the 4-position of the *N*-oxides had a dramatic effect on yield, stereoselectivity, and reaction rate. From their studies, 4-(3-phenylpropyl) pyridine *N*-oxide (P3NO) was identified as the optimal additive for the manganese–salen-catalyzed epoxidation by both stabilizing and activating the catalyst. The synthesis continued with the formation of the methyl oxazoline **69** by treatment with oleum in acetonitrile. Subsequent hydrolysis of the oxazoline and then crystallization with L-tartaric acid gave aminoindanol fragment **63** in 50% overall yield with greater than 99% enantiomeric excess.

To complete the synthesis of indinavir (Scheme 7.18), piperazine fragment **61** was coupled with triflate **62** in the presence of *i*-Pr$_2$NEt in 83% yield. Opening of the lactone with LiOH provided the hydroxyethylene isostere that was protected as a *tert*-butyldimethylsilyl ether. The acid moiety was coupled with (−)-*cis*-(1*S*,2*R*)-1-aminoindan-2-ol (**63**) to give amide intermediate **70**. Removal of both protecting groups with hydrochloric acid and subsequent *N*-alkylation with 3-picolyl chloride gave indinavir **60** in 10 total steps with an overall yield of 35%.[4]

The Sharpless asymmetric epoxidation [41,42] of allylic alcohols was a transformational reaction for organic chemists and led to Sharpless winning the Nobel Prize in chemistry, and serves as a versatile synthetic reaction for the organic chemist. The stereocontrol for the SAE is very predictable as illustrated in the stereochemistry model in Figure 7.11. With the allylic alcohol drawn as shown with the alcohol and small group (R$_1$) on the same side of the double bond, the use of the (−)-tartrate ligand results in delivery of the oxygen atom to the top face of the olefin. The

---

[4] *Note*: Although the Jacobsen epoxidation provides an efficient synthesis, it is not used in the manufacturing route.

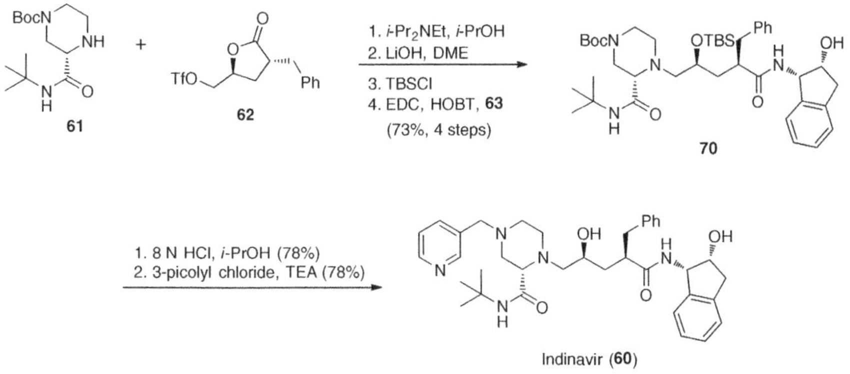

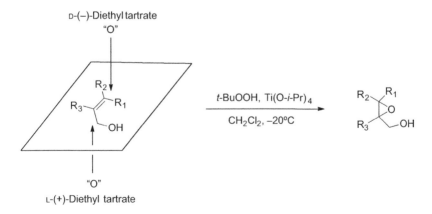

SCHEME 7.18 Synthesis of indinavir **60**.

(+)-tartrate ligand, conversely, results in epoxidation from the bottom face. The stereoselectivity for the epoxidation is excellent for achiral substrates with enantio-facial selectivity typically > 100:1. For substrates possessing chiral centers with a potential inherent facial selectivity themselves, the diastereoselectivity for the epoxidation can vary depending on a matched (same facial selectivity) or mismatched (opposing facial selectivity) case. This powerful reaction not only has been adopted as a routine reaction in academic and medicinal laboratories but also has found increasing use on large scale (> 500 g) and in pilot plants.

An excellent example highlighting the use of the SAE reaction on large scale is in the synthesis of SB-462795 (**71**) [43], a cathepsin K inhibitor. During the synthetic development of **71**, researchers at GlaxoSmithKline (GSK) noted that one of the main challenges in the synthesis is the azepanone ring bearing two stereocenters at C-4 and C-7. From their investigation, a route utilizing a ring-closing metathesis (RCM) between C-5 and C-6 proved to very efficient in constructing the ring system and their retrosynthetic analysis is shown in Figure 7.12. The azepane ring **72** is derived from dienoamino alcohol of the general structure **73** via a RCM and the C-7 stereocenter of

FIGURE 7.11 Sharpless asymmetric epoxidation stereocontrol model.

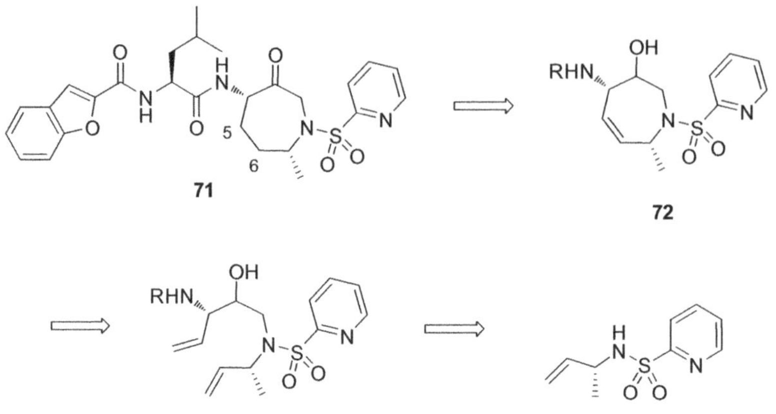

FIGURE 7.12   Retrosynthetic analysis of cathepsin K inhibitor **71**.

73 is derived from allyl sulfonamide **74**. Three different approaches utilizing a RCM were investigated with one being nonstereoselective and the other two consisting of either an aldol–RCM sequence or a Sharpless epoxidation–RCM sequence.

The nonstereoselective RCM route offered the advantage of potentially lowering the cost of goods and thus an extensive effort was undertaken. Unfortunately, the nonstereoselective nature of the route was too much to overcome due to the stereoisomeric mixture of later intermediates and the need for purification by flash chromatography. The difficulty in purification eventually outweighed the potential economic gain of not setting the C-3 and C-4 stereochemistry, and thus the researchers concluded that the two stereogenic centers would have to be carefully controlled in a route suitable for large-scale manufacturing. The two stereoselective routes are shown in Figure 7.13. In the first asymmetric route (Evans aldol–RCM), the azepanone moiety is derived from the C-3 epimeric amino alcohol **79**, which would arise from an Evans aldol condensation between aldehyde **86** and crotonate imide **85** to give the *syn* product **83**. In a subsequent Curtius reaction, the amino group would be installed with retention of configuration at C-4. For the second asymmetric route (Sharpless epoxidation–RCM), the azepanone moiety is derived from C-3 epimeric amino alcohol **80**. The C-3/C-4 stereocenters in **80** would be derived from epoxide **84**, which in turn would come from Sharpless epoxidation of divinyl carbinol **87**. The C-4 stereocenter will be inverted during the introduction of phthalimide via a Mitsunobu reaction.

Since the second route highlights the value of the SAE reaction and RCM reactions are covered in Chapter 5, only the second route in the synthesis of **71** is discussed in this chapter. The synthesis (Scheme 7.19) of the key epoxide intermediate begins with the catalytic SAE of divinyl carbinol **87** with diisopropyl tartrate as the ligand and cumene hydroperoxide as the oxidant to afford allylic epoxide **88** in 90% yield with excellent enantioselectivity (98% ee). In their epoxidation investigation and consistent with literature observations [44], cumene

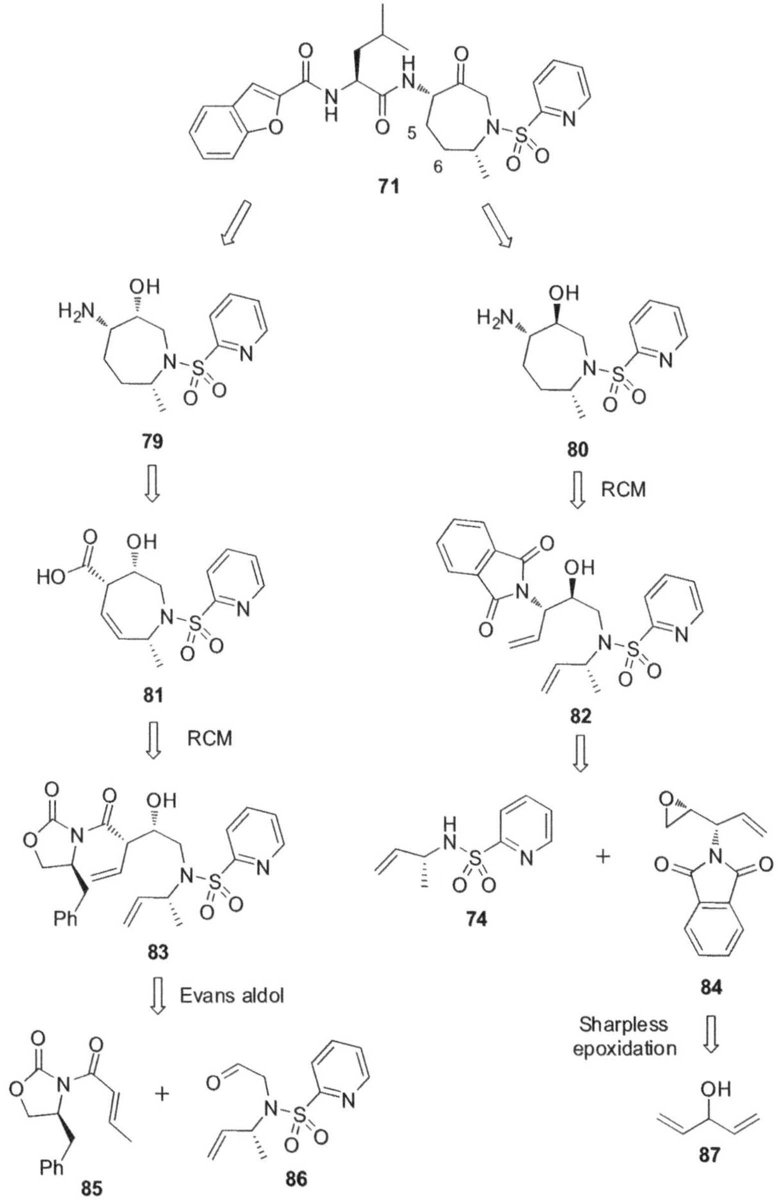

FIGURE 7.13   Retrosynthetic approaches of two asymmetric routes.

hydroperoxide provided a higher yield of the epoxide relative to *tert*-butyl hydroperoxide (75% yield, 97% ee). Exposure of **88** to phthalimide under Mitsunobu conditions inverted the C-4 stereocenter and provided the desired intermediate **84** in 85% yield. The isolation of the phthalimido epoxide **84** with high purity proved very challenging. The low melting point of **84** and high levels of reaction impurities

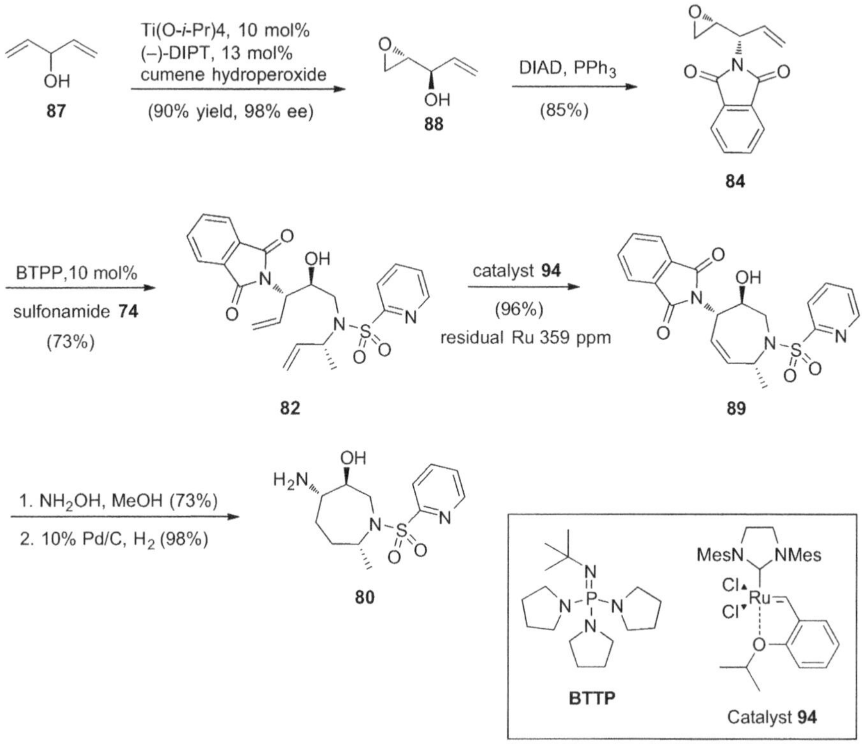

SCHEME 7.19 Preparation of amino alcohol **80**.

prevented the crystallization of **84**. To achieve the acceptable purity needed, batch chromatography purification was necessary and delivered > 200 kg of epoxide (in 60% yield from divinyl carbinol) for early-phase supplies.

Although suitable for early drug supplies, the synthetic sequence was not acceptable for commercial scale and therefore a two-step protocol to isolate epoxide **84** was developed. As shown in Scheme 7.20, reaction of epoxide **84** with toluene sulfonate under acidic conditions provided the more crystalline tosylate **90** that could be crystallized directly from the crude reaction mixtures after the Sharpless and Mitsunobu steps in excellent purity and yield. Epoxide **84** was regenerated in near-quantitative yield upon exposure of tosylate **90** to DBU in toluene. This two-step sequence provided a more scalable and cost-effective intermediate by eliminating the need for chromatographic purification of epoxide **84** and provided > 650 kg of the epoxide in 50% overall yield from divinyl carbinol. Base-catalyzed (10 mol%) coupling of epoxide **84** and sulfonamide **74** (Scheme 7.19) in the presence of the highly basic yet nonnucleophilic *tert*-butylimino-tri(pyrrolidino)phosphorane (BTPP) afforded the requisite RCM precursor **82** in 73% yield. Ring-closing metathesis of diene **82** with Hoveyda's second-generation catalyst **94** proved to be an exceptional reaction. The catalyst loading was only 0.5 mol% and provided the

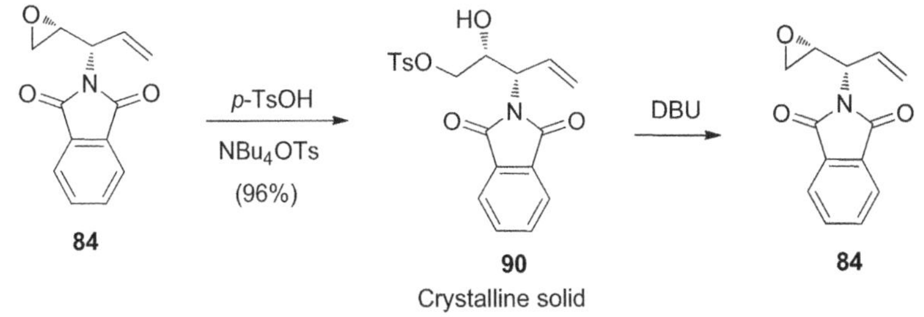

SCHEME 7.20 Epoxide ring-opening and ring-closing protocol.

desired azepane in 96% yield. Deprotection of the phthalimide with NH_2OH (73% yield) and subsequent hydrogenation (10% Pd/C, 98% yield) of the alkene provided amino alcohol **80**.

To complete the synthesis of **71**, the side chain acid fragment **92** was prepared from known acid **91** [45] and L-leucine as shown in Scheme 7.21 in 80% yield and coupled with amino alcohol **80**. The coupling of azepane amino alcohol **80** under most reaction conditions resulted in partial racemization (up to 10%) of the side chain stereocenter of **92**. Optimization of the amide coupling resulted in the use of EDCI with HOOBt (1 mol%) and provided the desired product **93** in 89% yield with >99% purity. The oxidation of alcohol **93** was extensively studied and the investigation led to the use of $DMSO/Ac_2O$ [46] to afford ketone **71** in 85% isolated yield. The process using amino alcohol **80** (Sharpless–RCM route) was successfully piloted on multikilogram scale, producing >200 kg of SB-462795 (**71**) that was suitable for clinical studies.

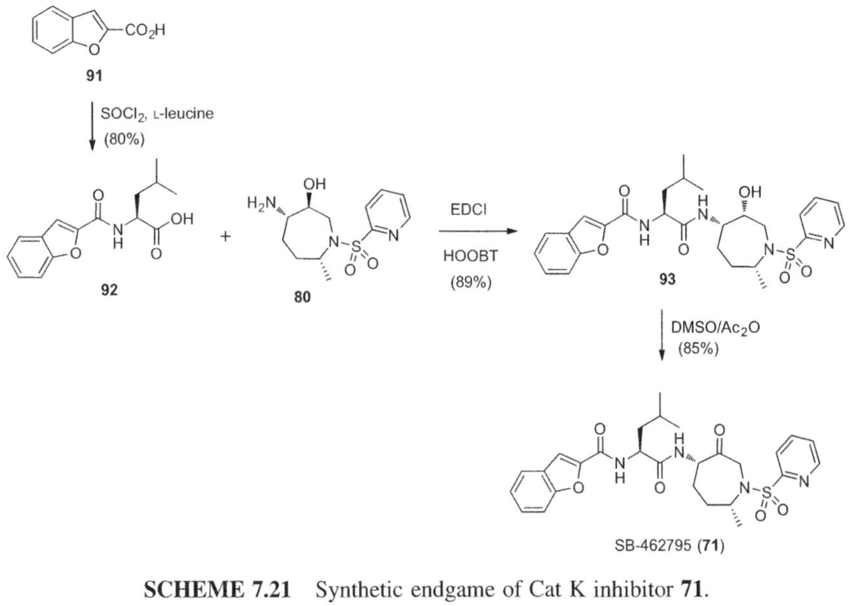

SCHEME 7.21 Synthetic endgame of Cat K inhibitor **71**.

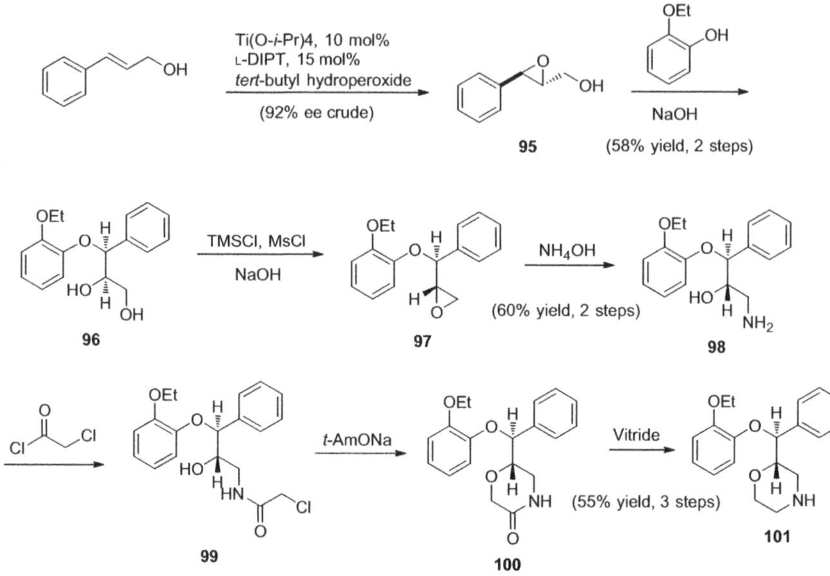

SCHEME 7.22   Preparation of (*S,S*)-reboxetine succinate **101**.

Another example to emphasize the high impact of the SAE reaction in drug development is Pfizer's synthesis of (*S,S*)-reboxetine succinate (**101**) [47], a drug in development for neuropathic pain as well as other indications [48]. There are published enantioselective syntheses in the literature [49] but (*S,S*)-reboxetine is usually prepared by resolution of the racemate as the (−)-mandelate salt [50]. Resolution is indeed a viable process in organic chemistry but inherently is inefficient due to a maximum product yield of 50%. Pfizer's asymmetric synthesis (Scheme 7.22) begins with the SAE (Ti(*i*-PrO)$_4$, L-DIPT, *tert*-butyl hydroperoxide) of cinnamyl alcohol to give epoxide **95** and establishing the two chiral centers early in the process. The enantiomeric excess of the crude epoxide was 92%.[5] Crude epoxide **95** was reacted with 2-ethoxyphenol and afforded intermediate **96** in 58% yield over two steps with >98% ee after crystallization. Diol **96** was converted to epoxide **97** upon exposure to TMSCl, MsCl, and NaOH. Opening of the epoxide with an ammonia source provided amino alcohol **98** in 60% yield over two steps. Acylation of **98** with 2-chloroacetyl chloride under Schotten–Baumann conditions (aqueous Na$_2$CO$_3$–toluene) afforded **99**, which upon exposure to sodium *tert*-amylate afforded the cyclized adduct **100**. The synthesis was completed with the Vitride reduction of the amide and subsequent succinate crystallization and gave **101** in 55% yield from amino alcohol **98**. The overall yield of (*S,S*)-reboxetine succinate was about 19% from cinnamyl alcohol, compared to <10% overall yield via resolution. Further-more, the asymmetric process reduces solvent use and waste generation by ∼50% compared to the resolution route.

_____

[5] Sharpless asymmetric epoxidation of cinnamyl alcohol is reported to give epoxide with >98% ee.

7.4.2 Dihydroxylation

The dihydroxylation of an alkene is a powerful chemical transformation since the resulting diol is in itself a functionality found in active pharmaceutical ingredients or the diol is a versatile synthetic intermediate. A variety of metal oxidants such as osmium(VIII), ruthenium(VIII), manganese(VII), and vanadium(V) have been developed for this transformation with osmium tetroxide being the reagent of choice. The osmium-mediated dihydroxylation of an alkene is stereospecific and delivers the two oxygen atoms from the same face of the alkene. Since most alkenes are prochiral, new stereocenters are often created. The critical dihydroxylation work of the Upjohn group in which N-methylmorpholine N-oxide was used as the cooxidant was a major breakthrough and led to the frequent use of the osmium-catalyzed dihydroxylation [51]. The process group at Pfizer nicely illustrated the utility of a racemic dihydroxylation protocol in the synthesis of their $\alpha 4\beta 2$ neuronal nicotinic acetylcholine receptor (nAChR) partial agonist varenicline **118**, which was approved in the United States for the treatment of smoking cessation in 2006 [52]. In the retrosynthetic analysis (Fig. 7.14), the key synthetic intermediate is the benzazepine core **125**, which in turn is derived from benzonorbornadiene **121**.

In the development of a process route, alternative approaches were investigated in the conversion of benzonorbornadiene **121** to the benzazepine core **125** [53]. The osmium dihydroxylation route (Scheme 7.23) outlined below was the initial frontrunner and provided large quantities of the required API. The known benzonorbornadiene **121** [54] was prepared from 1,2-dibromobenzene and cyclopentadiene via a benzyne Diels–Alder reaction in 89% yield. In optimizing the VanRheenen osmium dihydroxylation procedure of benzonorbornadiene **121** [55], researchers at Pfizer were able to reduce the osmium catalyst loading to 0.13–0.26 mol% in the presence of N-methylmorpholine N-oxide while maintaining an excellent yield of 89% of the diol **122**. It was noted that higher concentrations (0.5–1.5 M) of the reaction mixture resulted in the direct crystallization of diol **122** from the reaction mixture. It is also postulated that this crystallization phenomenon increases the reaction rate since crystallization presumably liberates active osmium from catalytically less active species allowing osmium reentry into the catalytic cycle and thereby driving the reaction to completion. Diol **122** was conveniently converted to the benzazepine core in a two-step oxidative cleavage/reductive amination procedure (Scheme 7.23). Exposure of diol **122** to $NaIO_4$ resulted in the *in situ* formation of dialdehyde **123** and subsequent addition of benzylamine and $NaBH(OAc)_3$ provided the benzazepine core adduct **124** in 84% yield over two steps. Benzylamine deprotection with hydrogenolysis (H_2, $Pd(OH)_2$) in the presence of HCl provided the key fragment

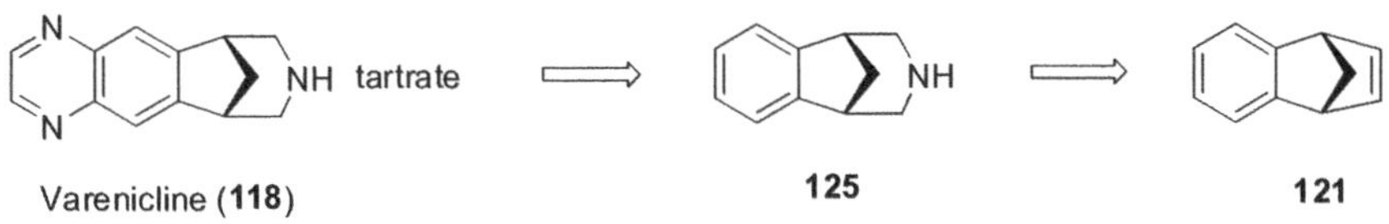

Varenicline (**118**)   **125**   **121**

FIGURE 7.14   Retrosynthetic analysis of varenicline (**118**).

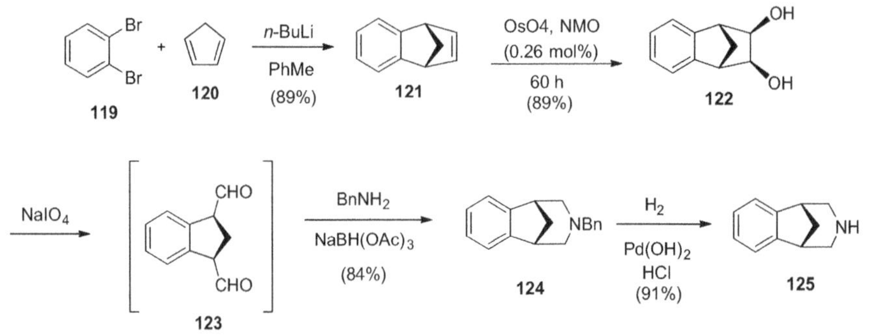

SCHEME 7.23   Synthesis of benzazepine core fragment **125**.

125 in 91% yield. This short and efficient synthetic sequence provided fragment **125** with an overall yield of 64–73% from the known benzonorbornadiene **121**.

To complete the synthesis of varenicline (Scheme 7.24), amine **125** was protected as the trifluoroacetamide with trifluoroacetic anhydride in 95% yield and followed by bis-nitration with triflic and nitric acid to give dinitro adduct **126**. Bis-nitro reduction of **126** by hydrogenation (H$_2$, Pd/C) gave the diamine intermediate **127**, which was immediately reacted with glyoxal in water and provided compound **128** in 85% yield over the last three steps. Simple deprotection of the amine under basic hydrolysis (2 N NaOH) and subsequent tartrate salt formation furnishes varenicline **118**.

To further expand the utility of the dihydroxylation reaction, Sharpless developed a catalytic asymmetric variant based upon the ligand acceleration work of Criegee [56]. In the early development of the Sharpless asymmetric dihydroxylation, pyridine was replaced with chiral nonracemic tertiary amines. The ligands of choice today are the cinchona alkaloid derivatives: dihydroquinine (DHQ) (**102**) and dihydroquinidine (DHQD) (**103**) (Fig. 7.15). In the asymmetric oxidation, the diastereomeric cinchona alkaloids exhibit opposite enantiofacial selectivity and thus essentially act as enantiomers.

The mechanism Sharpless proposed with two catalytic cycles for the dihydroxylation reaction is illustrated in Figure 7.16. The overall catalytic reaction begins with

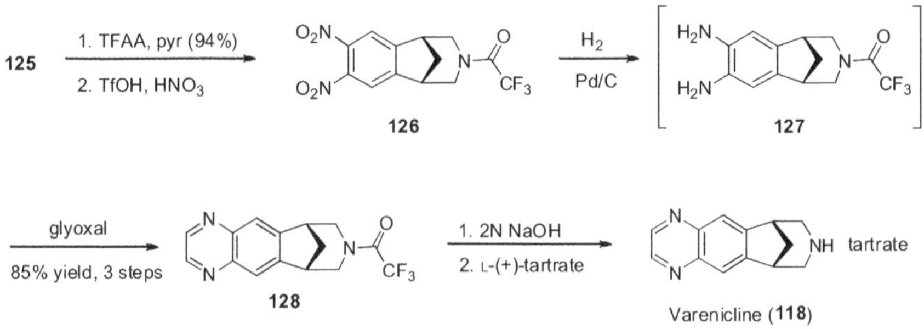

SCHEME 7.24   Endgame synthesis of varenicline (**118**).

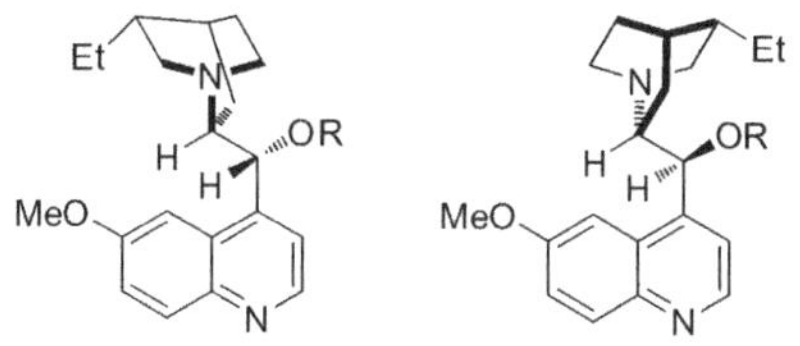

102　　R = alkyl, aryl　　**103**

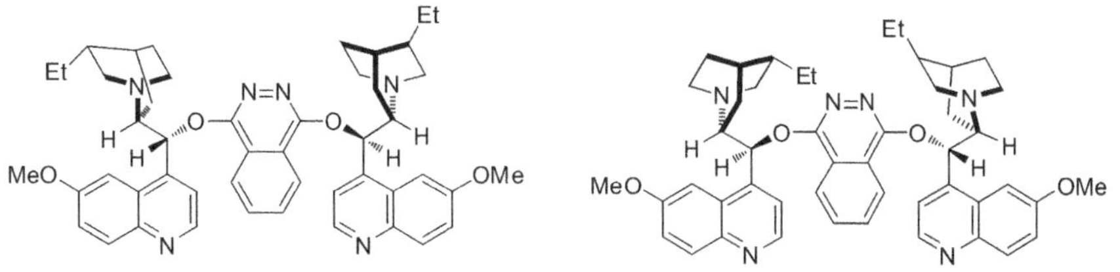

(DHQ)₂PHAL (**104**)
ligand used in AD-mix-α

(DHQD)₂PHAL (**105**)
ligand used in AD-mix-β

FIGURE 7.15　Sharpless dihydroxylation ligands.

unligated osmium species **106**, in the primary cycle. After coordination with the chiral ligand **L***, the osmium species react with alkene **107** to give osmium(VI) glycolate intermediate **108**. Oxidation of intermediate **108** by the stoichiometric cooxidant and loss of ligand provides intermediate **109**. Hydrolysis of **109** gives diol **110** and regenerates the unligated osmium species **106**. Alternatively, if the hydrolysis of intermediate **109** is slow, **109** may proceed to a secondary cycle where it reacts with another alkene molecule to form species **111**. Hydrolysis of **111** and ligand coordination leads back to intermediate **108** and diol **110**. It is postulated that the secondary cycle leads to lower enantioselectivity or formation of the opposite enantiomer, thereby degrading the ee of the primary cycle. With the understanding of the existence of two catalytic cycles in the reaction mechanism with only one of the cycles giving high enantioselectivity, several modifications were made to the reaction protocol to promote osmium ester hydrolysis and include the use of a biphasic reaction system and the use of organic sulfonamides as additives.

　　The scope of the Sharpless asymmetric dihydroxylation reaction is broad with the use of the phthalazine ligands (Fig. 7.15, R = aryl). The introduction of the aryl ether derivative ligands expanded the reaction to include four of the possible six classes of alkenes based on substitution pattern (Fig. 7.17). Monosubstituted (**112**), *gem*-disubstituted (**113**), *trans*-disubstituted (**114**), and trisubstituted (**116**) alkenes generally produce the corresponding 1,2-diols in high chemical yields with high enantioselectivity ranging from 79% to 99% ee. To further expand the usefulness and practicality of the dihydroxylation reaction, two AD-mix formulations labeled α and β were developed. The solid mixes contain trace amounts of the phthalazine ligand **104** (AD-mix-α) or **105** (AD-mix-β) and potassium osmate (total 0.6% by weight) blended with potassium ferricyanide and potassium carbonate (totaling 99.4%).

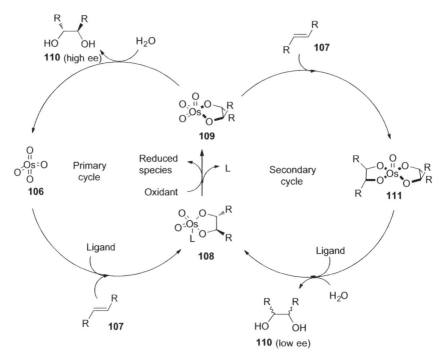

FIGURE 7.16   Sharpless dihydroxylation proposed catalytic cycles.

In the chemical industry, the general importance of dihydroxylation reactions is evident since both the Sharpless asymmetric dihydroxylation protocol and achiral dihydroxylation variants have been used on large scale to manufacture an API. The next example highlights the Sharpless asymmetric dihydroxylation reaction as the key chemical transformation. Often in the lead optimization phase of a drug discovery effort, several molecules are potential drug candidates and will require several hundred grams of the API for differential preclinical studies. In addition to the commonality of the biological target, quite often the molecules will share a common structural motif and thus require a large-scale synthesis of the common intermediate. In the development of one of their drug discovery programs, AstraZeneca was faced with this dilemma and required a robust and scalable route to α-hydroxycarboxylic acid **138** [57]. In evaluating the scale-up options, three scenarios were considered: (1) scale up the medicinal chemistry route as is, (2) modify the medicinal chemistry route, and (3) design a novel synthetic sequence. The third option was chosen and the

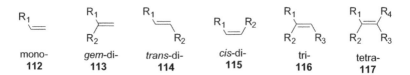

FIGURE 7.17   Alkene substitution patterns.

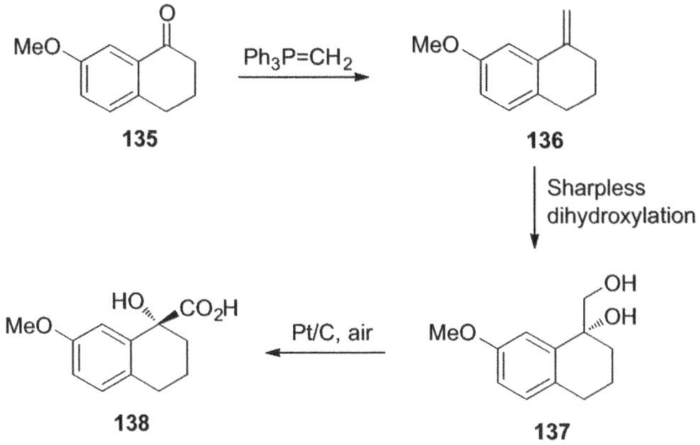

SCHEME 7.25   Preparation of α-hydroxycarboxylic acid **138**.

synthesis is shown in Scheme 7.25. Interestingly, the starting substrate 7-methoxy-1-tetralone was identical to the medicinal chemistry route. The synthesis begins with the Wittig olefination of ketone **135** with the phosphonium ylide $Ph_3P=CH_2$ to afford alkene **136** in excellent yields (97%). Subjection of **136** to the Sharpless asymmetric dihydroxylation reaction [(DHQD)$_2$-PHAL (**105**), $K_2OsO_4 \cdot 2H_2O$, $K_3Fe(CN)_6$)] provided diol **137** in good to excellent yields with outstanding enantiomeric purity (>98% ee). The loading of the osmium catalyst and phthalazine ligand was optimized to 0.05 and 0.1 mol%, respectively. Oxidation of the primary alcohol to the carboxylic acid **138** was cleanly achieved with a Pt/C catalyst in the presence of air in 82% yield. This short and efficient three-step synthesis provided the common intermediate **138** with high enantiomeric purity with levels of Os and Pt below the limits of detection (<5 ppm).

7.5   CONCLUSION

Despite the many challenges associated with running oxidation reactions, they are often needed and ran on manufacturing scale to synthesize an API. The examples in this chapter highlight the power of transition metals in reducing some of the risks associated with oxidation reactions and also demonstrate the importance of asymmetric oxidative catalysis. Without the use of transition metal-mediated oxidative catalysis, many of the synthetic routes would have increased step count, increased cost of goods, and/or racemic API with alternative synthetic approaches.

REFERENCES

[1] Dugger, R. W.; Ragan, J. A.; Brown Ripin, D. H. *Org. Process Res. Dev.* **2005**, *9*, 253.

[2] Dess, D. B.; Martin, J. C. *J. Am. Chem. Soc.* **1991**, *113*, 7277.

[3]   Caron, S.; Dugger, R. W.; Ruggeri, S. G.; Ragan, J. A.; Brown Ripin, D. H. *Chem. Rev.* **2006**, *106*, 2943.

[4]   Ley, S. V.; Norman, J.; Griffith, W. P.; Marsden, S. P. *Synthesis* **1994**, 639.

[5]   Watanabe, M.; Koike, H.; Ishiba, T.; Okada, T.; Seo, S.; Hirai, K. *Bioorg. Med. Chem. Lett.* **1997**, *5*, 437.

[6]   Giddings, S.; Mills, A. *J. Org. Chem.* **1998**, *53*, 1103.

[7]   (a) Pecka, J.; Stanek, J., Jr., ; Cerny, M. *Collect. Czech. Chem. Commun.* **1974**, *39*, 1192. (b) Cerny, M.; Pacak, J.; Stanek, J., Jr., *Carbohydr. Res.* **1970**, *15*, 379. (c) Cerny, M.; Kalvoda, L.; Pacak, J. *Collect. Czech. Chem. Commun.* **1968**, *33*, 1143.

[8]   (a) Palamidessi, G.; Bonanomi, M. *Farmaco Ed. Sci.* **1966**, *21*, 799. (b) Adachi, J.; Sato, N. *J. Org. Chem.* **1972**, *37*, 221. (c) Cheeseman, G. W. H.; Freeston, A. J.; Godwin, R. A.; Hough, T. L. *J. Chem. Soc., Perkin Trans. 1* **1975**, 1888.

[9]   Hicks, J. L.; Huang, C. C.; Goel, O. P.; Butler, D. E. *J. Label. Compd.* **1984**, *21*, 937.

[10]   Krapcho, A. P.; Maresch, M. J.; Gallagher, C. E.; Hacker, M. P.; Menta, E.; Oliva, A.; DiDomenico, R.; Da Re, G.; Spinelli, S. *J. Heterocycl. Chem.* **1995**, *32*, 1693.

[11]   Sugimura, T.; Paquette, L. A. *J. Am. Chem. Soc.* **1987**, *109*, 3017.

[12]   Sato, K.; Aoki, M.; Takagi, J.; Zimmermann, K.; Noyori, R. *Bull. Chem. Soc. Jpn.* **1999**, *72*, 2287.

[13]   Hida, T.; Nogusa, H. *Tetrahedron* **2009**, 270.

[14]   Hida, T.; Fukui, Y.; Kawata, K.; Kabaki, M.; Masui, T.; Fumoto, M.; Nogusa, H. *Org. Process Res. Dev.* **2010**, *14*, 289.

[15]   Bowden, K.; Heilbron, I. M.; Jones, E. R. H.; Weedon, B. C. L. *J. Chem. Soc.* **1946**, 39.

[16]   Carlsen, P. H. J.; Katsuki, T.; Martin, V. S.; Sharpless, K. B. *J. Org. Chem.* **1981**, *46*, 3936.

[17]   (a) Zhao, M.; Li, J.; Song, Z.; Desmond, R.; Tschaen, D. M.; Grabowski, E. J. J.; Reider, P. J. *Tetrahedron Lett.* **1998**, *39*, 5323. (b) Review of chromium-catalyzed oxidations: Muzart, J. *Chem. Rev.* **1992**, *92*, 113.

[18]   Gauthier, J. Y.; Chauret, N.; Cromlish, W.; Desmarais, S.; Duong, L. T.; Falgueyret, J.; Kimmel, D. B.; Lamontagne, S.; Leger, S.; LeRiche, T.; Li, C. S.; Masse, F.; McKay, D. J.; Nicoll-Griffith, D. A.; Oballa, R. A.; Palmer, J. T.; Percival, M. D.; Riendeau, D.; Robichaud, J.; Rodan, G. A.; Rodan, S. B.; Seto, C.; Therien, M.; Truong, V.; Venuti, M. C.; Wesolowski, G.; Young, R. N.; Zamboni, R.; Black, W. C. *Bioorg. Med. Chem. Lett.* **2008**, *18*, 923.

[19]   O'Shea, P. D.; Chen, C.; Gauvreau, D.; Gosselin, F.; Hughes, G.; Nadeau, C.; Volante, R. P. *J. Org. Chem.* **2009**, *74*, 1605.

[20]   Dauben, G. W.; Lorber, M.; Fulleron, D. S. *J. Org. Chem.* **1969**, *34*, 3587.

[21]   (a) Kimura, M.; Muto, T. *Chem. Pharm. Bull.* **1980**, *28*, 1836. (b) Prousa, R.; Schonecker, B. *J. Prakt. Chem.* **1991**, *5*, 333. (c) Ishii, Y.; Nakayama, K.; Takeno, M.; Sakaguchi, S.; Iwahama, T.; Nishiyama, Y. *J. Org. Chem.* **1995**, *60*, 3934.

[22]   (a) Parish, E. J.; Chitrakon, S.; Todd, K. L. *Org. Prep. Proceed. Int.* **1985**, *17*, 192. (b) Parrish, E. J.; Wei, T. *Synth. Commun.* **1987**, *17*, 1227. (c) Kumar, V.; Amann, A.; Ourisson, G.; Luu, B. *Synth. Commun.* **1987**, *17*, 1279. (d) Pearson, A. J.; Chen, Y. S.; Han, G. R.; Hsu, S. Y.; Ray, T. *J. Chem. Soc., Perkin Trans. 1* **1985**, 267.

[23]   Bakshi, R. K.; Patel, G. F.; Rasmusson, G. H.; Baginsky, W. F.; Cimis, G.; Ellsworth, K.; Chang, B.; Bull, H.; Tolman, R. L.; Harris, G. H. *J. Med. Chem.* **1994**, *37*, 3871.

[24]   Miller, R. A.; Li, W.; Humphrey, G. R. *Tetrahedron Lett.* **1996**, *37*, 3429.

[25] Murahashi, S. I.; Oda, Y.; Naota, T.; Kuwabara, T. *Tetrahedron Lett.* **1993**, *34*, 1299.

[26] Harre, M.; Haufe, R.; Nickisch, K.; Weinig, P.; Weinmann, H. *Org. Process Res. Dev.* **1998**, *2*, 100.

[27] Krysan, D. J.; Haight, A. R.; Menzia, J. A.; Welch, N. *Tetrahedron* **1994**, *50*, 6163.

[28] For a comprehensive listing of previous synthetic approaches see Baker, W. R.; Condon, S. L. *J. Org. Chem.* **1993**, *58*, 3277.

[29] (a) Corey, E. J.; Fuchs, P. L. *Tetrahedron Lett.* **1972**, *36*, 3769. (b) Piers, E.; Wong, T.; Ellis, K. A. *Can. J. Chem.* **1992**, *70*, 2058.

[30] Brougham, P.; Cooper, M. S.; Cummerson, D. A.; Heaney, H.; Thompson, N. *Synthesis* **1987**, 1015.

[31] Parikh, J. R.; Doering, W. v. E. *J. Am. Chem. Soc.* **1967**, *89*, 5505.

[32] Behrens, C. H.; Sharpless, K. B. *J. Org. Chem.* **1985**, *50*, 5696.

[33] Kornfeld, E. C.; Fornfeld, E. J.; Kline, G. B.; Mann, M. J.; Morrison, D. E.; Jones, R. G.; Woodward, R. B. *J. Am. Chem. Soc.* **1956**, *78*, 3087.

[34] Leanna, M. R.; Martinelli, M. J.; Varie, D. L.; Kress, T. J. *Tetrahedron Lett.* **1989**, *30*, 3935.

[35] (a) Russell, H. F.; Harris, B. J.; Hood, D. B.; Thompson, E. G.; Watkins, A. D.; Williams, R. D. *Org. Prep. Proced. Int.* **1985**, *17*, 391. (b) Johnson, H. E.; Crosby, D. G. *J. Org. Chem.* **1963**, *55*, 2794. (c) Borror, A. L.; Chinoporos, E.; Filosa, M. P.; Herchen, S. R.; Petersen, C. P.; Stern, C. A.; Onan, K. *J. Org. Chem.* **1988**, *53*, 2047.

[36] Srinivasan, K.; Michaud, P.; Kochi, J. K. *J. Am. Chem. Soc.* **1986**, *108*, 2309.

[37] Jacobsen, E. N.; Deng, L.; Furukawa, Y.; Martinez, L. E. *Tetrahedron* **1994**, *50*, 4323.

[38] Bell, D.; Davies, M. R.; Finney, F. J. L.; Green, G. R.; Kincey, P. M.; Mann, I. S. *Tetrahedron Lett.* **1996**, *37*, 3895.

[39] Zhu, J.; Prince, B. A.; Zhao, S. X.; Skonezny, P. M. *Tetrahedron Lett.* **2000**, *41*, 4011.

[40] (a) Singh, A. K.; Rao, M. N.; Simpson, J. H.; Li, W.-S.; Thornton, J. E.; Kuehner, D. E.; Kacsur, D. J. *Org. Process Res. Dev.* **2002**, *6*, 618. (b) Prasad, J. S.; Vu, T.; Totleben, M. J.; Crispino, G. A.; Kacsur, D. J.; Swaminathan, S.; Thornton, J. E.; Fritz, A.; Singh, A. K. *Org. Process Res. Dev.* **2003**, *7*, 821.

[41] Katsuki, T.; Sharpless, K. B. *J. Am. Chem. Soc.* **1980**, *102*, 5974.

[42] For reviews of the SAE, see the following references: (a) Tsutomu, K.; Martin, V. S. *Org. React.* **1996**, *48*, 1. (b) Pfenninger, A. *Synthesis* **1986**, *2*, 89.

[43] Wang, H.; Matsuhashi, H.; Doan, B. D.; Goodman, S. N.; Ouyang, X.; Clark, W. M., Jr., *Tetrahedron* **2009**, *65*, 6291.

[44] Romero, A.; Wong, C.-H. *J. Org. Chem.* **2000**, *65*, 8264.

[45] Fuson, R. C.; Kneisley, J. W.; Kaiser, E. W. *Org. Synth.* **1944**, *24*, 33.

[46] Albright, J. D.; Goldman, L. *J. Am. Chem. Soc.* **1965**, *87*, 4214.

[47] Henegar, K. E.; Cebula, M. *Org. Process Res. Dev.* **2007**, *11*, 354.

[48] (a) Hughes, B.; McKenzie, I.; Stoker, M. J. WO2006/000903 (2006). (b) Allen, A. J.; Hemrick-Luecke, S.; Sumner, C. R.; Wallace, O. B. WO2005/021095 (2005). (c) Wong, E. H. F.; Ahmed, S.; Marshall, R. C.; McArthur, R.; Taylor, D. P.; Birgerson, L.; Cetera, P. WO2001/001973 (2001).

[49] (a) Brenner, E.; Baldwin, R. M.; Tamagnan, G. *Org. Lett.* **2005**, *7*, 937. (b) Cebula, M.; Henegar, K. E. U.S. Patent 2005/187388 (2005).

[50] (a) Melloni, P.; Della Torre, A.; Lazzari, E.; Mazzini, G.; Meroni, M. *Tetrahedron* **1985**, *41*, 1393. (b) Prabhakaran, J.; Majo, V. J.; Mann, J. J.; Kumar, J. S. D. *Chirality* **2004**, *16*, 168.

[51] VanRheenen, V.; Kelly, R. C.; Cha, D. F. *Tetrahedron Lett.* **1976**, *17*, 1973.

[52] Coe, J. W.; Brooks, P. R.; Vetelino, M. G.; Wirtz, M. C.; Arnold, E. P.; Huang, J.; Sands, S. B.; Davis, T. I.; Lebel, L. A.; Fox, C. B.; Shrikande, A.; Heym, J. H.; Schaeffer, E.; Rollema, H.; Lu, Y.; Mansbach, R. S.; Chambers, L. K.; Rovetti, C. C.; Schulz, D. W.; Tingley, F. D., III; O'Neil, B. T. *J. Med. Chem.* **2005**, *48*, 3474.

[53] (a) Brooks, P. R.; Caron, S.; Coe, J. W.; Ng, K. K.; Singer, R. A.; Vazquez, E.; Vetelino, M. G.; Watson, H. H.; Whritenour, D. C.; Wirtz, M. C. *Synthesis* **2004**, 1755. (b) Singer, R. A.; McKinley, J. D.; Barbe, G.; Farlow, R. A. *Org. Lett.* **2004**, *6*, 2357. (c) Busch, F. R.; Hawkins, J. M.; Mustakis, L. G.; Sinay, T. G.; Watson, T. J. N.; Withbroe, G. J. WO2006090236 A1 (**2006**).

[54] (a) Wittig, G.; Knauss, E. *Chem. Ber.* **1958**, *91*, 895. (b) Coe, J. W.; Wirtz, M. C.; Bashore, C. G.; Candler, J. *Org. Lett.* **2004**, *6*, 158.

[55] VanRheenen, V.; Cha, D. Y.; Hartley, W. M. *Org. Synth.* **1988**, *6*, 342.

[56] (a) Hentges, S. G.; Sharpless, K. B. *J. Am. Chem. Soc.* **1980**, *102*, 4263. (b) Hentges, S. G.; Sharpless, K. B. *J. Org. Chem.* **1980**, *45*, 2257. (c) Berrisford, D. J.; Bolm, C.; Sharpless, K. B.; *Angew. Chem., Int. Ed. Engl.* **1995**, *34*, 1059.

[57] Ainge, D.; Ennis, D.; Gidlund, M.; Stefinovic, M.; Vas, L.-M. *Org. Process Res. Dev.* **2003**, *7*, 198.

8

INDUSTRIAL ASYMMETRIC HYDROGENATION

Hans-Ulrich Blaser

Applications of Transition Metal Catalysis in Drug Discovery and Development: An Industrial Perspective, First Edition. Edited by Matthew L. Crawley and Barry M. Trost.

© 2012 John Wiley & Sons, Inc. Published 2012 by John Wiley & Sons, Inc.

8.1   INTRODUCTION

Hydrogenation is without doubt the most important catalytic methodology for the manufacture of fine chemicals. For many years, most reactions were carried out with heterogeneous catalysts due to their exceptionally broad applicability for the chemo- and diastereoselective reduction of various functional groups, and since they are convenient to handle and catalyst separation is usually easy [1]. However, since regulations have demanded that chiral pharmaceuticals be produced as enantiomerically pure compounds, the asymmetric hydrogenation of C=C, C=O, and C=N functions has gained in importance and is now the most prevalent asymmetric catalytic method [1–4].

Progress in enantioselective hydrogenation has been summarized in several excellent reviews [5–8]. Industrial applications have extensively been described in Ref. 4 and two monographs on large-scale asymmetric catalysis [9,10] and several excellent up-to-date overviews [11–16] describing applications of enantioselective hydrogenation technology mainly in the pharmaceutical industry.

Due to its broad scope, asymmetric hydrogenation is the technology among transition metal-catalyzed processes with the most important industrial tradition. While the basic approach of selecting and implementing an efficient scalable route is quite classical for process development, there are a few special points concerning the selection of the chiral catalyst for which a generally accepted strategy has evolved over the years. For this reason, in this chapter not only relevant examples on industrial-scale hydrogenations will be examined but the underlying approach to process development will also be described in some detail. This general part will be illustrated with description of three detailed case histories for the manufacture of L-dopa, (*R*)-HPB ester (key intermediate for several ACE inhibitors), and sitagliptin.

8.2   THE ART OF PROCESS DEVELOPMENT

The choice of a development strategy that promises the best answer in the shortest time is the first decision at the start of every process development [17]. This strategy will depend on a number of considerations such as the goal of the development, the know-how of the investigators, the time frame, and the available manpower and equipment. In process development, there is usually a hierarchy of goals (or criteria) to be met. It is simply not possible to reach all the requirements for a technically useful process in one step. As depicted in Table 8.1, the catalyst selectivity (combined of course with an acceptable

TABLE 8.1 Catalyst Choice: Criteria and Requirements During Different Process Development Phases

Development Phase	Activities	Performance Criteria	Milestones	Ligand Sourcing Aspects
Feasibility phase of development	Screening	Selectivity, (activity)	Chemical feasibility	Availability in screening amounts (about 100 mg)
Bench-scale phase, demonstration of technical feasibility	Optimization, scale-up, catalyst handling quality risk analysis	Selectivity, activity, productivity	Technical feasibility	Catalyst supply <1 kg typically, quality, lead times
Pilot and production process	Scale-up, process adaptation to infrastructure, first production on scale	Selectivity, activity, productivity, recycling/ refining, metal removal, cost	Commercial feasibility	$>$Kilogram quantities of ligand and metal precursor, lead times, quality, metal refining

activity) is the first criterion—just as in academic research. But when a reasonable selectivity has been obtained, other criteria will become important: catalyst activity, productivity and stability, and catalyst separation (and maybe recycling). Then, questions such as the effect of substrate quality and the cost and availability of the chiral catalyst and other materials have to be addressed. The final process is usually a compromise since often not all of these requirements can be maximally fulfilled. It is useful to divide the development of a manufacturing process into different phases [18]; however, it is rarely possible to proceed in a linear fashion and very often one has to go back to an earlier phase in order to answer additional question before it is possible to go on.

8.2.1 Outlining and Assessing Synthetic Routes

This is of course the fundamental task of the development chemist of the pharma company with the goal to find the most economical and ecological technical synthesis in the time frame dictated by the development schedule. The starting point will be the synthesis developed in medicinal chemistry. However, in most cases, a new route has to be found and developed. When applying asymmetric catalysis, the maturity of a catalytic method (good procedures, reliability, known scope and limitations) is an important point for the synthetic chemist, since it is often difficult to transfer the results described for a particular substrate to even a close analog due to the high substrate specificity (low tolerance for structure variation even within a class of substrates). Finally, the question of patent protection has to be addressed here since many pharma companies are reluctant to use patented technology [19]. However, most technology providers have now devised accepted models for handling intellectual property rights [20].

The development time can be a hurdle, especially when the optimal catalyst has yet to be developed or no commercial catalyst is available for a particular substrate. When developing a process for a new chemical entity (NCE) in the pharmaceutical or agrochemical industry, time restraints can be severe. It is often more important to find a competitive process on time than an optimal process too late. So-called second-generation processes, for example, for chiral switches, for generic pharmaceuticals, or for the manufacture of other fine chemicals, have different requirements; here the time factor is usually not so important but a high-performance process is necessary.

When assessing proposed routes with enantioselective catalytic steps, the following criteria are important:

- chance of success for the key steps according to precedent (other closely related, efficient catalytic transformations);
- number and perceived difficulty of the noncatalytic steps;
- first estimation of costs for the overall synthesis.

8.2.2 Chemical Feasibility of the Enantioselective Catalytic Step

Once possible routes have been designed, the critical step, which is of course most of the time the catalytic step, must be tested experimentally. Here the major question is whether it is possible to find a chiral catalyst and the right reaction conditions to catalyze the desired transformation with satisfactory efficiency. The proper choice of the catalyst, that is, metal complex, chiral ligand or modifier, and in some cases additional additives, will have a dominant effect on catalyst performance. It is impossible to get good results with the wrong choice of the catalyst, no matter how well all other parameters are optimized.

Most effective homogeneous hydrogenation catalysts are complexes consisting of a central metal ion, one or more (chiral) ligands, and anions that are able to activate molecular hydrogen and to add the two H atoms to an acceptor substrate (see Fig. 8.1) [21]. The actual catalyst is prepared *in situ* from the ligand and a suitable metal precursor complex or a preformed and isolated ligand–metal (L–M) complex (see below) can be used. The nature of the ligand and sometimes the anion and of course the actual reaction conditions control the catalytic properties of a central metal atom. Experience has shown that low-valent Ru, Rh, and Ir complexes stabilized by tertiary (chiral) phosphorus ligands are the most active and versatile catalysts. Not surprisingly, most of the past as well as current research efforts are focused on this type of complexes.

From this it follows that the choice of the ligand is the most important element for the catalytic performance of a metal complex. The nature and structure of the ligand will influence how the metal center will react with H_2 and the substrate via a number of well-understood elementary steps such as oxidative addition, insertion, and reductive elimination. During these transformations, the ligand must be able to stabilize various oxidation states and coordination geometries. For the enantiose-lective hydrogenation, an impressively large number of chiral ligands are recorded in the literature that afford high enantioselectivities for a variety of catalytic reactions. However, if one has a closer look at the ligands really used by the synthetic organic chemist in academia, and even more so in industry, a very different picture emerges. Very few chiral ligands are applied on a regular basis for the synthesis of commer-cially relevant target molecules. Jacobsen once coined the term "privileged ligands" for these selected few. The following ligand classes that often have comparable performance profiles are known today: biaryl diphosphines, phospholanes, ferrocenyl-

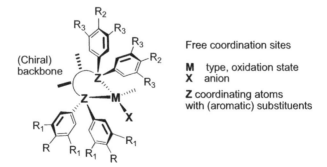

FIGURE 8.1   Design elements for chiral metal complexes for hydrogenation applications.

based diphosphines, and miscellaneous phosphines [22]. Commercially available ligand families both for screening and on production scale are depicted in Figure 8.2. Most of these ligand families are modular, that is, can be easily adapted to a particular substrate by the proper choice of the PR$_2$ moiety. Only one enantiomer of each ligand is depicted but in general both enantiomers are available even though in a few cases the prices might vary if the ligand is prepared from chiral pool material.

Despite the fact that heterogeneous catalysts would be easier to separate from the product, only two such catalytic systems have been developed up to now: Raney nickel modified with tartaric acid and Pt–alumina catalysts modified with cinchona alkaloids give high enantioselectivities but their application is essentially restricted to β-functionalized and α-functionalized ketones, respectively [23].

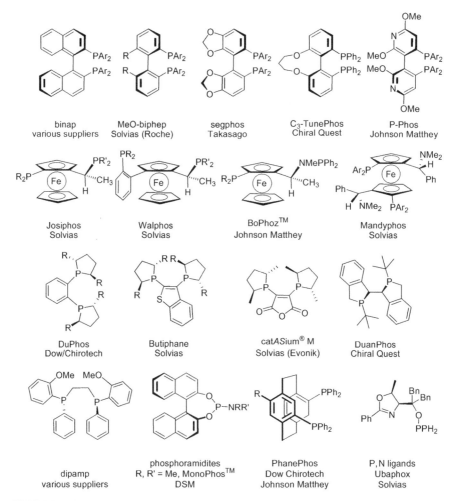

FIGURE 8.2 Structures and suppliers of commercially available ligand families for hydrogenation.

As most enantioselective catalysts are highly substrate specific, finding the right catalyst is the most crucial and time-consuming activity. In this phase, it especially important to cover a large experimental space (metal, ligand, solvent, additive, reaction conditions). This is where high-throughput screening comes in. While some pharma companies are able to do this in-house, this is the domain of technology providers such as Solvias [24] or DSM [25] who have the necessary expertise and personnel, the specialized equipment (screening robot, fast analytics, electronic data handling), and a large library of chiral ligands and metal complexes available.

8.2.3 Optimization and Scale-Up: Piloting the Catalytic Reaction

When a preliminary choice of the catalytic system has been made, the reaction conditions (H_2 pressure, temperature, and concentrations and ratios of reactants, catalyst, and auxiliary) are optimized. The most important criteria in this phase are chemoselectivity and enantioselectivity, catalyst productivity and activity, sensitivity of the catalytic system to the quality of all reagents, and the synthetic fit of the catalytic step into the complete manufacturing process. If the minimal requirements are not reached, one has to go back to finding a more suitable catalyst/auxiliary. Concerning the catalytic step, the following issues will have to be dealt with:

- *Ligand and Catalyst Handling.* It can often be problematic due to the high air and water sensitivity of organometallic compounds. On large scale, this can be an issue if no specialized equipment is available that allows handling under a nitrogen atmosphere. The sensitivity of the catalyst toward impurities (by-products in the starting material, oxygen, water, etc.) can usually be controlled with a strict purification protocol but cases have been reported where a change of the overall synthetic route of the substrate was necessary since a trace impurity in the substrate was a strong catalyst poison.

- *Availability of Ligands on Technical Scale.* As described below in the case of L-dopa, this was a major problem. When Knowles started his work, the most important part was the development of a suitable ligand. This situation continued well into the early 1990s when several companies started to sell chiral diphosphines. As mentioned above, the situation has changed since several classes of ligands are now available on scale [26]. But ligand supply can still be an issue when none of the established ligands give satisfactory results and an experimental ligand is chosen for development.

- *Catalyst Separation and Trace Metal Contamination.* In a survey [9a], catalyst separation was mentioned only once to have been a major obstacle when developing a homogeneous catalytic process. Distillation, product crystallization, and extraction were the techniques most frequently applied. It must be stressed that only in very rare cases are catalysts recycled, usually the noble metal residues are sent back to the catalyst producer. Residual metals in the product were more of an issue, especially for pharmaceutical applications and when the homogeneous catalyst was used late in the synthesis. Here, metal scavengers have shown to be an effective but not inexpensive solution [27]. Several companies such as Johnson Matthey, Evonik, or Phosphonics offer tailor-made scavengers for different metals and metal complexes.

- *Choice of Technology.* Most chiral chemicals are relatively small-scale products (1–1000 t/year for pharmaceuticals and 500–10,000 t/year for agrochemicals) and are usually produced in multipurpose batch equipment. Here, there will be only a brief discussion of the reactor choices for hydrogenation reactions that are usually carried out in the liquid phase. For a successful implementation, the following demands have to be met: (1) very good dispersion of the hydrogen gas (and a suspended catalyst for heterogeneous systems) in the reaction solution (efficient gas–liquid mixing and stirring); and (2) effective heat removal (reaction control) as well as safe handling of the sometimes pyrophoric and/or air-sensitive catalysts. In practice, two reactor types have proven to be capable of meeting these requirements as well as the need for high reliability in operation and ease of control: the stirred autoclave and the loop reactor. The loop reactor provides an efficient hydrogen dispersion and the heat exchanger surface is almost unlimited. It is especially useful when the space–time yield is very high (fast reaction, high substrate concentration) or when a low reaction temperature is required. The stirred autoclave is probably more versatile; it has an advantage when substrate slurries or viscous media have to be used or when the starting material is added continuously. In addition, it is usually easier to clean and less space and lower investment costs are required.

8.3 SELECTED CASE HISTORIES

In this section, three case histories are presented that illustrate the power of transition metal-catalyzed asymmetric hydrogenation in the synthesis of marketed drugs. These examples were chosen for three primary reasons: First, in all cases the authors give a great description of the state of the art when the process was developed as well as of the problems they had and the approaches that were chosen to resolve these problems. Second, the three examples show the scientific and technical progress in this area from the 1970s with the L-dopa process to the 1980–1990s for the (*R*)-HPB ester and the 2000s for the sitagliptin process. Finally, both homogeneous and heterogeneous processes are described.

8.3.1 L-Dopa (Monsanto, VEB Isis-Chemie)

The so-called L-dopa process is arguably the most famous transition metal-catalyzed enantioselective process. It was developed in the early 1970s by a team led by Knowles at Monsanto. This early success had a huge influence on the whole field of enantioselective catalysis and eventually resulted in Knowles receiving the Nobel Prize in 2001. Monsanto had been producing L-dopa, a drug for the relief of the Parkinson disease, on the scale of approximately 1 t/year for many years but the process is no longer in operation. A few years after Monsanto, the then East German

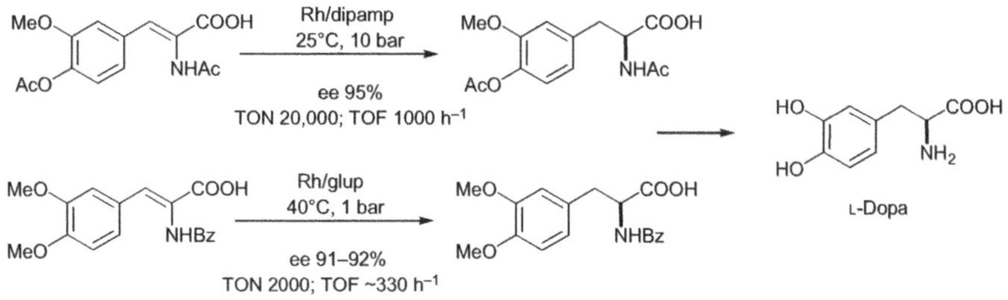

FIGURE 8.3 Monsanto (top route) and VEB Isis-Chemie (bottom route) L-dopa processes.

company VEB Isis-Chemie also developed a catalytic process and carried it out on a similar scale but also terminated the production after a few years. L-Dopa is now produced industrially using biotechnological methods.

The key step in the synthesis is the enantioselective hydrogenation of an enamide intermediate (Fig. 8.3). The stories surrounding the development of these early processes have been described in some detail by both Knowles [9b] and Selke [9c], the two principal investigators. Both teams had to start their endeavors when very little was known about homogeneous hydrogenation and even less on its enantio-selective version. While it was recognized that Rh complexes such as the Wilkinson catalyst were able to catalyze C=C bond hydrogenation, almost no chiral phosphines were known at the time. As a consequence, the search for suitable ligand and eventually the development of a technical synthesis was an integral part of the work.

The Monsanto process was carried out in a water/isopropanol mixture at relatively low temperature and pressure. Because the free ligand racemizes slowly, an isolated $[Rh(dipamp)(diene)]^+BF_4^-$ complex was used as the catalyst. With an ee value of 95%, turnover numbers (TONs) of 10,000–20,000, and turnover frequencies (TOFs) of $1000\,h^{-1}$, the catalyst performance is excellent, even for today's standards. It is therefore not surprising that for many years enamide hydrogenation was the standard test reaction for new ligands. One of the key factors for success was of course the dipamp ligand developed by Knowles and his team within an amazingly short time. Figure 8.4 shows the concept of the Monsanto scientists: (i) stereogenic phosphorous atom and (ii) bidentate structure.

The dipamp synthesis was later improved and Rh/dipamp complexes were later applied by NSC Technologies for the manufacture of several unnatural amino acids with good catalyst performances (ee 95–98%, TON 5000–20,000) [9d]. Further

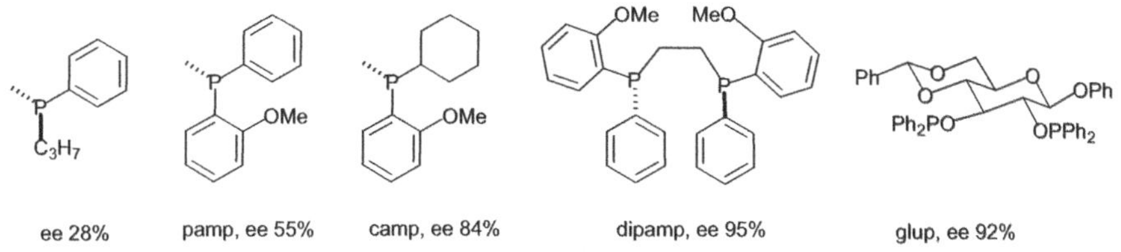

FIGURE 8.4 Ligands developed for the synthesis of L-dopa.

dipamp derivatives were synthesized by the group of Stephan [28] who found that replacing the methoxy by more bulky alkoxy groups leads to even more selective and efficient catalysts for the hydrogenation of activated alkenes.

The VEB Isis team chose a different approach by starting with a (cheap) chiral pool molecule for the construction of the glup ligand, also with two coordinating P atoms. The Rh/glup sulfate complex worked at 40°C/1 bar and with an enantios-electivity of 91–92% ee, TON of 2000, and TOF of about $330\,h^{-1}$ did not quite reach the performance of Rh/dipamp.

A very important feature of the Monsanto process is the fact that the reaction is started with a slurry of reactants and ends with a slurry of the pure product with close to 100% ee, allowing the separation of both the catalyst and the undesired racemate in one step. In both variants, critical issues were the quality of the starting material (enamide syntheses are often problematic) and especially the concentration of oxygen and peroxides in the reaction solution.

8.3.2 (R)-HPB Ester (Ciba-Geigy, Ciba SC/Solvias)

(R)-2-Hydroxy-4-phenylbutyric acid ethyl ester, the so-called (R)-HPB ester, is an important intermediate for the synthesis of several ACE inhibitors such as benazepril (Novartis) or cilazapril (Roche). Most approaches described in the literature are based on chemical or biochemical reduction methods starting from appropriate α-keto acid derivatives. Starting in the early 1980s, several teams in Ciba-Geigy tried different approaches to solve this important problem. As shown in Figure 8.5, three routes were found to be technically feasible. The biocatalytic reductions A and B were piloted and the heterogeneous hydrogenation C was introduced into production.

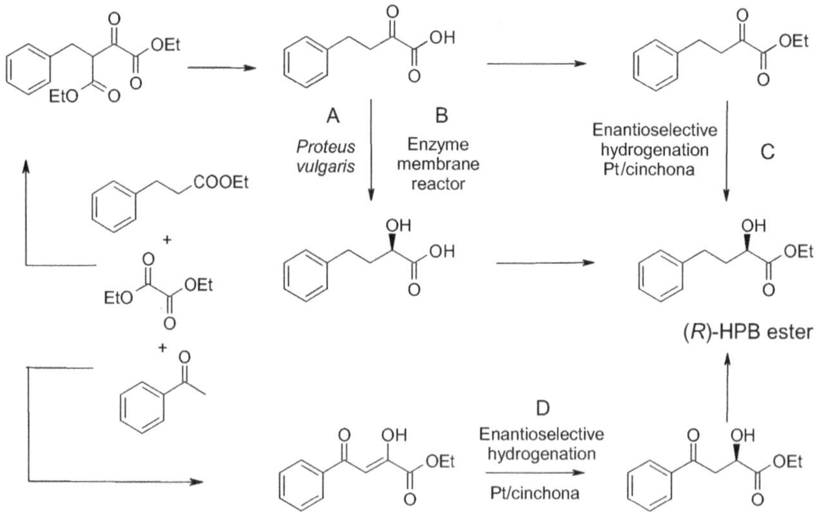

FIGURE 8.5 Catalytic routes to generate HPB ester successfully at Ciba-Geigy at the pilot and production scale.

The corresponding homogeneous hydrogenation using a Rh–norphos catalyst did not meet the requirements for a technical application. Some years later, Solvias in collaboration with Ciba SC developed a fourth variant D, again using a heterogeneous hydrogenation catalyst. The development of the two heterogeneous processes has been described in much detail [29]. Here, a short description of the most important findings and lessons is presented as well as an environmental assessment of the four different processes.

8.3.2.1 Route C: Heterogeneous Hydrogenation of Ethyl 2-Oxo-4-phenylbutyrate

The optimized process for the hydrogenation of the α-keto ester to the corresponding α-hydroxy derivative was carried out with Pt/Al$_2$O$_3$ catalyst and dihydrocinchonidine (HCd) as chiral modifier at rt and 60 bar H$_2$ pressure in toluene or AcOH (Fig. 8.6). Depending on the choice of modifier and solvent, ee's between 82% (HCd, toluene) and 92% (MeO-HCd, acetic acid) were obtained. This first-generation production process was run on a multiton scale.

When process development started, the Ciba-Geigy team already had worked for some time to reproduce results published by the group of Orito [30]. However, this turned out to be more difficult than expected and for this reason the development of a viable process for the HPB ester took more than a year. Even before the age of high-throughput screening, the obvious strategy was to screen for the best catalyst, modifier, and solvent, then to optimize relevant reaction parameters (p, T, concentrations, etc.), and finally, to scale up and solve relevant technical questions. Indeed, in the course of process development, more than 200 hydrogenation reactions were carried out. The most important results of this development work can be summarized as follows:

- *Catalyst.* 5% Pt/Al$_2$O$_3$ catalysts gave the best overall performance and the E 4759 from Engelhard was the final choice.
- *Modifier.* About 20 modifiers were tested; HCd (in toluene) and MeO-HCd (in AcOH) gave best results and were chosen for further development.
- *Solvent.* Jalett found that acetic acid was far superior to all classical solvents, allowing up to 92% ee for the HPB ester and 95% ee for ethyl pyruvate (then a

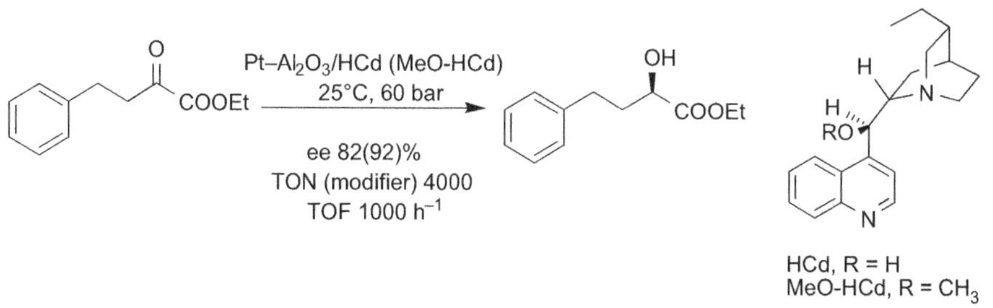

FIGURE 8.6 Heterogeneous hydrogenation of ethyl 2-oxo-4-phenylbutyrate.

new world record!). For technical reasons, toluene was chosen as solvent for the production process.

- *Reaction Conditions.* Best results (full conversion after 3–5 h, high yield, 80% ee) were obtained at 60 bar and rt with 0.5% (w/w) 5% Pt/Al_2O_3 (pretreated in H_2 at 400°C) and 0.03% (w/w) modifier.
- *Substrate Quality.* Enantioselective hydrogenation of α-keto esters proved to be exceptionally sensitive to the origin of the substrate.

After about 2 years, the production process was developed, patented, and scaled up resulting in a few hundred kilograms successfully produced in 1987 in a 500 L autoclave. The progress of the optimization can best be demonstrated by the variations in ee versus the experiment number in the different development phases (Fig. 8.7). The effect of various measures can be seen that led to improved enantioselectivities and a stable process.

8.3.2.2 Route D: Heterogeneous Hydrogenation of Ethyl 2,4-Dioxo-4-phenylbutyrate

A few years later, a new process for the (*R*)-HPB ester was developed by Solvias in collaboration with Ciba SC Life Science Molecules. After assessing a variety of synthetic routes, the one depicted in Figure 8.8 was chosen: Claisen condensation of cheap acetophenone and diethyl oxalate, followed by chemo- and enantioselective hydrogenation of the resulting diketo ester and hydrogenolysis to the HPB ester. Even though the 2,4-dioxo ester was a new substrate type, thanks to the significant experience of the development team for this kind of catalytic system it took only a few months to develop, scale up, and implement the new process. Key steps in the new process are undoubtedly the hydrogenation of the 2,4-dioxo ester with excellent chemoselectivity and satisfactory enantioselectivity and the successful enrichment to >99% ee via crystallization.

This second-generation process developed for Ciba LSM was run several times on a 10 kg scale. In addition to the success factors given above, the physical properties of

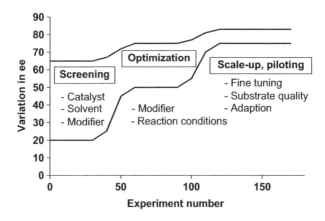

FIGURE 8.7   Different phases of HPB ester process development.

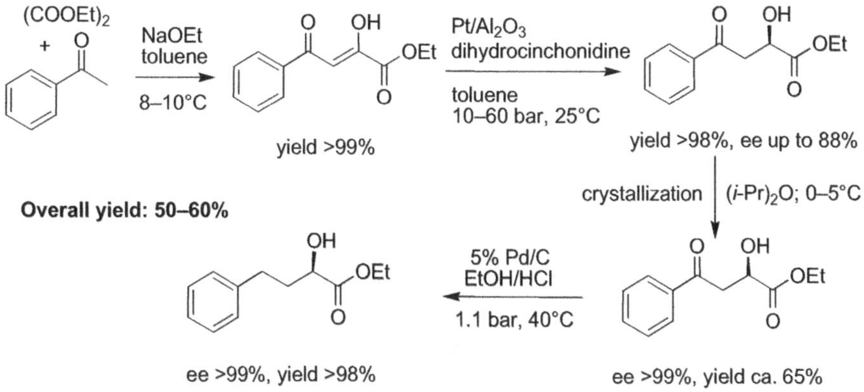

FIGURE 8.8 Solvias route to HPB ester.

the keto alcohol and the high chemoselectivity allowed enrichment to >99% ee in one crystallization, leading to a low-cost overall route. The critical issues were the substrate synthesis and quality. After enrichment to >99% ee, the keto group was removed hydrogenolytically with Pd/C without any racemization.

8.3.2.3 Comparison of Processes A–D

As already mentioned, two biocatalytic routes were also developed to the pilot stage by colleagues at Ciba-Geigy, namely the enantioselective reduction of the corresponding α-keto acid with immobilized *Proteus vulgaris* (route A in Fig. 8.5) and with D-LDH in a membrane reactor (route B in Fig. 8.5). It was therefore of interest to compare the four approaches. In collaboration with a team at ETH Zürich, the EATOS program (Environmental Assessment Tool for Organic Syntheses) was used in order to compare the mass consumption (kg input of raw materials for 1 kg of product) as well as other parameters [9e]. As shown in Figure 8.9, the new route D has the lowest overall mass consumption, even though ee and yield for the reduction step are the lowest. This is compensated by fewer steps, higher atom efficiency, and lower solvent consumption for synthesis and extraction. A rough estimation indicated that routes A, B, and C had similar production costs (keeping in mind that routes A and B were less developed) while route D had a clear cost advantage.

8.3.3 Sitagliptin (Merck)

The story of the development of the unprecedented asymmetric hydrogenation of an unprotected enamine as part of the manufacturing process for sitagliptin [31] has been described in detail in a recent account by Shultz and Krska [32]. In October 2006, sitagliptin was approved by the FDA as DPP-4 inhibitor for the treatment of type 2 diabetes and is now being marketed by Merck under the brand names Januvia and Janumet. As described by Shultz and Krska, the development work that started in 2002 was carried out under an immense time pressure. Since prior work had established an efficient three-step route to the triazole portion of the molecule,

FIGURE 8.9 Pilot processes for (R)-HPB ester: mass consumption (without water) for routes A–D.

efforts focused on accessing the β-amino acid functionality. As depicted in Figure 8.10, two approaches were evaluated, a substrate-controlled diastereoselective heterogeneous hydrogenation of an enamine/enamide bearing a chiral auxiliary and a catalyst-controlled asymmetric hydrogenation of an enamine/enamide.

For the substrate-controlled approach, the team identified (S)-phenylglycine amide as a suitable chiral auxiliary that gave remarkably high diastereoselectivity in the hydrogenation using Adam's catalyst. Hydrogenolytic removal of the chiral auxiliary using Pearlman's catalyst gave the free base of sitagliptin in excellent yield and purity. Key to the success of the diastereoselective hydrogenation was acid

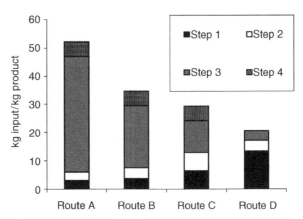

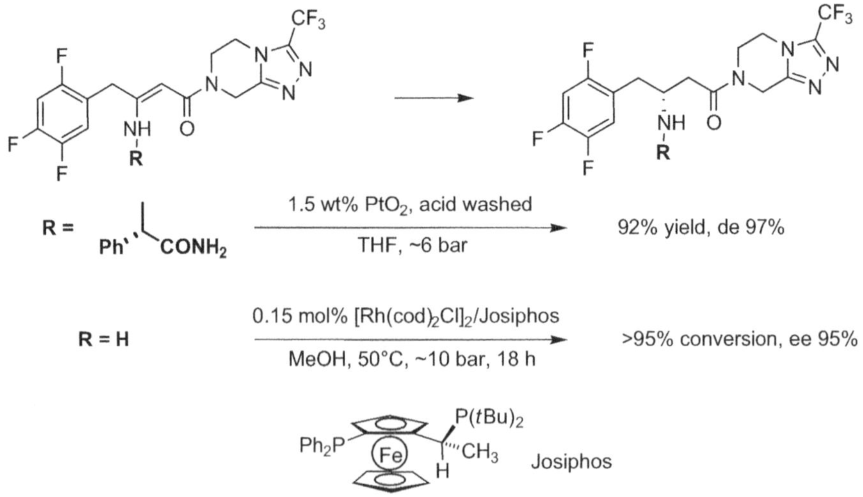

FIGURE 8.10 Substrate- and catalyst-controlled hydrogenation of sitagliptin (R = H) precursors.

washing of the Adam's catalyst, resulting in both high diastereoselectivity and activity.

For the catalyst-controlled approach, ample literature precedent suggested that asymmetric hydrogenation of a dehydro-β-amino acid derivative with a suitable N-protecting (and activating) group would be successful, provided a way could be found to synthesize the unsaturated substrate with acceptable control of olefin geometry. However, a protection/deprotection sequence required in such an approach would add unnecessary steps to the overall route as well as be less atom economic. Thus, the team decided to explore the asymmetric hydrogenation of the unprotected enamine, despite literature reports suggesting that substrates of this type were unreactive with typical asymmetric hydrogenation catalysts. The requisite enamine amide was easily obtained by treatment of the corresponding β-keto amide with NH_4OAc in methanol. Initial catalyst screening with a variety of Rh, Ru, and Ir complexes of traditional chiral bisphosphine ligands (e.g., binap) indeed showed low reactivity, low enantioselectivity, or both. However, the use of $[Rh(cod)Cl]_2$ in combination with newer ferrocene-based phosphine ligands such as ^{t}Bu-Josiphos gave promising reactivity and enantioselectivity (ee $\geq$ 90%).

At this point, the Catalysis Laboratory joined the effort to turn this promising lead into a practical, scalable process. Since at this point in time Merck had limited experience with scaling asymmetric hydrogenations and had no specific experience working with the Josiphos class of ligands, a strategic partnership was forged with Solvias AG, the catalyst supplier for Josiphos, which had world-renowned expertise in the field of asymmetric hydrogenation. In a joint effort, Merck and Solvias conducted hundreds of reactions over a very compressed period of 3 months in order to push the asymmetric hydrogenation reaction to the limits of its performance. Dozens of ligands and additives were screened, along with the usual parameters of solvent, temperature, and pressure.

The resulting optimized reaction is depicted in Figure 8.10. At a catalyst loading of 0.15 mol%, the reaction was complete in about 18 h delivering a product with 95% optical purity. The process including the associated chemistry for synthesizing the enamine amide and the downstream process for enantiopurity upgrade and final salt formation was subsequently run in the pilot plant on a hundreds of kilograms scale, less than 6 months after its discovery. Indeed, in the year following the development of the asymmetric hydrogenation route, over 1 t of sitagliptin was produced in this manner, and over 20 t has been produced up to 2007.

In a later publication [33], a reductive amination process was developed in collaboration with Takasago (see Fig. 8.11). A Ru-segphos catalyst is able to hydrogenate the enamine prepared *in situ* from the β-keto amide and ammonium salicylate. With ee's up to 99.5%, albeit at a modestly lower yield, this alternative process clearly has industrial potential. In addition to this chemical variant, a very elegant biochemical process was recently disclosed by a Codexis/Merck team [34] that not only provides very high enantioselectivities but also obviates the need for further purification. As described by the authors, starting from an enzyme that had the catalytic machinery to perform the desired chemistry but lacked any activity toward the prositagliptin ketone, various approaches such as substrate walking, modeling,

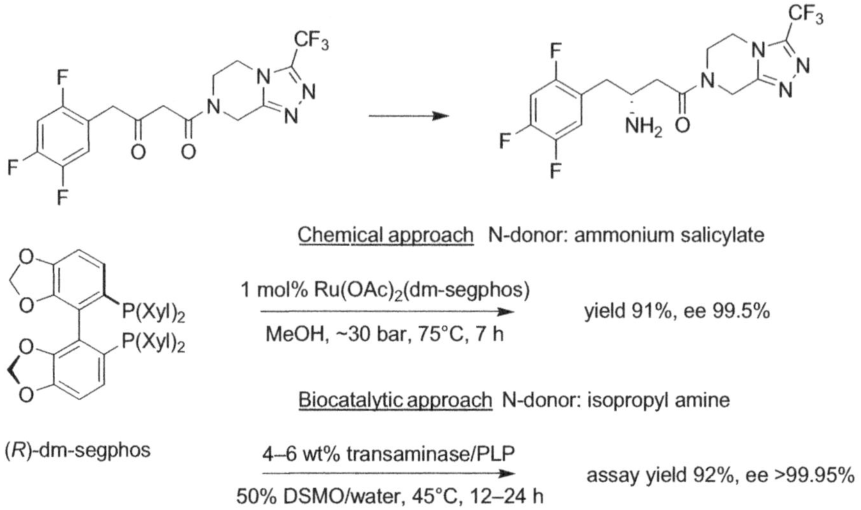

Chemical approach   N-donor: ammonium salicylate

1 mol% Ru(OAc)₂(dm-segphos)

$\xrightarrow{\hspace{3cm}}$ yield 91%, ee 99.5%

MeOH, ~30 bar, 75°C, 7 h

Biocatalytic approach   N-donor: isopropyl amine

(R)-dm-segphos   4–6 wt% transaminase/PLP

$\xrightarrow{\hspace{3cm}}$ assay yield 92%, ee >99.95%

50% DSMO/water, 45°C, 12–24 h

FIGURE 8.11   Direct reductive amination processes to sitagliptin.

and mutation were applied to create a transaminase with marginal activity for the synthesis of the chiral amine. This variant was then further engineered via directed evolution for practical application in a manufacturing setting. It is quite obvious that a massive R&D effort was needed in order to reach this very ambitious goal but it is also clear that this technology has the potential to replace the hydrogenation route used up to now.

8.4   SELECTED EFFECTIVE HYDROGENATION PROCESSES

In this section, a few pivotal and representative cases are briefly described where asymmetric hydrogenation has led to very cost-effective processes, either on a pilot scale or on a production scale for various applications (pharma, agro, fragrances). This can be either because the application of an asymmetric hydrogenation step has allowed a shorter, more effective process or because an extraordinarily effective enantioselective catalyst (high TON and TOF) has been developed. If available, some information on background and development histories as well as a short description of the process is given.

8.4.1   Production Processes for Penem Antibiotics and Ofloxacin Intermediates (Takasago)

An intermediate for penem antibiotics is produced by Takasago (Fig. 8.12) on a 50–120 t/year scale via an interesting hydrogenation reaction involving dynamic kinetic resolution of an existing stereogenic center with high enantio- and diastereoselectivities [35,36]. Later, it was reported that the newly developed biaryl

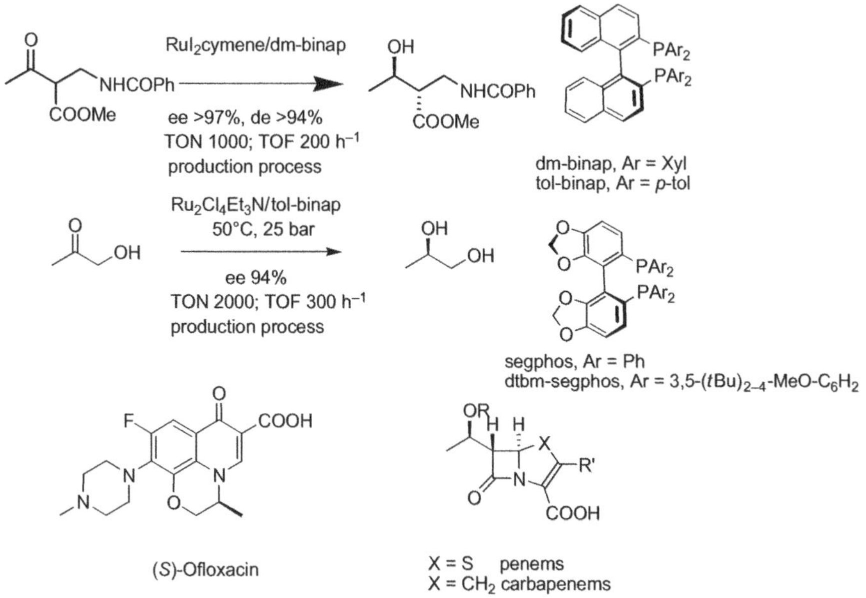

FIGURE 8.12 Production processes for penem antibiotics and ofloxacin intermediates.

diphosphine ligand dtbm-segphos can achieve even higher stereoselectivities with >99% ee and 99% de (TON and TOF not specified) [37,38]. (R)-1,2-Propanediol is an intermediate for (S)-ofloxacin, a bactericide originally sold as racemate. The (R)-diol is now produced by Takasago using a Ru-tol-binap catalyst on a 50 t/year scale. Recently, it was reported that segphos shows even better results achieving >98% ee and TON and TOF of 10,000 and ~1400 h^{-1}, respectively [25,37,38].

8.4.2 Pilot Processes for Building Blocks and Intermediates (Roche)

(S)-1,1,1-Trifluoro-2-propanol is the chiral building block in the synthesis of a variety of new type 1 glycine transporter (GlyT-1) inhibitors. Under optimized conditions, trifluoropropanol was obtained in 99.2% ee and 99.8% purity using [RuCl$_2$(S)-3,5-tBu-MeO-biphep)(R,R)-dpen)] at substrate-to-catalyst (S/C) ratio of up to 30,000 in the presence of 0.005 molar equivalents of sodium formate as base and 3 wt% of water (Fig. 8.13) [10a]. Special challenges were the development of a stable but still very active Ru-MeO-biphep-dpen complex and the fact that the reaction is run without solvent.

Several pilot processes were developed for a key intermediate for orlistat (drug against obesity) (Fig. 8.14). The use of a modified heterogeneous Raney nickel catalyst modified with tartaric acid allowed easy separation and recycling of catalyst. Crystallization gave product with an ee >99% [21b]. *Critical issues* were the catalyst preparation and a decrease of the ee with recycling (91 → 84%). Later, an alternative homogeneous process was developed and piloted using a Ru/MeO-biphep catalyst;

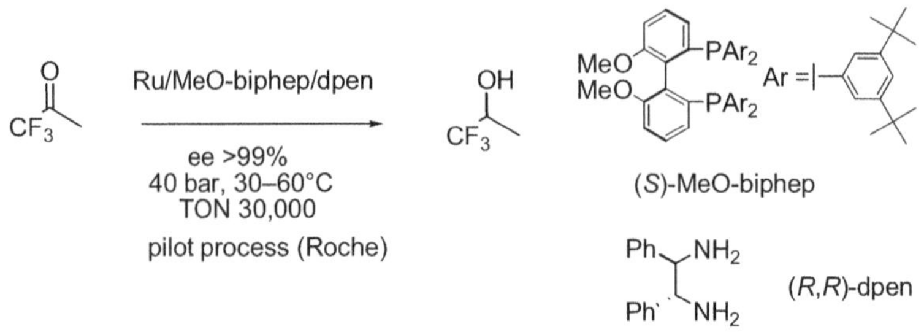

FIGURE 8.13   Pilot process for (S)-1,1,1-trifluoro-2-propanol.

ee was $> 99\%$, TON 50,000, and TOF 12,500 h^{-1} [39]. Both processes were applied to produce several tons of material.

A pilot process was developed for an intermediate of the NMDA 2B receptor antagonist Ro 67-8867 involving the hydrogenation/dynamic kinetic resolution of a cyclic α-amino ketone using an optimized Noyori procedure with a MeO-biphep ligand (Fig. 8.15) [40]. The ruthenium-catalyzed reaction was carried out on a 9 kg scale with excellent enantio- and diastereoselectivities and very high TON and TOF.

8.4.3   Pilot Processes for (R)-3,5-Bis-trifluoromethylphenyl Ethanol (Solvias/Rohner, Merck)

3,5-Bis-trifluoromethylphenyl ethanol (BTMP) is a building block for aprepitant, a NK-1 receptor antagonist. A very effective pilot process was developed by Solvias in collaboration with Rohner using a newly developed Ru/(phosphinoferrocenyl)oxazoline complex (Fig. 8.16) [41]. Even though the complex does not contain a NH bond, it shows very high activity for the hydrogenation of a variety of aryl ketones. Various reaction parameters and quality risk factors such as ligand structure, substrate quality, reaction conditions, and thermal safety were investigated. The reaction was carried out twice in a 4 m^3 autoclave on a 140 kg scale at 20 bar and 25°C with substrate-to-catalyst ratios of 20,000 with an enantiomeric excess of $> 95\%$. After crystallization, (R)-3,5-bis-trifluoromethylphenyl ethanol was obtained with an ee between 97.7% and 98.6% in 70% chemical yield. Merck has also developed a process for the transfer hydrogenation of 3,5-bis-trifluoromethyl acetophenone but with lower ee and TON [42].

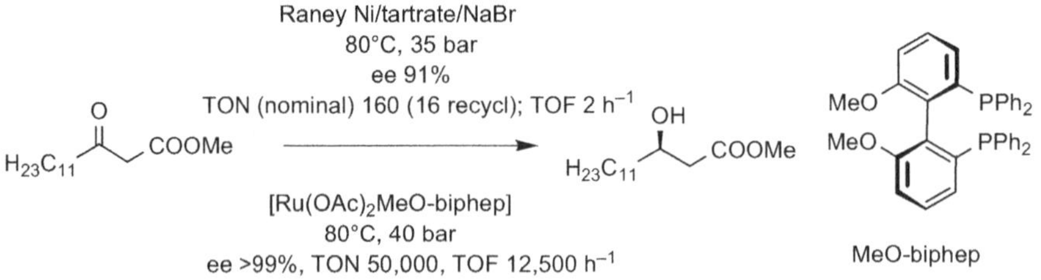

FIGURE 8.14   Pilot processes for orlistat intermediate.

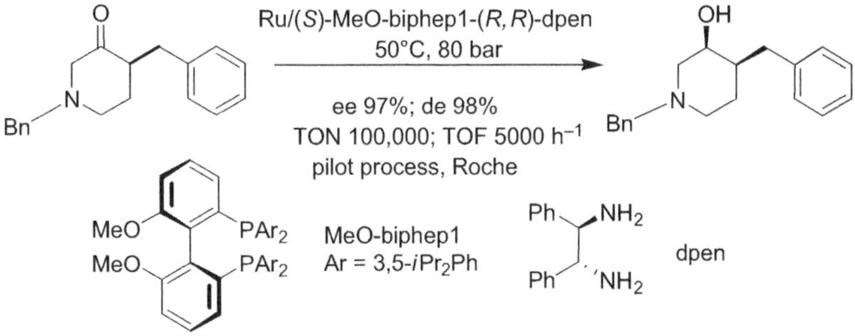

FIGURE 8.15 Pilot process for an Ro 67-8867 intermediate.

8.4.4 Pilot Processes for Taranabant Intermediates (Merck)

Two catalytic processes were developed and piloted on a 60–80 kg scale for the synthesis of CB-1R inverse agonist taranabant, a drug against obesity (Fig. 8.17). The hydrogenation of a tetrasubstituted enamide, a very difficult transformation, was accomplished using a Rh-Josiphos ligand in trifluoroethanol and accordingly required a relatively high catalyst loading [43]. The hydrogenation of a racemic aliphatic ketone with concomitant dynamic kinetic resolution (also a demanding reaction) was carried out with an optimized Noyori catalyst [44].

8.4.5 Bench-Scale Processes for Pregabalin Intermediate (Pfizer)

Several ligands were found to be successful for the Rh-catalyzed hydrogenation of a cyano-substituted α,β-unsaturated acid derivative, an intermediate for the anticonvulsant agent pregabalin (Fig. 8.18). While few details have been described concerning scale-up, it appears that multikilogram quantities have been prepared using the Rh-Me-DuPhos [45] as well as the Rh-TCFP [46] catalysts. Both catalysts achieve high enantioselectivities despite the fact that an E/Z mixture of the substrate is employed but the TCFP ligand achieves a much higher TON. At low catalyst loading, the Rh–L1 complex showed an unusual drop in enantioselectivity at lower pressures [47].

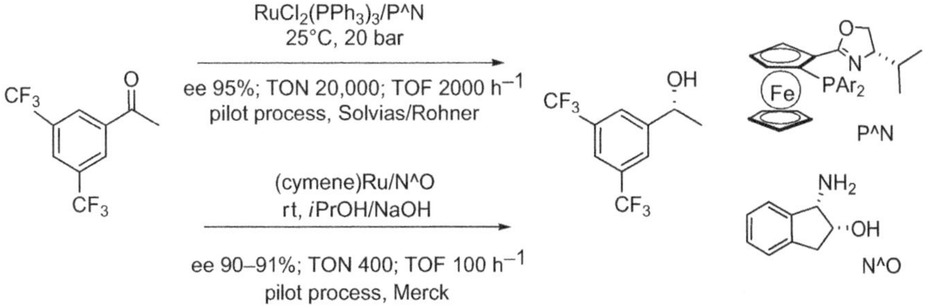

FIGURE 8.16 Pilot processes for (R)-3,5-bis-trifluoromethylphenyl ethanol.

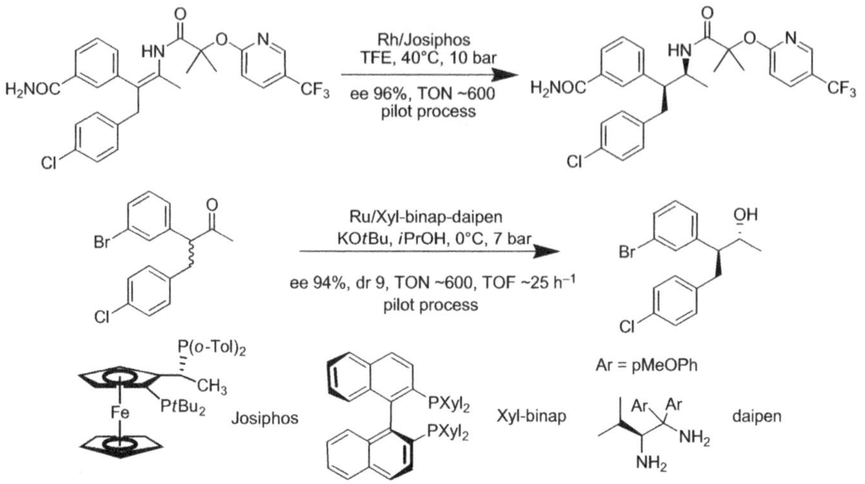

FIGURE 8.17 Pilot processes for taranabant intermediates.

8.4.6 Production Processes for Aliskiren Building Block (Solvias, DSM, BASF)

Aliskiren is a novel renin inhibitor, developed by Speedel and produced now on large scale by Novartis. While α,β-unsaturated carboxylic acids are usually hydrogenated using Ru catalysts, Rh complexes proved to be more suitable for this sterically hindered α,β-unsaturated acid (Fig. 8.19). Few details have been released for the first process developed by Solvias [48] using a Rh/Walphos catalyst with very good enantioselectivities. Recently, DSM has divulged results for a production process with a novel Rh/phosphoramidite/PPh$_3$ catalyst with somewhat lower enantioselectivity [10b] and BASF patented a process (and probably applies at production scale)

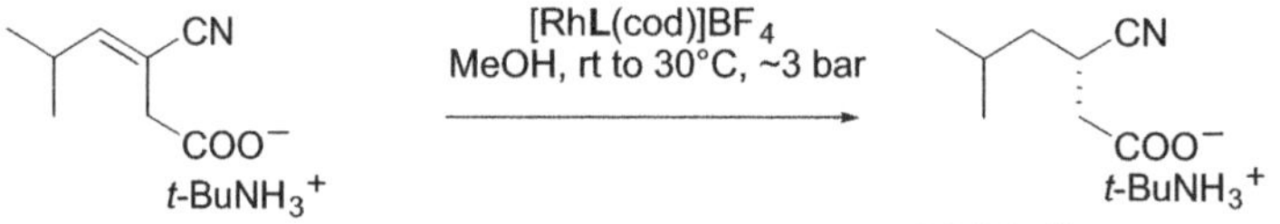

	ee (%)	TON	TOF (h^{-1})
Me-DuPhos	97.7	2700	700
L1 (35 bar)	97	1000	1000
TCFP	>99	27,000	700

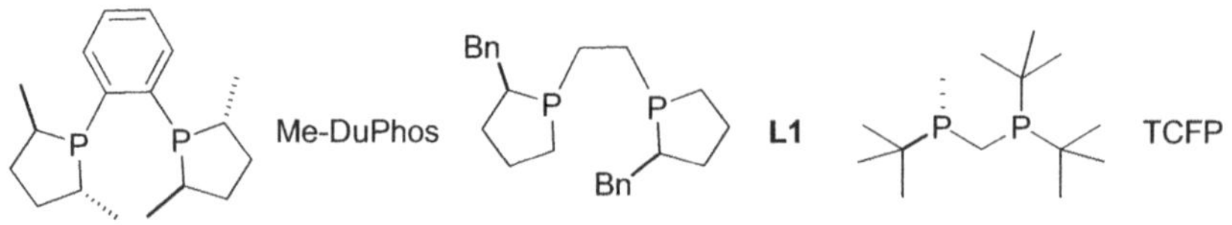

FIGURE 8.18 Bench-scale processes for pregabalin intermediate.

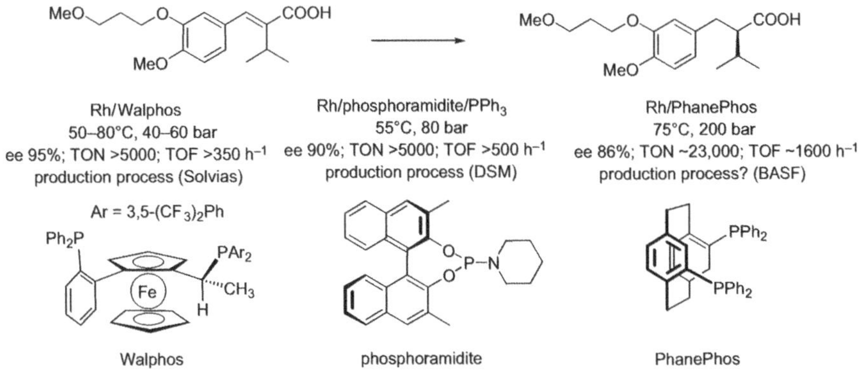

FIGURE 8.19 Production processes for aliskiren building block.

starting from a *E/Z* mixture of the unsaturated acid using a Rh/PhanePhos catalyst that requires very high pressures but still provides 86% ee [49].

8.4.7 Production Processes for Adrenaline and Phenylephrine (Boehringer Ingelheim)

The two production processes using α-amino ketone substrates were developed by Boehringer Ingelheim to improve existing resolution syntheses for adrenaline (R = OH) and phenylephrine (R = H) (Fig. 8.20) [50]. Unfortunately, few details are available but both processes are carried out on medium scale with a Rh/mccpm catalyst with very high TONs and TOFs, albeit with medium ee's of 88% that increase to >99% after precipitation of the free base.

8.4.8 Intermediates for Biotin Synthesis (Lonza, DSM/Solvias)

In the course of the development of a new technical synthesis for biotin (a water-soluble vitamin), the diastereoselective hydrogenation of a tetrasubstituted C=C bond turned out to be a key step. It is remarkable that with the exception of Rh/Josiphos catalysts (Fig. 8.21) [9f], homogeneous Rh complexes with most ligand classes achieved even lower diastereoselectivity than an achiral heterogeneous Rh/Al_2O_3 catalyst. The high effectiveness of the Rh catalysts with PPF-PtBu$_2$ as ligand was therefore even more surprising. The enantioselective hydrogenation (N-Bn

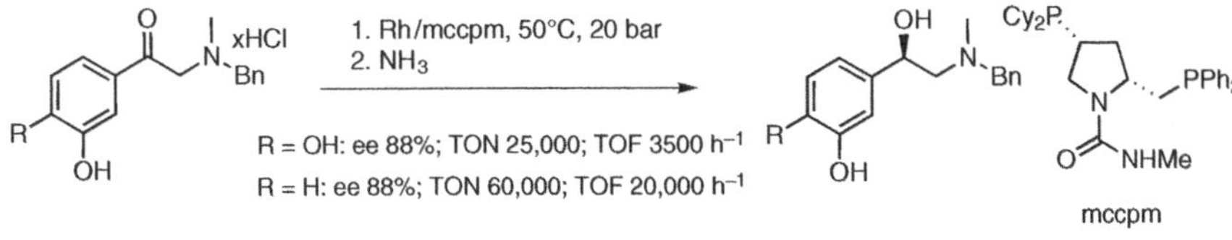

FIGURE 8.20 Production processes for adrenaline (R = OH) and phenylephrine (R = H).

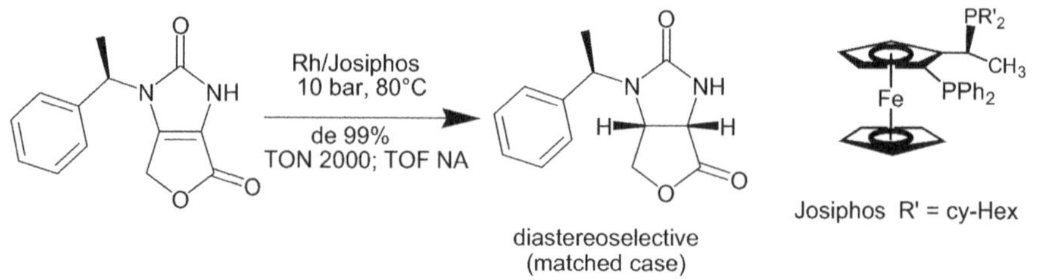

FIGURE 8.21 Production process for biotin intermediate.

instead of phenethyl) with Rh-Josiphos afforded the desired enantiomer with up to 90% ee. For the production process, the diastereoselective variant was chosen and for a few years several tons per year were manufactured.

A new biotin process based on the highly enantioselective hydrogenation of the easily accessible prochiral *meso*-anhydride building block catalyzed by an Ir–MeO-biphep catalyst was reported by a DSM/Solvias team (Fig. 8.22). It represents the first preparatively useful example of a catalytic asymmetric reduction of a cyclic anhydride to a chiral lactone. With a substrate-to-catalyst ratio of 5000, complete conversion, excellent yields, and ee values of >95% could be achieved. The process was piloted on a multikilogram scale, and production trials on the ton scale were run successfully.

8.4.9 Production Process for (S)-Metolachlor (Ciba-Geigy/Novartis/Solvias/Syngenta)

Metolachlor is the active ingredient of Dual®, one of the most important grass herbicides for use in maize and a number of other crops. In 1997, after years of intensive research, Dual Magnum® with a content of approximately 90% (1'S)-diastereomers and with the same biological effect at about 65% of the use rate was introduced in the United States and is now produced on a scale of >10,000 t/year. The key step is the Ir/Josiphos-catalyzed hydrogenation of an imine substrate (Fig. 8.23). This process is presently the largest known application of an asymmetric

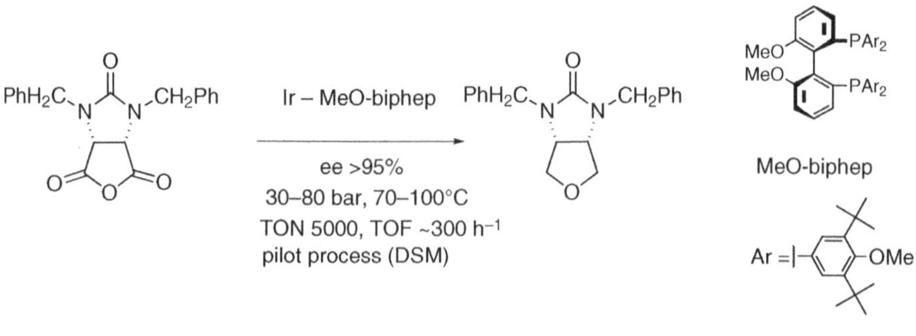

FIGURE 8.22 Pilot process for biotin intermediate.

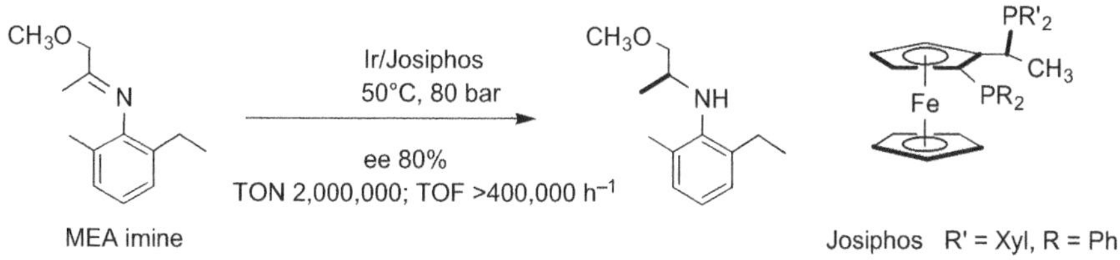

FIGURE 8.23 Production process for (S)-metolachlor intermediate.

catalytic hydrogenation and with turnover numbers of 2 million and turnover frequencies of up to 1.8 million is also the process with the highest catalyst efficiency [9g].

The search for a commercially viable process took many years [51]. Several approaches with Rh or Ir complexes using diphosphine ligands commercially available at the time were not successful. A critical breakthrough was achieved using Ir complexes of a new class of ferrocenyl-based ligands, now called Solvias Josiphos ligands. In the presence of acid and iodide ions, extremely active and productive catalysts were obtained. The optimized process operates at 80 bar hydrogen and 50°C with a catalyst generated *in situ* from [Ir(cod)Cl]$_2$ and the Josiphos ligand PPF-PXyl$_2$ (short name xyliphos) at a substrate-to-catalyst ratio of 2,000,000. Complete conversion is reached within 3–4 h, the initial TOF exceeds 1,800,000 h^{-1}, and enantioselectivity is approximately 80%.

8.4.10 Processes for Citronellol Production (Takasago, Roche)

One approach for the production of citronellol, which is also used as an intermediate for vitamin E, starts with geraniol (Fig. 8.24). This transformation requires a specific Ru precursor and is highly chemoselective. It is carried out by Takasago on 300 t/year scale [35,36]. Roche has reported a similar pilot process that works at 20°C and 60 bar using the same Ru precursor with MeO-biphep; the ee was 99%, TON 20,000, and TOF 1500 h^{-1} [21b].

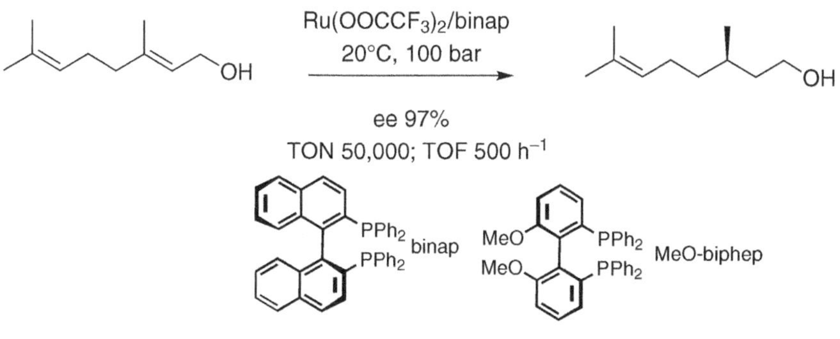

FIGURE 8.24 Production process for citronellol.

8.4.11 Production Process for (+)-cis-Methyl Dihydrojasmonate (Firmenich)

Methyl dihydrojasmonates are ubiquitous low-price perfume ingredients. Firmenich has established that (+)-cis-methyl dihydrojasmonate is the active stereoisomer, and has developed a Ru-catalyzed enantioselective process (Fig. 8.25) [52]. A novel Ru precursor [Ru(H)(cyclooctatriene)(ligand)]BF$_4$ and a new reaction system had to be developed because the classical Ru complexes and conditions for the hydrogenation of C=C bonds did not work. Besides the enantioselectivity, chemo- and cis-selectivity and activity problems (tetrasubstituted C=C) were solved on a very high level. A broad screening of Ru catalysts (partly in collaboration with Solvias) showed that selected Josiphos ligands and DuPhos satisfied the prerequisites. The production of the enriched isomer is presently carried out on a multiton/year scale.

8.5 CONCLUSIONS AND FUTURE DEVELOPMENTS

Is asymmetric hydrogenation a mature manufacturing technology? A somewhat personal definition of this term can be summarized as follows:

- Well-defined and widely known scope and limitations (selectivity, activity, productivity, functional group tolerance).
- Many existing technical applications; required equipment widely available.
- Routinely considered in route design during process development.
- Relevant catalysts and ligands are commercially available (including well-defined handling of IP issues) both in large numbers for screening and in technical quantities for production.

Considering this, the answer is "not yet"—with, maybe, the exception of the hydrogenation of a few privileged substrate types. Even though asymmetric hydrogenation is clearly the most advanced catalytic methodology, the results summarized

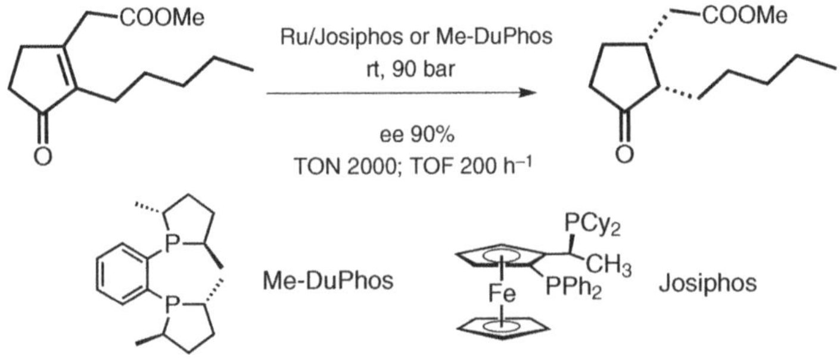

FIGURE 8.25 Production process for cis-methyl dihydrojasmonate.

in the preceding sections show that some of these requirements are "not quite fulfilled" or "not yet satisfactory." However, in the past few years, technical progress has accelerated and one of the most significant signs of this development is the growing number of companies active in this exciting area of chemical technology. The visibility of enantioselective hydrogenation as a superb manufacturing tool for chiral intermediates and active products will certainly be enhanced by scientific/technical publications but also by considerable marketing efforts.

Since the first processes have been implemented by Monsanto and Sumitomo in the early 1970s, the number of production processes has grown only slowly and comprises today (only) about 20–30 entries. Of these, about half are medium to large scale, all others are applied on a scale of a few tons per year or less and several of them are no longer in operation. On the positive side, many more processes developed to the pilot or bench scale are in principle ready for technical application.

There are several reasons for the rather slow progress. Maybe the most important one is the very high attrition rate for new chemical entities in the pharmaceutical industry (around 90% across all therapeutic areas), as well as the relatively low number of new drugs introduced in the past decade. Another reason is the fact that it takes more time (and money!) to find and develop a catalytic process compared to classical organic transformations. Even though it was shown that many enantioselective catalysts are very tolerant of functional groups, the scope and limitations of valuable enantioselective catalysts are not yet known sufficiently, making synthesis planning difficult. Last but not least, some chiral catalysts and ligands have not been readily available in large quantities and are not trivial to prepare, and some intellectual property issues are still difficult to deal with. Here it might help that patents for important ligands have expired (binap, tol-binap) or are near the end of their patent life (DuPhos, Josiphos) and that several ligand suppliers have developed solutions for handling intellectual property issues that are well accepted in the pharmaceutical industry.

Despite these hurdles, there are many signs that in the near future the industrial application of enantioselective catalytic technology will accelerate further and there are several points to strengthen this view. There is the usual time lag for any new technology to be used in actual production. However, there is hope that we are at the moment right at the beginning of the steep part of the classical S-shaped curve. More and more medium and small companies have expertise in developing catalytic syntheses and offer their services to companies who cannot or do not want to develop such processes in-house. In addition, several companies now offer various chiral ligands or catalysts in quantities required for large-scale processes.

On the chemical side, there is no doubt that new and more selective and active catalysts will be developed for ever more types of transformations. Hopefully, some of them will belong to the small elite group of privileged catalysts, able to tolerate significant structural variations without loss in catalyst performance. In addition, high-throughput experimentation will in many cases allow more tests to be carried out, thereby shortening the time needed for finding the right catalyst and improving the chances for success.

REFERENCES

[1] For a short overview see Blaser, H. U.; Steiner, H.; Studer, M. In *Transition Metals for Organic Synthesis*, 2nd ed.; Bolm, C.; Beller, M., Eds.; Wiley-VCH: Weinheim, 2004; Vol. 2, p. 125.

[2] Blaser, H. U.; Pugin, B.; Spindler, F. *J. Mol. Catal. A* **2005**, *231*, 1.

[3] For a recent update on industrial asymmetric catalysis see Blaser, H. U.; Pugin, B.; Spindler, F. In *Encyclopedia of Catalysis*, 2nd ed.; Horvath, I. T., Ed.; Wiley-Interscience, 2009.

[4] For a recent update on industrial asymmetric hydrogenations see Blaser, H. U.; Spindler, F.; Thommen, M. In *Handbook of Homogeneous Hydrogenation*; de Vries, J. G.; Elsevier, C. J., Eds.; Wiley-VCH, 2007; p. 1279.

[5] Blaser, H. U.; Malan, Ch.; Pugin, B.; Spindler, F.; Steiner, H.; Studer, M. *Adv. Synth. Catal.* **2003**, *345*, 103.

[6] Au-Yeung, T. T.-L.; Chan, A. S. C. *Coord. Chem. Rev.* **2004**, *248*, 2151.

[7] Tang, W.; Zhang, X. *Chem. Rev.* **2003**, *103*, 3029.

[8] Barbaro, P.; Bianchini, C.; Giambastiani, G.; Parisel, S. L. *Coord. Chem. Rev.* **2004**, *248*, 2131.

[9] (a) Blaser, H. U.; Schmidt, E. In *Large Scale Asymmetric Catalysis*; Blaser, H. U.; Schmidt, E., Eds.; Wiley-VCH: Weinheim, 2003; p. 1. (b) Knowles, W. S. In *Large Scale Asymmetric Catalysis*; Blaser, H. U.; Schmidt, E., Eds.; Wiley-VCH: Weinheim, 2003; p. 23. (c) Selke, R. In *Large Scale Asymmetric Catalysis*; Blaser, H. U.; Schmidt, E., Eds.; Wiley-VCH: Weinheim, 2003; p. 39. (d) Ager, D. J.; Lanemann, S. A. In *Large Scale Asymmetric Catalysis*; Blaser, H. U.; Schmidt, E., Eds.; Wiley-VCH: Weinheim, 2003; p. 257. (e) Studer, M.; Blaser, H. U.; Eissen, M.; Fauquex, P. F.; Hungerbühler, K.; Schmidt, E.; Sedelmeier, G. In *Large Scale Asymmetric Catalysis*; Blaser, H. U.; Schmidt, E., Eds.; Wiley-VCH: Weinheim, 2003; p. 91. (f) McGarrity, J. F.; Brieden, W.; Fuchs, R.; Mettler, H.-P.; Schmidt, B.; Werbitzky, O. In *Large Scale Asymmetric Catalysis*; Blaser, H. U.; Schmidt, E., Eds.; Wiley-VCH: Weinheim, 2003; p. 283. (g) Blaser, H. U.; Hanreich, R.; Schneider, H.-D.; Spindler, F.; Steinacher, B. In *Large Scale Asymmetric Catalysis*; Blaser, H. U.; Schmidt, E., Eds.; Wiley-VCH: Weinheim, 2003; p. 55.

[10] (a) Püntener, K.; Scalone, M. In *Asymmetric Catalysis on Industrial Scale: Challenges, Approaches and Solutions*; Blaser, H. U.; Federsel, H. J., Eds.; Wiley-VCH: Weinheim, 2010; p. 13. (b) Boogers, J. A. F.; Sartor, D.; Felfer, U.; Kotthaus, M.; Steinbauer, G.; Dielmans, B.; Lefort, L.; de Vries, A. H. M.; de Vries, J. G. In *Asymmetric Catalysis on Industrial Scale: Challenges, Approaches and Solutions*; Blaser, H. U.; Federsel, H. J., Eds.; Wiley-VCH: Weinheim, 2010; p. 127. (c) Klingler, F. D. In *Asymmetric Catalysis on Industrial Scale: Challenges, Approaches and Solutions*; Blaser, H. U.; Federsel, H. J., Eds.; Wiley-VCH: Weinheim, 2010; p. 171. (d) Bonrath, W.; Karge, R.; Netscher, T.; Roessler, F.; Spindler, F. In *Asymmetric Catalysis on Industrial Scale: Challenges, Approaches and Solutions*; Blaser, H. U.; Federsel, H. J., Eds.; Wiley-VCH: Weinheim, 2010; p. 27.

[11] Lennon, I. C.; Moran, P. H. *Curr. Opin. Drug Discov. Dev.* **2003**, *6*, 855.

[12] Lennon, I. C.; Pilkington, C. J. *Synthesis* **2003**, 1639.

[13] Iida, T.; Mase, T. *Curr. Opin. Drug Discov. Dev.* **2002**, *5*, 834.

[14] Ikunaka, M. *Chem. Eur. J.* **2003**, *9*, 379.

[15] Kumobayashi, H.; Miura, T.; Sayo, N.; Saito, T.; Zhang, X. *Synlett* **2001**, 1055.

[16] Shultz, C. S.; Krska, S. W. *Acc. Chem. Res.* **2007**, *40*, 1320.

[17] (a) Butters, M.; Catterick, D.; Craig, A.; Curzons, A.; Dale, D.; Gillmore, A.; Green, S. P.; Marziano, I.; Sherlock, J.-P.; White, W. *Chem. Rev.* **2006**, *106*, 3002. (b) Federsel, H.-J. *Acc. Chem. Res.* **2009**, *42*, 671. (c) Farina, V.; Reeves, J. T.; Senanayake, C. H.; Song, J. J. *Chem. Rev.* **2006**, *106*, 2734.

[18] Spindler, F.; Blaser, H. U. *Enantiomer* **1999**, *4*, 557.

[19] A very restrictive policy is described in Hawkins, J. M.; Watson, T. J. N. *Angew. Chem., Int. Ed.* **2004**, *43*, 3224.

[20] (a) Casey, G.; Johnson, N. B.; Lennon, I. C. *Chim. Oggi/Chem. Today* **2003**, *21*(12), 63. (b) Thommen, M.; Blaser, H. U. *Chim. Oggi/Chem. Today* **2003**, *21*(12), 6.

[21] (a) Brown, J. M. In *Comprehensive Asymmetric Catalysis*; Jacobsen, E. N.; Pfaltz, A.; Yamamoto, H., Eds.; Springer: Berlin, 1999; p. 121. (b) Schmid, R.; Scalone, M. In *Comprehensive Asymmetric Catalysis*; Jacobsen, E. N.; Pfaltz, A.; Yamamoto, H., Eds.; Springer: Berlin, 1999; p. 1439.

[22] Blaser, H. U.; Malan, Ch.; Pugin, B.; Spindler, F.; Steiner, H.; Studer, M. *Adv. Synth. Catal.* **2003**, *345*, 103.

[23] Studer, M.; Blaser, H. U.; Exner, C. *Adv. Synth. Catal.* **2003**, *345*, 45.

[24] Blaser, H. U.; Hoge, G.; Lotz, M.; Nettekoven, U.; Schnyder, A.; Spindler, F. *Chimia* **2008**, *62*, 476.

[25] de Vries, J. G.; Lefort, L. *Chem. Eur. J.* **2006**, *12*, 4722.

[26] Thommen, M.; Blaser, H. U. *Chim. Oggi/Chem. Today* **2003**, *21*(1–2), 27.

[27] Welch, C. J.; Albaneze-Walker, J.; Leonard, W. R.; Biba, M.; DaSilva, J.; Henderson, D.; Laing, B.; Mathre, D. J.; Spencer, S.; Bu, X.; Wang, T. *Org. Process Res. Dev.* **2005**, *9*, 198.

[28] Zupancic, B.; Mohar, B.; Stephan, M. *Org. Lett.* **2010**, *12*, 3022 and references cited therein.

[29] Blaser, H. U.; Studer, M. *Acc. Chem. Res.* **2007**, *40*, 1348.

[30] Orito, Y.; Imai, S.; Niwa, S. *J. Chem. Soc. Jpn.* **1979**, 1118; *J. Chem. Soc. Jpn.* **1980**, 670; *J. Chem. Soc. Jpn.* **1982**, 137.

[31] Hsiao, Y.; Rivera, N. R.; Rosner, T.; Krska, S. W.; Njolito, E.; Wang, F.; Sun, Y.; Armstrong, J. D.; Grabowski, E. J. J.; Tillyer, R. D.; Spindler, F.; Malan, C. *J. Am. Chem. Soc.* **2004**, *126*, 9918.

[32] Shultz, C. S.; Krska, S. W. *Acc. Chem. Res.* **2007**, *40*, 1320.

[33] Steinhuebel, D.; Sun, Y.; Matsumura, K.; Sayo, N.; Saito, T. *J. Am. Chem. Soc.* **2009**, *131*, 11316.

[34] Savile, C. K.; Janey, J. M.; Mundorff, E. C.; Moore, J. C.; Tam, S.; Jarvis, W. R.; Colbeck, J. C.; Krebber, A.; Fleitz, F. J.; Brands, J.; Devine, P. N.; Huisman, G. W.; Hughes, G. J. *Science* **2010**, *329*, 305.

[35] Kumobayashi, H.; *Recl. Trav. Chim. Pays-Bas* **1996**, *115*, 201.

[36] Akutagawa, S. *Appl. Catal.* **1995**, *128*, 171.

[37] Kumobayashi, H.; Miura, T.; Sayo, N.; Saito, T.; Zhang, X. *Synlett* **2001** 1055.

[38] Saito, M.; Yokozawa, T.; Ishizaki, T.; Moroi, T.; Sayo, N.; Miura, T.; Kumobayashi, H. *Adv. Synth. Catal.* **2001**, *343*, 264.

[39] (a) Birk, R.; Karpf, M.; Püntener, K.; Scalone, M.; Schwindt, M.; Zutter, U. *Chimia* **2006**, *60*, 561. (b) Schwindt, M. A.; Fleming, M. P.; Han, Y.-K.; Hodges, L. M.; Johnston, D. A.; Micheli, R. P.; Roberts, C. R.; Snyder, R.; Topping, R. J.; Püntener, K.; Scalone, M. *Org. Process Res. Dev.* **2007**, *11*, 524.

[40] Scalone, M.; Waldmeier, P. *Org. Process Res. Dev.* **2003**, *7*, 418.

[41] Naud, F.; Malan, C.; Spindler, F.; Rüggeberg, C.; Schmidt, A. T.; Blaser, H. U.; *Org. Process Res. Dev.* **2007**, *11*, 319.

[42] Hansen, K. B.; Chilenski, J. R.; Desmond, R.; Devine, P. N.; Grabowski, E. J. J.; Heid, R.; Kubryk, M.; Mathre, D. J.; Varsolona, R. *Tetrahedron: Asymmetry* **2003**, *14*, 3581.

[43] Shultz, C. S.; Krska, S. W. *Acc. Chem. Res.* **2007**, *40*, 1320.

[44] Chen, C.; Frey, L. F.; Shultz, S.; Wallace, D. J.; Marcantonio, K.; Payack, J. F.; Vazquez, E.; Springfield, S. A.; Zhou, G.; Liu, P.; Kieczykowski, G. R.; Kreska, S. W. *Org. Process Res. Dev.* **2007**, *11*, 616.

[45] Burk, M. J.; de Koning, P. D.; Grote, T. M.; Hoekstra, M. S.; Hoge, G.; Jennings, R. A.; Kissel, W. S.; Le, T. V.; Lennon, I. C.; Mulhern, T. A.; Ramsden, J. A.; Wade, R. A. *J. Org. Chem.* **2003**, *68*, 5731.

[46] Hoge, G.; Wu, H.-P.; Kissel, W. S.; Pflum, D. A.; Greene, D. J.; Bao, J. *J. Am. Chem. Soc.* **2004**, *126*, 5966.

[47] Hoge, G. *J. Am. Chem. Soc.* **2003**, *125*, 10219.

[48] Sturm, T.; Weissensteiner, W.; Spindler, F. *Adv. Synth. Catal.* **2003**, *345*, 160.

[49] Hettche, F.; Völkert, M.; Jäckel, C. WO2006097314 (2006), assigned to BASF.

[50] Klingler, F. D. *Acc. Chem. Res.* **2007**, *40*, 1367.

[51] For an account of this work see Blaser, H. U. *Adv. Synth. Catal.* **2002**, *344*, 17.

[52] Dobbs, D. A.; Vanhessche, K. P. M.; Brazi, E.; Rautenstrauch, V.; Lenoir, J.-Y.; Genet, J.-P.; Wiles, J.; Bergens, S. H. *Angew. Chem., Int. Ed.* **2000**, *39*, 1992.

INDEX

Applications of Transition Metal Catalysis in Drug Discovery and Development: An Industrial Perspective,
First Edition. Edited by Matthew L. Crawley and Barry M. Trost.
© 2012 John Wiley & Sons, Inc. Published 2012 by John Wiley & Sons, Inc.

Printed and bound by CPI Group (UK) Ltd, Croydon, CR0 4YY

16/04/2025

14658420-0002